The Ethological Roots of Culture

NATO ASI Series

Advanced Science Institutes Series

A Series presenting the results of activities sponsored by the NATO Science Committee, which aims at the dissemination of advanced scientific and technological knowledge, with a view to strengthening links between scientific communities.

The Series is published by an international board of publishers in conjunction with the NATO Scientific Affairs Division

A Life Sciences **B Physics**	Plenum Publishing Corporation London and New York
C Mathematical and Physical Sciences **D Behavioural and Social Sciences** **E Applied Sciences**	Kluwer Academic Publishers Dordrecht, Boston and London
F Computer and Systems Sciences **G Ecological Sciences** **H Cell Biology** **I Global Environmental Change**	Springer-Verlag Berlin, Heidelberg, New York, London, Paris and Tokyo

NATO-PCO-DATA BASE

The electronic index to the NATO ASI Series provides full bibliographical references (with keywords and/or abstracts) to more than 30000 contributions from international scientists published in all sections of the NATO ASI Series.
Access to the NATO-PCO-DATA BASE is possible in two ways:

– via online FILE 128 (NATO-PCO-DATA BASE) hosted by ESRIN, Via Galileo Galilei, I-00044 Frascati, Italy.

– via CD-ROM "NATO-PCO-DATA BASE" with user-friendly retrieval software in English, French and German (© WTV GmbH and DATAWARE Technologies Inc. 1989).

The CD-ROM can be ordered through any member of the Board of Publishers or through NATO-PCO, Overijse, Belgium.

Series D: Behavioural and Social Sciences - Vol. 78

The Ethological Roots of Culture

edited by

R. Allen Gardner

Beatrix T. Gardner

Center for Advanced Study,
University of Nevada,
Reno, Nevada, U.S.A.

Brunetto Chiarelli

Institute of Anthropology,
University of Florence,
Florence, Italy

and

Frans X. Plooij

Department of Psychology,
University of Groningen,
Groningen, The Netherlands

Kluwer Academic Publishers

Dordrecht / Boston / London

Published in cooperation with NATO Scientific Affairs Division

Proceedings of the NATO Advanced Study Institute on
The Ethological Roots of Culture
Cortona, Italy
June 21–July 3, 1992

A C.I.P. Catalogue record for this book is available from the Library of Congress.

ISBN 0-7923-3127-3

Published by Kluwer Academic Publishers,
P.O. Box 17, 3300 AA Dordrecht, The Netherlands.

Kluwer Academic Publishers incorporates the publishing programmes of
D. Reidel, Martinus Nijhoff, Dr W. Junk and MTP Press.

Sold and distributed in the U.S.A. and Canada
by Kluwer Academic Publishers,
101 Philip Drive, Norwell, MA 02061, U.S.A.

In all other countries, sold and distributed
by Kluwer Academic Publishers Group,
P.O. Box 322, 3300 AH Dordrecht, The Netherlands.

Printed on acid-free paper

Printed in the Netherlands

TABLE OF CONTENTS

STUDYING THE ETHOLOGICAL ROOTS OF CULTURE

R. Allen Gardner and Beatrix T. Gardner
Center for Advanced Studies
University of Nevada/296
Reno NV 89557

The NATO Advanced Study Institute on the Ethological Roots of Culture was held in Cortona, an ancient hill town in Tuscany.

Cortona Was Our Campus

From June 21 to July 3, 1992 a group of 60 scholars met to discuss recent discoveries on transmission of behavior in social groups and across the generations, in humans and other animals. Anthropologists, biologists and psychologists came to Cortona from 15 different countries, from as far away as India and Japan. Most came from Italy, of course, and the next largest contingent came from North America. The participants included 44 postdoctoral scholars; established scientists mingled with young researchers. Aris Poulianos, the President of the Anthropological Association of Greece and discoverer of hominid remains in the Petralona cave, was the most senior participant; Claire Pickford, who had just completed her undergraduate studies in the UK, the most junior. And Professor Poulianos, in particular, delighted in talking with the young graduate students and insisted they call him Ari.

Participants stayed in small hotels, in different parts of Cortona. Each morning, we gathered in the cloistered courtyard next to our meeting rooms. We waited for the caretaker to unlock the former chapel and bring in the audio-visual equipment: he was always rushing, and always a little late. At the end of the morning session, we walked a short distance up the Rua Guelfa, to the piazza in the center of town. There, we took lunch together, on the stone terrace of La Logetta, overlooking the piazza. All commercial activity in Cortona stops after lunch, and the conference took this break also. Sessions resumed later in the afternoon and lasted into the early evening. At dinner time, participants were on their own; usually, small groups set out for different restaurants within the town. The graduate students discovered a

R. A. Gardner et al. (eds.), The Ethological Roots of Culture, 1–13.

better way to dine: they bought bread and wine and other ingredients and took them to Cortona's park. Everyone contributed, and everyone shared. As the meeting went on, more and more of the participants came to join in this feast. The daytime views from the park are magnificent: the vast fertile Valdichiana below, with farms and olive groves and rows of cypress trees. At night, we strolled in the park amid the townspeople of Cortona; everyone seemed to be in the park then, from the very young to the very old.

Many participants remarked on the opportunity to meet freely in informal groups, to walk and talk together, or to stop for coffee and exchange ideas. One of the American students said it was like a college campus. Her comment seemed odd at first, but she had put it very well. The walled town of Cortona, with its ancient Etruscan walls, narrow cobbled streets, and handsome medieval buildings, had indeed become our college campus, more charming and more picturesque than any other we knew.

There was one behavioral theme that appeared repeatedly in one study session after another. This theme was that exploratory and innovative behavior is much more likely to emerge in times of plenty. The finding is clear in spite of all the old proverbs about necessity being the mother of invention, or all the Skinnerian dicta about deprivation and reinforcement, or all the principles of survival pressures driving evolutionary change. It emerged from studies of exploration and path breaking in mountain sheep, from studies of invention in monkeys, from studies of sign language in cross-fostered chimpanzees. Rich variations emerge from richly plentiful, even overplentiful conditions. That is when risk-taking replaces conservatism. This notion seemed particularly apt in Tuscany, so rich in the memory of one particular point in history. Over and over a visitor encounters the question of how so much genius in art, letters, and science could appear in so small a place and so brief a time. It seemed obvious after our exploration of the ethological roots of culture that this was also the period of greatest material wealth in the history of Tuscany.

Culture and Behavioral Biology

It is hardly likely that culture could arise from a unitary biological trait. It is much more reasonable to suppose that culture is the result of a complex of interacting traits that are involved in all aspects of human intelligence. It is also reasonable to suppose that, like other significant biological phenomena, the general principles that govern human intelligence are related to the general principles that govern the intelligence of all animals. This NATO ASI on the Ethological Roots of Culture joined in the search for these general biological principles.

Specifically, we looked at the transmission of behavior in social groups and across generations, as in: invention and spread of new habits in social groups of monkeys, development of tool use and communicative gestures in infant chimpanzees by observation and imitation of their mothers, development of local dialects of bird songs, transmission of predator recognition among birds, and traditional migratory routes of mountain sheep.

This NATO ASI, looked for patterns of variation, growth, and development rather than static modules. In her book, *A Calculating People*, P. C. Cohen (1982) traces the history of numeracy (a term meant as an analog to literacy) in England and America. She relates the rise of numeracy to growing interest in processes, trends and predictions (p 41-46).

> Before the seventeenth century in Europe, the order of the cosmos was dictated by classical and, specifically, Aristotelian systems of classification. All aspects of life and nature were comprehensible through the arrangements of categories meant to exhibit significant distinctions and to exhaust the possibilities of reality. The world was composed of four substances, the body had four humours, the life of man was framed by the seven stages of the aging process, and so on. The penchant for classification remained alive in the seventeenth century and found full expression in the Great Chain of Being, an idea that expressed the relations among all creatures on earth and in heaven by making explicit an assumed hierarchy of the natural and supernatural worlds. . . . Classification by categories is a reasonable method for ordering static things . . . (p. 44)

Cohen goes on to show how scientists and social leaders discovered that comparable measures over time reveal patterns of flux and change. Numerate scientists and citizens can study the motion of bodies and trends in births, deaths, trade, weather and a virtually unlimited world of variables. By studying variation over time, they could detect underlying functions and make reasonable predictions of future events. Eventually, this numerate strategy became commonplace in most natural sciences.

There has been a trend rather than a revolution from static categories to dynamic functions. The more successful branches of science have thrived as they gradually abandoned Aristotelian categories in favor of dynamic functions. The Aristotelian method survives to our time in tortuous academic debates over the precise difference between intentional and unintentional, conscious and unconscious, imitation and emulation, language and nonlanguage, human and nonhuman, natural and artificial, and so on.

Aristotelian logic applies rigorous rules of logic to sharp, either-or, yes-no categories to arrive at definitive conclusions. This is its great appeal. The weakness of Aristotelian logic is that carving nature in this fashion yields crude categories and distorted descriptions of nature. This in turn invalidates any conclusion of Aristotelian logic that rests on such crude categories which returns us to the confusion we had when we started. Fortunately, there are alternatives to Aristotelian logic that can deal with continuous rather than discrete natural functions. Moreover, these alternative logical systems apply straightforwardly to computer systems that perform quite mundane tasks, such as running subway trains and washing machines. Rigorous logical systems that deal with continuous, graded, hence "fuzzy," inputs and outputs perform their tasks more effectively and more efficiently than categorical systems. They are also easier to program and require less computation which makes them particularly suitable for robotic applications. These features should make continuous, dynamic logical systems even more attractive to behavioral biologists.

Behavioral biologists can learn from a time in history when chemists faced a similar choice between a hierarchy of Aristotelian categories and continuous, dynamic functions. According to a popular view, vitalism was defeated with the first synthesis of organic chemicals; in 1828 when Wöhler synthesized urea. Actually, the doctrine of a vital substance separating organic from inorganic matter dominated natural philosophy until quite recent times. The natural philosophers of Wöhler's time had much in common with their counterparts today. They argued, for example, that urea was a very lowly substance, perhaps not truly organic at all. They also complained that Wöhler's synthesis required unnatural amounts of heat. On their side, the pioneers of modern chemistry, were equally unimpressed. They had always assumed that there is only one kind of matter and one kind of chemical law. The synthesis of urea was applauded, it supported their position, but they were more interested in the new questions that now opened up before them. Forward looking nineteenth century chemists neither sought nor discovered a new form of alchemy by which inorganic could be transmuted into organic. What they sought were the basic laws that govern all matter (Brooke, 1968).

In these days of genetic engineering and artificial hearts, it is hard to believe that vitalism was an acceptable, even a dominant, position in chemistry and biology until recent times. Neither academic debate nor crucial experiments defeated vitalism. The debate was never won. No crucial experiment closed the gap between organic and inorganic. Instead, vitalism faded away as chemists learned to abandon categorical schemes in favor of dynamic variables and general laws.

In his report of the first synthesis of urea, Wöhler (1828/1959) barely mentioned the question of vitalism. Instead, he devoted his discussion to the finding that urea and the inorganic salt, ammonium cyanate, are both composed of the same proportions of nitrogen, carbon, hydrogen, and oxygen, yet have distinctly different physical and chemical properties. He cited a few other examples of what is now known as isomerism and suggested that a general law might be deduced from these cases. Wöhler was only 28 years old when he published that paper and he had a long and distinguished career ahead of him. Although he published more than three hundred articles touching on almost every significant aspect of nineteenth century chemistry during his extraordinarily active professional life, he rarely mentioned vitalism.

History vindicated Wöhler's judgment as to the relative significance of isomerism and vitalism. The pursuit of discoveries such as isomerism lead nineteenth century chemists to a whole new world of organic chemistry, leading beyond the synthesis of naturally occurring substances to the creation of entirely new chemicals. The debates over vitalism survive in the form of historical curiosities and cautionary tales.

The search for Aristotelian categories leads to endless academic debates over the wording of definitions and the ordering of hierarchies. The search for continuous functions leads to experiment and experiment leads to discovery. The philosopher and experimental psychologist Jerome Bruner put it very well in the following comment on topics closely related to this ASI.

> A third trend is also discernible: the bridging of gaps that before were not so much empty as they were filled with corrosive dogmatism. The gaps between prelinguistic communication and language proper as the child develops, the gap between gesture and word, between holophrases and sentences, between chimps signing and man talking, between sign languages and spoken ones, between the structure of action and the structure of language. I think that the renewal of interest in language as an interactive, communicative system has made these 'gaps' less like battlegrounds where one fights and dies for the uniqueness of man and more like unknown seas to be mapped (1978, p. viii).

Overview of this Volume

As directors of this ASI, we tried to influence the contributions and discussions to appeal to those who prefer the pursuit of new discoveries over the challenge of academic debate. Of course, with such an unruly and independent-minded group of participants, we were only partially successful in this, but we think that this influence emerges overall from a reading of this volume of contributions.

In the autumn of 1992 we based our graduate seminar on the contributions to this volume, then in the form of rough drafts. The references in each contribution usually provided generous suggestions for background reading and alternative views. The students, Elizabeth Ely and Matthew Swiergosz who had attended in Cortona, and Christine Appert, Jane Beatty, Norman Pecoraro, Alan Rees, and Patricia Simonet, who could only imagine Cortona, responded enthusiastically to the contributions and the background readings. They also contributed detailed notes and editorial suggestions for most of the contributions.

We organized our graduate seminar around the methods of the contributors -- field studies, laboratory studies, cross-fostered chimpanzees, human infant development, ethnography, and paleontology because adjacent papers in these sections seemed to talk most directly to each other. In the course of the seminar, we also found that discussions of articles in the separate sections easily evoked material from earlier and later sections.

PART I. FIELD STUDIES

The range of variation that is to be found among individuals of any given species is a fundamental fact of biology. Variations in traits that reflect variations in the gene pool are the source of evolutionary change. Descriptions of "the cat" or "the chimpanzee" or "the child" that fail to include the range of individual variation are necessarily distorted. The articles in this section document within-species variations in the behavior of macaques, chimpanzees, songbirds and social insects, and explore the conditions that give rise to them.

Early research on macaque communication focused on the 'norm' in an attempt to depict a standardized species-typical pattern. Anne Zeller finds considerable individual variation in the threat gestures of macaques. Differences in age, sex, and kin group of the sender yield variations in the facial gesture of threat. Macaques live in a complex social group and need to respond differentially to messages sent by different categories of individuals. In her focus on individual variation Zeller reveals the complexity and dimensionality of species-typical behavior and at the same time shows how these dimensions relate the behavior of different species.

In his field work with stumptail macaques, Dennis Rasmussen developed Pindex, a quantitative measure of the degree to which the range of an individual troop member resembles (central) or differs from (peripheral) that of the entire troop. Pindex values varied with age (infants most central, followed by juveniles, then adults) and across matrilines, and the highest ranking matrilineal subgroup was more central than the others. For adult individuals, Pindex values were consistent over the 2 years of the study.

Jane Goodall discovered so much about free-living chimpanzees and described them so well -- their communicative signals, hunting behavior, tool use and so on -- that the chimpanzees of Gombe seem to be the prototype of the wild chimpanzee. Bill McGrew reviews recent field work on behavioral differences between populations of wild chimpanzees. McGrew begins by describing his own astonishment when, on a visit from Gombe to Mahale, he saw pairs of chimpanzees grooming while clasping each other's uplifted arm. This striking grooming pattern had never been observed at Gombe, only 160 miles away. Longterm studies of wild chimpanzees, all across Africa reveal patterns of behavior that are habitual within some populations of chimpanzees but rare or absent in others such as hand clasp grooming and nut-cracking with stone or woood hammers. McGrew's rigorous scrutiny of the evidence and thorough analysis make us appreciate the complexities and multi-dimensionality of behavioral differences between populations.

Several of the contributors to this section, and to the next section as well, respond to the group that McGrew calls the "anti-culturalists," e.g., Galef, Tomasello, and Whiten. McGrew's direct response to the critics is respectful, thoughtful, and convincing. Yet it is the article as a whole, that is the more powerful response. McGrew and the other contributors to this section provide new data on cultural transmission in naturalistic settings, offer promising ways to proceed, reveal unexpected influential variables such as conditions of abundance, and suggest fresh models for cultural transmission, applicable to humans and other animals.

Kunio Watanabe reviews the extensive study of Japanese macaques on Koshima Island. In 1953, a year after the field workers began provisioning the troop with abundant supplies of sweet-potatoes, one juvenile female began to carry her sweet-potatoes to the nearby stream and wash them before eating them. This behavior gradually spread to other group members at first, to other juveniles, and then from mothers to infants, until it became fixed as a tradition of the group. Later, other innovative behaviors appeared, such as wheat washing, and swimming in the sea, all behaviors which had not been reported before or observed in other troops of Japanese macaques.

Watanabe's detailed ethological descriptions show variation and changes over the course of 40 years of observation. After 1972, abundant provisioning ended and Koshima monkeys received sweet potatoes only a few times a year. At the beginning of the studies, the provisioners put out soiled sweet potatoes because these were the cheapest on the market. After 1983, however, all sweet potatoes came to market already washed. Sweet potato washing persisted, and appeared in new generations, but the pattern changed from the original. The new pattern consisted of carrying and dipping sweet potatoes in the sea without the rubbing that once removed dirt. Perhaps dipping improves the taste, or perhaps dipping became a tradition.

Several varieties of wheat washing have appeared and spread in the troop. The wheat is gritty with sand when the field workers set it out for the monkeys who seem to find it much more palatable with the sand washed out. To wash the wheat, however, the monkeys must release handfuls in the water and then fish out the grains of wheat floating near the surface. More dominant monkeys discovered that they can grab wheat from lower ranking monkeys during this vulnerable moment. Lower ranking monkeys have developed safer counter-strategies for wheat washing and higher ranking monkeys have learned new ways of stealing. Watanabe also discusses factors that contribute to the emergence of innovation. Abundance rather than necessity is the mother of invention: new habits appear in monkey troops scattered throughout Japan after abundant provisioning. Moreover, it is the young monkeys, at play during a surplus of food and time, that are the innovators and also the first to take up the new habits of their playmates.

The last two articles in this section describe computer models devised by field biologists to simulate the transmission of behavior under wild conditions. Peter Slater's simulations are based on bird song, and what is known about the variations in structure of song both within and between populations. These simulations explore straightforward and attractively simple variables that influence social transmission. Slater's model assumes that learner chaffinches copy the songs of their neighbors, but they make errors. This simple model predicts the observed longevity of song types over generations, the size of dialect areas, and the persistence of dialect borders from season to season. Guy Theraulaz was concerned with swarm intelligence - collective behavior in which a coherent functional global pattern emerges. In this model, each individual responds to the activities of others, stimulating and informing them in return. Theraulaz shows that fundamental characteristics of insect societies, such as division of labor, networks of foraging trails, and nest architecture emerge from interactions with other individuals and with the local habitat. Specific patterns of behavior persist from generation to generation. Theraulaz brings the sad news that society breeds bureaucracy even in a colony of 20 primitive wasps.

PART II. LABORATORY STUDIES

Infants who live with their mothers, and indeed most young animals who live in a social group, repeatedly observe experienced conspecifics foraging, signaling, maintaining territories and following paths. In many cases, as in the field studies of chimpanzees by McGrew and of songbirds by Slater, the behavior patterns of the adults are transmitted to the naive young animals. In this section Paquette, Prato Previde, Suboski and Valsecchi describe laboratory studies of the transmission of foraging techniques and fear of predators. An impressive variety of species -- chimpanzees, hamsters, house mice, chicks, zebra fish, and freshwater snails -- transmit arbitrary patterns of behavior from individual to individual.

Emanuella Prato Previde and Paola Valsechhi show how hamster and mouse pups can acquire new foraging techniques from an experimentally trained adult female. Control pups spent the same amount of time in the apparatus, either alone or as a group or in the presence of an untrained adult female. Although a few of the control animals eventually used the new technique, the experimental subjects were far more likely to do so. Many interrelated factors combine to bring about behavior similarities between demonstrators and observers.

The social relationship between demonstrator and observer add a further dimension of complexity. In Prato Previde's study, the trained adult was the mother of the pups, and in Valsecchi's study she was an unrelated dam; both social situations were favorable for social transmission. In Paquette's study, a social group of four adolescent chimpanzees discovered how to use sticks to fish for honey, and all four gradually acquired skill in the fishing technique. But subordinate Merlin and Maya acquired the new foraging technique many days after dominant Spock and Sophie. Paquette's detailed description of the behavior of these individuals, shows how the social and affective relations between demonstrator and observer influence attention to the demonstrations as well as access to the tools and hence the acquisition of the new technique. In the case of the Japanese macaques, Watanabe mentions that older monkeys, especially older males, were the least likely to acquire the new habit of potato washing. Perhaps, the young and adolescent monkeys who were the innovators on Koshima Island were ineffective demonstrators for adult males.

Milton Suboski's work focuses on the stimulus rather than the response. Owls and hawks and other meat-eating birds, for example, prey on blackbirds and starlings and other smaller birds. The smaller birds can often fight off the predator, if they stick together and fly at their enemy from all directions. All of the defenders in the area join in the defense when they hear or see the the first bird begin the mobbing. In a series of classic experiments, Curio demonstrated that inexperienced blackbirds can learn to mob arbitrary targets. The experimenters housed groups of wild and cage-reared blackbirds in adjacent cages in such a way that they could show a stuffed owl to the wild birds while keeping the cage-reared birds from seeing the owl. What the cage-reared birds could see was an arbitrary object such as a stuffed specimen of a harmless honey-eating bird or a plastic bottle. When the experienced blackbirds called out and attempted to mob the owl, the inexperienced blackbirds followed them in calling out and attempting to mob the plastic bottle or the honey-eating bird.

Soon the inexperienced birds started mobbing the arbitrary object as soon as they saw it without waiting for the experienced birds to lead the attack. The experimenters then used the now trained, cage-reared blackbirds to train a fresh group of inexperienced blackbirds to mob an arbitrary object. They then let this second generation condition a third generation of inexperienced birds, and so on up to a sixth generation of blackbirds that learned to mob a plastic bottle by observing other blackbirds.

In his contribution, Suboski demonstrates similar principles in a variety of species. When they are injured, many social fish release a chemical that alarms other fish in the school. Experimenters can observe the alarm reactions of a school of fish confined to a laboratory fish tank by stimulating the live fish with the body fluids of a fish killed outside of the tank. The alarm reaction of zebra fish is fairly complex and involves swimming towards the source of the alarm chemical, infusion of the substance with tasting movements, followed by rapid darting movements as if trying to disperse and escape from the tank, followed by forming a compact group, dropping to the bottom of the tank, and swimming back and forth as close to the bottom as possible. In their natural habitat, the last maneuver would stir up mud and possibly hide the group from a predator.

Suboski and his associates first demonstrated that a small school of zebra fish would go through the full alarm reaction when their only stimulus was the sight of another school of zebra fish in an adjacent tank that was going through the alarm reaction after receiving a dose of the alarm chemical. The experimenters then conditioned a small school of zebra fish to react to an innocuous substance, called morpholine, when it was paired with the alarm chemical. After conditioning, morpholine alone induced the complete alarm reaction.

The experimenters next placed a second group of zebra fish in the same tank as the conditioned group and stimulated both with morpholine alone. When first stimulated with morpholine, the second group continued to swim normally, but they soon followed the conditioned fish in going through the alarm response. In this way the experimenters conditioned the second group to make the alarm response to morpholine whether or not the first group was present. Eventually, the second group (that had never experienced morpholine paired with the alarm chemical) conditioned a naive third group to make the alarm response to the innocuous morpholine. Suboski shows how a wide variety of species can acquire fairly elaborate and sustained patterns of response by exposure to a conspecific making the same response to an arbitrary stimulus.

PART III. CROSS-FOSTERED CHIMPANZEES

In cross-fostering, adults of one genetic stock rear the young of a different genetic stock. In its most dramatic form, the young and the parents belong to different species. Sign language studies of cross-fostered chimpanzees measure the amount of human culture transmitted to very young chimpanzees. Cross-fostered chimpanzees, like human children, only acquire human culture and signs gradually, in a process that takes many years. Allen and Beatrix Gardner describe the continuous, intensive observation and recording in their cross-fostering laboratory. The vocabularies of Washoe, Moja, Pili, Tatu, and Dar overlap with the vocabularies of human infants and the signs of the chimpanzees exhibit the fuzziness of natural language categories. The Gardners also compare the cross-fostering approach with more traditional modular approaches to the study of language-like behavior in nonhuman animals.

Project Washoe and its sequel with Moja, Pili, Tatu, and Dar, simulated with infant chimpanzees the conditions in which the language of human children develops gradually and piecemeal into the language of human parents. Beatrix and Allen Gardner trace patterns of growth and development in the early utterances of children and chimpanzees. The evidence for continuous processes and variables contradicts the yes-no, either-or Aristotelian logic of philosophical linguistics that has prevailed for so long. What the children and chimpanzees actually say supports, instead, a position that is much more compatible with modern natural science.

Washoe, Moja, Tatu, and Dar became young adults living in the Chimpanzee and Human Communication Institute in Ellensburg, Washington. They continued to sign and to participate in research. One of the most significant projects in which they have participated involved a drastic procedure. In order to see whether Washoe and the younger chimpanzees would transmit sign language to the infant Loulis, their human friends virtually stopped signing to them. Roger Fouts presents the findings of this 5-year-study, which demonstrate transmission of ASL to the next generation. In addition, Fouts describes Washoe's teaching methods for some of the signs that Loulis acquired.

Deborah Fouts introduced remote video recording, with no humans present, as a method for studying signed conversations among CHCI chimpanzees. Here, she reviews her own studies on Loulis' conversations with his mother and his peers, and subsequent studies that have used remote video recording to investigate private signing and imaginative play in Washoe, Moja, Tatu, Dar, and Loulis.

PART IV. INFANT DEVELOPMENT

Cultural anthropology and developmental psychology are closely allied, mutually reinforcing disciplines. A central issue in both disciplines is the transmission of behavior across generations. Frans Plooij takes the strong position that "It is in the daily customs and ritualized exchanges with the mother that culture is born." Busnel, Montagner and Plooij explore new directions within the field of mother-infant interactions.

When does acculturation begin? Marie Claire Busnel reviews recent studies that measure changes in fetal heart beat in response to acoustic stimuli, including speech and music. She presents evidence that the fetus is also receiving, remembering and integrating aspects of the culture of his group. An especially intriguing finding is the heart rate reaction to significant but subtle distinctions in speech. Busnel found different fetal heart rate reactions depending on whether the mother speaks to her fetus (e.g., *Bonjour Bébé*) or speaks to the experimenter (e.g., *Bounjour Madame*).

While mother-infant interactions are critical in acculturation, interactions with siblings and with peers are also influential. Hubert Montagner describes the complex social interactions between young children in his laboratory. His recent work shows influences from peers at an astonishingly early age. When 4-7 month old infants are brought face-to-face in special infant seats, they look at each other more than at their mothers. They are especially attentive to each other's face and hands. They touch hands frequently, suggesting a high level of coordination for 4-month-old humans. The infants even match vocalizations, match direction of gaze, and imitate rhythmic beating of the feet and hands as well as some actions with objects.

Frans Plooij's presents a model and a firm ground of evidence that integrate maturational and social variables. Conflicts with the mother stimulate the infant to develop new skills and new independence. In each period of conflict, mothers contribute to development by demanding that the infant reorganize its behavior according to its new potentials - never rejecting the infant as a whole, but only certain aspects of its behavior. Support for this model comes from Frans and Hetty Plooij's earlier observational studies of chimpanzee mother-infant pairs at the Gombe, and their more recent studies of human mother-infant pairs in the Netherlands, with questionnaires and direct observation.

PART V. ETHNOGRAPHIC AND HISTORICAL PATTERNS

This section contains reviews of cultural studies of humans, based on ethnographic and historical accounts. Traits that differ in separated human populations are regarded as cultural, while traits that show little variation across separate human populations are regarded as 'human nature' or biological givens. This seems simple enough, yet appropriate studies are challenging and rare. Linda Marchant and McGrew review methodological problems in the volume of research on human handedness that has appeared over the past 60 years. They show that even for a trait that is relatively easy to observe and measure, the existing data are so scanty and limited that right-handedness as a species-typical trait seems likely but not proven. They go on to outline the requirements for sound, ethological studies that remain to be done. Nonhuman primates must be studied in the same way, to determine whether right-handedness is specific to humans.

Michael Kirk Smith reviews olfactory communication in humans, particularly as related to reproductive activity and individual recognition. He includes a generous sample of the widely different responses to natural and synthetic odors that can be found across human groups, past and present. Sociologists, anthropologists, and expedition leaders have given vivid accounts of different human olfactory customs, such as perfuming rituals when welcoming strangers into the group, different uses of odor words and metaphors in human languages, and different human beliefs and practices regarding the suppression of body odors.

Bill and Claire Russell give us a lively description of postural and gestural signals that vary across human groups and change in the course of human history. Desmond Morris and his associates documented population differences in hand gestures, such as waving versus beckoning for summoning, and in head gestures, such as shaking versus tossing the head for 'no.' This ethnographic study was based on interviews with adult male informants from 40 localities in 25 countries, on the meaning and use of 22 human gestures. The Russells discuss their own work, on the Christian posture of prayer in Europe, as a case of historical change. The historical record for the praying figure extends back to the second century, and includes descriptions and prescriptions in written texts and portrayals in paintings and sculptures. These records show a marked change, that occurred between the 5th and 11th century, from praying while standing with outspread arms to praying while kneeling with hands together.

PART VI. PALEOANTHROPOLOGICAL PATTERNS

The contributors to this section review findings and hypotheses about the origins of human culture. In paleoanthropology, as Francesca Giusti points out, 'culture' is generally used as a synonym for the human 'way of life.' Chiarelli, Geist, Giusti and Slurink draw many intriguing connections between anatomy, behavior, and ecology. They discuss how bipedalism, tool use, hunting, male provisioning of young, a home base, and the use of language and other intelligent behaviors are interconnected, and how each is linked to reproductive fitness and survival in the environments encountered by early hominids. Their accounts weave together current evidence from the fossil record of hominid evolution, distributional evidence of hominid migrations, and the geological record of climatic changes. While selection pressures account for many aspects of the human 'way of life', intriguing puzzles remain. Geist, and Slurink also, describes the 'artful extensions' that humans have built around feeding, cleaning, body shelter, sexuality and other biologically essential activities. Cultural diversity stems from the different solutions of different human groups to common biological problems.

Accounts in terms of phylogenetic history, trace changes from nonhuman primates to hominids, and within hominids, from austrolopithecines to *H. sapiens*. Chiarelli discusses the dramatic increase in hominid brain size in the last 2 or 3 million years. Val Geist describes changes in foraging, from surface foraging and digging for tubers and roots in Australopithecines, to the beginning of hunting in *H. habilis* and *H. erectus*, and then to advanced hunting, in *H. sapiens*. Advanced hunters intercepted migratory ungulates by predicting yearly migrations with calendars, lured prey to a group of hunters that dispatched them with spears, and cured the excess meat into stores for later use. Francesca Giusti insists that modern accounts must be in terms of change and variation, rather than of linear progression from one type to another. She documents the various ways that tool use, meat consumption, male provisioning of young and other key features of the human way of life, are combined and recombined in different hominid species, and in different populations within the our own species.

References

Brooke, J.H. (1968). Wöhler's urea, and its vital force? – A verdict from the chemists. *Ambix*, *15*, 84-114.

Bruner, J. (1978). Foreword to A. Lock (Ed.), *Action gesture and symbol: The emergence of language*, (pp. vii-viii). New York: Academic Press.

Cohen, P.C. (1982). *A calculating people*. Chicago: Univeristy of Chicago Press.

Wöhler, F. (1828/1959). On the artificial production of urea. In W.C. Dampier & M. Dampier (Eds.), *Readings in the literature of science*, (pp. 213-216). New York: Harper. (Reprinted from *Annalen der Physik und Chemie*, 1828).

EVIDENCE OF STRUCTURE IN MACAQUE COMMUNICATION

A.C. Zeller
Department of Anthropology
University of Waterloo
Waterloo, Ontario
N2L 3G1
Canada

Abstract

Analysis of filmed facial gestures in three species of macaques reveals differential frequencies of component use in categories divided by age, sex and kingroup. This differential use pattern permits the attribution of some principles underlying human language structure to non-human primate communication codes, supporting the gradualist approach to language development

Introduction

When humans listen to other humans talking, several levels of information are transmitted, especially if the individuals know each other well. However, in spite of a considerable amount of non-verbal information transfer, most of the meaning is transmitted through the vocal-auditory channel. The vocal aspects of information include the message, the identity of the speaker, and information about the social status of the speaker. While the message is usually coded in the meaning of the words, individual recognition often occurs through paralinguistic aspects, such as tone of voice, turn of phrase, syllable emphasis or speed of articulation. All of these idiosyncratic features allow us to tell from voice alone who is on the other end of the telephone saying 'hello'. We can also use paralinguistic factors to decipher indicators of social class from spoken communication. These extend from fairly minor indicators such as vowel length or word choice, to clear markers of social background such as accent, dialect or language. Social categorization for monkeys is a little different than it is for humans. The analogy with human social class would probably be rank order among monkeys based on a familial membership and lifetime achievement. These are factors that influence ranking positions in monkeys. In addition, the factors of age, gender and matriline affiliation are very important for each animal in establishing its relationships with other members of the group. Not all interactions in a group are based on rank relationships because friendships, and individual antagonism, as well as kin aiding all affect how individuals interact. For example, females may protect their friend's infants from aggression, and most adults treat male and female infants differently.

Inappropriate responses to communicative signals by young animals are often treated tolerantly by adults until the youngsters have had the opportunity to learn the communication system and the social and individual identity of their group mates.

R. A. Gardner et al. (eds.), The Ethological Roots of Culture, 15–39.

These observations suggest that macaque communication systems are much more complex than simple indicators of emotional states.

Three levels of complexity are currently under investigation. Indications of referential function are suggested by research on coos (Green 1975), alarm calls, (Struhsaker 1967; Seyfarth et al 1980) and grunts (Seyfarth and Cheney 1984). In fact there is evidence that such referential function can also be revealled in vocalizations of some lemurs and New World Monkeys. (Cheney and Seyfarth 1990). Recognition of individuals by acoustic cues alone has been discovered in many monkeys (eg. macaques, baboons and vervets) as well as apes (Bauer and Philip 1983). In some cases individual vocal productions are accurately assessed as belonging to troop members as outsiders (Cheney and Seyfarth 1990). Another aspect of complexity in primate communication systems is the possibility that variability seen in the morphological structure of macaque facial gestures is not random, but transmits information about the social situation of the sender; in particular indicators of age, sex and kin group. The three types of facial gestures investigated in this study in three species of macaque do seem to incorporate such indicators. This ability to code social markers for fairly abstract concepts such as age and kin group reveals a meta-level of complexity in information transfer which parallels some aspects of information gained through paraliguistic channels in human communication. As Bickerton suggests (1990) the ability to categorize both objects and concepts is an important basis for language ability.

Young rhesus monkey distress screams can indicate to their mothers whether they are being attacked by kin or non-kin individuals. (Gouzoules et al. 1984). Dasser (1988) has also revealed behavioural evidence of the ability to recognize relationship concepts such as mother-offspring, sisters, and non-kin among macaques. My research is focused on ascertaining whether there are communicative indicators of age, sex and kin group categories. The facial gesture system of the three species of macaque under study does seem to reveal such indicators.

The purpose of this paper is to utilize the results of this research on facial gestures to investigate whether there is evidence that some of the complex characteristics considered to be unique to human language, are in fact present in primate communication systems. If most of these characteristics can be recognized in monkey communications, it provides a phylogentic support for an argument that the advanced attributes of human language rest on a base derived from prehuman forms and are not necessarily confined to modern humans. If monkeys can be demonstrated to utilize these communicative features then it seems reasonable to argue that apes, who are more closely related to us than monkeys would have the ability to use them also.

By this I do not mean that human language developed from current primate communication but that complex communicative features which are shared between these related taxa, may have developed in parallel, in socially complex forms. In other words, rather than arguing that humans have a totally unique verbal system based on a set of abilities which are only manifested in human spoken language, there is a possibility that humans share some aspects of the foundations of their complex language system with non-human primates.

A number of linguists, but principally (Hockett 1958, 1960; see also, Davis 1978) have organized a set of criteria which form a basis for human language. These criteria include Generalization, Interchangeability, Duality of Pattern, Cultural Transmission Specialization, Creativity, Prevarication, Arbitrariness, Displacement and Semanticity. Some of these criteria have been recognized as attributes of animal communication but most are considered only to occur in human language. If simple levels of most of these markers can be demonstrated to occur in a monkey communication system, it would provide support for the argument that the jump to human language may not be as qualitative as we have previously thought. I have acquired evidence for all of these criteria except Displacement and Semanticity. Since Semanticity practically depends on the use of words it would be impossible to find in a non-linguistic communication system. In order to make the argument about potential for ape use of a simple level of language (Proto-language, Bickerton 1990) more forceful, I have chosen to examine the visual communication systems of three species of macaques, M. sylvanus the barbary, M. fuscata the Japanese, and M. fascicularis the longtail macaque. If simple occurrences of language defining criteria can be demonstrated in monkey level communication, then ape level communication must surely be as complex. This work supports a gradualist approach for the evolution of language ability. My viewpoint regards language as a set of principles, rather than depending on performance criteria (Langacker 1968) and is in opposition to the idea that language is about representation rather than communication (Bickerton 1990).

Methodology

My research procedure involved watching and filming eleven free ranging adult and sub adult M. sylvanus from Gibraltar, 18 M. fuscata from Arashiyama West, in Texas and 12 M. fascicularis macaques from Monkey Jungle in Florida. Current work on communication recognizes the importance of studying a free ranging intact social group, rather than dyads of caged primates. (eg. Cheney and Seyfarth 1990). I also analyzed whole sequences - here called units - rather than focusing on the "critical instant" in an episode of communication. I chose threat behaviour for the major analysis because of its frequency and ease of recognition, but analysis of friendly approach and fear/submission behaviour provided a comparison with this data for M. sylvanus.

Communication is clearly a two way phenomenon involving a sender and a receiver with constant potential for each individual to modify the behaviour of the other. However, due to the detailed nature of the observations and the free ranging condition of the animals, film could only be gathered on one individual at a time, and the focus was always on the sender. Future research emphasizing the more interactive aspects of the facial gesture system may become possible with more sophisticated recording devices and in situations where the animals are more habituated to the presence of cameras and operators. Only a relatively small proportion of the film exposed could be analyzed due to the stringent quality standards required in order to see facial movements clearly.

The use of film as a recording mechanism was an extremely important aspect of this study. Most interactions were completed in a few seconds and some in less than one second. In order to see what happened, let alone describe it in detail, a filmed record was necessary. The constant frame speed also permitted an accurate control of elapsed time. Previous communication research by Hinde and Rowell (1962) among others had focused on the "norm" in an attempt to depict a standardized pattern and any variation seen was discounted. My study endevoured to record and analyse exactly how each animal made a threat. Considerable variation was present in the form, or shape, of each individual's threat gesture when it was analyzed in terms of the elements that composed it. These elements, called components, were specific movements of certain parts of the face and head such as eyebrows raised or ears forward. The differences between the manifestations of a threat were evident both between different animals and within the repeated threat behaviour of one individual. I chose to study component use in terms of the variability that occurred between the different animals.

The actual analysis concerned the movements of 34 aspects of the face and head; movement of the eyes, eyebrows, ears, lips, nostrils, and others. Of these precisely defined movements, sixteen had not previously been described for the genus Macaca. Lists of components used were made for each unit of threat performed by each animal. I recorded which components were used (eg. nostrils flared, eyebrows raised) and in how many frames of film they occurred. (For detailed methods see Zeller 1978, 1980, 1985, 1986) (Tables 1, 2 & 3). I compared the component use patterns in the 3 species, and within the age, sex and kin group classes in 3 ways. The first focus of research assessed the consistency of component use. This involved ascertaining which components were used in all, or all but one of the units by all animals in each classification. In this way the basic pattern of components used by each category could be identified and compared. The consistently used components were called Constant components, and those used more intermittently were called Variable components (Table 4). In the second stage of analysis I recorded for each individual the number of units in which each component occurred. This provided the frequency of occurrence for that component and was called Unit Use. I also recorded for each unit the number of frames in which a particular component occurred. This provided the rate at which each component was used and was called Percentage of Overall Use. Thus I had two measures with which to compare individuals and classes - frequency and rate. The animals were grouped into categories by AGE, SEX, and (for M. sylvanus) KIN GROUP. and the various groups were compared, eg. Males or Females, or Kingroup O to Kingroup W.

Thirdly, I assessed the degree of similarity and difference within and between categories by comparison to the Group Mean. In this procedure, the overall occurrence of a component for each individual in a class, such as female M. sylvanus was compared with the mean level of occurrence for the group. If the rate was 5% above or below the Group Mean it was classed as prevalent or infrequent respectively. This allowed me to compare use patterns of males to females within a species. Direct comparisons between species were not undertaken at this stage of analysis. In order to confirm the reliability of scoring the use or disuse of the various components in the frames of film, 10 units for M. sylvanus and 10 units for M. fuscata were independantly assessed by another observer who I had trained. The concordance in judgement was good with 93% of the components being scored identically. Most of the frames of film (each with 34 components) which differed did so in only one component, so that only 5.2% of the frames showed scoring differences. Because of this high level of concordance I did not use a second observer when analysing the M. fascicularis film.

TABLE 1. List of Threat components in descending order of use for M. sylvanus

Components	Frequency of 67	Frequency of Units(%)	Overall Frequency(%)
1. Eyes staring	65	97.01	51.9
2. Eyebrows raised	64	95.52	46.4
3. Piloerection	62	92.53	80.0
4. Eyebrows forward	61	91.04	37.4
5. Ears back	58	86.56	52.9
6. Mouth open threat	56	83.58	20.2
7.* Mouth stretch	55	82.09	14.4
8. Nostrils flared	52	77.61	22.9
9. Head raised	52	77.61	27.8
10. Eyelid flicker	50	74.63	20.5
11. Head forward	50	74.63	40.0
12.* Lower jaw protrude	44	65.67	10.8
13. Ears forward	43	64.18	24.9
14. Mouth closed	40	59.70	6.9
15.* Mouth forward	30	44.78	10.7
16.* Mouth protrude	26	38.80	7.9
17.* Upper jaw protrude	25	37.31	9.5
18. Mouth open grimace	17	25.37	2.1
19.* Skin around eyes back	11	16.41	1.5
20. Incisors show	8	11.94	1.5
21.* Upper lip puffy	7	10.45	0.7
22.* Mouth puffy	6	8.96	0.5
23. Tongue protrude	6	8.96	0.5
24. Lower canine	5	7.46	1.0
25. Tongue show	5	7.46	0.7
26.* Rear crouch	3	4.48	0.5
27.* Mouth open wider right	3	4.48	0.2
28.* Mandrill mouth	2	2.99	0.1
29.* Mouth open wider left	2	2.99	0.1
30. Upper canine	2	2.99	0.3
31.* Lower jaw displaced right	1	1.49	0.1
32.* Body crouch	1	1.49	0.6
33.* Upper lip back	0	0	0
34.* Lower jaw displaced left	0	0	0

* The components indicated by asterisks are the 16 components discovered by detailed observation of the data.

Breaks in the column emphasize major differences in the frequency of use. Note differences between species.

TABLE 2. List of Threat components in descending order of use for <u>M. fuscata</u>

Components	Frequency of 69	Frequency of Units(%)	Overall Frequency(%)
1. Eyes staring	69	100	81.3
2. Piloerection	67	97.1	81.3
3. Mouth closed	67	97.1	54.1
4. Nostrils flared	62	89.8	55.0
5.* Mouth stretch	61	88.4	82.1
6. Mouth open threat	57	82.6	42.1
7. Eyelid flicker	54	78.2	10.2
8. Mouth forward	54	78.2	28.9
9. Eyebrows raised	51	73.9	50.1
10. Eyebrows forward	51	73.9	33.0
11. Ears forward	50	72.4	39.7
12. Ears back	48	69.5	36.2
13. Head forward	48	69.5	48.4
14.* Lower jaw protrude	33	47.8	23.5
15.* Upper jaw protrude	30	43.4	10.1
16.* Mouth protrude	29	40.2	10.7
17.* Lower jaw displaced right	21	30.4	4.5
18. Head raised	17	24.6	5.6
19.* Mouth puffy	17	24.6	8.6
20.* Skin around eyes back	17	24.6	6.1
21.* Upper lip puffy	15	21.7	2.3
22.* Mouth open wider left	14	20.2	1.5
23.* Lower jaw displaced left	12	17.3	1.5
24. Incisors show	11	15.9	4.7
25.* Mouth open wider right	9	13.0	1.9
26. Mouth open grimace	7	10.1	1.2
27. Lower canine	5	7.2	.5
28.* Body crouch	5	7.2	1.3
29. Upper canine	4	5.7	.5
30.* Rear crouch	3	4.3	1.5
31.* Upper lip back	3	4.3	.2
32. Tongue show	2	2.8	.2
33. Tongue protrude	2	2.8	.04
34.* Mandrill mouth	0	0	0

* The components indicated by asterisks are the 16 components discovered by detailed observation of the data.

Breaks in the column emphasize major differences in the frequency of use. Note differences between species.

TABLE 3. List of Threat components in descending order of use for M. fascicularis

Components	Frequency of 84	Frequency of Units(%)	Overall Frequency(%)
1. Mouth open threat	80	93.0	51.2
2.* Mouth stretch	78	90.6	76.3
3. Mouth closed	78	90.6	37.5
4. Eyebrows raised	76	88.3	68.0
5. Nostrils flared	74	86.0	71.5
6. Piloerection	73	84.8	82.6
7. Eyelid flicker	72	83.7	16.1
8. Eyes staring	71	82.5	75.2
9. Ears back	67	77.9	64
10. Head raised	58	67.4	36.1
11. Head forward	56	65.1	42.9
12. Eyebrows forward	44	51.1	29.4
13. Ears forward	40	46.5	30.1
14.* Upper jaw protrude	40	46.5	25.1
15. Mouth open grimace	33	38.3	7.0
16.* Skin around eyes back	26	30.2	17.0
17.* Mouth protrude	25	29.0	14.2
18. Tongue show	23	26.7	2.2
19. Incisors show	18	20.9	4.1
20. Tongue protrude	16	18.9	1.7
21.* Upper lip puffy	13	15.1	3.2
22. Mouth forward	11	12.7	7.2
23. Upper canine	9	10.4	1.9
24. Lower canine	9	10.4	1.6
25.* Mouth open wider left	6	6.9	1.5
26.* Lower jaw protrude	6	6.9	3.1
27.* Mouth puffy	4	4.6	2.7
28.* Rear crouch	3	3.4	1.5
29.* Lower jaw displaced right	3	3.4	.45
30.* Mouth open wider right	2	2.3	.18
31.* Body crouch	2	2.3	.96
32.* Mandrill mouth	0	0	0
33.* Upper lip back	0	0	0
34.* Lower jaw displaced left	0	0	0

* The components indicated by asterisks are the 16 components discovered by detailed observation of the data.

Breaks in the column emphasize major differences in the frequency of use. Note differences between species.

TABLE 4. Constant components used by each species

Component	sylvanus	fuscata	fascicularis
Eyes staring	+	+	
Eyebrows raised	+		
Eyebrows forward	+		
Piloerection	+	+	
Mouth closed		+	
Mouth open threat			+
Mouth stretch			+

Results

Basically, my findings for this study include eight major points.

1. In all 3 species of macaques there is a marked overlap of components used in threat, but the pattern is not the same.

2. In threat, friendly and fearful interactions in Macaca sylvanus, there is almost complete overlap of the components used although the proportions differ. (Table 5).

3. The third discovery concerned the non-occurrence of individual idiosyncrasy in the use of components. Unexpectedly, I found that there was almost no individual use of components. This differed from a subjective impression that some animals had a monopoly on particular components and used them consistently and exclusively. In terms of the model this indicates that individual distortion was not a significant factor in producing variation in the signal.

4. Class based differentiation of the signal was the fourth feature to be examined. Discrimination within each of the three classes of AGE, SEX, and KIN GROUP,was carried out by overall use (rate) (Table 6). The most notable feature of this analysis was the clarity of distinction within the KIN GROUP class in M. sylvanus threat (Table 7). The non-kin individual present was also clearly indicated.

5. Differences in component use occur between males and females in all three species, although the pattern of difference is not the same. Thus there is not a 'male' or 'female' pattern of threat in macaques. (Table 8).

6. Differences in component use occur between the three age classes; old adult, young adult and sub adult in all three species. However only the sub adult category shows marked similarity in pattern with three components used at prevalent rates and five used at infrequent rates in the two species in which sub adults were studied. (Table 9).

TABLE 5. List of components used in Threat, Friendly approach and Fear response arranged in descending order of use in Threat by M. sylvanus

Components	% use in Threat	% use in Friendly	% use in Fear
1. Eyes staring	97.01	82.60	82.35
2. Eyebrows raised	95.52	82.60	88.23
3. Piloerection	92.53	91.30	88.23
4. Eyebrows forward	91.04	78.26	88.23
5. Ears back	86.56	78.26	82.35
6. Mouth open threat	83.58	82.60	70.58
7. Mouth stretch	82.09	65.21	70.58
8. Nostrils flared	77.61	73.91	58.82
9. Head raised	77.61	78.26	88.23
10. Eyelid flicker	74.63	60.86	76.47
11. Head forward	74.63	73.91	76.47
12. Lower jaw protrude	65.67	56.52	52.94
13. Ears forward	64.18	60.86	88.23
14. Mouth closed	59.70	47.82	41.17
15. Mouth forward	44.78	21.73	23.52
16. Mouth protrude	38.80	13.04	5.88
17. Upper jaw protrude	37.31	9.09	23.52
18. Mouth open grimace	25.37	39.13	82.33
19. Skin around eyes back	16.41	13.04	11.76
20. Incisors show	11.94	26.08	47.05
21. Upper lip puffy	10.45	4.34	0
22. Mouth puffy	8.96	4.34	11.76
23. Tongue protrude	8.96	26.08	5.88
24. Lower canine	7.46	17.39	17.64
25. Tongue show	7.46	26.08	29.41
26. Rear crouch	4.48	0	17.64
27. Mouth open wider right	4.48	4.34	5.88
28. Mandrill mouth	2.99	0	5.88
29. Mouth open wider left	2.99	4.34	0
30. Upper canine	2.99	8.69	23.52
31. Lower jaw displaced right	1.49	4.34	5.88
32. Body crouch	1.49	4.34	11.76
33. Upper lip back	0	8.69	5.88
34. Lower jaw displaced left	0	0	0

Components 1-4 were classified as Constant Components over all animals for threat. Components 2-3 were classified as Constant Components over all animals for friendly interactions.

There is a distinct break between the frequencies of the first 14 components and the last 19 in both threat and friendly behaviour.

TABLE 6. Overall use anaylsis of threat behaviour in M. sylvanus

Components	AGE category			SEX category	
	OA	YA	SA	M	F
Nostrils Flared	+				
Eyes Staring	+				
Head Raised	+				
Lower Jaw Protrude	+				
Ears Back			+	+	
Mouth Closed			+	+	
Rear Crouch			+		
Eyebrows Forward	+				
Mouth Open Gimace			+		

OA = Old Adult, **YA** = Young Adult, **SA** = Subadult

TABLE 7. Separation by kin group on basis of overall use threat behaviour in M. sylvanus

Component	Kin Group				Individual
	W	B	M	O	Af
Head Raised	+				+
Lower Jaw Protrude	+				
Nostrils Flared	+				
Eyelid Flicker	+				
Ears Forward		+			
Upper Jaw Protrude		+			
Head Forward		+		+	
Mouth Stretch		+			
Ears Back			+		
Mouth Protrude			+		
Mouth Forward			+		
Mouth Closed				+	
Eyes Staring					+
Mouth Open Grimace					+
Incisors Show					+
Rear Crouch					+
Mouth Open Threat					+

TABLE 8. Comparison of SEX Classes by Prevalent and Infrequent component use. (Prevalent=5% higher than group mean; Infrequent=5% lower than group mean)

	M. sylvanus		M. fuscata		M. fascicularis	
	Male	Female	Male	Female	Male	Female
Mouth open threat			+	-		+
Mouth open grimace						-
Mouth closed	+		-			+
Nostrils flare	-					
Eyelid flicker	-					-
Eyebrows raised	-			-		+
Eyebrows forward			+	-		-
Eyes staring	-					
Ears back	+	-				
Ears forward			+	-		+
Mouth forward			+			+
Mouth puffy			-			-
Mouth stretch			+	-		
Skin around eyes back			-			+

TABLE 9. Comparison of AGE Classes by Prevalent and Infrequent component use.(Prevalent=5% higher than group mean; Infrequent=5% lower than group mean)

	M. sylvanus		*M. fuscata*		*M. fascicularis*		
	Old	Subadult	Old	Young	Old	Young	Subadult
Mouth open threat	+			-	+	+	-
Mouth closed	+	+		+	+	-	+
Nostrils flare		-		-			-
Eyelid flicker		-			+		
Eyebrows raised		-		+	-	-	-
Eyebrows forward		-		-	+	+	+
Eyes staring		-			+		-
Ears back	-	+		+	-	-	+
Ears forward					-	+	
Head forward	+			+			+
Head raised	+						-
Mouth stretch				-			-
Lower jaw protrude				-			+
Upper jaw protrude		-		+	+		-
Mouth forward		-					-
Mouth puffy		-				-	+
Rear crouch		+					+
Skin around eyes back							-

7. The seventh result occurred when the kin groups in M. sylvanus were assessed in terms of prevalent component use for Threat, Friendly and Fearful interactions. (Table 10). In all 3 cases the kin groups were clearly distinguished, as well as the non-kin individuals. This emphasizes the importance of kin group as a preeminent social focus which is clearly marked in the communication code. This discovery suggests that kin affiliation is a major factor in social organization. The importance of matrilines in influencing behaviour such as intervention, aid seeking, aunt behaviour, and the organization of respect hierarchies has long been recognized but has always been considered to rest on individual recognition. While I am sure individual recognition is a very important aspect of organizing kin based behaviour, the discovery that there are recognition signals built into the communication system would help explain how outside animals who come into the group could learn who belonged to which kin line. It also indicates that component use in threat gestures (and probably in most other communicative patterns) is largely learned by the young animal from its kin.

8. The eighth and final result was the surprising discovery that there appears to be differential attribution of meaning to mouth and non-mouth components. This was derived from the discovery that those components that occurred at 90% or higher levels of frequency were mainly non-mouth components, whereas those that occurred less often (variable) were very frequently mouth components. This held true for threat, friendly and fearful interactions in M. sylvanus ($P > .01$). (Table 11). When comparisons were made between species for categories of Male, Female, Old Adult, and Young Adult, the breakdown of those components which occurred at similar frequencies (Frequency Index $< .1$, Hohmann 1991) and those used at different frequencies (Frequency Index $> .5$) also indicated that non-mouth components were attributes of similarity, and mouth components correlated with differences between species at a significance level of $P < .05$. (Table 12, 13, 14, 15). This division between mouth and non-mouth components means that the choice of components to be included in a gesture was not random. The most consistently used components involved non-mouth areas, while the components that added the major part of the variability to the system came from one small but active region of the face ie. the mouth. Differentiation of this sort indicates that these two divisions of the face add different types of information to the gesture. This indicates a more complex level of information coding than has heretofore been established for primate communication. Both message and class indicators can be sent simultaneously using different parts of the face. It also means that random irregularity or noise did not contribute substantially to the variability.

TABLE 10. Separation by Kin Group on the basis of overall use prevalent components - M. sylvanus

Threat	Kin Group				(Non kin)
	W	B	M	O	A[1]
Nostils flare	+				
Head raise	+				
Lower jaw protrude	+				
Eyelid flicker	+				
Ears forward		+			
Upper jaw protrude		+			
Mouth stretch		+			
Mouth protrude			+		
Mouth forward			+		
Mouth closed				+	
Eyes stare					+
Mouth open grimace					+
Incisor show					+
Rear crouch					+
Mouth open threat					+

Friendly	Kin Group				
	W	B	O	M[1]	A[1]
Mouth stretch	+				
Eyes stare	+				
Insicors show	+				
Upper canine	+				
Nostrils flare	+				
Ears forward	+				
Eyebrows raise		+			
Upper jaw protrude		+			
Lower jaw protrude			+		
Low canine				+	
Eyelid ficker				+	
Ears back					+
Head forward					+
Mouth closed					+

Fear	Kin Group		
	W	B	O
Nostrils flare	+		
Eyebrows forward	+		
Head forward	+		
Head raised	+		
Piloerection	+		
Lower jaw protrude	+		
Ears back		+	
Incisors show			+
Upper canine			+
Lower canine			+
Tongue show			+
Eyelid flicker			+
Upper jaw protrude			+

TABLE 11. Anaylsis of Constant and Variable component use by attribution as Mouth and Non-mouth location, in *Macaca sylvanus*

Threat	Constant Component	Variable Component
Mouth	6	98
Non-mouth	39	58
df=1	X^2=34.17	P<.001

Friendly	Constant Component	Variable Component
Mouth	17	51
Non-mouth	45	45
df=1	X^2=9.55	P<.001

Fear	Constant Component	Variable Component
Mouth	35	26
Non-mouth	13	30
df=1	X^2=7.48	P<.01

TABLE 12. Mouth vs Non-mouth Components used at similar or different frequencies by three species of Male macaques

	Similar		Different
Mouth	4		25
Non-mouth	10		7
df=1		$X^2 = 10.39$	
		$P < .05$	

TABLE 13. Mouth vs Non-mouth Components used at similar or different frequencies by three species of Female macaques

	Similar		Different
Mouth	2		5
Non-mouth	8		8
df=1		$X^2 = 5.99$	
		$P < .05$	

TABLE 14. Mouth vs Non-mouth Components used at similar or different frequencies by three species of Old adult macaques

	Similar		Different
Mouth	7		23
Non-mouth	14		3
df=1		$X^2 = 15.59$	
		$P < .05$	

TABLE 15. Mouth vs Non-mouth Components used at similar or different frequencies by three species of Young adult macaques

	Similar		Different
Mouth	4		20
Non-mouth	12		6
df=1		$X^2 = 10.83$	
		$P < .05$	

Discussion

These results provide some information which helps to assess the potential of non-human primates to utilize complex concepts which have previously been attributed only to human language. Some of these are attributes which are being examined in chimpanzee use of human invented languages. I feel that even a rudimentary demonstration of these criteria in an inter-monkey communication system supports claims that apes have the background to understand and use such concepts in their two way communication. Eight major criteria which have been attributed as key properties of language are demonstrated in this analysis of macaque communication patterns.

1. Generalization requires the individual to take salient common properties in a system and utilize them to recognize or demarcate a particular category. The differential use of components by KIN GROUP and the differentiation by AGE and SEX would indicate that the ability to categorize exists for primates.

2. Interchangeability is the aspect of communication which allows senders and receivers to exchange roles. This is an important aspect of a language system even though exact messages may be culturally constrained (eg. servant speaking to master). Within the same species of macaque the content aspects of the messages can be used by both and it is the social rather than physical constraints which govern which components are used by each individual as class markers.

3. Duality of pattern usually suggests that particular message elements can be recombined to produce different types of messages. Since the list of components which are used to indicate threat (Table 5) are also almost coincident with the components used in Friendly approach and Fear responses, the components themselves would appear to have little meaning, and the frequency, intensitiy, and combination would appear to differentiate the messages. In other words the same lists of components are combined and recombined to make different messages. Also, the same components seem to be used as markers for different kin groups, but mainly in different message situations. For example, kin group O uses high levels of lower jaw protrude as a marker in Friendly interactions, while kin group W, used the same component at high levels in Fear responses. Thus the components themselves do not seem to have meanings attached to them, but the important indicator is derived from their pattern of use.

4. Cultural transmission is indicated by the ability to learn the whole range of components used by the group and especially by learning which components are utilized more frequently in an animal's own kin group. Infants seem to learn mainly from their mothers. This is analogous to building up bird dialects except that among primates the different patterns of component use are completely comprehensible within the group.

5. Specialization refers to the degree with which a message must be interpreted by the receiver, as opposed to the message itself producing physical results (such as a punch in the face to indicate anger). Some primate communicative episodes are not specialized, but many of them stand for a non-physical referent. The physical effect of 'ears back' or 'eyelids flicker' on a receiver is non-existent, but the combination of components provides a particular message with both content and social referents.

6. Creativity is demonstrated by primates' ability to select and combine components to express not only the message aspect of the gesture, but also indicate membership in AGE and SEX classes as well as in KIN GROUP. Of the thirty-four possible components some are chosen and used within a circumscribed range.

The frequency of components used by any individual in the group runs from 7 to 22, with modal numbers between 13 and 18 for the different individuals. I have termed this attribute creativity rather than productivity, because it is unlikely that primates can construct large numbers of new messages from known components. They can learn the pattern of new members coming into the group but it seems unlikely that they would utilize the new pattern themselves.

7. The attribute of prevarication is based on observational data. Prevarication generally implies an attempt to deceive. This has been claimed for symbol using apes (eg. Savage-Rumbaugh and McDonald 1988). One particular example in the Gibraltar macaques involved a typical behaviour for this group. Gibraltar macaque males carry and care for young black infants. This pattern is organized in such a way that the adult male tends the infant when it is very young. When more infants are born the sub-adult males share the baby-tending behaviour. However, the sub-adults acquire the babies from the adult male, and return them to him, since the females are very reluctant to give their infants up to a sub-adult. The first time a sub-adult manages to carry and hold an infant he usually must overcome some resistance from the mother. In this particular case one sub-adult male was approaching the adult male who had the infant. Behind the male sat the mother who was watching the sub-adult. From the mother's perspective the adult male appeared to be threatening the sub-adult with some intensity, using considerable amounts of body and head movement. From the front view, however, the male's mouth was very slack and his eyes were not staring intently. The sub-adult did not run away as he normally would but merely leaned his body away. To the mother the sub-adult appeared to be receiving a severe threat and she did not get up and chase him herself. This sequence ended with the sub-adult getting close enough to snatch the infant and run, completing his first successful attempt to hold a baby. The mother then chased him and the male followed and intervened between them. This sequence demonstrates an attempt by the adult male to deceive the mother by appearing to direct a more severe threat towards the sub-adult than he was actually performing.

8. The concept of arbitrariness involves the use of non-representational aspect to refer to phenomena. Rumbaugh (1977) argues that this must be separable in space and time from the event being referred to. This separation is beyond our powers to ascertain, from an observational stand point, but the use of a nonrepresentational signal can be inferred from close observations of actions and responses. The idea of an intention movement is common in ethological literature, in which a small part of the movement stands for the whole. (Hinde 1974; 71). An escalation of threat sequence begins with a mild look, graduating to eyes stare, eyebrows raise and lower, piloerection, and mouth stretched while the nostrils flare. Then the mouth opens, the body leans forward, the hand slaps the ground, and the head lunges. If the threat continues, the body lunges and then the animal will chase and occasionally grab and bite, which is full scale aggression. This sequence approaches in some ways the idea of a biological cline - where the two ends work in different ways but are completely connected by a finely graded series. Eyes stare, eyebrows raise and lower, and piloerection are not iconic for chase and bite but they are certainly responded to as if the end result were a real possibility. The receiver's response, or lack of responses, is different in different categories of individuals. Also, the complexities of social interaction, such as baboons attacking the sisters or mothers of the monkeys they have recently been in conflict with (Strum 1990), are undoubtly furthered by receiving clear reminders of who is related to whom.

The major advances that signing chimpanzees make over this level of communication is the ability to demonstrate displacemnt and semanticity. A gradualist viewpoint on the study of communication suggests that the specialized features of language have their root in the communicative patterns of our closest living relatives. Chimpanzees and other apes who use human introduced systems demonstrate the mental ability to categorize, refer to the past, play with words and other complexities of communication which have not yet been found for monkeys (this volume). However the increasing evidence for the complexity of macaque communication systems suggests that the mental abilities underlying some of the criteria used to define human language exist in the brains of our primate relatives. Study of the biological roots of language is an important aspect of studying the biological roots of culture.

Conclusion

The conclusions of this study indicate that communication patterns, at least in these three macaque species, exist at a more complex level than mere transmitters of a single message. This level of complexity suggests that what may appear as redundancy in primates, the simultaneous use of vocal and facial gestures, may indeed represent a layering of information. I am suggesting that perhaps monkeys use vocal channels for individual signature markers and some message aspects, while they use facial signals for additional messages and category markers such as SEX, AGE, KIN GROUP and perhaps species. (Kingdon 1980). Humans have developed verbal communication to such a level of finesse that they can load one channel with a wide variety of levels of meaning. But as early primate groups developed in complexity the value of social as well as external information in a message would have increased. It seems quite possible that different elements of the same message may have been divided between different channels, giving the appearance of redundancy when in fact a meta-message was being transmitted. Efficient operation of such a complex code system would require a substantial level of learning by young animals who could use their knowledge to categorize their social companions as well as to communicate successfully. Indeed, close study of the variability expressed between animals and within and between classes of animals, such as the various kin groups, gives us access to social information about how the group sees itself divided. The discovery that the different regions of the face, divided as mouth and non-mouth, reflect attribution of consistency and variability, indicates a whole meta-level of information in gestural communication. For what other parts of the body can similar dichotomies be found?

From these discoveries we can argue that there is a glimmering of some criteria considered necessary for the presence and use of human language. If it is accepted that monkeys have these types of capabilities in their communication systems, it will be much easier to admit that apes may have the necessary mental abilities to use a simple form of human language.

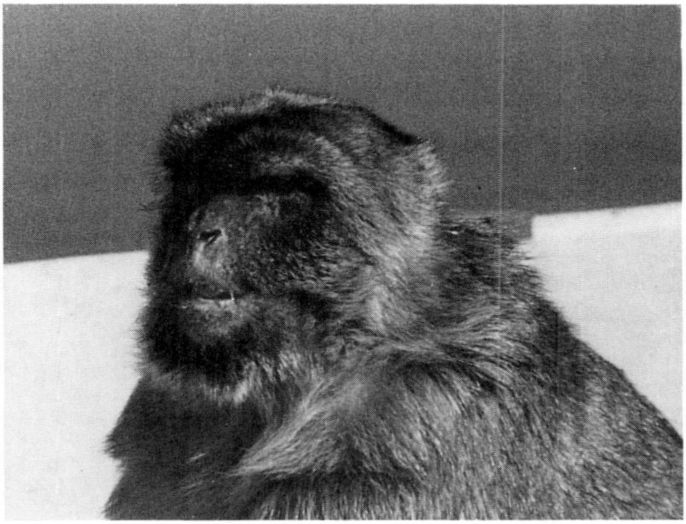

Figure 1 & 2. Kin Group B - Threat. Young Adult Male in Fig. 2 is the son of Old Adult Female in Fig. 1.

Figure 3 & 4. Kin Group W - Threat. Young Adult Female in Fig. 4 is daughter of Old Adult Female in Fig. 3. Note similarities within Kin Group, particularly Nostrils flare, Eyebrows lowered, Piloerection, and Mouth stretch.

Figure 5. Subadult Male - Grimace. Note components: Eyebrows raised, Ears back, Mouth open grimace, Skin around eyes back, Incisors show, Nostrils flare, Upper jaw protrude.

Figure 6, 7, 8 & 9. Comparison of Threats in Three Species of Macaques. Fig. 6, Subadult, M. fuscata; Fig. 7, Adult Female, M. fuscata; Fig. 8, Adult Male, M. fascicularis; Fig. 9, Adult Male, M. sylvanus, Note particularly the difference between the adult male M. fascicularis and M. sylvanus.

References Cited

Bauer, H.R. and Philip, M. (1983) Facial and vocal individual recognition in the common chimpanzess. Psychological Record 33: 161-170.

Bickerton, D. (1990) Lanugage and Species. Chicago. University of Chicago Press.

Cheney, D.L. and Seyfarth, R.M. (1982) Recognition of individuals within and between groups of free-ranging vervet monkeys. American Zoologist 22:519-529.

Cheney, D.L. and Seyfarth, R.M. (1990) How Monkeys See the World. Chicago. Chicago University Press.

Davis, F. (1978) Eloquent Animals. Coward, McCann and Geghegan Inc. New York.

Dasser, V. (1988) Mapping social concepts in Monkeys. In: Machiavellian Intelligence: Social expertise and the evolution of intellect in Monkeys, Apes and Humans. Ed. R.W. Byrne and A. Whiten. Oxford. Oxford University Press. Pp. 85-93.

Green, S. (1975) Variation of vocal patterns with social situation in the Japanese Monkey (Macaca fuscata): A Field study. Primate Behavior, 4. Academic Press, New York. Pp. 1-102.

Gouzoules, S., Gouzloules, H. and Marler, P. (1984) Rhesus monkey (Macaca mulatta) screams: Representational signalling in the recruitment of agonistic aid. Animal Behavior 32:182-193.

Hinde, R.A. (1974) Biological Basis of Human Social Behavior. McGraw-Hill, New York.

Hinde, R.A. and Powell, T.E. (1962) Communication by postures and facial expressions in the rhesus monkey (Macaca mulatta). Proceeding Zool. Society of London 138:1-21.

Hockett, C.F. (1958) A Course in Modern Linguistics. McMillan Co. New York.

Hockett, C.F. (1960) Logical considerations in the study of animal communication. In: Animal Sounds and Communication. Eds. W.E. Langon and W.N. Tavolga. American Institute of Biological Sciences, Washington. Pp. 392-430.

Hohmann, G. (1991) Comparative Analysis of Age - and Sex - Specific Patterns of vocal Behaviour in Four Species of Old World Monkeys. Folia Primatologica 56:133-156.

Kingdon, J.S. (1980) The role of visual signals and face patterns in African forest monkeys (quenons) of the genus Cercopithecus. Trans. Zool. Society of London 35:425-475.

Langacker, R.W. (1968) Language and Its Structure. San Francisco. Harcourt Brace and World Inc.

Rumbaugh, D. (ed.) (1977) Logical considerations in the study of animal communication. In: Animal Sounds and Communication. Eds. W.E. Langon and W.N. Tavolga. American Institue of Biological Sciences, Washington. Pp. 392-430.

Savage-Rumbaugh, S, and McDonald, K. (1988) Deception and social manipulation in symbol-using apes. In: Machiavellian Intellegence: Social expertise and the evolution of intellect in Monkeys, Apes and Humans. Ed. R.W. Byrne and A. Whiten. Oxford. Oxford University Press. Pp. 224-237.

Seyfarth, R.M. and Cheney, D.L. (1984) The acoustic features of vervet monkey grunts. J. Acoust. Soc. Am. 75:1623-28.

Seyfarth, R.M., Cheney, D.L. and Marler, P. (1980) Monkey responses to three different alarm calls: Evidence for predator classification and Semantic Communication. Science 210:801-3.

Struhsaker, T.T. (1967) Auditory communication among vervet monkeys (Cercopithecus aethiops). In: Social Communication among primates. Ed. S.A. Altmann. Chicago. University of Chicago Press. Pp. 281-324.

Strum, S. (1990) Almost Human. Norton. New York.

Zeller, A.C. (1978) Film analysis as a field technique: a study in communication in Macaca sylvanus of Gibraltar. Unpublished Ph.D thesis. University of Toronto.

Zeller, A.C. (1980) Analysis of primate facial gestures as a contribution to Linguists. Paper in Linguistics: International Journal of Human Communication 13(4):565-606.

Zeller, A.C. (1985) Component Patterns in Gesture Formation in Macaca sylvanus of Gibraltar. Canadian Review of Physical Anthropology 4(2):35-42.

Zeller, A.C. (1986) Comparison of Component Patterns in Threatening and Friendly Gestures in Macaca sylvanus of Gilbraltar. Current Perspectives in Primate Social Dynamics. Eds. D.M. Taieb and F.A. King. New York. van Nostrand Reinhold Press. Pp. 487-504

THE CENTRAL-PERIPHERAL STRUCTURE OF THE TANAXPILLO COLONY OF STUMPTAIL MACAQUES (*Macaca arctoides*)

D. R. RASMUSSEN
Animal Behavior Research Institute, Madison, WI 53715
E. RIORDAN
University of Washington, Seattle, Washington 98195
M. FARRINGTON
California State University, Long Beach, CA 90840
E. KELLY
University of Pennsylvania, Philadelphia, PA 19104
J. NACHMAN
University of Wisconsin, Stevens Point, WI 54481
S. FERNANDEZ
Columbia University, New York, NY 10027
A. CHURCHILL
University of Pennsylvania, Philadelphia, PA 19104

ABSTRACT. We developed a peripherality index, pindex, as a measure of the placement of nonhuman primates within the central-peripheral structure of their group. We used pindex to measure the central-peripheral structure of the troop of stumptail macaques, *Macaca arctoides*, provisioned and maintained on Tanaxpillo Island in Lake Catemaco, Mexico. We based our analyses on 446 hours of focal samples on individuals. We found infants to be the most central troop members, followed by juveniles and then adults. Matrilineal subgroups had similar values of pindex. Central individuals spent more time eating the clumped and provisioned food. Individuals who were peripheral were more likely to die the following year. Spatial position is associated with many aspects of the macaques' society and ecology. Position within the spatial structure of the troop may be transmitted between generations within matrilines. Our results show study of the central-peripheral structure of nonhuman primate groups may add to our knowledge of the biological roots of human culture.

1. Introduction

If we are to gain a comparative understanding of the biological roots of culture, we must describe nonhuman animal societies and determine how these societies are influenced by environmental and social factors. Scientific evaluation of the biological roots of culture also requires replicable and quantitative measures. We focus here on the measure, evaluation and correlates of the central-peripheral structure of nonhuman primate social organization.

A central-peripheral troop spatial structure has been described in provisioned troops of Japanese macaques (*Macaca fuscata*) and other species of nonhuman primates for over 20 years (Table 1). Individuals of higher agonistic rank, or the young and vulnerable troop members, are often reported to be near their group's center. Lower ranking individuals, or the adult males most capable of defending the group, are reported to be more peripheral.

R. A. Gardner et al. (eds.), The Ethological Roots of Culture, 41–60.

Table 1. Some reports on spatial patterns of nonhuman primate groups.

Species	Structure*	Evidence	Reference
C. nigrivittatus	1	quantitative	Robinson, 1981
M. arctoides	1	qualitative	Lopez-Lujan et al., (1989)
M. fuscata	1	qualitative	Itani, 1954, Sugiyama & Ohsawa, 1982, Casey & Clark, 1976
M. fuscata	2	quantitative	Wada & Matsuzawa, 1986
M. fuscata	5	quantitative	Rasmussen & Rasmussen, 1979
M. mulatta	1	qualitative	Southwick et al., 1965
M. sinica	5	qualitative	Dittus, 1979
P. cynocephalus	1	qualitative	Washburn & DeVore, 1961
P. cynocephalus	3	quantitative	Collins, 1984
P. ursinus	4	qualitative	Stoltz & Saayman, 1970
P. ursinus	4	quantitative	Anderson 1981, Byrne et al., 1989
T. gelada	5	qualitative	Crook, 1966

M. = Macaca P. = Papio T.= Theropithecus C. = Cebus

* **Structure:** Spatial pattern reported

Types of spatial structure:
1. High ranking adult males and females and offspring of high ranking females in center with individuals of decreasing agonistic rank more peripheral. Most peripheral troop members are low ranking adult and subadult males.
2. Adult females and infants in center. Males dispersed randomly.
3. Adult females and their infants in clusters toward the middle of group. Males more at edges than females and more often in front edge during troop movement. Agonistic rank of mothers positively and significantly associated with time in clusters. Tendency for estrous females and consorts to be on periphery.
4. Adult females and their offspring cluster around adult males.
5. Resource related differences in patterns of use of space strongly associated with environmental resources.

Members of a matriline may have similar positions within the central-peripheral structure (Sugiyama & Ohsawa, 1982). Position within the central-peripheral structure has also been linked to access to resources, such as provisioned food, and survival (Sugiyama & Ohsawa, 1982). The central-peripheral structure can be strongly influenced by resource distribution. For example, shade and logs used while resting along the edges of a corral have been shown partially to turn the central-peripheral structure of a troop of Japanese macaques inside out: higher ranking individuals were frequently found using these resources on the corral's periphery (Rasmussen, 1973; Rasmussen & Rasmussen, 1979).

Published information on the central-peripheral structure of nonhuman primates indicates: (1) The spatial structure may exist within provisioned and nonprovisioned groups. (2) Related individuals may share similar positions. (3) Position in the central-peripheral structure may be transmitted between generations. (4) Position within the central-peripheral structure may influence reproductive success. And (5) the structure can be strongly influenced by ecological factors. We feel these characteristics of the central-peripheral structure warrant its further examination as an aspect of the social organization of nonhuman primates that may shed light on the biological bases of human cultural differences.

The central-peripheral structure has neither gained the recognition nor attention of other aspects of social organization such as agonistic rank. A reason why descriptions of the central-peripheral spatial structure are not an integral part of most studies may be the absence of broadly accepted quantitative methods for evaluation of the structure. As with agonistic rank (Bernstein, 1981), some scientists also remain unconvinced of its importance. Others feel spatial structure may be partially due to the use of provisioned food (Gartlan, 1968). We agree that providing food in piles may be a factor giving rise to central-peripheral structure (Rasmussen & Rasmussen, 1979). Yet the resources we have created and their distribution influence human cultural differences. There are also strong theoretical reasons to believe spatial structure will exist when food is not provisioned (Hamilton, 1971; Busse, 1984).

Here we describe the central-peripheral spatial structure found within a provisioned troop of stumptail macaques, *M. arctoides*. We first describe a peripherality index, pindex, we developed to assess the position of every troop member within the central-peripheral structure. We then describe the results of analyses conducted to gauge the accuracy, consistency, and utility of pindex. We then examine the relationship between pindex and several factors that other studies suggest may be linked with position in the central-peripheral structure. We describe the relationships between individuals' positions in the troop and their age and sex. Pindex is then related to troop members' use of provisioned and natural foods. We then examine the relationship between position in the central-peripheral structure and mortality. Finally, we describe the relationship between an individual's position in the structure and the position of other members of their matriline.

2. Method

2.1. SUBJECTS AND STUDY SITE

Stumptails live on Tanaxpillo Island in Lake Catemaco, Veracruz State, Mexico. A submerged area forms a rock bridge approximately 15 m in length and divides the island. The macaques frequently wade and swim across the bridge. Trees and cliffs located primarily on the island's perimeter provide shade.

The troop moves around the island as a unit (Rasmussen, 1991, Rasmussen, et al., 1992). Perhaps the need for individuals to maintain proximity for visual and auditory contact is a major reason why the troop has retained its integrity as a mobile unit. Several features of the island make such contact between members either impossible or difficult at distances greater than 10 to 20 m: Tanaxpillo Island is elongate. It is about 295 m long and has maximum width of 55 m. At many points it is less than 15 m wide. There is a ridge running the length of the island from East to West. Monkeys on the north side of the island are often out of visual and auditory contact with those on the south side. There are also several cliff grottoes, shrubs and trees.

The soil on the volcanic rock forming the island is thin and the macaques use

much of the vegetation either as a substrate or as food. Estrada and his colleagues (Estrada & Estrada, 1976, 1977; Estrada et al. 1978) have described the naturally occurring plants and animals on Totogochillo Island including those eaten by the macaques. The flora and fauna on Totogochillo Island are very similar to those on Tanaxpillo. The natural food is enhanced with approximately 11 liters of commercial pellet food and .05 cubic meters of fruits and vegetables every day.

Some troop members occasionally swam the channel to Tanaxpi island. This is a larger island occupied by a family that farms all but the southern tip (the island East of Tanaxpillo Island in Fig. 1 of Estrada & Estrada, 1976). Tanaxpi was apparently less desirable to troop members since the farmers' dogs chased the monkeys. Three adult males (Bruno, Blass and Miguel) spent most of their time in an all male group on the southern tip of Tanaxpi Island in 1988. We did not count these males as troop members. We collectively refer to the troop and the all male group as the Tanaxpillo colony.

2.2. THE DEMOGRAPHIC COMPOSITION OF THE TANAXPILLO COLONY

There were 45 individuals in the Tanaxpillo colony alive at some point during the period spanned by collection of data in 1988 and 1989 (Table 2). Four troop members were among those introduced in 1974 onto Totogochillo Island (Estrada & Estrada, 1976). The rest were descendants of the initial troop members. In 1979 (Rodriguez-Luna, personal communication) the troop was moved to the nearby Tanaxpillo Island (the most southern island in Figure 1 of Estrada & Estrada, 1976).

The youngest female observed with an infant was 4.4 years old. Only one male younger than this age, Martin, was observed to ejaculate and he was only once observed to ejaculate. This age was therefore used to separate adults from nonadults by their ability to reproduce. Individuals were not classified as infants if they were older than 200 days of age when we started data collection, at this age nursing ceased to contribute much to food intake and individuals were seldom carried by their mothers.

During data collection in 1988 there were 38 to 39 members of the Tanaxpillo colony. During data collection in 1989 there were 35 members of the Tanaxpillo troop. Between 1988 and 1989 there was a loss of 10 and a gain of 7 colony members. The decline in numbers follows a trend: In 1984-1986 there were 57 members in the troop (Rodriguez-Luna et al, 1986) and by June 1991 there were less than 30.

2.3. OBSERVATION METHODS, CONDITIONS AND SCHEDULE

We could approach within 1-5 m of the macaques when the study was initiated (Rasmussen, 1991). Thirty-three scientists and undergraduates collected the data in 1988 and 35 scientists and undergraduates collected the data in 1989. This number of personnel allowed the collection of a relatively large and representative sample of the location of focal subjects throughout all daylight hours. In 1988 and in 1989 data were collected on 32 days between June 26 and August 14. We collected data on a focal subject in teams of 3 to 6 observers.

Observation teams were composed of undergraduates and orchestrated by a graduate student or scientist. One or two teams of observers collected data during each of three shifts on the island. The first shift was from 6 am to 10 am. The second was from 10 am to 2 pm. And the third was from 2 pm to 6 pm.

An observation team conducted 14 min sampling sessions on individual focal subjects. Members of the team had defined roles: The principal observer took primary responsibility for keeping the focal subject in sight and called

observations to the data recorder for entry into a lap-top computer. Assistants watched the focal subject from different locations and verified calls made by the principal observer. One assistant carried a map of the island with identification numbers of the 5 x 5 m quadrats mapped and painted on the island (Rasmussen, 1991) and called the quadrat in which the subject was located to the data recorder. Data collection roles were rotated to prevent observer fatigue. Scientists or graduate students participated in all roles, monitored data accuracy and directed team efforts. At least one scientist or graduate student was present on the island when data were collected.

Table 2. ID code, name, sex/age class, troop membership status and birth dates of the member of the Tanaxpillo colony in 1988 and 1989. Individuals of each sex listed by descending age.

ID Code	Name	Sex/Age1 Class 88 89	Troop2 Status 88 89	Birth Date	ID Code	Name	Sex/Age Class 88 89	Troop Status 88 89	Birth Date
Females:					**Males:**				
PI	Pierre	AF AF	X D	07/15/74	CH	Chimuelo	AM AM	X D	04/13/72
LO	Lolita	AF AF	X X	03/01/77	PA	Pancho	AM AM	X X	07/24/73
MN	Mirna	AF AF	X D	11/22/78	TI	Tigre	AM AM	X X	12/20/74
UC	Ucilia	AF AF	X D	05/24/80	BR	Bruno	AM AM	P X	07/08/79
SI	Sigrid	AF AF	X X	09/16/80	WI	Wilfrido	AM AM	X X	10/02/80
EN	Enriqueta	AF AF	X D	10/02/80	BL	Blas	AM AM	P D	04/15/82
PD	Piedad	AF AF	X X	03/03/82	FL	Florencio	AM AM	X E	05/09/82
LU	Lupita	AF AF	X X	03/13/83	MG	Miguel	AM AM	P D	05/22/82
FN	Florinda	AF AF	X X	04/13/83	MI	Millo	JM AM	X D	02/21/84
BI	Blonie	AF AF	X X	10/03/83	MA	Martin	JM AM	X X	03/24/84
BA	Blenda	AF AF	X X	10/22/83	UL	Ulises	JM AM	X X	06/11/84
CA	Carolina	JF AF	X X	06/13/84	MS	Martes	JM JM	X X	02/01/86
FA	Flora	JF AF	X X	02/06/85	UR	Urano	JM JM	X X	03/17/86
LC	Lucrecia	JF JF	X D	04/25/85	MO	Milliano	IM JM	X X	04/20/88
TS	Tess	JF JF	X X	06/19/85	SY	Sylvio	IM JM	X X	06/14/88
MV	Minerva	JF JF	X X	04/06/86	FY	Floyd	IM JM	X X	07/16/88
CL	Columbia	JF JF	X X	05/08/86	FK	Flojo	N IM	N X	06/12/89
BS	Brasilia	JF JF	X X	06/28/86					
LE	Leonor	JF JF	X X	09/15/86					
LR	Lorena	JF JF	X X	12/01/86					
PL	Pili	JF JF	X X	01/29/87					
PR	Pilar	JF JF	X X	03/15/87					
US	Ursula	IF JF	X X	04/01/88					
LQ	Loquita	IF JF	X X	07/07/88					
LL	Lulu	IF N	N X	12/13/89					
PN	Pinata	N IF	N X	12/15/89					
BK	Blonda	N IF	N X	06/08/89					
SA	Sylvia	N IF	N X	06/08/89					

1. AF = adult female; AM = adult male; JF = juvenile female; JM = juvenile male; IF = infant female; IM =infant female; N = not yet born.
2. X = present in troop; D = died; E = emigrated; P = peripheral; N = not born
D? disappeared on mainland, presumed dead.

Subjects could be clearly seen from most points on the island and were observed from a distance of 5 to 15 m. Sampling sessions were divided into 2 min intervals. Instantaneous samples on the quadrat number in which focal subjects were located were made at the beginning of each 2 min interval. We used a 15 sec warning tone prior to the onset of a 2 min interval to alert observers to an upcoming instantaneous sample. If the subject's location could not be determined at the instant of data collection, the location of the subject was not recorded for that 2 min sample.

Focal subjects were sequentially selected from a list derived by sampling each age-sex category without replacement. If a subject could not be located after a 10 min search, the next subject on the list was observed. After the sampling session, a search was made for the missed subject and, if located, it was observed. The list was started again from the top when observations were completed on the last subject.

2.4. DATA BASE

The 1988 data base consists of 273.2 hours of 2 min samples (8196 samples) of the location of individual focal subjects. The 1989 data base consists 173.03 hours of 2 min samples of location (5191 samples). A video tape of the observation methods and troop are available (Ordonez et al., 1991).

The number of samples per subject varied (1988: mean=204.9 2 min samples, SD=45.08 1989: mean=148.31 2 min samples, SD=23.89). Some variation was due to the death or birth of subjects. Some variation was also due to differences in the difficulty in locating subjects.

2.5. PINDEX

The ideal data for the analysis of spatial structure would be simultaneous maps of the location of all group members. Such data might be obtained from overhead pictures in areas without cover. Given the exceptionally difficulty in obtaining the ideal data, two broad types of less ideal data may be used for the description of spatial structure: First, the geographic locations of individuals (Rasmussen, 1973). Second, the location of individuals relative to the group's mobile boundaries (Collins 1984). Appropriate choice of either geographic or relative location is partly contingent on the size of the range of the group relative to the size of the group's mobile boundaries.

When housed in an enclosure, a group's mobile boundaries may be equivalent to its range (Figure 1A). When this is true, all variation in the use of geographical space is variation among individuals since the group cannot move as a unit. The geographical location of individuals is therefore excellent for assessing spatial structure (Rasmussen, 1973).

When in a natural environment, the mobile boundaries of a group are often small relative to its range (Figure 1C). In this case individual variation in use of geographical space could be overwhelmed by the group's variation. The spatial structure within the group therefore must be assessed relative to the mobile boundaries of the group (Robinson 1981; Collins 1984; Wada & Matsuzawa, 1986) or with measurements of distances between group members (Kummer, 1968).

When there is distinctive variation in both the individual's and the group's use of space (Figure 1B) either location or relative position within the mobile boundaries of a troop could be used for description of spatial structure. The size of the mobile boundaries of Tanaxpillo troop compared to the area of the island (6750 m^2) approximates this intermediate condition.

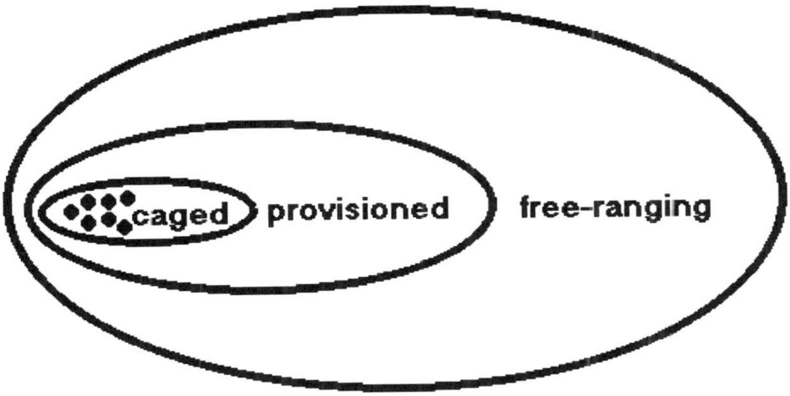

Figure 1. Influence of relative size of the range of a group to the group's mobile boundaries on relative variation in individual's use of space. Black circles represent monkeys, borders of the elipses represent boundaries of the group's range.

We chose to use geographic location to assess individuals' position within the central-peripheral spatial structure since Tanaxpillo Island is relatively small. Individual variation in use of geographic space therefore could be detected. Individual preferences for locations could be quantitatively assessed since we used a fine-grained grid of 5 x 5 meter quadrats.

Greater accuracy and consistency in the assessment of geographic location was another reason for our choice of this variable for the measure of the colony's central-peripheral structure. Geographic location may be assessed on individually observed focal animals without reference to the sometimes very difficult to determine simultaneous distribution of all individuals in the group.

We developed a peripherality index, pindex, to measure the placement of individuals within the central-peripheral spatial structure. Pindex is calculated on the focal samples of the quadrat locations of individual subjects. Pindex assesses degree of dissimilarity between an individuals' use of quadrats and the proportion of time spent by all troop members in quadrats. The more dissimilar an individual's use of quadrats is to that of the troop, the larger is the value of pindex.

Pindex is a chi-square like statistic. The numerator is the sum of the squared deviations of the proportion of focal samples of a subject in quadrats from the expected proportion of samples in quadrats. The denominator is the expected proportion of samples in quadrats. The observed proportion of samples of a subject in a quadrat is the number of focal samples of the subject in that quadrat divided by all focal samples of that subject's location. The unweighted mean proportion of samples of all colony members in a quadrat is used as the expected proportion of samples of the subject in that quadrat.

$$\mathrm{PINDEX} = \frac{\sum_{j=1}^{T} (p_{ij} - P_j)^2}{P_j}$$

where:

N = the number of subjects. N = 40 in 1988, N = 35 in 1989

T = the number of quadrats in which troop members were sampled. T = 227 in 1988, T= 232 in 1989

p_{ij} = the proportion of all 2 minute samples of the location (l) of subject i in quadrat j

$$= \frac{l_{ij}}{\sum_{j=1}^{T} l_{ij}}$$

P_j = the unweighted mean proportion of all samples of the subjects in quadrat j

$$= \frac{\sum_{i=1}^{N} p_{ij}}{N}$$

The larger the magnitude of pindex the more divergent is the individual's pattern of use of quadrats (Fig. 2). Most individuals are sometimes in the center of the troop and sometimes on the periphery. Pindex provides an overall evaluation of the degree of peripherality.

2.6. FEEDING

We defined feeding as placing an edible object in the mouth. We collected a 1/0 sample on each of 12 natural and 12 provisioned food categories. If, for example, the focal subject ate leaves and provisioned monkey chow during a 2 min interval, a 1 was recorded for feeding on leaves and a 1 for feeding on monkey chow and 0s for the other food types. We calculated the proportion of time (number of 2 minute intervals) feeding on each food type for every subject. These proportions were used to calculate the mean proportion of time feeding on provisioned foods and on natural foods.

As in educational testing (Anastasi, 1968; Suen & Ary, 1989) our analyses focus on covariation of variables, not values of individual variables. Our analyses center on changes in time spent feeding as a function of range use rather than on either the true frequency or duration of feeding. When intervals are 2 minutes or less in duration (Simpson & Simpson, 1977), 1/0 samples strongly correlate with both duration and frequency (Martin & Bateson, 1993). 1/0, frequency, and duration samples should therefore have similar relationships with position in the central-peripheral structure.

We could neither reliably count the number of items eaten in every food category nor keep track of the duration of time feeding on each food type. Instantaneous sampling would miss many food types eaten during the day. We therefore felt 1/0 sampling was the appropriate method for our study.

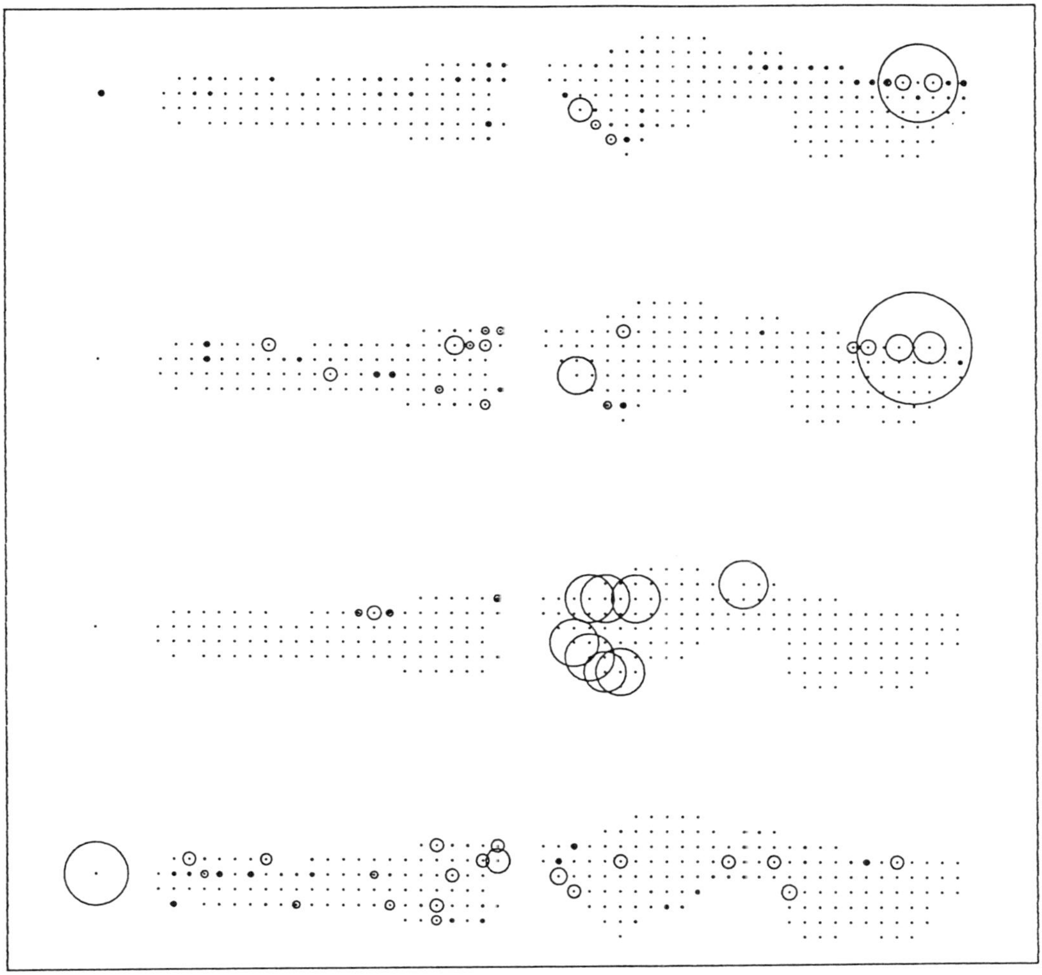

Figure 2. These diagrams illustrate variation in use of space assessed by pindex in 1988. The diameter of the circles is directly related to the proportion of time spent in quadrats. The points are centers of quadrats. Tanaxpi island is indicated by the single point to the west of Tanaxpi Island. 2A illustrates the unweighted mean proportion of time spent in quadrats by all members of the Tanaxpillo colony. 2B is the pattern of use of quadrats by Pilar, the juvenile female offspring of the highest ranking adult female in the group, Piedad. Pilar was the most central colony member. 2C is the pattern of range use of Miguel. Miguel was one of three bachelor males who did not move with the troop and who spent most of his time on Tanaxpi Island. Pindex evaluated Miguel as the most peripheral colony member.

3. Results

3.1. SAMPLING BIASES AND THEIR STATISTICAL CONTROL

We conducted analyses to determine if variation in the number of samples collected subjects influenced pindex. The individuals' pindex values were regressed on the number of samples. The regression equations were run separately for the 1988 and the 1989 data. Examination of the residuals indicated that they were positively skewed. Logarithmic transformation of pindex resulted in distributions of residuals that closely approached normality.

The values of pindex from 1988 and from 1989 were significantly correlated with the number of samples collected on subjects. In 1988 there was a negative correlation of -.49 ($P < .0013$) between pindex and number of intervals collected on subjects. In 1989 there was a negative correlation of -.42 ($P < .0128$) between pindex and the number of intervals collected on subjects. Fewer samples were collected on the more peripheral individuals.

Fewer samples were collected on peripherals since they were often away from the rest of the troop and so were more difficult to locate. Peripherals also tended to select locations where they were difficult for observers to find such as behind rocks, in dense foliage or in tree tops. Some had favorite "hiding places" where observers learned to find them, such as a cliff grottoes covered with vines and bushes (see also Rasmussen, 1973; Rasmussen & Rasmussen, 1979). The covariation between values of pindex and number of samples collected on subjects partially validates pindex since it indicates peripheral individuals were often in places away from other troop members.

We control for variation in number of intervals collected on subjects in all our analyses. The residuals from the regression of pindex on number of intervals is used in our descriptive analyses. The first variable entered in all our inferential multiple regression analyses is the number of intervals collected on subjects.

Statistical control of samples collected on individuals may therefore remove some of the variation pindex was designed to assess. Fortunately, values of pindex and the residualized logarithmic transformations of pindex were strongly correlated. For the 1988 data the correlation between the residualized logarithmic transformation of pindex and raw values was +.87 ($P < .0001$). For the 1989 data the correlation was +.91 ($P < .0001$). Both the original, and the residualized values of pindex, therefore place individuals in very similar relative positions within the central-peripheral structure of the colony.

3.2. QUALITATIVE ESTIMATES

Primatologists often use qualitative observations to estimate the degree to which monkeys are central or peripheral (Table 1). These estimates are associated with quantitative variables, such as differential mortality of offspring (Sugiyama & Ohsawa, 1982). We therefore felt that if pindex measures peripherality, it should provide results somewhat consistent with our qualitative evaluations of peripherality. We knew the colony members who were peripheral since they were often difficult to locate for sampling. Their more isolated positions made their peripheral position more distinctive than subtle differences between individuals toward the center of the troop. As a first step toward evaluation of the utility of pindex, we therefore compared individuals evaluated to be most peripheral by pindex with our qualitative impressions of the most peripheral individuals.

In 1988, three adult males, Miguel, Bruno and Blas spent most of their time on Tanaxpi island and we seldom saw them within the mobile boundaries of the troop. Of these three, Blas was most often observed within the troop's mobile boundaries. All three males were very peripheral to the center of mass of the troop. Pindex evaluated Miguel as the most peripheral colony member followed by Bruno. Blas was evaluated as the fourth most peripheral adult male.

A multiple regression analysis revealed these males were significantly more peripheral than other colony members. Division of individuals into troop or members of the 3 male group accounted for 17.4 % of the variation in the logarithmic transformation of pindex ($P=.002$; $F=11$; $DF=1/37$) after entry of the number of 2 minute intervals collected on troop members.

Ucilia was a crippled adult female often observed on the periphery of the troop. Pindex evaluated Ucilia as the most peripheral adult female and the third most peripheral member of the Tanaxpillo colony. Wilfredo appeared to be the most peripheral of the adult males who moved with the troop. Pindex evaluated Wilfredo as the fifth most peripheral adult male.

In 1989 the adult males evaluated to be the most peripheral by pindex were again those estimated to be most peripheral by observers. Bruno was evaluated as the most peripheral adult male by the observers and pindex. Bruno was the sole survivor from the all male group. After a series of agonistic encounters with troop members that left several dead, Bruno became a peripheral member of the troop. Wilfredo was the other adult male most often observed on the troop's periphery. He was evaluated by pindex to be the second most peripheral of the adult males.

Not all of our qualitative estimates were consistent with evaluations by pindex. The evaluation of Chimuelo as the fourth most peripheral troop member by pindex was a surprise. Chimuelo was an adult male who was the brother of the two adult males (Tigre and Pancho) we felt were the most central adult male troop members. Although evaluated as peripheral by pindex, Chimuelo often engaged in affiliative behavior and formed alliances with his brothers. These are characteristics of a high ranking male. However, Chimuelo was crippled from an old injury and seldom fed with the rest of the troop. Chimuelo may have not fed with the troop to avoid the aggression associated with competition for this highly sought after resource (Jolly & Rasmussen, 1990). This may account for his pindex score.

Piedad was the female most observers felt to be of highest agonistic rank in 1988 and 1989. Surprisingly pindex placed Piedad as the third most peripheral of the adult females in 1989. In 1989 Piedad had an infant female, Pinata, and spent a great deal of time with her daughter away from the center of the troop. Piedad's care of Pinata may have been partially responsible for her peripherality as assessed by pindex. The matriline to which Piedad belonged also suffered the highest mortality rate of all matrilines (see below). The decreased number of matrilineal allies may also have increased Piedad's peripherality.

3.3. CONSISTENCY OF THE CENTRAL-PERIPHERAL STRUCTURE

Another way of assessing the validity of pindex is to examine the correlation between 1988 and 1989 pindex values. If the values are strongly correlated this indicates the central-peripheral organization of the Tanaxpillo colony remained relatively constant between years. Correlated pindex values across years would also indicate that pindex, and the sampling methods necessary for its calculation, tended to assess individuals in the same way across years.

There were 30 members of the Tanaxpillo colony who were present in both 1988 and 1989. The correlation between values of pindex in 1988 and 1989 for these 30 was +.36 ($P=.0493$). We expected values of pindex for adults would be more strongly correlated across years since positions of the nonadults in the spatial structure of the troop could change as they matured. This expectation was confirmed: there was a correlation of +.67 ($P<.001$) for the 15 individuals who were adults in 1989 and present in 1988. For the 9 females who were adults in 1989 and present in 1988 there was a very strong +.82 ($P<.001$) correlation between values of pindex in 1988 and 1989 (Fig. 3). For the 6 males who were adults in 1989 and present in 1988 there was a stronger correlation of +.90 ($P=.0138$) between values of pindex in 1988 and 1989 (Fig 3). These correlations are particularly strong when the number of deaths among the adults is considered along with the changes in the social organization of the troop that accompanied these deaths. There was no correlation between the values of pindex in 1988 and 1989 ($r=-.03$, $P=.93$) for the 15 individuals who were not adults in 1989 and present in 1988.

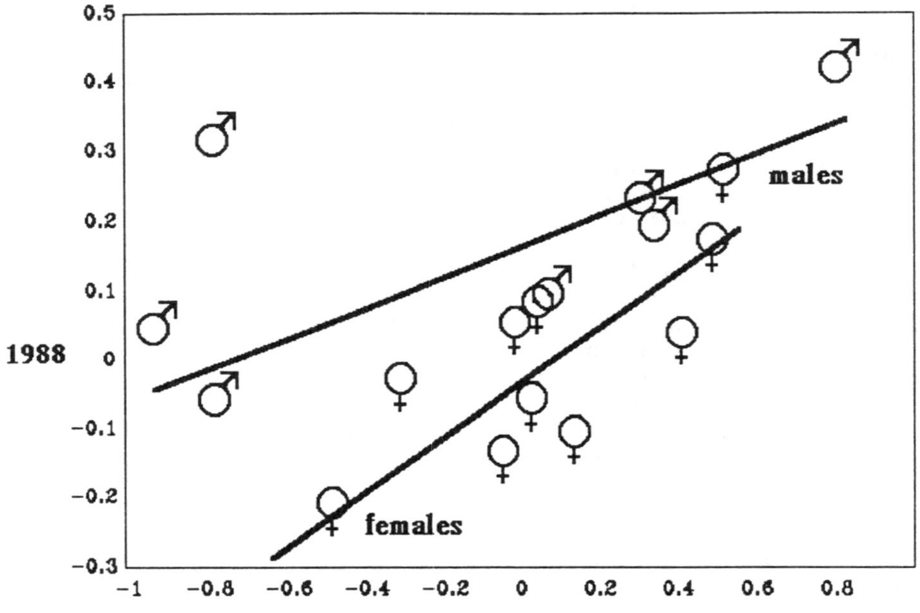

Figure 3. Relationship between residualized values of pindex in 1988 and 1989 for adult females ($r=+.82$, $P=.0068$) and adult males ($r=.90$, $P=.0138$).

3.4. AGE, SEX AND THE CENTRAL-PERIPHERAL STRUCTURE

Our observations led us to believe that the most central troop members were infants and juveniles. These troop members appeared to form a social focus of the troop (Fig. 4). We calculated the mean values of pindex for each age/sex class of troop members for our initial examination of the data. We found that nonadults tended to be more central than the adults in both 1988 (nonadult mean = -.11, adult mean = +.12) and 1989 (nonadult mean= -.05, adult mean =+.06). Infant females were the most central troop members in both 1988 (mean = -.24) and 1989 (mean = -.12).

We used multiple regression to examine the relationship between age, sex and peripherality. With multiple regression, we could determine the association between pindex and age (in days) of individuals as well as the interaction between their age and sex. After entry of number of samples, age and sex were entered simultaneously into the equation followed by the interaction between age and sex.

Females were coded as 1 and males as 0.

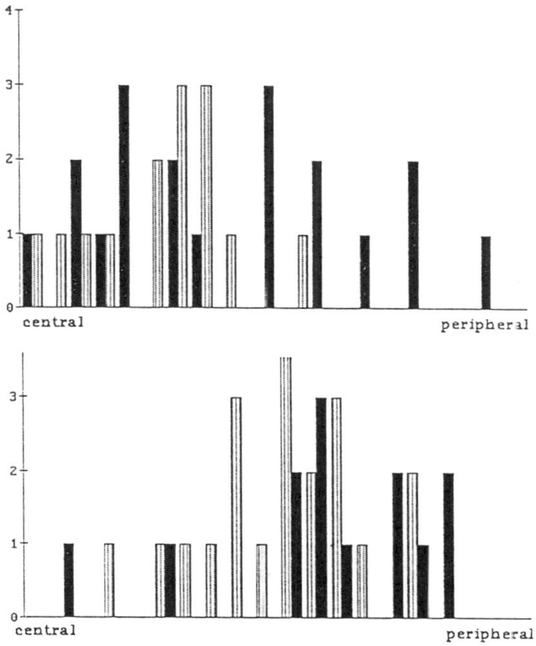

Figure 4. Positions of adults and nonadults in the central-peripheral structure. Black columns are adults, shaded columns are nonadults. Heights of columns indicate number of individuals for a given value of pindex. Note nonadults tended to be more central than adults in 1988 (A) data and 1989 (B).

In 1988 age and sex together accounted for 19% of the variance in pindex ($F=6.11$, $DF=2/36$, $P<.01$). The partial correlations show that age, not sex, was the significant correlate of individuals' peripherality. There was a large and significant positive partial correlation between age and pindex ($pr=+.43$, $P<.001$). There was a weak and nonsignificant partial correlation between pindex and sex ($pr=-.24$, $P=.1502$). The negative partial correlation indicates males tended to be more peripheral than females. The age by sex interaction was not significant and low in magnitude (R^2 Change=.01, $F=.65$, $P=.43$). Both males and females therefore had similar tendencies to be more peripheral with increased age. Like our descriptive analysis, the regression analysis therefore indicated that the most central group members tended to be youngest.

In 1989 neither age nor sex were associated with a significant proportion of variation in the values of pindex (R^2 Change=.02, $F=.31$, $DF=2/31$, $P=.73$). The age by sex interaction accounted for a significant 12.38% of the variation ($F=5.41$, $DF=1/30$, $P<.03$). Females tended to be more peripheral with increased age (Fig. 5A) as they did in 1988. Unlike in 1988, older males tended to be more central in 1989 (Fig. 5B). The change in the direction of the relationship between pindex and age for males was not due to changes in pindex in the males who were adults in 1989 (Fig 3). The change was due major alterations in the adult male composition of the colony and the low correlation between the pindex values in 1988 and 1989 for the males who were not adults in 1989. Three older

and peripheral adult males died or disappeared: Miguel, Blass, and Chimuello. After the death of these males, the oldest remaining in the troop were also central males.

3.5. ACCESS TO FOOD AND THE CENTRAL-PERIPHERAL STRUCTURE

Previous studies have shown more central troop members may have greater access to resources such as provisioned food (Rasmussen & Rasmussen, 1979; Sugiyama & Ohsawa, 1982). We also observed that more central members of the Tanaxpillo colony had greater access to the provisioned food (Ordonez et al., 1991). We therefore conducted multiple regression analyses of the relationship between pindex and the rate at which troop members ate provisioned or natural food in 1988 and 1989.

In our initial regression analysis of the 1988 data we first entered number of 2 minute intervals collected on subjects to control for variation in sample size. The mean rate of eating provisioned food and the mean rate of eating natural food were then entered simultaneously. Together these variables account for 12% of the variance in pindex ($F=3.43$, $DF=2/36$, $P<.04$).

Nearly all the variance in pindex associated with rates of eating provisioned or natural food was associated with mean rate of eating provisioned food ($pr^2=15\%$, $T=-2.52$, $P<.02$, Fig. 5). Only .08% of the unique variance in pindex was associated with the unique variance in the rate of eating natural food ($T=-.17$, $P=.87$). Both of the partial correlations were negative: individuals who were more peripheral tended to spend less time eating both natural and provisioned food.

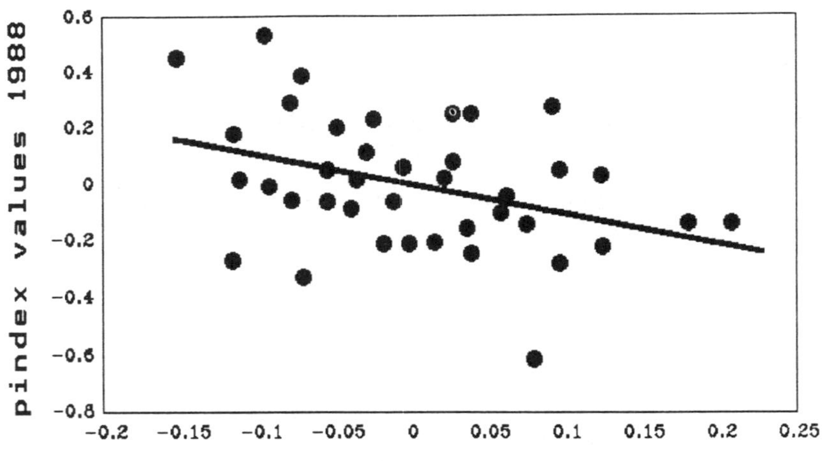

Figure 5. Graph of the partial correlation between individuals' degree of peripherality as assessed by pindex in 1988 and the proportion of time they spent eating provisioned food.

We have shown that older individuals tended to be more peripheral in 1988. This relationship led us to question whether the association we found between rate of eating provisioned food and pindex may have been due to the correlation between age and rate of feeding. It might be suggested, for example, that older individuals spent less time eating provisioned food and were more peripheral and

this caused a spurious correlation between pindex and rate of eating provisioned food.

We therefore conducted a hierarchical multiple regression analysis in which the number of 2 minute intervals collected on subjects and the age of subjects were entered before the mean rate of eating provisioned food. We found the rate of eating provisioned food was associated with pindex after control of the variance associated with age (R^2 Change=.073, F=5, DF=1/36, P<.03). There was no significant interaction between rate of eating provisioned food and age.

The partial correlation between pindex and rate of eating provisioned food was -.35 (T=-2.24, P<.03). This result shows the correlation between rate of eating provisioned food and pindex is not a passive consequence of age. Mean rate of eating provisioned food was associated with nearly as much unique variance in pindex as age (age: pr=.42, T=2.75, P<.01). Both variables were associated with a unique and significant proportion of variation in pindex.

Unlike for the 1988 data, the initial regression analysis of the 1989 data did not indicate rates of eating natural and provisioned food were associated with pindex (R^2 Change=.10, F=2.02, DF=2/31, P=.15). We thought these unexpected results might be due to the complex relationships we found between age, sex and pindex in 1989. We therefore conducted a multiple regression analysis with age, sex and the interaction between age and sex entered before rates of eating provisioned or natural food.

The rate of eating natural and provisioned food was not related to pindex (R^2 Change = .06, F=1.45, DF=2/28, P=.25). Although not significant, the partial correlation between pindex and rate of eating provisioned food was negative as in 1988 (pr^2=.02, T=-.63, P=.54). Individuals who spent more time eating provisioned food did, when the other variables were held constant, tend to have a more central position in the troop. Also, as in 1988, the rate of eating provisioned food accounted for more variation in pindex than did rate of eating natural food (pr^2=.0004, T=.08, P=.93).

The interactions between feeding, age, and sex were significant (R^2 Change=.29, F=5.17, DF=4/23, P<.001). None of the individual components of the interactions was significant. We therefore defer interpretation of these components.

3.6. SURVIVAL AND THE CENTRAL-PERIPHERAL STRUCTURE

There was a striking association between mortality and peripherality as evaluated by pindex in 1988 (Fig. 6). Five of the seven adults who died between onset of data collection in 1988 and the onset of data collection in 1989 were among the eight most peripheral in 1988.

We conducted an inferential multiple regression analysis on these data. Colony members were separated into those who lived throughout the study period and those who were alive in 1988 but who died by the end of data collection in 1989. After entry of the number of intervals collected on subjects, this dichotomy accounted for 13.97% of the variation in values of pindex in 1988 (F=8.33, DF 1/37, P=.0065). Individuals who died were therefore likely to be more peripheral in 1988.

There were 6 matrilines with living members in Tanaxpillo colony. Each was descended from one of the females who were in the troop when it was formed in 1974 (Fig. 7). There was considerable variation in the proportion of deaths occurring in matrilines. There was no mortality in the 10 members of the Flor matriline (#4) present in 1988. It is of interest that members of the Flor matriline had the highest mean value of pindex in 1988 and that this was the only matriline in which no deaths occurred during this study. The Blondie and Flaca matrilines

each had 1 death among the 6 individuals present in 1988 or 17% mortality. The Blanca and Jarocha matrilines each had three members in 1988 and each had one death, or 33% mortality. The Margarita matriline was the largest with 12 members in 1988. Five of these 12 members died during the time spanned by this study, the mortality rate was therefore a staggering 42%.

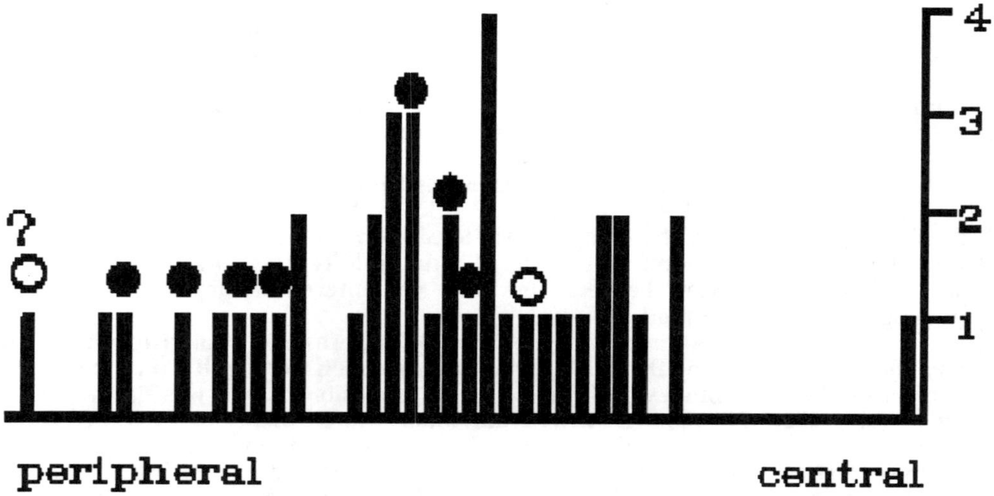

Figure 6. Histogram of values of Pindex in 1988. Values of pindex decrease from left to right. Height of columns indicate number of individuals at a given value of pindex. Black circles above columns indicate a death by the end of this study in 1989. The two open circles indicate an adult male with this pindex value emigrated from the troop. The open circle with a question mark above it is for an adult male, Miguel, who left the troop and who was never observed again. He is most likely dead. Note the much higher mortality among the macaques in the peripheral tail of the distribution.

3.7. MATRILINES POSITION IN THE CENTRAL-PERIPHERAL STRUCTURE

In some matrilines members shared similar positions in the central-peripheral structure. For example, the males most observers felt to be highest in agonistic rank and most central, Tigre and Pancho, belonged to the Jarocha matriline. Sugiyama & Ohsawa (1982) report members of matrilines occupy similar positions in the central-peripheral structure of Japanese macaques. These scientists grouped matrilines by position in the estimated central-peripheral structure and found central matrilines had higher birthrates when food was provisioned.

We have shown that age and sex are significantly associated with placement in the central-peripheral structure. We therefore conducted multiple regression analyses that controlled for matriline age and sex composition. The analyses used were the same as those previously described with number of intervals collected on subjects, age and sex, and the sex by age interaction entered hierarchically. After these variables, matrilineal membership was entered into the equation. Membership in matrilines was effects coded (Cohen & Cohen, 1983).

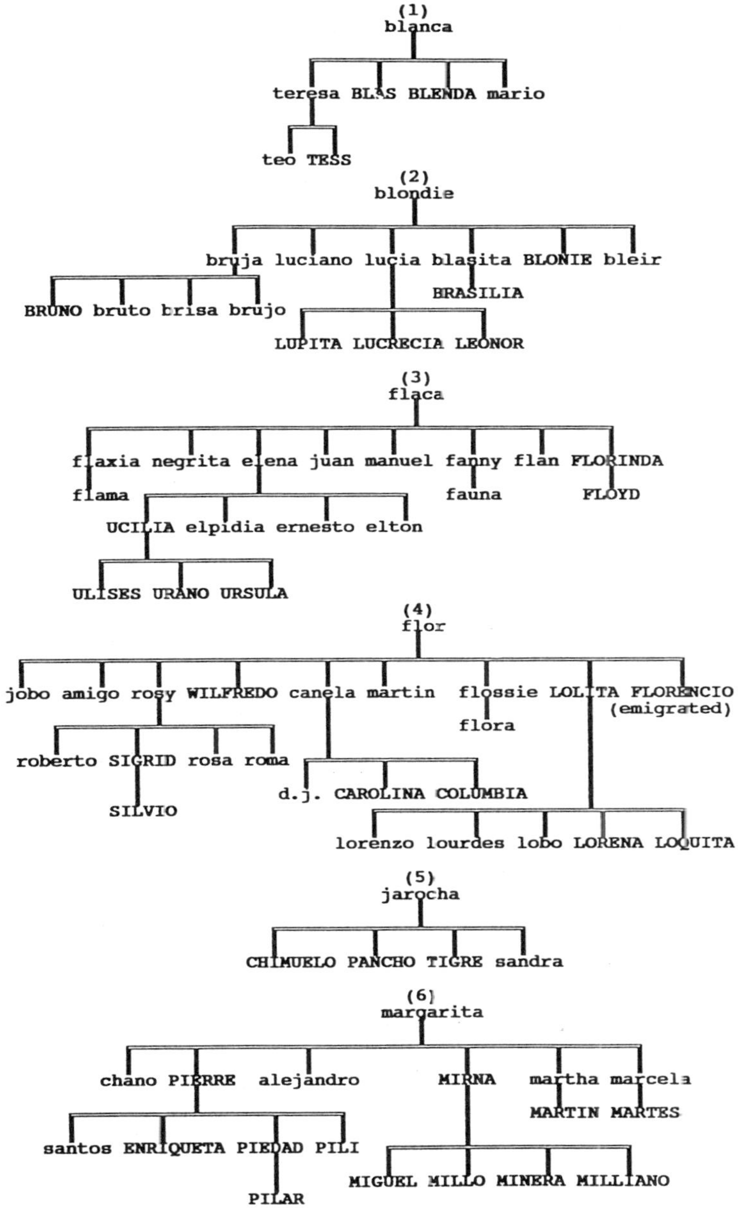

Figure 7. Matrilines with living descendants in the Tanaxpillo colony during data collection in 1988 or 1989. Individuals with all letters in their name in upper case were alive in 1988 and 1989. Individuals with names in all lower case were not alive during data collection in 1988 or 1989. Individuals with first letter in name in upper case and the rest in lower case died between 1988 and the end of data collection in 1989. Individuals with first letter of name in lower case and the rest of their name in upper case letters were born after the end of data collection in 1988 and were alive in 1989.

Matrilineal membership in 1988 did not account for a significant proportion of variation in pindex (P=.11; F=1.98; DF=5/30). In 1989 membership in matrilines was associated with a large and significant 28.58% of the variation in pindex (P=.0143; F=3.57; DF=5/25). Matrilineal membership was associated with greater variation in pindex than sex, age, or the sex by age interaction. Examination of the individual matrilines indicated significant partial correlation between pindex and membership in the Jarocha matriline (pr= -.62, P=.001), The Jarocha matriline was more central than other matrilines. Membership in the Flor matriline (pr=+.45, P=.018) and the Blanca matriline (pr=+.42, P=.028) were associated with a more peripheral position in the troop's spatial structure.

4. Discussion

Our results illustrate the importance of knowledge of the central-peripheral structure of the colonies' society for understanding its complex and multifaceted social organization. Our results also suggest knowledge of the central-peripheral structure may be of use for understanding differences between troops and perhaps for learning more about the biological bases of culture.

Comparison of our impressions of individuals' placement within the central-peripheral structure with the quantitative evaluations of pindex suggest pindex does assess aspects of the central-peripheral structure similar to those assessed in qualitative studies. This "observational" validity has also been confirmed in a study conducted on a captive group of chimpanzees (James-Aldridge & Gorena, 1992). We do caution that an individual's pindex value is but a single number and, as such, cannot possibly summarize all aspects of his or her position within the troop (Sade et al., 1988). As with agonistic rank (Hinde & Datta, 1981), knowledge of the central-peripheral structure is only useful to the extent that it is associated with other aspects of social organization. And our analyses have shown the central-peripheral structure is associated with basic biological aspects of the social organization of the Tanaxpillo colony.

Our results suggest, for example, that a central position confers a selective advantage. The evidence for this is as follows: There has been a steady decrease in the size of the troop since at least 1984. A central position within the troop is associated with higher access to the high quality resource of provisioned food. Peripherality and mortality were positively associated. A more central position in the troop therefore appears to be associated with greater access to resources and lower mortality. There therefore appears to be selection for characteristics associated with a more central position in the troop.

Position in the central-peripheral structure is a trait that is sometimes significantly associated with matrilineal membership. This suggests position in the central-peripheral structure may be partially acquired across generations. The transmission of characteristics associated with position in the central-peripheral structure may lead to characteristics of matrilines. Environmental characteristics could be greater strength of central group members due to greater access to the provisioned food. Genetic characteristics could be biological bases of competitiveness (Jolly & Rasmussen, 1992) associated with a more central position in the troop.

ACKNOWLEDGEMENTS. Logistic support was provided by Dr. J. Fa, L. Marsh, E. Rodriguez-Luna, D. Canales, staff of the primate field station of the University of Veracruz, and 75 student assistants. P. Dubois and Dr. S. Sholl wrote programs for data collection and management. L. Nash and K. Crooks helped compile information on matrilines. Dr. Newell-Morris, Mr. R. Gorena and the editors provided useful comments. Support was provided by Animal Behavior Research Institute, SFS, the Wisconsin Regional Primate Research Center, the Patronato Pro-Universidad Veracruzana, A.C., NIMH NRSA 1 F32 MH09419-01 RERA, NIH, NSF Grants 880414 and 890080, NATO, APA, and Florida State University. This is publication 21 of the Animal Behavior Research Institute.

References

Anastasi, A. (1968). *Psychological Testing, 3rd edition.* The MacMillan Co./London.

Anderson, C. M. (1981). Subtrooping in a Chacma baboon (*Papio ursinus*) population. *Primates*, 22, 445-458.

Bernstein, I.S. (1981). Dominance: the baby and the bathwater. *Behavioral and Brain Sciences*, 4, 419-429.

Busse, C. D. (1984). Spatial structure of chacma baboon groups. *International Journal of Primatology*, 5, 246-261.

Byrne, R.W., Whiten, A., & Henzi, S.P. (1989). Social relationships of mountain baboons: Leadership and affiliation in a non-female-bonded monkey. *American Journal of Primatology*, 18, 191-207.

Casey, D. E., & Clark, T. W. (1976). Some spacing relations among the central males of a transplanted troop of Japanese macaques (Arashiyama West). *Primates*, 17, 433-450.

Cohen, J., & Cohen, P. (1983). *Applied multiple regression/correlation analysis for the behavioral sciences*, 2nd edition. Lawrence Erlbaum/Hillsdale, N.J.

Collins, D. A. (1984). Spatial pattern in a troop of yellow baboons (*Papio cynocephalus*) in Tanzania. *Animal Behaviour*, 32, 536-553.

Crook, J.H. (1966). Gelada baboon herd structure and movement: a comparative report. *Symposia of the Zoological Society of London*, 18, 237-258.

Dittus, W. P. (1979). The evolution of behaviors regulating density and age-specific sex ratios in a primate population. *Behaviour*, 69, 265-302.

Estrada, A., & Estrada, R. (1976). Establishment of a free-ranging colony of stumptail macaques (*Macaca arctoides*): relations to the ecology I. *Primates*, 17, 337-355.

Estrada, A., & Estrada, R. (1977). Patterns of predation in a free-ranging troop of stumptail macaques (*Macaca arctoides*): Relations to the ecology II. *Primates*, 18, 633-646.

Estrada, A., Estrada, R., & Manzolillo, D. (1978). Further data on predation by free-ranging stumptail macaques (*Macaca arctoides*). *Primates*, 19, 401-407.

Gartlan, J. S. (1968). Structure and function in primate society. *Folia Primatologica*, 8, 89-120.

Hamilton, W. D. (1971). Geometry for the selfish herd. *Journal of Theoretical Biology*, 31, 295-311.

Hinde, R. A., & Datta, S. (1981). Dominance, an intervening variable. *Behavioral and Brain Sciences*, 4, 442.

Itani, J. (1954). Japanese monkeys in Takasakiyama. In: Imanishi, K. (ed.), Nihon Dobutsuki II, Kobunsha, Tokyo.

James-Aldridge, V., & Gorena, R. L. (1992). Pindex. *International Journal of Psychology*, 27, 432.

Jolly, D. W., & Rasmussen D. R. (1990). Differences in competitive rates when feeding on natural or provisioned food and when not feeding in the Catemaco troop of stumptail macaques. *American Journal of Primatology*, 20, 202.

Jolly, D. W. & Rasmussen, D. R. (1992). Provisioning food in piles: breeding a more competitive monkey? Evidence from the Tanaxpillo troop of stumptail macaques in Mexico. *American Journal of Primatology*, 27, 36.

Kummer, H. (1968). *Social organization of Hamadryas baboons*. University of Chicago Press, Chicago.

Lopez-Lujan, A. X., Ramirez Ochoa I., Mayagoitia L., Mondragon-Ceballos R. (1989). Sex differences in intra-group spacing behaviour in stumptailed macaques (*Macaca arctoides*). *Folia Primatologica*, 52, 102-108.

Martin, P., & Bateson, P. (1993). *Measuring behavior: An introductory guide* (2nd ed.). Cambridge: Cambridge University Press.

Ordonez, D., Guss, L., Nash, A. & Rasmussen, D. R. (1991). Observation, sampling, and data collection methods used in a two year research project on the social ecology of stumptail macaques. *American Journal of Primatology*, 24, 125.

Rasmussen, D. R. (1973). Spacing in adult male Japanese macaques. *Primate News*, 11,11-15.

Rasmussen, D. R. (1991). Observer influence on range use of *Macaca arctoides* After 14 years of observation? *Laboratory Primate Newsletter*, 30, 6-11.

Rasmussen, D. R., Biggs, R. & Gorena, R. (1992). Wind direction is associated with daily variation in use of an island by a troop of stumptail macaques (*Macaca arctoides*). *Laboratory Primate Newsletter*, 31, 12-14.

Rasmussen, D. R., & Rasmussen, K. L. 1979. Social ecology of adult males in a confined troop of Japanese macaques (*Macaca fuscata*). *Animal Behaviour*, 27, 434-445.

Robinson, J. G. (1981). Spatial structure in foraging groups of wedge-capped capuchin monkeys (*Cebus nigrivittatus*) *Animal Behaviour*, 29, 1036-1056.

Rodriguez-Luna, E., Carrillo-Castilla, P., Aguilera-Rayes, U., Barney-Guillarmo, H. (1986). Changes in the dominance hierarchy of a group of macaques (*Macaca arctoides*) living in free-ranging: 1984-1986. *Primate Report*, 14, 229-230.

Sade, D. S. Altmann M., Loy J. Hausfater G., & Breuggeman J. A. (1988). Sociometrics of *Macaca mulatta*: II. Decoupling centrality and dominance in rhesus monkey social networks. *American Journal of Physical Anthropology*, 77, 409-425.

Simpson, M. J. A. & Simpson, A. E. (1977). One-zero and scan methods for sampling behaviour. *Animal Behaviour*, 25, 726-731.

Southwick, C. H., Beg, M. A., & Siddiqi, M. R. (1965). Rhesus monkeys in North India. In: DeVore, I. (ed.), *Primate behavior: field studies of monkeys and apes*. Holt, Rinehart & Winston/New York. pp. 111-159.

Stoltz, L. P., & Saayman, G. S. (1970). Ecology and behaviour of baboons in the Northern Transvaal. *Annals of the Transvaal Museum*, 26, 99-143.

Suen, H. K. & Ary, D. (1989). *Analyzing quantitative behavioral observational data*. Lawrence Erlbaum Associates/Hillsdale, New Jersey.

Sugiyama, Y. & Ohsawa, H. (1982). Population dynamics of Japanese monkeys with special reference to the effect of artificial feeding. *Folia Primatologica*, 39, 238-263.

Wada, K., & Matsuzawa, T. (1986). A new approach to evaluating troop deployment in wild Japanese monkeys. *International Journal of Primatology*, 7, 1-16.

Washburn, S. L., & DeVore, I. (1961). The social life of baboons. *Scientific American*, 204, 62-71.

CULTURAL IMPLICATIONS OF DIFFERENCES BETWEEN POPULATIONS OF FREE-RANGING CHIMPANZEES IN AFRICA

W.C. McGrew
Department of Sociology & Anthropology
Miami University
Oxford, OH 45056, U.S.A.

ABSTRACT

Behavioral differences exist among populations of wild chimpanzees (Pan troglodytes, Pongidae) across Africa. Deciding if these are cross-cultural differences demands careful interpretation and rigorous scrutiny. Even standard patterns like social grooming may show variation in detail between neighboring populations. Unique patterns of tool-use, such as using hammers to crack nuts, also vary across far western Africa. At least 34 populations of free-ranging chimpanzees show tool use, and 13 show at least one habitual pattern of tool-use. Regional differences exist among the three subspecies or geographical races of chimpanzees. Convincing, empirical demonstration of social customs and traditions in non-human species is problematical, and even something so simple in principle as innovation is hard to show in practice. However, culture need not depend on imitation, teaching, or language, either in humans or in other species.

R. A. Gardner et al. (eds.), The Ethological Roots of Culture, 61–79.

1. INTRODUCTION

Can the concept of culture be applied validly to another species? This paper first reports (as a case study) a kind of grooming shown by wild chimpanzees which seems to be a truly social custom. The example serves to introduce the practical pitfalls and potentials of seeking to answer the question posed. Next, the paper focuses on a type of tool-use, hammer-and-anvil, which varies across populations and has important archaeological implications. Broadening further, an exhaustive catalogue of habitual tool-use across all known field-studies is presented. Finally, the evidence of regional and local patterns of tool-use by wild chimpanzees is assessed. The paper then turns to mechanisms of cultural processes, especially innovation, before ending with responses to recent criticisms by the "anti-culturalists". If concepts such as culture are to help us understand the behavior of our nearest relations, we must avoid simplistic and sloppy extrapolation.

Two long-term field studies of wild chimpanzees have proceeded in parallel in western Tanzania, and most of the published knowledge of the natural behavior of individual chimpanzees comes from these. Goodall's (1986) research group in the Gombe National Park has focussed on the Kasakela community of chimpanzees. The project begun by the African Primate Expedition at Kasoje in the Mahale Mountains, initially under the direction of Itani and later of Nishida (1968, 1990), focussed first on K Group, then later on M Group.

For many reasons, these two longitudinal studies are ideal candidates for comparative studies. First, both are of the eastern subspecies of chimpanzee. This should reduce the chances that any differences found between them are genetic. Second, both are relic populations of what was once part of a continuous cline, since disrupted in modern times by deforestation. In straight-line distance the two study sites are only about 170 kilometers apart, and it is unlikely that the two populations have had time to differentiate markedly through genetic drift. Third, they occupy similar types of habitat: mixed forest and woodland on the rugged escarpment of the Great Rift which forms the eastern shore of Lake Tanganyika. Fourth, both studies have focussed on one or two communities or groups in which all members are individually recognizable and well-known. Fifth, both have been studied using similar methods initially based on provisioning, although it has since been abandoned. This has enabled near-continuous human monitoring of all known chimpanzee activities at very close-range. The differences between Gombe and Kasoje are far fewer than those between either of them and any other long-term study site.

2. CASE STUDY: GROOMING

Grooming is the co-ordinated fine manipulation, sometimes linked with the use of lips or tongue, and close inspection of the body surface of the self or of another individual. In many kinds of primates, including humans, grooming has at least two distinct but compatible functions: On one level it is hygienic, serving to remove ecto-parasited, extraneous matter, and body products. On another level, social grooming is an intimate interaction between two friends or relations. Normal chimpanzee grooming shows a species-typical form which is the same for all known wild populations. Variants on the standard form can be seen both in the wild and in captivity, but these are idiosyncratic elaborations.

Caroline Tutin and I made a brief study of chimpanzees at Kasoje in 1975. To our surprise, we witnessed a behavioral pattern which at that time had not been described for chimpanzees. This we called the grooming-hand-clasp. It always occurred at the beginning of or during an otherwise normal bout of social grooming. Each of the participants simultaneously extended an arm overhead and then either or both grasped the other's wrist or hand, or both clasped the other's hand. Meanwhile, the opposite hand was used to groom the other individual's underarm area revealed by the upraised limb. In doing so, the two chimpanzees sat facing one another on the ground in a symmetrical configuration. Either both raised their right arms and groomed with their left, or vice versa. The effect was striking (for details see McGrew & Tutin, 1978):

(1) Participants engaged in mutual grooming only in pairs.

(2) Only adults and adolescents performed the pattern.

(3) Performance of the behaviour was evenly distributed in the group, suggesting that it was a regular activity, and not idiosyncratic.

(4) The pairs tended to be of mixed-sex rather than of same-sex composition.

(5) At an average rate of once every 2.4 hr, the frequencies exceeded those for tool-use, predation, food-sharing, and almost all sexual and agonistic behavior.

Otherwise, the chimpanzees of Kasoje showed social grooming that was typical of that seen in other wild populations of the same and of different geographical races of chimpanzees, as well as that shown by captive groups.

Against this must be set the total absence of the grooming-hand-clasp among the chimpanzees of Gombe over thousands of hours of observation over more than 30 years. It is simply unknown there. The two nearest behavioral patterns known at Gombe are: (a) brief use of an upraised arm in grooming invitation, but only one individual does this at a time and the upraised arm is not touched by the other (Goodall, 1968) and (b) more prolonged grasping of branches overhead while engaged in social grooming.

Only one other report of this behavioral pattern has come from other field studies of apes. In the Kibale forest, Ghiglieri (1984) found that it occurred in 38% of non-maternal grooming sessions shown by the Kanyawara chimpanzees in his secondary study-area. Most intriguingly, it was entirely absent from the grooming repertoire of the neighboring Ngogo chimpanzees who lived 10 kilometres to the southeast.

What are the origins of these differences between neighboring communities of chimpanzees? It is possible that the behavioral differences seen between Gombe and Kasoje and between Kanyawara and Ngogo reflect differences in genotype. However, there seems not to have been enough time for genetic drift to account for the contrast. Alternatively, it is possible that the chimpanzees at Kasoje and Kanyawara are responding individually to a selection pressure present in their environment but absent at Gombe and at Ngogo. However, it is hard to see how differences in habitat could account for the presence or absence of the grooming-hand-clasp. It is conceivable that Kasoje's chimpanzees suffer from (say) a bothersome axillary parasite which requires more frequent attention, but there is no evidence to suggest this. This leaves by exclusion the third possibility that the grooming-hand-clasp of Kasoje's chimpanzees is some sort of social custom.

3. HAMMERS AND ANVILS

Stone tools have always been of special significance to prehistorians interested in the evolutionary origins of technology. Lithic artifacts preserve well, and so the earliest known tools in the archaeological record have been interpreted as pebble hammers. Given this, what can be said about non-human and especially chimpanzee use of hammers and anvils?

[These apparently simple objects are vexedly hard to define with logical consistency. Hammers (sensu strictu) are here considered to be tools that are used in the hand to apply explosive percussive force to a resting goal-object. The result is that the goal-object is fractured to reveal its contents. Anvils (sensu strictu) are here defined as fixed objects, usually in the substrate, which support goal-objects to be hammered. Thus, anvils are not tools because they are not handled and remain stationery. In reality, goal-objects may be struck directly against a hammer, an anvil, one another, or the substrate. Also, anvils may be moved, although between and not during blows].

Apart from chimpanzees, the only primates to use hammers or anvils habitually and spontaneously are capuchin monkeys (Cebus apella). This occurs both in nature (Izawa & Mizuno, 1977) and in captivity (Visalberghi, 1990), but there is a notable difference between these two conditions. All object-smashing confirmed for wild capuchins is anvil-use, while captive capuchins show anvil-use, hammer-use, and hammer-and-anvil use.

Wild chimpanzees have been reported to use hammers and anvils at seven main sites: Bossou, Cape Palmas, Mt. Kanton, "Liberia", Sapo, Tai, and Tiwai Island. (Kortlandt & Holzhaus, 1987, reported a number of other sub-sites). Free-ranging chimpanzees after release have been seen to do so at a further three places: Abuko, Assirik, and Bassa Islands. (Table 1 gives details of references). However, behavioral data on the tools being used are available from only three of these, Bassa, Bossou, and Tai. Other reports are anecdotal, second-hand, or circumstantial and this may lead to confusion unless criteria for inference from indirect data are explicitly stated.

4. CATALOGUING TOOL-KITS

By the early 1970s, it was clear that wild chimpanzees at various sites in Africa had different repertoires of tool-use. Goodall (1973) produced the first catalog of tool-use by free-ranging chimpanzees; it included 10 sites, and all but Gombe's data were based on short-term studies or single sightings. Teleki (1974) followed with a list of 12 sites, of which only five were common to Goodall's catalog of a year earlier! The most extensive published catalog is that of Beck (1980), who compiled findings from 20 sites across Africa. More recently, Goodall (1986) produced another list, but this had only 16 populations of wild (but not released and free-ranging) chimpanzees. All of the previous efforts are now dated.

Table 1 lists 34 populations or groups of free-ranging chimpanzees in Africa that have shown some kind of tool-use, if a tool is defined as a moveable, inanimate object used to facilitate acquisition of a goal (cf Beck, 1980). This minimal measure shows that the far western sub-species has 15 records, the central-western has 13, and the eastern has only six.

No previous catalog of tool-kits has sought to distinguish between <u>habitual</u> versus <u>rare, idiosyncratic, or questionable</u> tool-use by chimpanzees. Thus habitual use is here restricted to patterns shown repeatedly by several members of a group. It excludes a single instance by one individual, a single instance by several individuals, several instances by only one individual, and all instances of insufficient data. Cases by released chimpanzees are also omitted, because they may have been influenced or even shaped by their human caretakers. This is <u>not</u> to say that single cases are useless. On the contrary, it takes only one example to show a capacity. Many of the instances classed as non-habitual may merely be patterns awaiting more evidence. Finally, many of the older records are anecdotal because such natural history "notes" were adequate by then-current standards of scientific reporting.

TABLE 1: African study-sites of free-ranging chimpanzees at which tool-use has been recorded (For full details of references, see McGrew, 1992.)

Site	Country	Subspecies	Major Sources
Abuko (r)	The Gambia	v	Brewer 1978; Goodall 1973
Assirik	Senegal	v	McBeath & McGrew 1982; McGrew et al. 1979a
Assirik (r)	Senegal	v	Brewer 1978, 1982
Ayamiken	Equat. Guinea	t	Jones & Sabater Pí 1969, 1971
Baboon (r)	The Gambia	v	Brewer & McGrew 1990
Banco	Ivory Coast	v	Hladik & Viroben 1974
Bassa (r)	Liberia	v	Hannah & McGrew 1987
Belinga	Gabon	t	McGrew & Rogers 1983
Beni	Zaire	t	Kortlandt 1965
Bossou	Guinea	v	Albrecht & Dunnett, 1971; Sugiyama 1981, 1990; Sugiyama & Koman 1979, 1988
Budongo	Uganda	s	Sugiyama 1969
"Cameroon"	Cameroon	t	Merfield & Miller 1956
Campo	Cameroon	t	Sugiyama 1985
Cape Palmas	Gabon	t	Savage & Wyman 1844
Dipikar	Equat. Guinea	t	Jones & Sabater Pí 1969, 1971
Filabanga	Tanzania	s	Itani & Suzuki 1967
Gombe	Tanzania	s	Goodall 1964, 1968, 1970, 1973, 1986; McGrew 1974, 1977, 1979; Teleki 1974
Ipassa (r)	Gabon	t	Hladik 1973
Kanka Sili	Guinea	v	Albrecht & Dunnett 1971
Kanton	Liberia	v	Kortlandt & Holzhaus 1987
Kasakati	Tanzania	s	Izawa & Itani 1966; Suzuki 1966
Kasoje (Mahale)	Tanzania	s	McGrew & Collins 1985; Nishida 1977, 1980b; Nishida & Hiraiwa 1982; Nishida & Uehara 1980; Uehara 1982
Kibale	Uganda	s	Ghiglieri 1984, 1988; Wrangham et al. 1994
Kahuzi-Biega	Zaire	s	Yamagiwa et al. 1988
"Liberia"	Liberia	v	Beatty 1951
Lopé	Gabon	t	Tutin & Wrogemann 1993
Mayang	Equat. Guinea	t	Gonzalez-Kirchner & Sainz de la Maza 1992
Mbomo	Congo	t	Fay & Carroll 1990
Ndakan	Cent. Afr. Rep.	t	Fay & Carroll 1990
Ngoubunga	Cent. Afr. Rep.	t	Fay & Carroll 1990
Okorobiko	Equat. Guinea	t	Jones & Sabater Pí 1969, 1971; Sabater Pí 1974
Sapo	Liberia	v	Anderson et al. 1983
Tai	Ivory Coast	v	Boesch & Boesch 1990; Rahm 1971
Tenkere	Sierra Leone	v	Alp 1993
Tiwai	Sierra Leone	v	Whitesides 1985
West Cameroon	Cameroon	t	Struhsaker & Hunkeler 1971

(r) = released populations
s = eastern subspecies
t = central western subspecies
v = far western subspecies.

Table 2 presents a stricter catalog, limited to habitual tool-use shown only by wild chimpanzees. Only 13 populations showed a total of 49 habitual tool-use patterns that met the above criteria, and within these, the range was from one to 11 per site. Six of the sites showed but one pattern (Campo, Kanton, Kibale, Okorobiko, Sapo, Tiwai), but like Lopé or Assirik (N=2) all were places where the subjects of study were totally unhabituated or only minimally tolerant of close-range observation. Overall, a clear positive correlation emerges between degree of habituation or length of study and number of identified patterns of habitual tool-use, which suggests that the results are incomplete for most populations.

No pattern of tool-use even comes close to being universal. The most widespread, the use of a hammer to crack open nuts, is known at only five sites. The inter-related agonistic or anti-predatory patterns of weapon-use (flail-club-missile) occur in five populations, but one of these, Kanka Sili, was experimentally induced. More impressive are the five occurrences of ant-dipping, especially as this pattern was found from the wettest forest site at Bossou to the driest savanna site at Assirik, and at points in between (Alp, 1993).

The functional nature of the habitual tool-use is instructive: Most (26 of 49) are subsistence activities of acquiring or processing food, especially social insects (N=13) or nuts (N=5). Only one each relates to meat, water, or other plant foods. As noted above, the weapon-use total (N=13) may be inflated by its make-up of related patterns. The 10 remaining habitual types are split between self-directed (leaf-napkin, self-tickle) and apparently ritualized communicative signals (play-start, leaf-groom, leaf-clip).

5. REGIONAL AND LOCAL PATTERNS

If the 1960s can be characterized as the initial period of descriptive ethnography of chimpanzees, then the 1970s can be thought of as the decade of gross, regional comparisons, followed by the 1980s, which produced finer-grained, local comparisons. As discussed above, until Goodall (1968) showed that Gombe's chimpanzees had a tool-kit (although she did not use the term), all previous accounts were anecdotes. Such minimal ethnography continues to fill in gaps, especially in populations whose habituation is incomplete. For example, ant-dipping at Bossou was first seen only in 1987 (Sugiyama et al., 1988), though the chimpanzees had been studied since 1976.

The first attempt to compare different populations of chimpanzees was by Struhsaker & Hunkeler (1971). (See Table 3). They set a pattern of using the geographical races or sub-species of chimpanzees as a basis for regional comparison. They hypothesized that far western chimpanzees were hammer-users to smash nuts while eastern and central-western chimpanzees were termite-"fishers". The latter entailed lumping all forms of tool-use to get

TABLE 2: Habitual patterns of tool-use of wild chimpanzees (X = present)
(See McGrew, 1992, for references)

Pattern	Gombe	Bossou	Mahale	Tai	Kanka Sili	Lopé	Assirik	Others
Termite-fish	X		X				X	
Ant-dip	X	X		X			X	Tenkere
Honey-dip	X			X		X		
Leaf-sponge	X	X	X					
Leaf-napkin	X							Kibale
Stick-flail	X	X	X		X			
Stick-club	X	?X	X	X	X			
Missile-throw	X	X	X	X	X			
Self-tickle	X							
Play-start	X		X					
Leaf-groom	X		X					
Ant-fish			X			X		
Leaf-clip		X	X					
Gum-gouge		X						
Nut-hammer		X		X				Kanton, Sapo, Tiwai
Marrow-pick				X				
Bee-probe				X				
Branch-haul		X						
Termite-dig								Campo Okorobiko
Total	11	8	9	7	3	2	2	

TABLE 3: Hypothesized regional differences in tool-use across wild chimpanzees
(See McGrew, 1992, for references).

East (P.t.s.)	Central West (P.t.t.)	Far West (P.t.v.)	Source
-----------------Termite fishers--------------		Nut smashers	Struhsaker & Hunkeler 1971
----------------Ant/termite fishers------------		Nut smashers	Nishida 1973
Foliage industry	Stick industry	Stone industry	Sabater Pí 1974
Ant/termite probers	Termite probers	Pounders	Teleki 1974
Termite fishers	Termite diggers	Termite fishers	McGrew et al. 1979
----------------Non-nut crackers----------------		Nut crackers	McGrew 1992

termites. Nishida (1973) followed the same dichotomous distinction, but expanded fishing to include ants as well as termites, given his findings on Componotus-eating at Kasoje.

Teleki (1974) refined this scheme into the first three-way one, whereby far western chimpanzees pounded with hammers, central-western chimpanzees probed for termites (but not ants), and eastern chimpanzees probed for both ants and termites. This distinction was untenable even before it appeared, as Hladik (1973) had reported extensive ant-fishing at Ipassa. At the same time, Sabater Pí (1974) presented another three-way distinction, based on stones (far west), sticks (central west), and foliage (east). Already published data belied this too, as Goodall's (e.g. 1968) reports of ant-dipping with sticks were well-established. Both Teleki's and Sabater Pí's comparisons ignored Goodall's (1973) report of Brewer's observations of wild-born, rehabilitated chimpanzees in The Gambia probing for termites, thus making far wester chimpanzees more than nut-crackers.

McGrew et al. (1979) presented a new kind of regional comparison. First, they held prey constant (Macrotermes termites) and concentrated on one kind of tool-use (probes of vegetation). Second, they presented detailed and systematic comparisons of features of tool-use, such as dimensions of tools. Third, they related differences in tool-use to differences in the ecology of prey and of raw materials, such as patterns of rainfall and termites' mound-building. Finally, in addition to hypotheses, they gave explicit ways in which these could be falsified. The result was a scheme whereby forest-living (central-western) chimpanzees were termite-diggers and savanna- (far western) and woodland-living (eastern) chimpanzees were termite-fishers. These hypotheses were later falsified by a report of termite-fishing in the forest at Belinga in the central-west region.

So, what is the current state of regional comparisons? Some differences still persist: Use of hammers of stone or wood to crack open hard-shelled containers still has been reported only in the far west beyond the Dahomey Gap (McGrew 1992). Use of sticks as picks to perforate or to dig up termites' mounds remains unique to the central-western chimpanzees of the equatorial forests, as apparently does the use of the brush-stick (Sugiyama, 1985). Use of medicinal plants in self-treatment of illness has been recorded so far only from the eastern subspecies of the chimpanzee (Huffman & Seifu, 1989). All other apparent regional distinctions have fallen by the wayside, though many await further study.

The overall picture of variety in chimpanzee tool-use has seen a pendular change. From a position of generalizing about the species as recently as 25 years ago, the tendency now is to emphasise differences, not similarities (Boesch & Boesch, 1990; Sugiyama, 1990). Contrasts are well known across regional races, populations, and communities or groups. However, establishing the existence of differences says nothing in itself about their origins, cultural or otherwise (McGrew, 1992).

6. NON-HUMAN CULTURE?

To what extent do species other than humans possess something akin to humans culture? One problem lies in the area of <u>definition</u>. For example, it is all very well to produce theoretically satisfying definitions, but if these cannot be translated into operational ones, then empirical testing is impossible. This paper is not about definition, but it assumes that satisfactory definitions exist or can be devised. The crucial next step is to assess on the basis of comparison of data the usefulness of the comparative method for understanding the origins of culture.

Before going on, it seems advisable to review briefly other approaches to chimpanzee culture which have been advanced. The earliest and still most common of these was for the observer to describe the ingenious and sometimes astonishing acts of a single population of apes. These patterns were seen as being the results of individual learning, even if their origins could not be specified. Such results emerged in the 1960s from the wild, especially from long-term studies such as those of the Kasakela community at Gombe and of K group at Kasoje.

The problem with all such studies is that with a sample size of one, be it an individual, group or population, there is no variance. One can always argue that a unique set of conditions has prompted an idiosyncratic response, whether ontogenetically or phylogenetically programmed. If only one population of chimpanzees preys upon a painfully stinging kind of ant, then this may reflect the unusual make-up of a marginal habitat where subsistence is harsh (McGrew, 1983). That is, extraordinary circumstances may prompt a similarly extraordinary response. There is no need in these cases to call upon any social factor, which is assumed to be a necessary, if not sufficient, condition for culture.

A more useful approach is that of limited comparison. This depends on having at least two appropriate sets of data, and such raw material for chimpanzees became available from the 1970s onwards. The most common paired comparison has been that of two neighboring populations such as Gombe versus Kasoje (McGrew & Tutin, 1978; Nishida <u>et al</u>., 1983), although a few others have gone further afield.

There are several drawbacks to such limited comparisons. One also follows from the constraints of small sample size: An apparently notable contrast may merely reflect an anomaly in one of the two sets compared. Another is that selective comparisons of certain tasks may emphasize a few differences at the expense of even more similarities between them. Conversely, approximate and general rather than systematic and precise comparisons may obscure differences which lurk behind vague generalizations of similarity. Most problematic, however, is that comparisons of the subjects may really be comparisons of the scientists, especially if their methods of data collection differ. For example, if in comparing diets, one set of data is based on fecal sampling (which is biassed against food-items of only soft tissue that are fully digested) and the other is based on direct observation (which is biassed against inconspicuous events, except in totally observable subjects), it would not be surprising to find apparent but false differences. This is not an idle complaint; it applies to most possible pairings of data from long-term studies of chimpanzees.

Put another way, the ideal two-way comparison involves a close match-up, with all or as many as possible of the independent variables being controlled for. Contrasts found in two sets of subjects will be more convincing if done by the same investigators using the same methods, etc. This is rarely achieved. Another approach is to randomize out these variables by using many sets of data, on the assumption that differences in technique, or definition, or sampling, or precision, or any other noise in the system, will pale to insignificance. Any phenomenon which survives this treatment is likely to be a robust one. For example, if all wild chimpanzees build arboreal sleeping-platforms by inter-weaving the terminal branches of trees, regardless of when and where studied, for how long, by whom, using whatever system of recording, then it seems safe to conclude that one is dealing with a universal aspect of chimpanzee nature. The rub is that many sets of data are needed for this sort of exercise, and this really became possible only in the 1980s, as fieldwork, especially in West Africa, advanced (McGrew, 1992).

7. CULTURED CHIMPANZEES?

Given that differences exist across populations of wild chimpanzees, are these cultural? Galef (1976, p.77) stipulated three means by which such a difference might occur: (a) genetically transmitted propensities which are virtually independent of environmental influence affecting their expression in ontogeny; (b) similarly structured transactions between individuals and their environment in one community as opposed to different transactions in another community; (c) transmission of behavioral patterns through social learning from one individual to another, according to the norms of the community in which they live.

Abundant examples now exist of the social transmission of acquired behavior in groups of primates. These have been demonstrated in quantitative observations and experiments both in laboratory and field (see review in Nishida, 1987).

Contrasts in behavior between groups need not be cultural. Differences can be innate, just as similarities can be learned, to use the old-fashioned, nature-nurture terms. Neither does complexity nor variance in behavioral patterns tell us whether they are of cultural or idiosyncratic origin. The simplest pattern may be acquired by imitation, and phenotypic variety may reflect genotypic variation, as in pleiotropy. The only way to be confident of the cultural nature of an act is to see it being done for the first time by an individual and then passed on to others (McGrew, 1992).

8. INNOVATION

There seem to be a least four ways that innovation could happen (cf. Kummer & Goodall, 1985): The clearest would be the spontaneous invention (in a constant environment) of a new pattern that was then copied by others. This could occur in an intact group and would be recognized by the novelty of the act set against a history of negative data. A second way is by diffusion, in which an already-skilled performer joins a naive group and shows them (inadvertently or not) the pattern. A third way is when the new pattern is prompted by environmental change. An existing habit may be altered by natural forces or the group may occupy a new habitat and its members adapt to the change. A fourth way is really a special case of the third, when the agent is intervention. Sympatric humans (or other species) may intentionally or accidentally shape the behavior of the chimpanzees. Any of these four types of innovation, if followed by dissemination, could change the technology of a group of apes.

Taking these in reverse order, there is much evidence of the effects of human influence, especially as most behavioral data come from provisioned subjects. Both the spread of the use of levers to open banana distribution boxes, and the famous example of Mike and the paraffin tins, are well-known from Gombe (Goodall, 1973). More hypothetical, but equally inadvertent, is Kortlandt's (1986) view that the wild chimpanzees of Bossou learned to crack palm nuts from the local humans. He stated that the work-sites are identical in appearance, distinguishable only by their locations. (Whatever the similarities, equally plausible would be the opposite conclusion, that the humans originally learned the pattern from the apes). Less disputable are deliberate human interventions, such as Brewer's (1978) teaching her rehabilitant chimpanzees to use stone as hammers to smash open *Afzelia* pods, a custom never seen in wild apes.

There seem to be no cases of natural environmental changes being seen to cause innovation in tool-use by chimpanzees. The closest case may be the use of hammers to crack open oil palm nuts, in that the wide range of feral oil palms is a recent development on an evolutionary time-scale. It is not known how the oil palm spread so widely, but it was probably a combination of natural seed dispersal and unnatural human horticulture, though chimpanzees may have contributed.

Kasoje's chimpanzees present several well recorded examples of dietary innovations, all of them cases of the apes making use of feral cultigens such as mangoes originally planted but then left behind by departed villagers (Takasaki, 1983). This was occupation by the apes of an empty but hardly natural 'niche'. The simplest explanation for the lack of innovation in response to environmental change is that the time-scale of field primatology is too short to pick up all but the most catastrophic of natural changes, none of which has yet been documented.

The arrival of tool-use by immigration has probably occurred at Kasoje, in that termite-fishing females from K Group have moved to M Group (Takahata, 1982). Uptake of the pattern by the residents, that is, diffusion, remains to be seen. A more striking case was the dramatic spread of palm nut cracking on one of the Bassa Islands, as reported by Hannah & McGrew (1987). For 12 weeks before the arrival of an adult female to join the group of rehabilitates on the island, no signs of tool-use were seen. Within hours of arrival, she began to use stones to crack nuts, and within a month nine of the 13 chimpanzees were also cracking nuts. Strictly speaking, this does not show social learning, as the other chimpanzees may already have known the pattern and only have been prompted by her enthusiastic performance to recall it. However, this seems far-fetched.

Finally, there is no known case of spontaneous invention of a tool-use pattern, followed by dissemination, in the pure sense (cf. Kummer & Goodall, 1985). That is, all known cases can be alternatively explained by the previous three ways. Perhaps this should not be too disappointing. How many studies, however long-term, of traditional peoples in situ have ever reported spontaneous invention of new tools? Or, if the Acheulean hand-axe persisted virtually unchanged for hundreds of thousands of years, why should we expect to be lucky enough to see notable changes in the material culture of apes in three decades?

9. CULTURE DENIED?

In a thoughtful and provocative chapter, Tomasello (1990) tackled in detail the problem of whether or not chimpanzee tool-use is cultural. His conclusions are skeptical, if not totally negative. (Others have posed similar questions, e.g. Whiten, 1989; Galef, 1990). Based on his work with captive chimpanzees (Tomasello et al., 1987) he argued that there is no convincing experimental evidence of chimpanzee imitation. This, plus a similar lack of data on chimpanzee teaching, persuaded him that all of chimpanzee behavior can be explained without invoking culture. This seems compelling, especially as he concluded that all potentially impressive cases of tool-use practice in the wild have been independently and individually invented by chimpanzees in the laboratory. Thus, candidates like nut-cracking, leaf sponging, termite-fishing, ant-dipping, etc. can be discounted. How solid is his case?

Failure to find true imitation under contrived and impoverished laboratory conditions is hardly surprising. Asking immature apes to learn from a brief exposure to a strange adult when they are evolved to learn from their mothers over years is asking a lot. Failure to find imitation in the field, when in principle it can never conclusively be proven under uncontrolled conditions, is illogical. A similar argument probably applies to failure to find teaching, though until this phenomenon is clearly defined in operational terms, its confirmation or otherwise will be messy. (Boesch, 1991, presented suggestive anecdotal evidence for teaching in nut-cracking, but there are alternative, non-teaching explanations for most or all of what he provided). Thus the question of whether or not chimpanzees show imitation or teaching remains open.

But even if chimpanzees never imitated or taught, this does not require us to deny them culture. As Tomasello remarks, many behavioral differences across human populations persist without such mechanisms. Wynn (1993) gave several examples of skilled human operations that are acquired "only" by observational learning. What (small) proportion of human knowledge is transmitted by intentional instruction? Interesting as the detailed mechanisms of information transfer may be to experimental psychologists, they play little part in discussions of culture in ethnology or prehistory.

Tomasello's (1990) claim that all notable tool-use practices of wild chimpanzees have been invented by chimpanzees individually in the laboratory is simply wrong. Sumita et al. (1985) trained zoo chimpanzees to crack domesticated nut species. Kitahara-Frisch & Novrikoshi's (1982) single subject never learned to leaf-sponge, though she did dip for juice with some items of vegetation. No experiment has ever simulated any of the delicate probing tasks such as termite-fishing shown by wild chimpanzees, and all of the purported examples listed by Tomasello (e.g. Menzel et al., 1970) are simple reaching tasks that are readily done by monkeys (Beck, 1973). However, even if captive chimpanzees did spontaneously show all of these patterns (and under such conducive circumstances this would hardly be surprising), it need not say anything about chimpanzee culture in the wild.

As Tomasello rightly points out, the challenge is to tackle empirically chimpanzee culture in nature, or at least in naturalistic, free-ranging settings. There are several promising ways to proceed. One is to perform field experiments in which a crucial variable such as exposure to particular objects and others is manipulated. Hannah & McGrew's (1987) report of the spread of nut-cracking was not such an experiment, but it could easily have been. Another is to be more precise in operationalising key points to be looked for in field settings, as Russon & Galdikas (1993) have done for imitation in rehabilitated orangutans. Finally, and probably most demandingly, new theoretical approaches, such as King's (1991) emphasis on social information transfer, rather than limited (and outmoded?) ideas such as imitation or stimulus enhancement, may prove useful.

Tomasello (1990) has rightly called to task those who use terms like culture and imitation lightly, but their applicability to chimpanzee life has hardly been tested. Whether culture occurs without imitation, or imitation without culture remains to be seen. (Heyes, 1993).

So are chimpanzees cultural or not? In commenting on McGrew & Tutin's (1978) original report, Washburn & Benedict (1979) thought not. For them, even if another species satisfied all eight conditions, it could not be granted cultural status unless it also had language. Language may be the most efficient means of natural communication yet devised and human beings may be the only naturally linguistic creatures (both debateable points), but why should language be essential to culture? Both living pre-verbal humans and non-verbal non-humans readily learn new behavioral patterns without it. Moreover, if language were a necessary and sufficient condition for the emergence of culture, then we would probably have to deny culture to evolutionarily pre-linguistic creatures ancestral (Noble & Davidson, 1991) to anatomically modern humans. This would require denying culture to those responsible for the Shanidar flower burial (Leroi-Gourhan, 1975), which seems excessive.

It may be that, in the end, human culture is unique to humans, just as chimpanzee culture is unique to chimpanzees. But if we want to know about australopithecine culture, or early Homo culture, we might well be advised to tackle it from as many directions as possible, including that of our African ape sibling species.

REFERENCES

Alp, R. (1993) Meat eating and ant dipping by wild chimpanzees in Sierra Leone. Primates, 34: 463-468.

Beck, B.B. (1980) Animal Tool Behavior: The Use and Manufacture of Tools by Animals. New York: Garland STPM Press.

Boesch, C. (1991) Teaching among wild chimpanzees. Animal Behaviour, 41: 530-532.

Boesch, C. & Boesch, H. (1990) Tool use and tool making in wild chimpanzees. Folia primatologica, 54: 86-99.

Brewer, S. (1978) The Chimps of Mt. Asserik. New York: Alfred A. Knopf.

Byrne, R. & Whiten, A. (eds.) (1988) Machiavellian Intelligence: Social Expertise and the Evolution of Intellect in Monkeys, Apes, and Humans. Oxford: Clarendon Press.

Galef, B.G. (1976) Social transmission of acquired behavior: A discussion of tradition and social learning in vertegrates. In: Advances in the Study of Behavior, Volume 6, Rosenblatt, J.S., Hinde, R.A., Shaw, E., & Beer, C. (eds), New York: Academic Press, pp 77-100.

Galef, B.G. (1992) The question of animal culture. Human Nature, 3: 157-178.

Ghiglieri, M.P. (1984) The Chimpanzees of Kibale Forest. New York: Columbia University Press.

Gonzalez-Kirchner, J.P. & Sainz de la Maza, M. (1992) Sticks used by wild chimpanzees: A new locality in Rio Muni. Folia Primatologica, 58: 99-102.

Goodall, J.v.L. (1968) The behaviour of free-ranging chimpanzees in the Gombe Stream Reserve. Animal Behaviour Monographs, 1: 161-311.

Goodall, J.v.L. (1973) Cultural elements in a chimpanzee community. In: Precultural Primate Behavior, Menzel, E.W. (ed.), Basel: S. Karger, pp. 144-184.

Goodall, J. (1986) The Chimpanzees of Gombe. Cambridge, Mass.: Harvard University Press.

Hall, E.T. (1959) The Silent Language. Greenwich: Fawcett, pp 192.

Hannah, A.C. & McGrew, W.C. (1987) Chimpanzees using stones to crack open oil palm nuts in Liberia. Primates, 28: 31-46.

Heyes, C.M. (1993) Imitation, culture and cognition. Animal Behaviour, 46: 999-1010.

Hladik, C.M. (1973) Alimentation et activite d'un groupe de chimpanzes reinstroduits en foret gabonaise. Terre et la Vie, 27: 343-413.

Huffman, M.A. and Seifu, M. (1989) Observations on the illness and consumption of a possibly medicinal plant Vernonia amygdalina (Del.) by a wild chimpanzee in the Mahale Mountains National Park, Tanzania. Primates, 30: 51-63.

Izawa, K. & Mizuno, A. (1977) Palm-fruit cracking behaviour of wild black-capped capuchin (Cebus apella). Primates, 18: 773-793.

King, B.J. (1991) Social information transfer in monkeys, apes, and hominids. Yearbook of Physical Anthropology, 34: 97-115.

Kitahara-Frisch, J. & Norikoshi, K. (1982) Spontaneous sponge-making in captive chimpanzees. Journal of Human Evolution, 11: 41-47.

Kortlandt, A. (1986) The use of stone tools by wild-living chimpanzees and earliest hominids. Journal of Human Evolution, 15: 77-132.

Kortlandt, A. & Holzhaus, E. (1987) New data on the use of stone tools by chimpanzees in Guinea and Liberia. Primates, 28: 473-496.

Kummer, H. & Goodall, J. (1985) Conditions of innovative behaviour in primates. Philosophical Transactions of the Royal Society of London, B308: 203-214.

Leroi-Gourhan, A. (1975) The flowers found with Shanidar IV, a Neanderthal burial in Iraq. Science, 190: 562-563.

McGrew, W.C. (1983) Animal foods in the diet of wild chimpanzees: Why cross-cultural variation? Journal of Ethology, 1: 46-61.

McGrew, W.C. (1992) Chimpanzee Material Culture: Implication for Human Culture. Cambridge: Cambridge University Press.

McGrew, W.C. & Tutin, C.E.G. (1978) Evidence for a social custom in wild chimpanzees? Man, 13: 234-251.

McGrew, W.C. Tutin, C.E.G. & Baldwin, P.J. (1979) Chimpanzees tools and termites: Cross-cultural comparisons of Senegal, Tanzania and Rio Muni. Man, 14: 185-214.

Menzel, E.W., Davenport, C. & Rogers, R. (1970) The development of tool-using in wild-born chimpanzees. Folia primatologica, 12: 273-283.

Nishida, T. (1973) The ant-gathering behaviour by the use of tools among wild chimpanzees of the Mahali Mountains. Journal of Human Evolution, 2: 357-370.

Nishida, T. (1987) Local traditions and cultural transmission. In: Primate Societies, Smuts, B.B., Cheney, D.L., Seyfarth, R.M., Wrangham, R.W. & Struhsaker, T.T. (eds.), Chicago: University of Chicago Press, pp. 462-474.

Nishida, T. (ed.) (1990) The Chimpanzees of the Mahale Mountains. Tokyo: University of Tokyo Press.

Nishida, T., Wrangham, R., Goodall, J., & Uehara, S. (1983) Local differences in plant-feeding habits of chimpanzees between the Mahale Mountains and Gombe National Park, Tanzania. Journal of Human Evolution, 12: 467-480.

Noble, W. & Davidson, I. (1991) The evolutionary emergence of modern human behaviour: language and its archaeology. Man, 26: 223-253.

Russon, A.E. & Galdikas, B.M.F. (1993) Imitation in free-ranging rehabilitant orangutans (Pongo pygmaeus). Journal of Comparative Psychology, 107: 147-161.

Sabater Pí, J. (1974) An elementary industry of the chimpanzees in the Okorobiko Mountains, Rio Muni (Republic of Equatorial Guinea), West Africa. Primates, 15: 351-364.

Struhsaker, T.T. & Hunkeler, P. (1971) Evidence of tool-using by chimpanzees in the Ivory Coast. Folia primatologica, 15: 212-219.

Sugiyama, Y. (1985) The brush-stick of chimpanzees found in south-west Cameroon and their cultural characteristics. Primates, 26: 361-374.

Sugiyama, Y. (1993) Local variation of tool and tool behaviour among wild chimpanzee populations. In The Use of Tools by Human and Non-human Primates, Berthelet, J. & Chavaillon J. (eds.), Oxford: Oxford University Press.

Sugiyama, Y., Koman, J. & Sow, B. (1988) Ant-catching wands of wild chimpanzees at Bossou, Guinea. Folia primatologica, 51: 56-60.

Sumita, K., Kitahara-Frisch, J. & Norikoshi, K. (1985) The acquisition of stone-tool use in captive chimpanzees. Primates, 26: 168-181.

Takahata, Y. (1982) Termite-fishing observed in M group chimpanzees. Mahale Mountains Chimpanzee Research Project Ecological Report No. 18.

Takasaki, H. (1983) Mahale chimpanzees taste mangoes--toward acquisition of a new food item? Primates, 24: 273-275.

Teleki, G. (1974) Chimpanzee subsistence technology: materials and skills. Journal of Human Evolution, 3: 575-594.

Tomasello, M. (1990) Cultural transmission in the tool use and communicatory signalling of chimpanzees? In: Language and Intelligence in Monkeys and Apes, Parker, S.T. & Gibson, K.R. (eds.) Cambridge: Cambridge University Press, pp. 274-311.

Tomasello, M., Davis-Dasilva, M., Camak, L. and Bard, K. (1987) Observational learning of tool-use by young chimpanzees. Human Evolution, 2: 175-183.

Tutin, C.E.G. & Wrogemann, D. (1993) Tool-use by chimpanzees (Pan t. troglodytes) in the Lopé Reserve, Gabon.

Visalberghi, E. (1990) Tool use in Cebus. Folia primatologica, 54: 146-154.

Washburn, S.L. & Benedict, B. (1979) Non-human primate culture. Man, 14: 163-164.

Whiten, A. (1989) Transmission mechanisms in primate cultural evolution. Trends in Ecology and Evolution, 4: 61-62.

Wrangham, R.W., de Waal, F.B.M., & McGrew, W.C. (1994) Introduction: The challenge of behavioral diversity. In: Chimpanzee Cultures, Wrangham, R.W., McGrew, W.C., de Waal, F.B.M., & Heltne, P.G. (eds.), Cambridge, Mass: Harvard University Press.

Wynn, T. (1993) Layers of thinking in tool behavior. In: Tools, Langauge and Cognition in Human Evolution, Gibson, K.R. & Ingold, T. (eds.), Cambridge University Press, pp. 389-406.

Yamagiwa, J., Yumoto, T., Ndunda, M. & Maruhashi, T. (1988) Evidence of tool-use by chimpanzees (Pan troglodytes schweinfurthii) for digging a bee nest in the Kahuzi-Biega National Park, Zaire. Primates, 29: 405-411.

Precultural Behavior of Japanese Macaques: Longitudinal Studies of the Koshima Troops

Kunio Watanabe
Primate Research Institute,
Kyoto University,
Inuyama, Aichi, 484 Japan

Abstract. Longitudinal changes in potato and wheat washing behaviors by Japanese macaques on Koshima Island were analyzed along with those of the surrounding situations during the last forty years. Potato washing behavior was maintained by most monkeys even after 20 years had past since artificial food supplies were greatly decreased. Wheat washing behavior has developed in variety, but was diminishing in frequency. The wheat washing behavior was more stereotypic and differed by individual, which may indicate that the inventive behavior must be fixed in the own behavioral repertoire of the performers. The new habit could spread over the group members through two different processes; 1) from a young initiator to old monkeys through kinship and play mates, and 2) from old initiator(s) to the other members. The attitude of monkeys together with surrounding situations may contribute to the process and rapidity of propagation of the new habits. The possible behavioral mechanisms for transmitting the habits are discussed.

1. Introduction

Japanese macaques on Koshima Island were provisioned beginning in 1952. Since then, all individuals in the group were identified and observations on them have been carried out continuously by many researchers. In 1953, one year after the commencement of successful provisioning, it was found that a juvenile female washed the soiled sweet-potatoes in the nearby stream before eating them (Kawai, 1965; Kawamura, 1954; Mito, 1971). She may have noticed that it would be much more comfortable for her to eat sweet potatoes after washing, as soil covering the potatoes could be easily cleaned off in the water. The behavior had gradually spreads to the other group members. The newly-acquired behavior, thus, was transmitted to the other group members and then was fixed as a tradition of the group. Further, Koshima monkeys subsequently participated in other innovative behaviors such as wheat washing, and swimming in the sea, which had not been reported before or in the other areas. This has made Koshima monkeys famous throughout the world as the only species other than humans to perform "cultural behavior." The cultural meanings of the different food habits between

R. A. Gardner et al. (eds.), The Ethological Roots of Culture, 81–94.

groups were vigorously studied in Japan (e.g., Kawamura, 1965). Nevertheless, most studies of the cultural behaviors of Japanese macaques were made prior to 1965, and new information has been limited (e.g., Huffman, 1984; Watanabe, 1989).

During these last forty years, the situation surrounding the Koshima monkeys has inevitably changed and several new generations have been born. In 1952,when artificial provisioning was started, the Koshima group consisted of 22 monkeys. The artificial provisioning was rather infrequent and only small amounts of food (usually sweet potatoes and wheat grains) were given by the researchers who visited the island and/or by some sympathetic local peoples during the first period. Monkeys moved around freely over the island searching for foods and other necessities. After 1964 an abundant supply of food was provided for them every day. After 1972, however, the provisions given to the monkeys were drastically restricted because of a change in management policy (Mori, 1979; Watanabe et al., 1992). Since abundant food supplies influenced the reproduction of Koshima monkeys, it was expected that a population increase on the island would cause several difficult problems for the management of the monkeys. The food given to them, therefore, was limited only for the use of scientific studies, to approximately about 3 kg of wheat grains, two or three times a week. Further, sweet-potatoes were only provided several times a year after 1972. It is noteworthy that the limitation of artificial food supplies brought on several deleterious effects on the reproduction of Koshima monkeys, although the population size of the island stabilized at around one hundred (Mori, 1979; Watanabe and Mori, 1991; Watanabe et al., 1992). These can be summarized as a decrease in birth ratio, higher infant mortality, delayed primiparous age, and an increase in the death of younger females (Fig. 1). The monkeys apparently began to develop slowly, taking more time to reach maturation, and even the size of adult monkeys was smaller.

The aim of this paper are twofold. First, to describe the longitudinal changes in "precultural behavior" during the last forty years in accordance with such changes in the surrounding situation. And second, to discuss the possible factors contributing to the creation of such innovative behavior and their propagation among group members. Most of our observation has been rather anecdotal, with systematic surveys being made only in 1961-62 by Kawai (1965) and in 1983-84 by ourselves (Kawai et al., 1992). This anecdotal information comes from former and present staff of the Koshima Field Station of Primate Research Institute, which I joined in 1982.

2. Potato-Washing Behavior

According to Mito (1971), the young juvenile female started this behavior by rolling a sweat-potato in the shallow water before eating it. The stream was not very deep and the soil could be cleaned off after being rubbed in the sand in the water. Mito (1971) impressively described this female, named "Imo" (potato in Japanese), as playfully repeating this behavior during the first day. As time passed, Imo began to go deeper into the water and to wash a potato by grasping it in the water with one hand and brushing it with the other (Kawai, 1965; Mito, 1971). The behavior spread to other monkeys: first to her play mates, to her mother and sisters, and then their friends. Older monkeys never tried this new method of cleaning potatoes and stuck to conventional methods; i. e., stroking the soil off with the fur of one arm while grasping it with the other, rubbing the soil off with a rock, or brushing it between their hands. Such behaviors are commonly seen

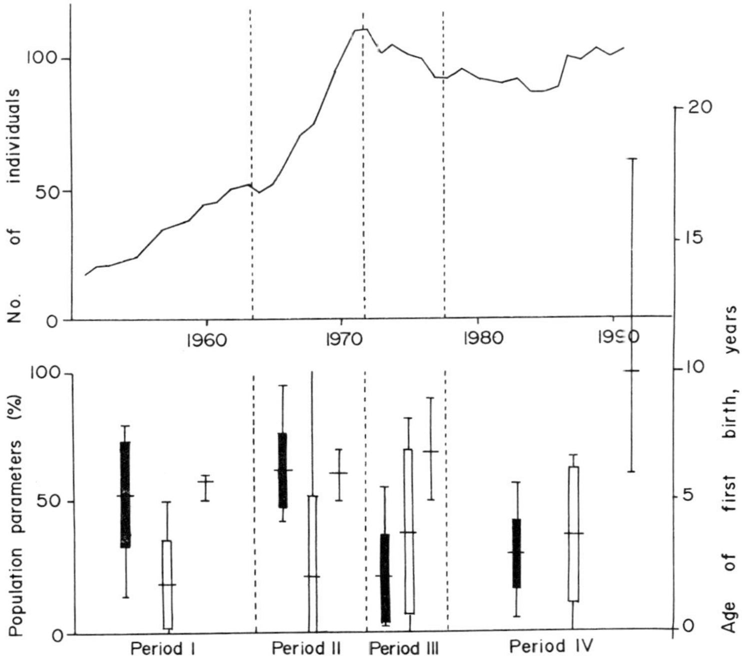

Fig. 1. Progressive changes in population parameters of Koshima macaques. The change in the total population is shown at the top. The birth rate of adult females of 5 years of age or more ((left solid bar), infant mortality within the first year of life (central open bar) and age at first birth (right bar) are indicated for each period: Period I, from 1952 to 1963; Period II, from 1964 to 1971; Period III, from 1972 to 1977; and Period IV, from 1978 to 1991. The bars show the overall ranges within the respective periods, means (horizontal short line), and one standard deviation (thick sections).

in Japanese macaques when foodstuffs are covered with soil. Older monkeys did not use the water to clean potatoes. During 1961-62, Kawai (1965) surveyed the propagation of sweet-potato washing behavior and found that 36 out of 49 monkeys of two years old or more had acquired this behavior. Further, most monkeys of less than eleven years old had acquired the behavior, while a large portion of older monkeys, especially males, had not, even though ten years had past since Imo first started the behavior. Monkeys must acquire the new habit during the young ages.

Kawai (1965) noted at that time that the potato washing behavior was apparently changing to "seasoning" from the original brushing-type behavior. The potatoes had first been washed solely by fresh water, but many monkeys had begun to preferably eat after washing, or just dipping, them into sea water. Koshima monkeys had learned that the sweet-potato could be more tasty when flavored by salt contained in the sea water. Monkeys born after 1955 may have been more inclined to seasoning during 1961-62.

During the summers of 1983 and 84, we made a similar survey on the sweet-potato and wheat washing behaviors (Kawai et al., 1992). At this time, the soiled sweet-potatoes were no longer available since potatoes appeared on the market only after cleaning the soil off. We recorded 2,060 examples, by 58 individuals, in which monkeys carried sweet-potatoes and washed and/or dipped them in the water (Fig. 2). The most frequently observed was "*Seasoning*" (50%), where monkeys just dipped the potato, holding with one hand, into the sea water repeatedly while eating it (Fig. 3). Further, it was often observed that monkeys dipped and fiddled with the potato in the sea water (*Rinsing*; 33%). The greatest number of the present potato washing behavior, thus, were presumably done only for the salty flavor. It is surprising that most Koshima monkeys, including very young animals, still engaged in the "potato washing behavior," even after 10 years had past since food supplies for them had been severely restricted. In other words, obtaining sweet potatoes only several times a year was sufficient for Koshima monkeys to maintain the behavior.

The original type of potato washing behavior, "*Brushing*," was also still seen though it was not very frequent (1%). "*Rubbing*" potatoes on the rocks or sand (5%), and between the hands (12%) in the water, could be variations of the original type of behavior. It should be noted that brushing or rubbing could be commonly seen when Japanese macaques cleaned other soiled foods. What the monkeys acquired was the use of water, or later, sea water, and the habit of carrying potatoes to the place where they could use water. All of these behavioral types were distributed throughout all group members and no monkeys seemed to engage persistently in any one specialized behavior.

Information on this behavior for other years is anecdotal. However, we are quite sure that most Koshima monkeys have continuously performed potato washing behavior during these forty years and that, at the present time, they continue to show similar behaviors. It is often misleadingly understood that Koshima monkeys wash all potatoes before eating them. This is not true. Instead, they tend to do it rather opportunistically. During the surveys in 1983-84, eleven monkeys were never observed to wash potatoes. All of them were the subordinate monkeys among group members and/or males wandering around the group. The sea shore or the stream near the feeding ground may not provide enough suitable places for all group members to wash potatoes. It is very common that many monkeys eat directly at the feeding place without washing them. We, therefore, need to conduct examinations in order to

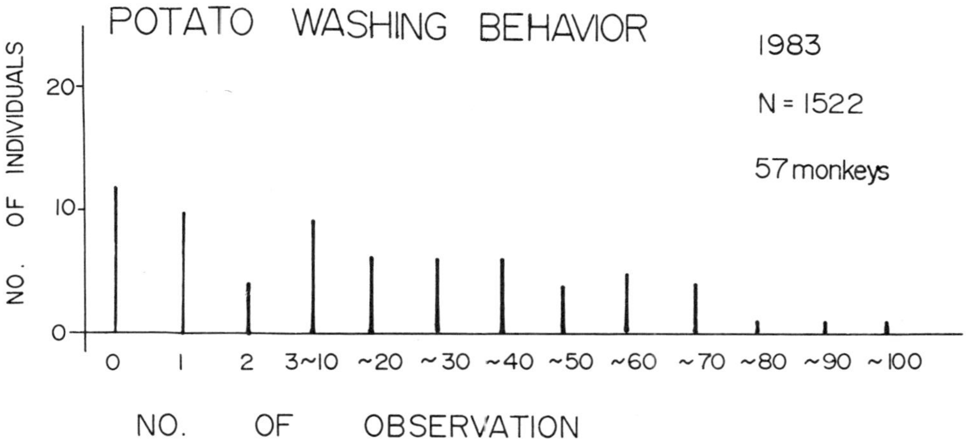

Fig. 2. Observed frequency of the potato washing behavior by individuals in 1983-84.

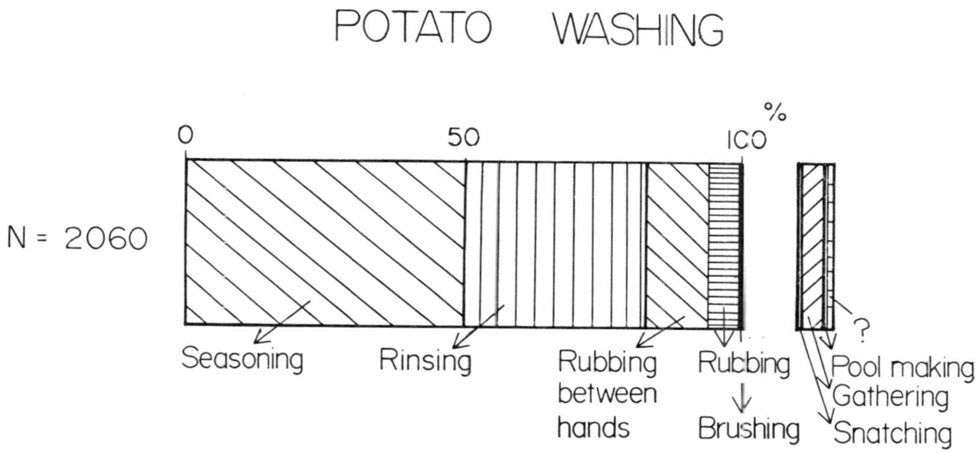

Fig. 3. Frequency of each variation of the potato washing behavior observed in 1983-84. The relative frequencies of *Pool Making*, *Gathering* and *Snatching* behavior recorded are also indicated on the right.

delineate the actual state of the behavior in the group.

3. Wheat-Washing Behavior

Observation of the wheat washing behavior may provide much more interesting implications since the obvious changes occurred after the food supplies for the monkeys had been decreased in 1972. In August 1962, 39% of the 49 monkeys excluding infants of one year old or less, had acquired this behavior (Kawai, 1965). The wheat washing behavior was also invented by the same juvenile female, "Imo," in 1956. Wheat grains may intermingle with sand when it is given on the sandy beach on the island. Monkeys gather wheat grains mixed with sand in their hand(s), carry it to the nearby water, and put it into the water. This makes the wheat grains separate from the sand and monkeys can conveniently eat it. This wheat washing behavior was also transmitted through a similar process as with potato washing behavior, and became fixed among the group members (Kawai, 1965).

The original behavior then, was to carry wheat grains intermingled with sand in their hand and to put them into the water (*Throwing*). This behavior has, however, a serious defect. When monkeys put their grains of wheat into the water, they can not protect it from the nearby dominant animals. In fact, some dominant monkeys who did not acquire the washing behavior, became specialized snatchers of wheat grains. They might wait near the water for other monkeys to bring the sand-wheat mixture and to put it in the water. The newly-invented countermeasure was for monkeys to keep grains in their hand(s) and drop a little bit at a time, repeatedly, from the sand mass grasped in their forefingers (*Removing*). The monkeys could, then, successfully avoid snatching monkeys since they could quickly flee carrying most of the wheat grains. This method was initiated by a four years old female, in 1970 when abundant provisions were still being given for them.

By 1974-75, Koshima monkeys had adopted several types of wheat washing behavior (Mori, pers. comm.). The three dominant adult males were devoted to snatching grains which other monkeys put into the water. Many young adult females had learned the removing method, and used it properly. Some other monkeys might grasp the sand-wheat mixture and shake their hand in the water. In this manner, the sand was washed away and only the grains remained in the palm (*Screening*). Others simply walked into the water, grasping the mixture in their hand(s), which produced a similar result (*Walking*). Mito (1971) described that some adult females had begun to throw wheat grains which were near the water, directly into the water by sweeping them with a hand (*Sweeping*). This method is much more efficient than hand carrying when they are close to the water. Carrying grain in their tiny hand(s) is not efficient since the number of wheat grains which monkeys can carry at one time is very limited. Further, some wheat grains always float away with the stream or the waves in the sea. Some monkeys often wait for floating grains downstream and/or in the sea where wheat grains concentrate after floating on the waves (*Gathering*). Thus, the wheat washing behavior developed into several variations as the artificial provisioning continued. Each individual in the group would adopt their own method in accordance with their social status and the situation.

In 1983-84 when we surveyed the wheat washing behavior, the situation was changing. We recorded 2,237 examples of wheat washing behavior (Fig. 4 & 5). This figure excludes cases of *Snatching*, *Sweeping* and *Gathering*, since these were difficult to assess precisely in the crowded situation and we might need different measures to deal

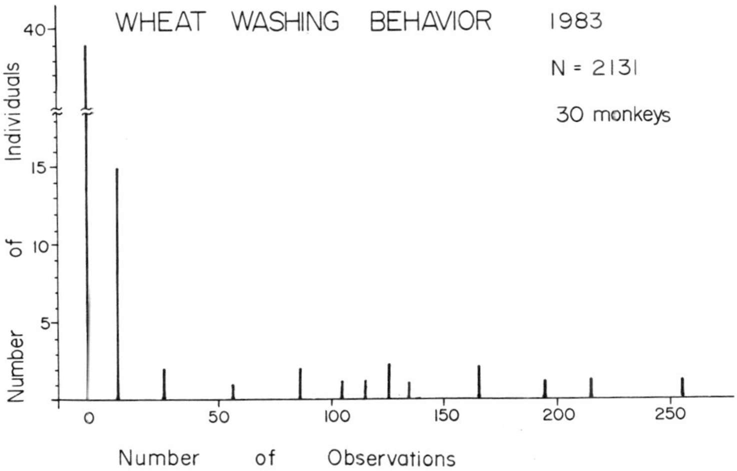

Fig. 4. Observed frequency of the wheat washing behavior by individuals observed in 1983-84.

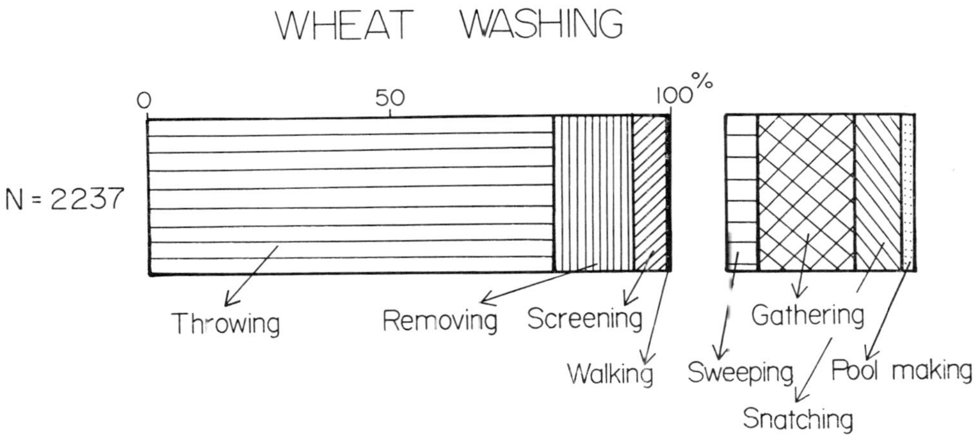

Fig. 5. Frequency of each variation of the wheat washing behavior observed in 1983-84.

with them. Thus, this figure includes only examples when monkeys gathered wheat grains in their hand(s) and carried it to the water. Of these 97% were performed only by 13 monkeys, and 99% by 15 monkeys. A shared characteristic was that most of these monkeys were born before 1971 (13 individuals) or else were dominant females in the group (two individuals)(Fig. 6). Most of these old monkeys have died one by one.

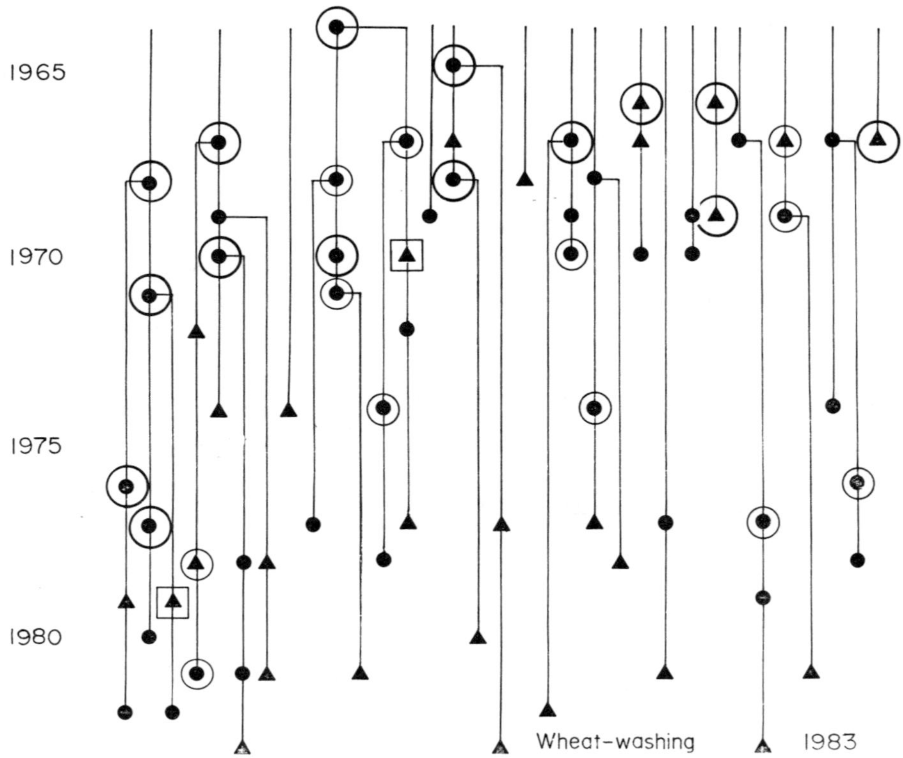

Fig. 6. Animals who performed wheat washing behavior and the genealogy of Koshima monkeys in 1983. Females are shown by solid circles and males by solid triangles. Large open circles indicate the 15 monkeys who frequently performed the behavior and small open circles those who performed occasionally. Two males often made a pool, indicated by squares. The lineages are roughly arranged by the dominance hierarchy from left to right.

However, one young female, born in 1985, began to wash wheat grains from 1989. In 1992, we can see only four individuals who often wash wheat grains among the 74 individuals in the main group.

In the survey of 1983-84, we could find another variation of the behavior. When the wheat grains were given on the beach which was still wet at low tide, monkeys dug out the sand and made small pools by the oozing water. They then swept nearby grains into the pool and collected them. This behavior was often difficult to discriminate from a hollow which could be made unintentionally through the monkeys' stirring the sand in order to take grains, especially when the beach was submerged. However, one adult male and one four years old male used to make a pool by forcibly digging the sand. Their intentions were, therefore, quite obvious in these cases.

Table 1. The variation of wheat washing behavior by individuals.

	Throwing	Removing	Screening	Walking	Total
Males					
Geshi	85				85
Geba	147		1		148
Noso	211		7		218
Bohze	24	46	11		81
Females					
Ume	147				147
Uri	132	3	5		140
Utsugi	212	4	78	11	305
Cha	23		1		24
Kurumi	230		11		241
Kuri	85	48	15	2	150
Isu	180			1	181
Icho	90	2	4		96
Sansho	114	1			115
Zakuro	22	235	1	1	259
Sakaki	21	1			22

The type of wheat washing behavior tended to be fixed according to the individual (Table 1). Of the 15 monkeys who frequently performed wheat washing behavior, 11 engaged almost exclusively in *Throwing*. The great majority of *Removing* was done only by three individuals. One adult female, Zakuro, performed 69% of the *Removing* in the group, and it accounted for 91% of her washing behavior. *Screening* and *Walking* were done primarily by another adult female, Utsugi, (57% and 69%, respectively). Further, left or right handedness was clearly apparent when Koshima monkeys carried the sand-wheat mixture though not when they carry sweet-potatoes (Watanabe and Kawai, 1993).

Japanese macaques do not usually part spontaneously with food which they once have in the hand when other animals are nearby. Wheat washing behavior may force it on them. Some monkeys learned to do so only bit by bit, or to make a pool where they rake in wheat grains. In these respects, wheat washing behaviors are more innovative than potato washing (Nishida, 1987). That is, monkeys need to add a more novel behavior to their own repertoire than just the use of water when they engaged in the wheat washing behavior. This may be the reason that wheat washing behaviors tend to become more stereotypic and individual-specific than the potato washing behavior.

4. Propagation and Surrounding Conditions.

The diminishing wheat washing behavior suggests several interesting features of the precultural behaviors of Koshima monkeys. First, the more novel a behavior is, the more difficult it is to transmit to other group members. Sweet potatoes have been given to them only occasionally since 1972, while wheat grains have been provided for two or three times a week. Nevertheless, most group members still engage in the potato washing behavior. The easiness of the behavior may also be reflected in the elasticity of potato washing behavior. The pattern of wheat washing behavior, on the other hand, was fixed in each individual to some extent. There were several monkeys who only occasionally engaged in the wheat washing behavior. When we surveyed in 1983-84, 15 individuals, other than 15 monkeys who frequently performed wheat washing behavior, were seen carrying sand-wheat mixture in their hands to wash, although the frequencies were very low (range 1-6, N=26). These monkeys may have previously acquired the wheat washing behavior but did not incorporate it into their daily lives. Further, it has been found that some other monkeys also occasionally began to wash wheat grains. In 1988, a three years old female was often observed to rake wheat grains into the water from a distance of one meter. Another adolescent female, born in 1982, showed a similar behavior and also made a pool in 1989 to rake in grains. A juvenile female born in 1984, sometimes, made pools. In 1992, a three years old male showed *Throwing* in order to wash grains. Additional examples were also occasionally reported by the staff of the Koshima Field Station. Thus, many monkeys showed behavior apparently intended to wash wheat grains in the water. However, they still appeared inexperienced. Most of these examples may not have been fixed in the behavioral repertoire of their performers and disappeared over time. The novel behavior must be fixed as a more stereotypic pattern before it can be maintained by the performer.

Changes in the surrounding situation may be even more important. Monkeys can carry only a small number of wheat grains in their hand(s) and therefore, they must repeatedly carry it many times. Further, they will often be robbed by the dominant monkeys when they part with it in the water. Therefore, the wheat washing behavior is much more time consuming and wasteful than potato washing behavior. A sweet potato has a big mass and monkeys may be satisfied after eating only two or three peaces, which a monkey can usually carry very easily for a long distance. Such easiness and the lower cost of potato washing behavior must function to assist in maintaining this behavior by koshima monkeys, despite the greatly decreased opportunity to learn it. In order to take more wheat grains, monkeys can pick up the grains one by one and eat it directly without any other work. The wheat grains are not usually intermingled with sand just after they have been scattered on the beach. Therefore, all monkeys engage in taking it directly following provisioning, and, only after they have consumed a fairly large portion, do several monkeys begin to wash grains. The teeth of Koshima monkeys appear to wear out relatively sooner than monkeys in the other monkey parks in Japan (Hamada, pers. comm.). This would contribute to greater comfort for those monkeys who do wash wheat grains. When the amount of food decreased, however, it was quite natural that wasteful work was diminished.

Many studies on the cultural behavior of Japanese macaques have indicated that there are two patterns of transmittal of the newly acquired behaviors among the group members. One is that the behavior is transmitted through kin relationship and/or play mates. Characteristic of this pattern is that young monkeys initiate the behavior and then older monkeys learn it. The new behavior will be propagated rather progressively and usually requires time. This pattern of transmission could be found in the precultural behavior of Koshima monkeys (Kawai, 1965; Kawamura, 1954; Mito, 1971), the acquisition of new food items (Itani, 1958) and of Stone-play (Huffman, 1984). The other type of transmission is that adult monkeys might first acquire the new food items, and then young monkeys followed (Azuma, 1968; Kawamura, 1959, 1965; Watanabe, 1989; Yamada, 1958). In the later cases, most group members often accepted new food items very rapidly. The difference between these two patterns of the invention and propagation can be analyzed in terms of the surrounding situation and the attitude of the monkeys.

It has been reported that monkeys of several different groups recently started to play with stones (Huffman, 1984). The monkeys may gather, carry, cuddle, rub, roll in hands, or clack them, making loud noises. These groups are scatteredly distributed through Japan and interchanges of individuals are absolutely impossible. Included are the groups of Arashiyama, Takasakiyama, Takagoyama (Huffman, 1984), and Gagyuzan (Fukuda, pers. comm.). The common feature is that these groups have all been provisioned sufficiently for a long time. Huffman (1984) properly inferred that the habit was no longer propagated in the Takagoyama group once provisioning had ceased. Being attracted by abundant foods, the provisioned group would become stationary near the feeding ground and the monkeys would have a surplus of time. The initial acquisition of new habits by young monkeys and the progressive propagation might be a reflection of such abundant free time in their lives. Young monkeys may first acquire the new habit through playing during a possible time surplus and other monkeys follow them. Since the new habit is an outcome of abundance in life, the habit will not necessarily be transmitted rapidly to the other monkeys. The habit, however, may spread gradually among the group members as long as the common situation prevails.

The other type of propagation is characterized by the eventual acquisition of a new habit by many adult monkeys. Koshima monkeys learned to eat raw fish around 1980 (Watanabe, 1989). Fish eating was first reported just sporadically among adult males on the island. However, the habit did not spread to most group members. In 1982, about three years after the first observation, raw fish eating was observed in four adult females in the group. It was the turning point for the habit to spread among the group members. Most group members then began to take raw fish until 1986. It seemed that raw fish were not to the monkeys' taste since they often ignored it when other foods were available. Koshima monkeys can obtain natural foods on the island freely. However, the population density on the island, one hundred monkeys / 30 ha, is about ten times higher than that on Yakushima Island where similar forest occurs (Maruhashi, 1980). Such an over-crowded situation or, to be more precise, a paucity of foods, forced Koshima monkeys to adopt other food items into their repertoire. All Koshima monkeys, thus, might have been ready to accept new food items. Once several group members had started to eat raw fish, the other monkeys might just follow. When such an attitude toward new food items was thus prevailing, they could be easily accepted by most group members. Yamada (1958) and Azuma (1968) reported the rapid propagation of new food habits, and that most group members began to eat them within several days. These observations were made in the groups just after provisioning was started. It is expected, therefore, that the monkeys could readily accept new food items on these occasions.

5. The Possible Mechanism by which Monkeys Acquired New Habits.

It is difficult to make detailed observations in the field on the mechanisms by which a new habit is transmitted to other monkeys. It is, especially, impossible to record all interactions between monkeys scattered on the beach. However, it is likely that monkeys gradually learn potato and wheat washing behaviors. Most Koshima monkeys apparently had had the experience of eating foods floating in the stream or in the sea which might have been carried and put into the water by other individuals prior to the time they acquired washing behaviors (*Gathering*). Further, it is likely that most monkeys have eventually experienced the sweeping of wheat grains into the water. This is because when the beach is submerged on rainy days or at low tide many hollows can be made while monkeys stir the sand, which results in a similar effect to *Sweeping* or *Pool-making*. It is a fact that we have never seen the typical example of imitation or observational learning for these behaviors. That is, monkeys do not acquire these habits just from a few observations of the others' behavior and then mimicking it. Further, the behavior patterns acquired do not seem related to those of relatives. The behavior pattern of carrying foods to the water may appear individually and abruptly during continual provisioning. These all suggest that most of these behaviors could be acquired by individual learning in the specialized environment of Koshima and/or local enhancement (Galef, 1976; Fragaszy and Visalbergi, 1990). It is, however, very usual that new habits spread soon after initiations by a limited number of monkeys among the group. Some mechanisms undoubtedly function for transmitting the new habits.

Higuchi (1992) studied the propagating process of a new habit in the Koshima group using operant conditioning. During a total of 90 sessions (one session was one hour), 34 monkeys learned to push a panel to get a soy bean. He classified the monkeys into four types of acquiring a new habit by the frequencies of trials and their observations of others' successful behavior: 1) local enhancement and trials, both frequencies were high; 2) individual learning, a few observations and many trials; 3) observational learning, many observations and a few trials; and 4) rapid learning, both few. The results indicated that most monkeys acquired the habit by local enhancement and through 'trial and error,' while the observational learning was rare. The instrument for experiment attracted many monkeys and they watched as a monkey successfully took soy beans. However, the behavior of pushing a panel could not be acquired only by observations and monkeys always needed trials for acquisition. The ability of Japanese macaques to imitate other's behavior is, therefore, quite limited if it exists at all.

Higuchi (1992), on the other hand, described the possibility of imitation found in infant monkeys, where infants often repeated the behavior of their mothers when pushing the panel repeatedly or just after she finished it. It was, however, not observational learning as infants rarely took soy beans.

Yamada (1958) and Azuma (1965) described the behaviors when monkeys began to accept new food items such as wheat grains. Both emphasized that monkeys watched the face of the initiator eating the novel food from very close distances. It seemed that monkeys were interested in the eating behavior of the other, not the food. They might pick up, smell and gnaw the food, while often looking at the initiator and then, start to eat it. Infants took several grains but the habit was very unclear until the mother started to eat the new food. It is not clear whether these behavior indicate true imitation performed by Japanese macaques since the desire for the food was apparently stimulating them. Kawamura (1965) reported the example where all monkeys of Minoo-B group started chewing on woods just after seeing five males in a cage persistently gnawing wooden frame. Wood chewed on were similar with the material of the cage. Social facilitation and/or true imitation, thus, may contribute in part to the propagation of new habits. In order to clarify this, more detailed analysis must be obtained on the types of behavior and the developing stages of a monkeys behavior.

Acknowledgments

I thank R. A. Gardner, A. B. Chiarelli, B. T. Gardner and F. X. Plooij for inviting me to the NATO Advanced Study Institute. M. Kawai, A. Mori, S. Mito, F. Kanchi and N. Yamaguchi provide me with their unpublished data. I am indebted to E. Ingmanson, M. A. Huffman, Y. Higuchi and M. Setoguchi for their kind helps and critical reading of the manuscript.

References

Azuma, S.(1968) 'Acquisition and propagation of food habit in a troop of Japanese monkeys'. in C. R. Carpenter (ed.), Social Regulators of Behavior in Primates, Bucknell University, Lewisburg, pp.284-292.

Fragaszy, D. M. and Visalberghi, E. (1990) 'Social processes affecting the appearance of innovative behaviors in capuchin monkeys', Folia Primatol. 54, 155-165.

Galef, B.G. (1976) 'Social transmission of acquired behavior: a discussion of tradition and social learning in vertebrates', Advances in the study of behavior 6, 77-100.

Higuchi, Y. (1992) The Cultural Behavior of Japanese Macaques, Kawashima Shobo, Tokyo. (in Japanese)

Huffman, M. A. (1984) 'Stone-play of *Macaca fuscata* in Arashiyama B troop: transmission of a non-adaptive behavior', J. Hum. Evol. 13, 725-735.

Itani, J. (1958) 'On the acquisition and propagation of a new food habit in the natural group of the Japanese monkey at Takasaki Yama', Primates 1, 84-98. (in Japanese with English summary)

--------- and Nishimura, A. (1973) 'The study of infrahuman culture in Japan: a review', E. W. Menzel (ed.) Precultural Primate Behavior, Karger, Baser, pp.26-50.

Kawai, M. (1965) 'Newly-acquired pre-cultural behavior of the natural troop of Japanese monkeys on Koshima Islet', Primates 6, 1-30.

---------, Watanabe, K. and Mori, A. (1992) 'Precultural behaviors observed in free-ranging Japanese monkeys on Koshima Islet over the past 25 years', Primate Report 32, 143-153.

Kawamura, S. (1954) 'A new type of action expressed in feeding behavior of Japanese monkeys in the wild', Seibutsu-Shinka 2, 11-13. (in Japanese)

------------ (1959) 'The process of sub-human culture propagation among Japanese macaques', Primates 2, 43-60.

------------ (1965) 'Sub-culture among Japanese macaques', In S. Kawamura and J. Itani (eds.), Monkeys and Apes: Sociological Studies, Chuokoronsha, Tokyo, pp.237-289. (in Japanese)

Maruhashi, T. (1980) 'Feeding behavior and diet of the Japanese monkey (*Macaca fuscata yakui*) on Yakushima Island, Japan', Primates 21, 141-160.

Mito, S. (1971) Japanese Monkeys in Koshima: Twenty-five Years of Observation, Popula-sha, Tokyo. (in Japanese)

Mori, A. (1979) 'Analysis of population changes by measurement of body weight in the Koshima troop of Japanese monkeys', Primates 20, 371-397.

Nishida, T. (1987) 'Local traditions and cultural transmission', In B. B. Smuts, D. L. Cheney, R. M. Seyfarth, R. W. Wrangham and T. T. Struhsaker (eds.) Primate Society, The university of Chicago Press, Chicago, pp.462-474.

Watanabe, K. (1989) 'Fish: a new addition to the diet of Koshima monkeys', Folia Primatol. 52, 124-131.

----------- and Mori, A. (1991) 'Prolonged effects of developmental retardation on the prospective reproduction concerning to food-shortage observed in Japanese macaques on Koshima Island', In A. Ehara, T. Kimura, O. Takenaka and M. Iwamoto (eds.), Primatology Today, Elsevier, Amsterdam, pp.393-396.

-----------, ------- and Kawai, M. (1992) 'Characteristic features of the reproduction of Koshima monkeys, *Macaca fuscata fuscata*: a summary of thirty-four years of observation', Primates 33, 1-32.

----------- and Kawai, M. (1993) 'Lateralized hand use observed in the precultural behavior of the Koshima monkeys (*Macaca fuscata*)', In J. P. Ward and W. D. Hopkins (eds.), Primate Laterality: Current Behavioral Evidence of Primate Asymmetries, Springer-Verlag. New York, pp.183-192.

Yamada, M. (1958) 'A case of acculturation in a subhuman society of Japanese monkeys', Primates 1, 30-46. (in Japanese with English summary)

BIRD SONG LEARNING: A MODEL OF CULTURAL TRANSMISSION?

P.J.B.SLATER AND J.M.WILLIAMS[1]
School of Biological and Medical Sciences,
University of St Andrews,
Fife KY16 9TS, U.K.

[1] *Present address:*
Seabirds Team,
Joint Nature Conservation Committee,
16-17 Rubislaw Terrace,
Aberdeen AB1 1XE, U.K.

ABSTRACT. The learning of bird song is a classic example of cultural transmission among animals, yet the extent to which it shares characteristics with other cultural phenomena has been questioned. We discuss this and review features of the song learning process, together with its consequences, such as changes in song with time and in space. Finally, we describe simple computer simulations of the song learning process in which the effect of altering several different variables was assessed. This simple model gives surprisingly complex results: some points that it raises may be relevant to other cultural processes whether or not one accepts song learning as being one of their number.

1. Introduction

It is well known that many birds learn details of their vocalizations from other individuals. The phenomenon has probably arisen independently several times as it occurs in groups which are not at all closely related to each other, such as the parrots (Psittaciformes: Todt 1975) and hummingbirds (Apodiformes: Baptista and Schuchmann 1990) as well as, most strikingly, in the Passeriformes. This order is split into the Oscines, or songbirds, and the Sub-Oscines. Both of these sub-orders produce the rather long and complex vocalizations that we call "song", but they differ markedly in the extent to which learning from others plays a role in its ontogeny. In the songbirds, all species studied to date learn their songs, and learning also plays a part in the development of many of their simpler calls (Slater 1989). However, there is no evidence for such an influence in Sub-Oscines: even young birds exposed only to the rather different songs of closely related species (Kroodsma 1984, 1989), or deafened so that they cannot hear any sounds (Kroodsma and Konishi 1991), produce normal species-specific songs. Because song learning is widespread among the songbirds, and because this group includes the great majority of the birds that produce elaborate and complex vocalizations, this article will concentrate on them.

The mechanism of song learning is often referred to as "cultural transmission" (Jenkins 1978, Petrinovich 1988) or as "imitation" (Mundinger 1970, Aoki 1989), and the changes in time that result from it are labelled as "cultural evolution" (Slater and Ince 1979, Lynch and Baker 1986, Trainer 1989). The extent to which these processes provide an example of culture, as these words imply, is an issue which we must consider before going any further. The answer to the question depends very much on how one defines terms. In a general sense the transmission of song must be an example of cultural transmission,

R. A. Gardner et al. (eds.), The Ethological Roots of Culture, 95–106.

because one animal learns very precise and detailed information from another. On the other hand, it can be argued that the similarity between song learning and other cultural phenomena breaks down at a more detailed level. For example, in song transmission most of the novelty appears to be due to random copying error and may not be adaptive in any way, as we will discuss in the next section, whereas other cultural processes involve the spread of novel adaptive behaviour patterns through a population. Song learning "just happens to be from other individuals, and this sets the scene for cultural evolution" (Slater 1986).

Another contrast is drawn by Whiten and Ham (1992) in their recent survey of imitation in animals. They adopt the definition of Thorndike (1898) that imitation is "learning to do an act from seeing it done." Thorndike argued that it was reasonable to differentiate between the vocal imitation of birds and the visual imitation of mammals, because the former was a specialization rather than indicative of a generalized imitative ability. Things have changed somewhat since Thorndike's time. There is probably now better experimental evidence for visual learning in birds than in any mammalian group other than the apes (see Lefebvre and Palameta 1988, Palameta 1989). Pepperberg (e.g. 1990) has also shown that the vocal learning capacity of a parrot can be used to reveal a range of unsuspected cognitive abilities. But the basic point, that song learning itself is a more restricted and specialized ability than the visual imitation of apes and humans, seems nevertheless a good one.

Whiten and Ham (1992) draw another distinction between imitation in the visual and auditory modalities. They argue, following Palameta (1989), that auditory copying allows an animal to compare its own output with that of the animal it imitates in the same modality, whereas this is not so in the case of visual imitation. The latter has more demanding "computational requirements" they suggest, because the animal must adopt for itself the model's representation of the act. This is a rather more difficult argument. In both cases the subject must match its own pattern of muscular contractions to that shown by a model, the output of which is all that it can perceive. In both cases a process of trial and error may help it to match that output, despite differences between its perception of its own actions and those of its model. While the view of a model performing an action may be very different from that of the animal when it does it itself, presumably sounds heard from the outside world are also very different in quality from those the animal hears itself producing. It will be interesting to see how such discussions develop: the chapter by Byrne (in press) provides a useful perspective.

Various reasons have thus been put forward as to why song learning in birds may be viewed as different from other cultural processes. In the next section, we will review what is known about song copying and the consequences that it has for the structure of song both within a population and between populations. At the very least, this will point to an interesting analogy with cultural transmission in other animals. The final section will rely on no more than this analogy. It will describe computer simulations of cultural transmission which are based on bird song but may have a wider relevance. Here we will exploit the ambiguity of the word "model" in our title: whether or not song learning is a good example of culture, the model we have developed of it may shed light on similar processes in other animals. The key message that we derive from these simulations is that remarkably simple processes may lead to very complex outcomes. This will come as no surprise to an evolutionist, but it may perhaps have something to say to those who study culture in animals such as ourselves, the complexity of which may seem daunting.

2. Cultural Transmission of Bird Song

2.1 THE PROCESS OF SONG COPYING

Beyond the fact that young songbirds learn the details of the songs that they sing from other individuals, it is hard to generalize. If we just look at a few basic features of song learning, such as at what age song is learnt, from whom and how accurately, almost every species that has been studied shows a unique combination of characteristics (Slater 1989). A few examples will give the flavour of this diversity.

In considering the timing of learning we must distinguish between memorization and production. In some species this distinction is clear as the two processes do not overlap in time. Song production is normally limited to adult males and, as testosterone titres rise, the song produced develops from variable and unstructured subsong, through a period of practice to the typically stereotyped and clearly organised full song. The song that is produced may be a perfect copy of one that was only heard many months earlier, long before the young bird began to sing itself (e.g. Marler and Peters 1987, Slater and Ince 1982). However, in other cases memorization and production overlap. There are many cases where the songs that young males produce match those of the neighbours alongside whom they set up their territories (e.g. Payne 1981). In such cases the young males often learn the song in both senses simultaneously, so that they match the songs they produce to those that they hear rather than to their memory of sounds heard earlier. Another possibility is that the young birds do not memorize the sounds that they hear at all, but start off with a large repertoire of different sounds from which they then select those that best match their neighbours. This is what Marler (1990) refers to as "action-based learning". Some species are known to have far more elements in subsong than in full song and those that are retained may be the matching ones (Marler and Peters 1982). A particularly striking example of this comes, not from males which match their neighbours, but from cowbirds (Molothrus ater), in which young males develop songs matched to the preferences of the females with which they are housed (West and King 1988).

Song production is thus normally limited to adult males in breeding condition, but memorization may take place either as they come into breeding condition or at an earlier stage. There is some evidence of birds learning from their fathers in the period before they become independent (e.g. Grant 1984, Zann 1990), but this appears not to be widespread. The influence of territorial neighbours is more common, and there are several reports of birds changing song in adulthood, especially where they move from one area to another in which the birds that surround them sing differently (e.g. Payne 1985, Nurnberger et al. 1989).

The timing of learning is linked to the question of from whom birds learn. Birds that memorize songs as juveniles are likely to learn those of their father and other males singing in his neighbourhood. Birds that memorize later are most likely to copy those of those that surround them when they set up territories. A theme of increasing importance in recent studies of song learning has been that of social relationships, such as that in the cowbird example mentioned above. Early experiments often used tape recorders to train young males, and the subjects often failed to learn from songs to which they were exposed in this way. It is now clear that males of many species are more prone to learn from those with whom they can interact, visually and vocally, than from songs which they simply hear (e.g. Petrinovich and Baptista 1987). In a striking example, Adret (in prep.) has found that young male zebra finches (Taeniopygia guttata), which will not normally learn from tapes or individuals with which they cannot interact, will learn from tape recordings which they can present to themselves by key pressing in a Skinner box. Their control over the situation seems to be an adequate substitute for interaction with a live tutor.

That learning can be extremely accurate has been demonstrated in many experiments in the laboratory, where sonagraphic comparisons between tutors and tutees has frequently shown very precise matching. In line with this, field data have shown that males of many species may sing songs which match those surrounding them to a fine degree of detail. But the extent of such matching varies between species. Similar songs may be subtly different between the birds in a neighbourhood. Young birds trained in the laboratory may produce copies which fail to match those of their tutors in various respects.

Inaccuracy generates novelty. As well as failing to copy the songs that they hear with complete fidelity, novelty can arise because birds copy several different individuals. Many song birds have a repertoire of different songs or phrases. At an extreme this may run to several hundred different themes. In such cases, even where only a few song types are involved, it is unusual for them all to be copied from the same individual. Thus one bird in a neighbourhood may have songs A and B, another B and C and a third A and C (e.g. Slater et al. 1980, McGregor and Krebs 1982). When one takes repertoires into account, the capacity for the generation of novelty is very great. If songs are learnt independently from different neighbours, and each bird has 6 or 8 song types, it would be quite unusual for two individuals to share a repertoire.

2.2 CHANGES IN TIME

That song learning is not always totally accurate is clear from the fact that the song types in an area are often related to each other in form. One of them may be a combination of two others, starting like one but finishing like a second. Two may be exactly the same except that a section present in one is missing in the other (e.g. Slater and Ince 1979). Such differences may be attributed to "cultural mutations", mistakes or inaccuracies that have crept in during the copying process.

Because most songbirds mature at one year of age, and as many as 50% of adults may not survive from one year to the next, it is feasible to trace the songs present in a population through time to look for such changes. Jenkins (1978) first did this in the saddleback (Philesturnus carunculatus), a rare New Zealand passerine a small group of which had been introduced for conservation reasons to the offshore island where he worked. He found that it was usually possible to pinpoint who learnt from whom because of the close similarity of their songs, but that this copying sometimes involved inaccuracies on the part of the young bird. While the similarity of its song to one of its neighbours made it clear who its tutor had been, an added element or a missing one meant that its song was unique to itself. This new song type might then spread to other birds in the population in subsequent cultural generations.

Later studies confirm that the songs in an area change through time, ones that are copied inaccurately leading to new types, while others that are not copied at all go extinct. Thirty-five song types were recorded from chaffinches (Fringilla coelebs) in an English wood in 1978 and only three types were found to be in common with a sample of 22 that had been recorded there 18 years earlier (Ince et al. 1980). Indigo buntings (Passerina cyanea) have a limited number of stereotyped elements, but the sequence of these making up a song leads to many different song types in a population. Groups of birds share sequences, but the frequency of particular patterns rises and falls. Payne et al. (1981) estimate that the half-life of a sequence of three element types averages 3.8 years. Evidence from both these species suggests that these changes are simply due to random errors in copying rather than having adaptive significance in themselves (Lynch et al. 1989, Payne et al. 1988). In the chaffinch it has been estimated that at least 85% of copying is accurate (Slater et al. 1980): it is hard to see what benefit errors might give their possessors when they occur at such a low rate.

The extent to which song changes with time seems to vary quite considerably between species, with some reports of changes from year to year, and others of songs in an area remaining much the same over long periods. The conclusions depend to some degree on the level of detail at which the analysis is carried out: fine changes will obviously accumulate more rapidly than massive ones. But the changes will also depend on a number of factors concerning the process of song learning. For example, the following points may be important:

1. Can individuals change their songs from year to year? In some species there is rapid turnover of songs within the population, and individuals appear to learn new songs each year. In bobolinks (Dolichonyx oryzivorus), for example, Avery and Oring (1977) found that only the two commonest songs in a population persisted from one year to the next. Trainer (1989) found a population of yellow-rumped caciques (Cacicus cela) to change their songs gradually in concert with one another. All males in a breeding colony shared their repertoire of 5-8 song types, but 78% of the songs present each year were new.

2. Do birds learn from one or several individuals? In the chaffinch, males appear to learn each song from one individual, so that songs are generally passed from one animal to another, rather than the characteristics of several tutors being blended to produce the song of the young bird. On the other hand, blending of characteristics may occur in other species, and this is a conservative influence. Thielcke (1987) found treecreeper (Certhia brachydactyla) song variants in a population to be the same in 1983 as they had been in 1960, and argued that this was because young birds "take an average" when learning. This difference is very much the same as the difference between "particulate" as opposed to "blending" inheritance: genes are particulate and this form of inheritance allows variation to persist; if inheritance worked by the blending of parental characteristics, then variation would rapidly disappear.

Thus, song learning with a low rate of inaccuracy will lead to changes in time. If learning is restricted to one period of life, or if it involves blending the characteristics of more than one tutor, changes will tend to be slow. If, on the other hand, birds learn afresh each year, and learn each song from a single tutor, then more rapid turnover of the song types within the population would be expected.

All the changes with time discussed so far are likely to stem from random copying errors. However, in some cases the changes involved may be adaptive. For example, Lehtonen (1983) has argued that the songs of great tits (Parus major) in Finland have become simpler over the past few decades as an adaptation to a noisier environment. Where systematic changes have been found in the frequency of certain song types (e.g. Gibbs 1990), it also seems likely that this is because some types are for some reason copied in preference to others. However, cases like this are exceptional: many of the changes with time that have been found in song do not appear to require adaptive explanations.

2.3 CHANGES IN SPACE

Male birds tend not to disperse far between the site where they hatch and the site where they breed. As a result, whether they copy their songs as fledglings or as young adults, they normally sing songs typical of the area. But their songs may differ quite markedly from those of birds elsewhere. Once again, this is probably simply because differences between localities build up through copying errors, rather than because geographical variation has any adaptive significance. The idea that local differences in song act as barriers to dispersal, and thus help to ensure that birds only mate with individuals well adapted to their own area, while an attractive one, has not been supported by most studies. This idea was largely put forward to account for cases where there are sharp dialect boundaries (Nottebohm 1969). These have been found in several species with very simple songs, all the birds in an area singing the same, with a sudden switch to a different song when one crosses the boundary to another area (Baptista 1975, Bjerke and Bjerke 1981, McGregor 1980). Such dialect boundaries remain perplexing and also controversial (see Baker and Cunningham 1985 and the associated commentaries): in some cases there is doubt about whether they really exist; in other cases they may do so only as a passing relic of different groups of birds recolonising an area that had become suitable to them; in yet other cases they may exist because groups of birds singing differently are separated by a geographical barrier. Such explanations would account for their existence without the need to propose that they are in any way adaptive.

While sharp dialect boundaries are rare, geographical variation is common. This may simply involve gradual change in the songs present as one moves from one place to another, with no clear discontinuities. The word dialect may not always be appropriate to describe such variations. It implies that all the birds in one area share a feature or features in their song which individuals from other areas lack. In species which have more than one or two song types, it is more common for a whole variety of types to be present in each area, with no song characteristic which easily allows distinction between that area and others.

One reason why birds with more than one or two song types present a more complex picture, is that individuals in these species do not normally build up their repertoires by copying just one tutor. Young male white-crowned sparrows (Zonotrichia leucophrys) probably copy their single song type from one individual rather than blending the characteristics of several. In the corn bunting (Miliaria calandra), males have two song types and these appear to be transmitted as pairs: some birds sing A and B, some C and D, but the other possible combinations of these four types are very seldom found (McGregor 1980). The whole repertoire appears to be a cultural unit. But this is exceptional. In most species where individuals have more than one song type all possible combinations of types are found within different repertoires (e.g. Slater et al. 1980). Rather than learning all their songs from the same tutor, young males appear to pick up different songs from the different individuals to which they are exposed. This leads the distribution of song types to be a complex mosaic.

3. A Useful Model for Culture?

In modelling the cultural transmission of song, we have used computer simulations in which a number of variables could be modified. Initially, the case in which each individual had only a single song type was studied by Goodfellow and Slater (1986). The individuals were set up in a matrix of "territories" within the computer so that each of them, except for those along the border of the matrix, had four neighbours. We examined the effect of different mortality rates, and different song copying error rates, on how the songs present in the population changed over time. Thus, at the start of each breeding season, each territory had a probability of being empty determined by the mortality rate. It was then occupied by a new bird, which copied its song from one of its four neighbours. The probability of this copying being inaccurate, so that a new song type was created, was determined by the copying error rate.

Taking realistic annual mortality (30-40%) and copying error rates (15%) suggested by our work on chaffinches, this simulation produced small groups of birds sharing a song type, not dissimilar to those that have been found for some species in the wild (e.g. Payne 1981). However, the likelihood of large dialect areas, with 100 or more birds sharing a type, being generated in this way seems very low: groups of this size would only arise if miscopying was extremely rare. For example, the mean group size only rose to above 10 birds, even when mortality was low (30%), if copying was accurate on over 98% of occasions.

While our simulations could not generate large dialect areas, they suggested that the borders between these might persist once they arose. This might occur if two populations of birds with very different songs spread into a new area so that their territories came to border on each other, as might happen after changes in habitat through burning or agricultural cycles. A matrix was set up with all birds on one side singing one song and all those on the other singing another. After 20 years the border between the two areas was in much the same place, though it shifted 2-3 territories in one direction or another in some trials. Again, however, this assumes very low (<2%) copying error rates.

This simple initial simulation assumed that each bird had four neighbours, which is near the mean figure for chaffinches. It also assumed that each had only a single song type, as is the case with some bird species, of which the white-crowned sparrow is the best known and studied. However, most species have more than one song type: in chaffinches, individuals can have anything from one to six, and most have two to four. Williams and Slater (1990) investigated the possible influence of these variables with another series of simulations. In these mortality was kept constant (at 40%) as was the copying error rate (15%). But repertoire size was varied from one to six in different runs, and birds had either two, four or eight neighbours from which they could copy. New birds built up their repertoires by choosing songs at random from a list of those possessed by their neighbours, the main constraint being that the same song could not be learnt twice. (The notion that a song might be learnt twice may seem a curious one. However, Slater (1983) has argued that this may well occur in chaffinches. Data on both the relative frequencies of song types in their output, and on the temporal pattern of singing, are more easily interpreted in some birds if it is assumed that the same song type is represented twice in the repertoire. We would not expect it to happen very frequently, however, and simulations in which such double learning was permitted did not yield markedly different results).

The simulation was run for "100 years" to achieve an equilibrium and various measures were then extracted from the resulting matrix: (1) The mean longevity of those song types that had both appeared and become extinct during the course of the simulation; (2) The number of birds, out of the total of 169 in the matrix, possessing the commonest song type; (3) The mean number of birds sharing a song; (4) The mean proportion of the repertoire shared between neighbours (measured by the Jaccard similarity index); and (5) A fragmentation index to measure the tendency for songs shared by two or more birds to occur in groups of birds that were not on adjacent territories.

Table 1 summarizes how repertoire size and number of neighbours affect these five measures. Although repertoire size did not affect two of the measures, it was clear that both it and number of neighbours did have a profound influence on various aspects of the distribution of song types within the population.

TABLE 1. Summary of results of simulations in which numbers of neighbours and repertoire sizes were varied (from Williams and Slater 1991).

	Effects of Increasing	
	Repertoire Size	Number of Neighbours
Mean Song Longevity	RISE	FALL
Number with Commonest Song	RISE	RISE
Mean Number with a Song	NO CHANGE	RISE
Similarity between Neighbours	FALL	FALL
Fragmentation Index	NO CHANGE	MORE WITH 4 AND 8 THAN WITH 2

Our final simulations, described by Williams and Slater (1991), explored the importance of different possible "learning rules". As mentioned above, the rule used by Williams and Slater (1990) was that all neighbour songs were listed and each one to be learnt was chosen at random, with the single constraint that the same song could not be learnt twice. However, if a song type was in the repertoire of three neighbours, it was three times as likely to be learnt as one that only a single neighbour sang. This rule will make common songs more likely to be copied than rare ones. While it could well be the rule that wild birds use, this is not necessarily so. We therefore explored three other learning rules: (1) Each song type has an equal probability of being learnt, so that the songs of neighbours are only added to the list if they are not on it already; (2) The song that is learnt is that shared by most neighbours, or chosen at random from amongst these; (3) The song that is learnt is that shared by fewest neighbours, or chosen at random from amongst these. In making these tests, for simplicity, we fixed error rate at 15%, mortality at 40%, number of neighbours at 4, and repertoire size at 1.

The first of these changes did not have a marked impact on the results of the simulations, but the other two had striking effects. If the commonest song was learnt, mean song longevity was much reduced, whereas learning of the rarest song led songs to persist for much longer. At first sight this appears rather strange. However, it is probably because learning of the commonest song makes it very difficult for a new song type to spread into the population whereas, when the rarest is learnt, a song which is just about to die out becomes chosen for learning and so tends to persist. Where birds chose the commonest song type for learning, the number of individuals which could share a song rose quite sharply, so that 25-30 birds could sing the same type. Indeed, if error rate was reduced to 2%, groups of over 100 birds might arise with the same type: far more than in the original simulations where the song to be learnt was chosen at random. It is therefore possible that, if this learning rule is that employed by wild birds, dialect areas in which many birds share a song type, such as those that have been described for white-crowned sparrows (Baker and Thompson 1985), might arise.

We do not yet know which of these learning rules is the most realistic for birds in the wild. However, it may well be that, at least in some species, individuals do not just select a random neighbour for copying. Gibbs (1990) found the commonest song amongst a population of the Darwin's finch Geospiza fortis to become rarer over several years, whereas the rarer songs increased in frequency. Birds with rarer songs seemed, for some reason, to do rather better, both in terms of longevity and of breeding success. This suggests that there may be some advantage to learning the rarer song.

Two main points emerge from our simulations. First, it is clear that copying processes which are normally accurate but which generate errors on a small proportion of occasions can produce patterns of song distribution not dissimilar to those observed amongst birds in the wild. Second, it is also clear that a variety of factors, such as the numbers of birds available to be copied, repertoire size and the learning rules involved in copying can have a profound influence on various aspects of song sharing and distribution amongst birds. We have made very few assumptions. The simulations are based mainly on the idea that young birds learn their songs from neighbours when they set up their territories for the first time but that, on a small proportion of occasions, song copying is not accurate so that new types emerge. What is interesting is that this rather simple model can generate a variety of varied and complex outcomes.

Whether or not one accepts the transmission of bird song from one generation to the next as an example of culture, it does illustrate how simple rules may give rise to complex outcomes. Faced by the complexity of culture, perhaps we should not be shy in searching for underlying simplicity.

4. Acknowledgements

Work by the authors described in this article has been assisted by a studentship to J.M.W. from the University of St Andrews, together with grants to P.J.B.S. from the Science and Engineering Research Council and the Royal Society. We are grateful to Dr Richard Byrne for his comments on the manuscript.

5. References

Aoki, K. (1989) 'A sexual-selection model for the evolution of imitative learning of song in polygynous birds', Am. Nat. 134, 599-612.

Avery, M. and Oring, L.W. (1977) 'Song dialects in the bobolink (Dolichonyx oryzivorus)', Condor 79, 113-118.

Baker, M. C. and Cunningham, M. A. (1985) 'The biology of bird-song dialects', Behav. Brain Sci. 8, 85-133.

Baker, M. C. and Thompson, D. B. (1985) 'Song dialects of white-crowned sparrows: Historical processes inferred from patterns of geographic variation', Condor 87, 127-141.

Baptista, L. F. (1975) 'Song dialects and demes in sedentary populations of the white-crowned sparrow (Zonotrichia leucophrys nuttalli)', Publ. Zool. Univ. Calif. 105, 1-52.

Baptista, L. F. and Schuchmann, K. (1990) 'Song learning in the Anna hummingbird (Calypte anna)', Ethology 84, 15-26.

Bjerke, T. K. and Bjerke, T. H. (1981) 'Song dialects in the redwing, Turdus iliacus', Ornis Scand. 12, 40-50.

Byrne, R.W. (In press) 'The evolution of intelligence', in P.J.B.Slater and T.R.Halliday (eds.), Behaviour and Evolution, Cambridge University Press, Cambridge.

Gibbs, H. L. (1990) 'Cultural evolution of male song types in Darwin's medium ground finches, Geospiza fortis', Anim. Behav. 39, 253-263.

Goodfellow, D. J. and Slater, P. J. B. (1986) 'A model of bird song dialects', Anim. Behav. 34, 1579-1580.

Grant, B. R. (1984) 'The significance of song variation in a population of Darwin's finches', Behaviour 89, 90-116.

Ince, S. A., Slater, P. J. B. and Weismann, C. (1980) 'Changes with time in the songs of a population of chaffinches', Condor 82, 285-290.

Jenkins, P. F. (1978) 'Cultural transmission of song patterns and dialect development in a free-living bird population', Anim. Behav. 26, 50-78.

Kroodsma, D. E. (1984) 'Songs of the alder flycatcher (Empidonax alnorum) and willow flycatcher (Empidonax traillii) are innate', Auk 101, 13-24.

Kroodsma, D. E. (1989) 'Male eastern phoebes (Sayomis phoebe; Tyrannidae,Passeriformes) fail to imitate songs', J. Comp. Psychol. 103, 227-232.

Kroodsma, D. E. and Konishi, M. (1991) 'A suboscine bird (eastern phoebe, Sayornis phoebe) develops normal song without auditory feedback', Anim. Behav. 42, 477-487.

Lefebvre, L. and Palameta, B. (1988) 'Mechanisms, ecology, and population diffusion of socially learned, food-finding behavior in feral pigeons', in T.Zentall and B.Galef (eds.), Social Learning: Psychological and Biological Perspectives, Erlbaum, Hillsdale, New Jersey, pp. 141-164.

Lehtonen, L. (1983) 'The changing song patterns of the great tit Parus major', Ornis Fennica 60, 16-21.

Lynch, A. and Baker, A. J. (1986) 'Congrence of morphometric and cultural evolution in Atlantic island chaffinch populations', Can. J. Zool. 64, 1576-1580.

Lynch, A., Plunkett, G. M., Baker, A. J. and Jenkins, P. F. (1989) 'A model of cultural evolution of chaffinch song derived with the meme concept', Am. Nat. 133, 634-653.

Marler, P. (1990) 'Song learning: the interface between behaviour and neuroethology', Phil. Trans. R. Soc. Lond. B 329, 109-114.

Marler, P. and Peters, S. (1982) 'Developmental overproduction and selective attrition: new processes in the epigenesis of bird song', Dev. Psychobiol. 15, 369-378.

Marler, P. and Peters, S. (1987) 'A sensitive period for song acquisition in the song sparrow, Melospiza melodia: a case of age-limited learning', Ethology 76, 89-100.

McGregor, P. K. (1980) 'Song dialects in the corn bunting (Emberiza calandra)', Z. Tierpsychol. 54, 285-297.

McGregor, P. K. and Krebs, J. R. (1982) 'Song types in a population of great tits (Parus major): their distribution, abundance and acquisition by individuals', Behaviour 79, 126-152.

Mundinger, P. C. (1970) 'Vocal imitation and individual recognition of finch calls', Science 168, 480-482.

Nottebohm, F. (1969) 'The song of the chingolo (Zonotrichia capensis) in Argentina: description and evaluation of a system of dialects', Condor 71, 299-315.

Nurnberger, F., Siebold, D. and Bergmann, H.-H. (1989) 'Annual changes of learned behaviour - variation of song pattern in free-living chaffinches, Fringilla coelebs, during the breeding season', Bioacoustics 1, 273-286.

Palameta, B. (1989) 'The importance of socially transmitted information in the acquisition of novel feeding skills by pigeons and canaries', Ph.D. Thesis, University of Cambridge.

Payne, R. B. (1981) 'Song learning and social interaction in indigo buntings', Anim. Behav. 29, 688-697.

Payne, R. B. (1985) 'Behavioral continuity and change in local song populations of village indigobirds Vidua chalybeata', Z. Tierpsychol. 70, 1-44.

Payne, R. B., Payne, L. L. and Doehlert, S. M. (1988) 'Biological and cultural success of song memes in indigo buntings', Ecology 69, 104-117.

Payne, R. B., Thompson, W. L., Fiala, K. L. and Sweany , L.L.(1981) 'Local song traditions in indigo buntings: cultural transmission of behavior patterns across generations', Behaviour 77, 199-221.

Pepperberg, I. M. (1990) 'Some cognitive capacities of an African grey parrot', Adv. Study Behav. 19, 357-409.

Petrinovich, L. (1988) 'Individual stability, local variability and the cultural transmission of song in white-crowned sparrows (Zonotrichia leucophrys nuttalli)', Behaviour 107, 208-240.

Petrinovich, L. and Baptista, L. F. (1987) 'Song development in the white-crowned sparrow: modification of learned song', Anim. Behav. 35, 961-974.

Slater, P. J. B. (1983) 'Sequences of song in chaffinches', Anim. Behav. 31, 272-281.

Slater, P. J. B. (1986) 'The cultural transmission of bird song', Trends Ecol. Evol. 1, 94-97.

Slater, P. J. B. (1989) 'Bird song learning: causes and consequences', Ethol. Ecol. Evol. 1, 19-46.

Slater, P. J. B. and Ince, S. A. (1979) 'Cultural evolution in chaffinch song', Behaviour 71, 146-166.

Slater, P. J. B. and Ince, S. A. (1982) 'Song development in chaffinches: what is learnt and when?', Ibis 124, 21-26.

Slater, P. J. B., Ince, S. A. and Colgan, P. W. (1980) 'Chaffinch song types: their frequencies in the population and distribution between the repertoires of different individuals', Behaviour 75, 207-218.

Thielcke, G. (1987) 'Langjährige Dialektkonstanz beim Gartenbaumläufer (Certhia brachydactyla)', J. Orn. 128, 171-180.

Thorndike, E. L. (1898) 'Animal intelligence: An experimental study of the associative process in animals', Psychol. Rev. Monogr. 2 (8), 551-553.

Todt, D. (1975) 'Social learning of vocal patterns and modes of their application in grey parrots (Psittacus erithacus)', Z. Tierpsychol. 39, 178-188.

Trainer, J. M. (1989) 'Cultural evolution in song dialects of yellow-rumped caciques in Panama', Ethology 80, 190-204.

West, M. J. and King, A. P. (1988) 'Female visual displays affect the development of male song in the cowbird', Nature 334, 244-246.

Whiten, A. and Ham, R. (1992) 'On the nature and evolution of imitation in the animal kingdom: Reappraisal of a century of research', Adv. Study Behav. 21, 239-283

Williams, J. M. and Slater, P. J. B. (1990) 'Modelling bird song dialects: the influence of repertoire size and numbers of neighbours', J. theoret. Biol. 145, 487-496.

Williams, J. M. and Slater, P. J. B. (1991) 'Simulation studies of song learning in birds', in J-A.Meyer and S.Wilson (eds.), Simulation of Adaptive Behavior, M.I.T.Press, pp. 281-287.

Zann, R. (1990) 'Song and call learning in wild zebra finches in south-east Australia', Anim. Behav. 40, 811-828.

SWARM INTELLIGENCE IN SOCIAL INSECTS AND THE EMERGENCE OF CULTURAL SWARM PATTERNS

G. THERAULAZ

C.N.R.S. U.R.A. 664
C.R.B.C., Université Paul Sabatier
118, route de Narbonne
31062 Toulouse
France
&
Santa Fe Institute
1660 Old Pecos Trail, Suite A
Santa Fe, NM 87501
U.S.A.

J.-L. DENEUBOURG

Université Libre de Bruxelles,
Service de Chimie Physique
C.P. 231, Boulevard du Triomphe
1050 Bruxelles
Belgium

KEYWORDS / ABSTRACT : Swarm Intelligence / Self-organization / Social Insects / Collective Problem Solving / Artificial Life / Emergent functionality. This paper focuses on the reciprocal causality which exists between the organization of social insects' colonies (behavioral differentiation, spatio-temporal patterns of activities, collective decisions, ...) and the structure of the environment as a consequence of the actions of the colony. This constitutes a basis to explore different kinds of phenomena : how to achieve a collective task assignment without any centralized organizer; how a collective representation emerges at the level of the colony without any symbolic representation at the individual level; how to collectively build a complex architecture without a blueprint. To characterize these collective behaviors in which a coherent functional global pattern emerges through simple local agents interacting with each other and with their local near environment we introduced the concept of Swarm Intelligence. Swarm Intelligence applies to animal or even artificial groups that exhibit intelligent behavior at a group level, mainly through achieving a distributed problem solving, well beyond that shown by its individual members. In swarms the structure of the social organization or the architecture of the nests produced is perpetuated from one generation to another. This creates a process that enables a collective structure to be transmitted without any true cultural tradition.

1. A Distributed Perception-Action System

For a long time insect's societies have been considered as a super-organism (WHEELER, 1911; EMERSON, 1939, 1952) or a somewhat like a collective brain, but this metaphor loses the main characteristic of these societies (see BONABEAU and THERAULAZ, 1994 for a complete historical overview). In fact, what is particularly striking is that its components are separated from each other and that each individual is mixed with the material components of the environment where it is moving about. It is a kind of brain having its cells mixed in with the problem to solve. Then the society is able to sense, process and act on the information coming from the environment in which it is scattered but in a distributed way, both the perception and the action taking place at the individual level. This fact constitutes the essential property of the swarm activity which

R. A. Gardner et al. (eds.), The Ethological Roots of Culture, 107–123.

accounts for its specific behavioral, and architectural patterns. The collective performance results from the integration of the myriads of individuals' activities, with each individual both processing information produced by the activities of others and stimulating and informing them in their turn. And for the observer "the whole behaves as a unit in the same way as if there was a coordinating agent "virtually" present in its center" (VARELA, 1991).

We would like to emphasize the reciprocal causality that exists between the organization of the colony in its various aspects (behavioral differentiation, spatio-temporal patterns of activities, collective decisions, ...) and the structure of the environment as a consequence of the collective action of the individuals of the colony. This reciprocal causality is able to explain both the specific patterns of colonies' organization as well as the architectures that are built by the colony. This mutual process could have also affected the behavioral evolution in social insects in order to produce specific kinds of organizations as well as specific kinds of architectures. Moreover, such a mutual process seems to be largely widespread. Indeed, we would like to point out that similar kinds of phenomena appear with human collective behaviors leading to specific patterns.

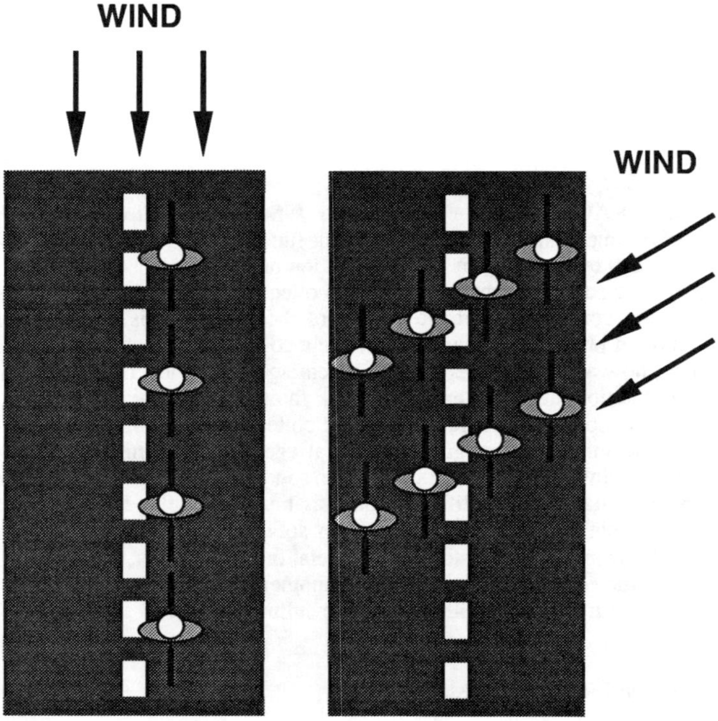

Figure 1. Example of field effect : the collective pattern of the squad in a cycle racing is a function of the direction of the wind.

One of the best examples appears in cycle racing. When the squad is faced with wind it takes on a specific pattern which depends on the direction of the wind (see figure 1). When an individual adjusts his position in the wind in some way, he creates a nearby zone which is less exposed to the wind. In so doing he encourages the other cyclists to take up this position. The fact is that this kind of pattern emerges either in a team in which all the individuals are cooperating together or in a group where every individual is running for himself. Indeed such a structure is able to emerge just as easily by "trial and error". This structure can be obtained with the help of a learning process or the individuals can rediscover the right position without any learning process. This kind of field effect that leads to specific structures can be seen in other situations. Thus, in BRITTANY, some villages spontaneously adopt an arrow-shape structure in order to minimize the effect of the wind (see GRILLOT, 1975). Some of these patterns are then frozen by the cultural process, and appear in some places where they lose their original functionality.

In order to characterize this kind of collective behavior in which a coherent functional global pattern emerges through simple local agents interacting with each other and with their local near environment, we have introduced the concept of *Swarm Intelligence* (THERAULAZ *et al.*, 1990b, THERAULAZ and GERVET, 1992). Swarm Intelligence applies to animal or even artificial groups that exhibit intelligent behavior at a group level well beyond that shown by its individual members. The dynamic organization of a social insect's colony is a striking example of this notion, whose efficiency is reflected in their widespread ecological success. It is now evident that a colony's collective performance depends as much on the interactions between the individuals and with their environment as on an individual complexity. In spite of the workers generally limited behavioral repertoire and learning capacity, and their inherently random character, the colony is a well-organized and stable structure that nevertheless responds adaptatively to the unpredictable nature of its environment and changes in its internal composition due to births and deaths.

Section 1 defines some of the concepts of Swarm Intelligence, while section 2 presents various aspects of this distributed intelligence and focuses at each time on the role played by the reciprocal causality we introduced above in the genesis of specific patterns. Section 3 outlines the role played by the environment and tries to establish the distinction between true and false cultural swarm patterns.

2. Fundamentals of Swarm Intelligence

These last years, the interest in the adaptability of social systems in insects has been renewed (see DENEUBOURG & GOSS, 1989 for an overview). Indeed, these non-linear systems can display a large variety of rich and even very complex behaviors, even though the constituent individual behavior is paradoxically very simple and has a strong random component. Local behavioral and environmental constraints and local information control the behavior of each individual. The resulting collective patterns characterize either the differentiation and spatio-temporal organization of the insects or the kinds of structures which are produced. One particular approach is to analyze the performance of the social group as a form of collective problem solving. In this kind of analysis, the question is to determine what are the characteristics at the level of the different elements which govern the efficiency of the solving process and how the environmental factors regulate the form of the solution adopted by the colony. In order to do this, the studies combine a detailed biological analysis of individual behavior with numerical simulations to link this level with the colony's global behavior. The concepts we introduce thereafter are aiming to describe the phenomenology of the dynamical organization of swarms as well as the formal causes of their collective behavior.

2.1. SWARM

We define a swarm as a set of (mobile) agents which are liable to communicate directly or indirectly (by acting on their local environment) with each other, and which collectively carry out a distributed solution to a problem (THERAULAZ *et al.*, 1990b ; THERAULAZ & GERVET, 1992). In this sense we refer to *emergent functionality* (STEELS, 1990, 1991) or *functional self-organization* (ARON *et al.*, 1990), since this emerges from swarm's internal dynamics and its interaction with the environment. The swarm functioning induces both the genesis of functional collective patterns which characterize the differentiation and spatio-temporal organization of the agents of the swarm and the parallel organization of the material elements in the environment upon which each agent has an action. We will see that close connections exist between these different patterns and the specific categories of problem which are able to be solved by a swarm. Swarm intelligence can then be considered as a property of non-intelligent agents exhibiting collectively intelligent behavior.

2.2. PROBLEM AND COLLECTIVE PROBLEM SOLVING

In the framework of swarm functioning the concept of a problem can be defined as a kind of description of the position of a biological or artificial agent, where a functional outcome is described as a goal even though some parameters having the possibility to evolve with time are described as constraints. One can consider the problem to be set when the goal, the constraints and the lawful procedure to move from an initial state S_0 to a final state S_f take into account the swarm and the environment in which the swarm is scattered. It is worth noting that this definition not only applies to a swarm but also to a single agent. The swarm is characterized by the collective resolution of the problem. Depending on whether an artificial or biological system is considered, the description of the problem to be solved will take a different look :

- when we consider an artificial system, the problem can be conceived before the design of the swarm whose local elementary behavioral rules will bring the system to solve this problem in a given environment;
- but when we consider a biological system, the specification of the problem is equivalent to identifying a specific biological function (*e.g.* : the building behavior, the task assignment, ...).

The solution of the problem can be considered in both cases as a particular state of the swarm environment system through which the functional outcome looked for is reached. As a rule, a great number of solutions can exist for a given problem, meaning that a given goal is compatible with several states of the system constituted by the swarm and its environment. Thus the "collective resolution of the problem" lies in the structural co-dependency process between the swarm and its environment in which the functional outcome described as a goal is reached.

Taking an anthill as an example ; observation shows different elements constituting the brood (eggs, larvae and pupae) are sorted and aggregated into piles of the same type by the workers (SENDOVA & FRANKS, in prep.). The sorting can be finer and discriminate several larval instars. In this example the goal is sorting the different elements which constitutes the brood; the problem is how to achieve this ? The variation in the final number of aggregates we obtain, such as three (eggs, larvae, pupae), four (eggs, small larvae, big larvae, pupae) or more, represents different solutions to the problem. The resolution is the process by which the swarm reaches one of these solutions (see DENEUBOURG *et al.,* 1991 for further information about a particular example of sorting algorithm).

2.3. WHAT KINDS OF PROBLEMS CAN BE SOLVED BY A SWARM ?

Solving a problem with a swarm amounts to a morphogenetic process leading to a form which is the solution of the problem. Such a process involves both a structuring of the group of agents and the environment in which the swarm moves. If one specifies the different kind of problems a swarm is able to solve, one identifies the elements this structuring process is acting on. Indeed, in a swarm the structure of the environment and the organization of the group of agents mold each other. Both make up the double sides of the same structuring process. The problem to be solved may still turn on one face or the other : organizing the environment or specifying and organizing in space and time the individual activities of each agent.

Numerous studies have dealt with these two classes of problems. Problems such as task allocation, coordination, collecting material inspired by social insects or other social groups can support similar discussions and speculations. Different examples of this decentralized and collective intelligence have been discussed, including building behavior (DENEUBOURG, 1977 ; BELIC *et al.*, 1986 ; SKARKA *et al.*, 1990), collective choice (PASTEELS *et al.* 1987 ; BECKERS *et al.*, 1990 ; SEELEY *et al.*, 1991 ; CAMAZINE & SNEYD, 1991), the formation of trail networks (ARON *et al.* 1990), sorting (CAMAZINE, 1991 ; CAMAZINE *et al.*, 1990 ; DENEUBOURG *et al.* 1991), collective exploration (DENEUBOURG *et al.*, 1989 ; FRANKS 1989 ; FRANKS *et al.* 1991) and dynamical division of labor (DENEUBOURG *et al.* 1987 ; CORBARA *et al.* 1991 ; THERAULAZ *et al.*, 1991b) synchronization and the generation of oscillations (FRANKS *et al.* 1990 ; GOSS & DENEUBOURG, 1988 ; KRAFFT & PASQUET, 1991).When the problem turns on the structuring of the environment, the swarm changes the structure characterizing a set of objects spread over the environment with handling operations. Different algorithms have been elaborated to enable an artificial swarm to sort different types of objects. The algorithm used by DENEUBOURG *et al.* was inspired from the processes used in ant colonies to sort their brood (DENEUBOURG *et al.* 1990b, 1991). When we simulate a swarm of artificial ants with simple elementary and reshaping properties reproducing the behavior of ants, the swarm's activity is coordinated and we obtain the sorting of two or more classes of objects. But the problem may also turn on the structuring of the behavioral units of the swarm to realize a spatio-temporal distribution (BECKERS *et al.* 1990 ; DENEUBOURG *et al.*, 1989, 1990a ; FRANKS *et al.*, 1989 ; GOSS *et al.*, 1989) or a functional specialization in the agents' activities (THERAULAZ *et al.*, 1990a, 1990b, 1991a, 1991b).

3. Embodiments of Swarm Dynamics

We shall now examine several examples of swarm dynamics in which the reciprocal interaction between the structure of the colony and the organization of the environment accounts for the specific patterns produced. To do this, we examine three self-organizing models with each one focusing on a particular aspect of the swarm's activity.

3.1. A COLLECTIVE TASK ASSIGNMENT WITHOUT ANY CENTRALIZED ORGANIZER

Division of labor occurs in many eusocial insects' societies. At each time, all the individuals of the colony must perform a certain number of tasks which vary in space and time and which depend on the internal needs of the colony as well as on the particular environmental conditions. Then the question is : how the colony is able to allocate the tasks to be done and to coordinate the activities of each individual. In other words how is it possible to relate local individual interactions, their effect onto the individuals in which they occur and how on this basis is it possible to explain the organization of the great behavioral forms which appear in the colony. This occurs in particular in a primitively eusocial wasp species called *Polistes dominulus* Christ (THERAULAZ *et al.*, 1991a). These wasps have little differentiation and no morphological differences between castes or predetermined control of activities depending on age or on any other known physiological predetermination. Here again, the integration of the individuals' activity depends essentially on the interactions occurring between the colony members and with their immediate environment.

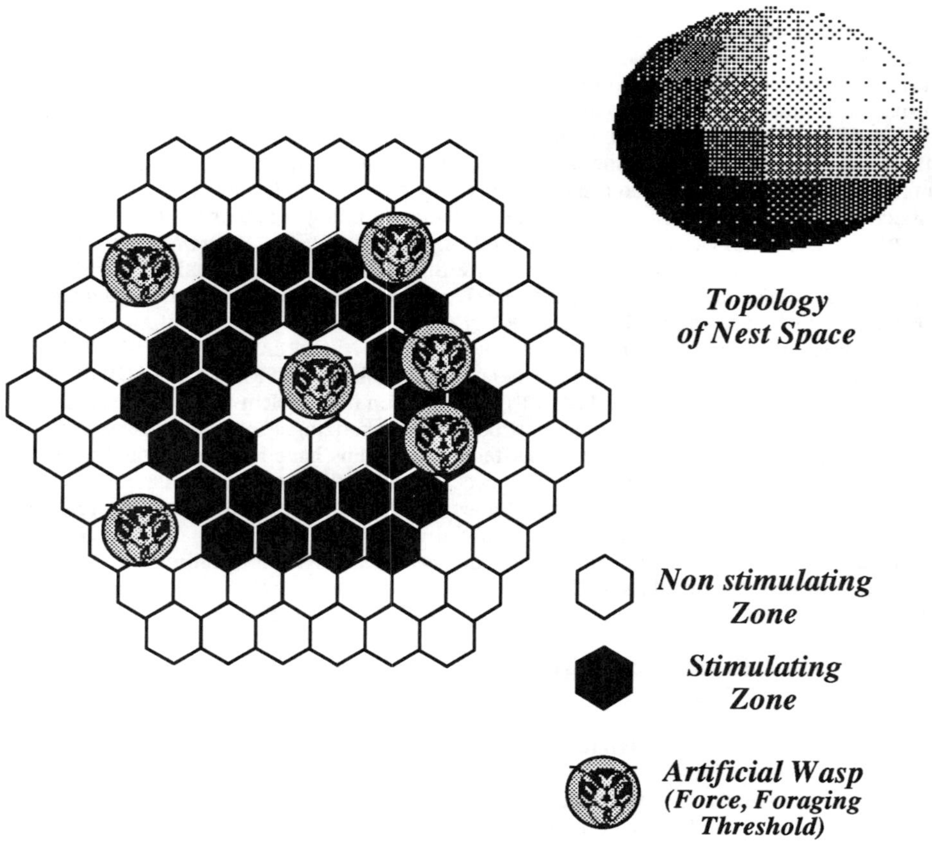

Figure 2. Structure of the nest space

A double morphogenetic process occurs. First, each individual acquires its own behavioral profile, related to the different tasks in and outside the nest. Second, the ensemble of these profiles constitutes the way in which the colony has divided the different tasks to be performed amongst the different members. In the colonies, one can distinguish three main tasks : egg laying (always performed by just one individual : the queen or the α-individual), foraging and brood care, and finally doing nothing. The individuals involved in this later "task" constitute a sort of reserve labor pool that can be more or less mobilized in response to the overall work load. Furthermore, each of these three tasks is oriented towards a different part of the nest; the egg-layer remains in a zone where there are cells ready to receive eggs, the foragers/brood careers' activity is concentrated towards zones occupied by the larvae, and the inactive wasps are found at the edge and the back of the nest.

Traditional approach has been to consider the queen or the α-individual as the main central organizer of the colony through changes of its own level of activity. However, the implementation of the interactions which control the individual behavior and the simulation of artificial colonies reveals that the task assignment, the parallel spatial distribution process and the hierarchical organization of the colony emerge out the interactions occurring both between individuals and with their environment. Two types of interactions control individual behavior in *Polistes dominulus* Christ wasp colonies. One is a hierarchical type of interaction which appears in dominance scenes. When two wasps meet, they carry out a ritualized fight in which one wins (the dominant) and the other loses and is pushed toward the edge of the nest (the subordinate). The more subordinate the individual is, the less likely it is to show dominance behavior. Moreover, the higher the probability that an individual will dominate in a hierarchical interaction, the greater its propensity will be to frequently dominates its subordinates. In this way, each wasp reinforces its probability of dominating through the number of times it is successful in hierarchical interactions and each subordination has the opposite effect. These relationships extend throughout the society as a whole and organize it along a linear hierarchy. The second type of interaction is a trophic type which controls the relationships between the individuals and the environment, particularly with the brood, which in turn triggers some types of activities towards the brood, such as brood-tending and/or collecting prey and the subsequent feeding of larvae. Each individual is characterized by differential reactivity threshold to the various brood components (the eggs, small and large larvae and pupae), which may evolve along the time in function of the individual's activity. In fact, there is a parallel self-facilitating process which causes the increase in the amount of time spent foraging by individuals which have already foraged previously (THERAULAZ *et al.*, 1991a).

3.1.1. *The elementary behavioral rules.* In the model each wasp is represented by a unit having the ability to interact both with similar units and with some characteristics of the environment. The model describes the activity of n wasps, whose number remains constant with time where dominance-subordination exist, and integrates two local action rules. Each individual is characterized by two variables, the force F which intervenes during random encounters within the nest, and by a response threshold σ to larval stimulations located in a specific zone of the brood (see figure 2). At the beginning of the simulation, the n individuals of the colony are identical. The behavioral forms which characterize the colony are then generated by the combined action of two autocatalytic processes (dominance and foraging). The force value of each individual varies with each combat. Every time it wins or loses its force F increases or decreases along with an intrinsic coefficient of reinforcement which is common to all the individuals. Moreover its probability of winning or losing is a function of its own and rival force. Thus each time a wasp wins it increases its force and its probability of winning its next combat. This positive feedback rapidly separates a group of initially equal wasps into a dominance hierarchy with dominants (high force individuals) and subordinates (low force ones). The foraging threshold evolves along a similar process. There is again a positive feedback mechanism, since each time a wasp forages it increases its probability of foraging, and each time it doesn't it decreases its probability of foraging the next time. Again this can rapidly separate a group of initially identical wasps into two subgroups of foragers and non-foragers.

3.1.2. *The colony' behavioral patterns.* The system is characterized by one or more stable configurations (or attractors), corresponding to the force, foraging threshold and average position of the initially equals individuals. For a given set of parameters values, the system spontaneously evolves towards one of these configurations. But one cannot predict what any particular individual history will be. Figure 3 presents a simulation of a colony of 8 individuals. Three behavioral categories of individuals are generated which correspond exactly to those appearing

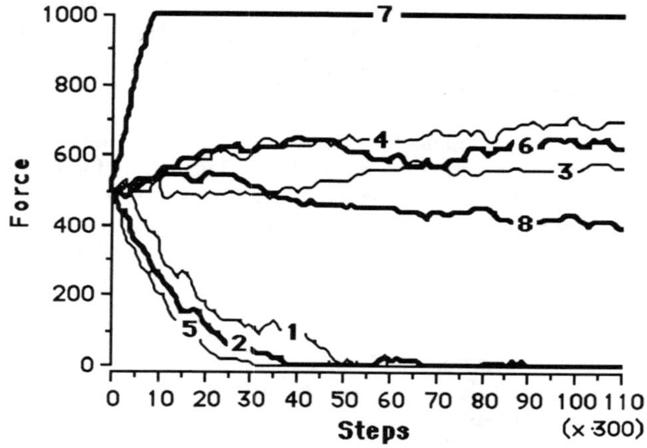
1000
800
600
400
200
0
Force
7
4
6
3
8
5
2
1
0 10 20 30 40 50 60 70 80 90 100 110
Steps
(x 300)

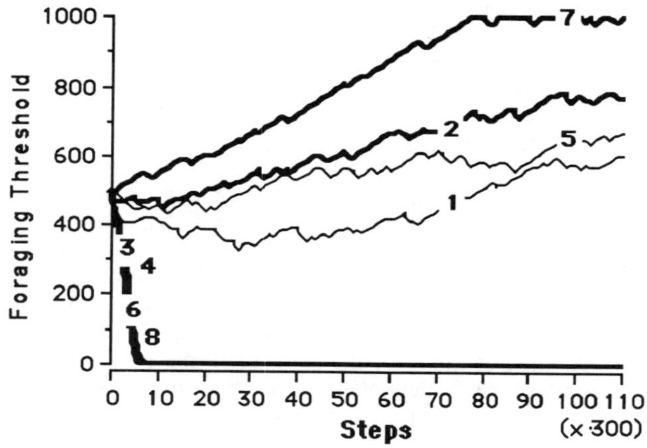
1000
800
600
400
200
0
Foraging Threshold
7
2
5
1
3
4
6
8
0 10 20 30 40 50 60 70 80 90 100 110
Steps
(x 300)

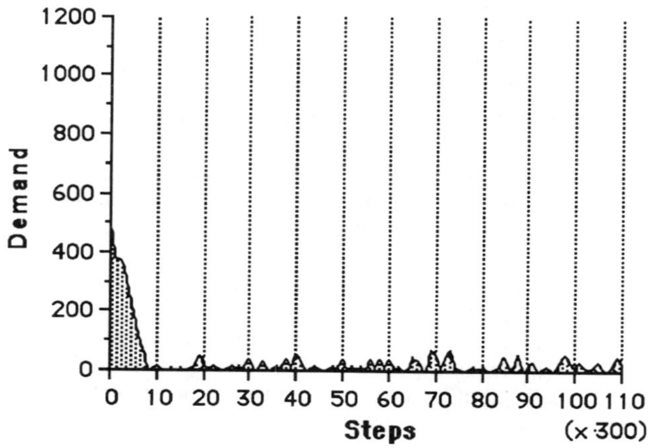
1200
1000
800
600
400
200
0
Demand
0 10 20 30 40 50 60 70 80 90 100 110
Steps
(x 300)

Figure 3. Collective behavioral organization in a wasp nest. The model describes the parallel evolution of the behavioral variables (Force and Foraging Threshold) in a colony of 8 individuals. For each individual the coefficient of reinforcement of the force and the foraging threshold is 10, and the forgeting coefficient for the foraging is 0.03.

---

in real colonies. The first behavioral form corresponds to a unique individual, whose force increases rapidly to a maximum, and whose foraging threshold increases to a maximum. This alpha-individual therefore remains on the nest, involved in many encounters and always dominates the others. The second behavioral form characterizes a group of individuals whose foraging threshold rapidly decreases to a minimum. They are the active foragers who control the level of the brood's hunger which drops to fluctuate around a low level as they take up their foraging role and bring food. As they are often outside the nest they can maintain an intermediate level of force. And the last behavioral form characterizes individuals with intermediate or high foraging threshold. So they forage occasionally, remain most of the time on the back of the nest, they are frequently dominated and have a low force.

Here we can see that the collective pattern that is generated reflects the coupling between the structure of the nest space and the elementary behavioral rule which characterize the individuals. The group function is generated by a double feedback system controlling individual behavior : Positive feedback amplifies small fluctuations among the competing wasps and has morphogenetic effects. In the case of the force value, direct competition among individuals leads to the hyperactivity of a single wasp up to the biologically possible limit. The more indirect nature of the competitive interactions between the wasps explains why the number of foragers varies as a function of the intensity of the larval demand and the foraging time. The cyclical relationship which exists between brood satisfaction and foraging activity maintains a regular level of social demand which regulates the number of foragers (*i.e.* the wasps which have a minimum foraging threshold). Moreover, the structure of the nest space accounts for the spatial distribution of the three behavioral groups of wasps.

3.2. A COLLECTIVE REPRESENTATION WITH REPRESENTATIONLESS INDIVIDUALS

Two of the main features of the communications used by social insects in order to self-organize are randomness and positive feed-back. A standard example is given by food recruitment in ant colonies (see figure 4). When a new food source is discovered by an ant, it lays down a transient chemical trail. This pheromonal marking promotes in an indirect manner the other ants of the colony to follow this trail towards the place where the food was discovered. These ants, after feeding at the source, will reinforce the trail when they return to the nest. In so doing they themselves change from recruits to recruiters. This kind of communication enables the amplification of a discovery but also the selection of information through the competition which might exist between several discoveries. Indeed, with the help of this autocatalytic logic and the use of spatial constraints, the strength of the "near source" signal increases faster than the strength of the "far source" signal. The latter will be eliminated afterwards. In this way a food source close to the nest will be selected more rapidly than a food source far away from the nest. This collective response of the society, showing in this special case a temporal structuring of the individuals' activity, is the solution to the problem in hand : in this example the exploitation of the closest food-source.

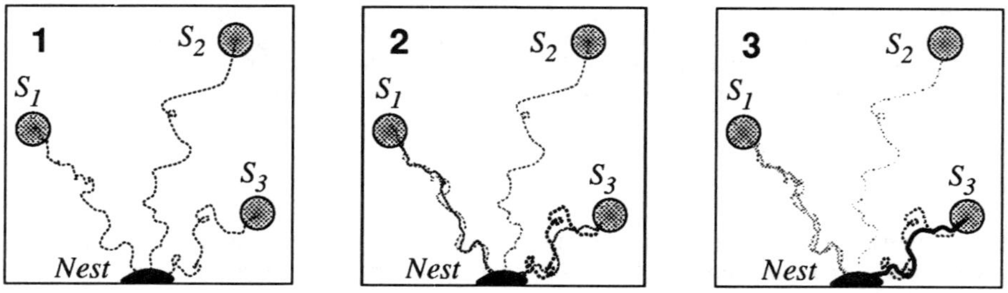

Figure 4. The exploitation of the closest food source shows how small random fluctuations can be amplified by the colony. S_1, S_2 an S_3 are the food sources ; the thickness of the curves is a function of the stream of individuals following the trail.

Through this example we put in place all the elements which give the collective problem solving in insect societies its original features. Even though none of the individuals in the colony is informed of all the possible ways to find a solution, and none have an explicitly preprogrammed solution, the colony as a whole converges towards an adaptive solution. We can thus see that without any particular spatial coding the colony solves the problem using only an autocatalytic logic and the geometrical and physical constraints of the environment in which it moves.

3.3. A COLLECTIVE CONSTRUCTION WITHOUT A BLUEPRINT : ARCHITECTURES WITHOUT ARCHITECTS

One question that has been for a long time set is about the way the insects collectively produced their nest architecture. Here again it seems that something like a blueprint is embedded somewhere in the brain of each insect. Indeed, as we shall see the architecture of the nests reflect the dynamics of the local behavioral interactions between the insects and their near environment. Figure 5 shows some aspects of the great diversity of nest architecture we observe in wasps. The variation of nest design extends from one cell per comb with an elongate form (a) to larger single combs (b) and multiple stacked combs with a varying number of cells per comb (c). Combs are suspended either to the substrate or from the rim of the cells of the upper comb. Then the question is : why do these structures have the form they have ? To answer this question we build a model in which the environment is a lattice divided in n x m cells which can be empty (0) or full (1). Moreover the behavior of each wasp is idealized with local behavioral rules.

3.3.1. *The elementary behavioral rules.* Two types of behavioral rules can be tested. The first which is called stigmergic is inspired by different biological observations. Our first goal with such a model is to examine the power and the limit of given rules, rather than to fit theoretical and experimental results. GRASSE introduced the concept of stigmergy in 1959 to explain the collective genesis of the termites' nest (GRASSE, 1959). The basic idea is that no direct interactions are necessary to coordinate the work of a group, but that the interactions between the nest and the workers are enough. The working termites modify their environment, providing new stimuli. These new stimuli induce new behavioral responses which in their turn modify the environment. With this succession of stimulus-reaction, the society is able to produce a structure. It is the work itself which assumes the coordination of the workers' activities. So in the

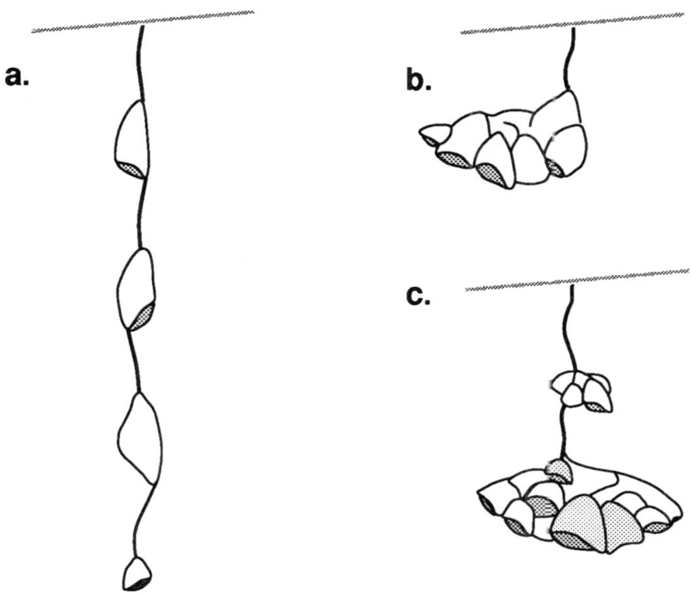

Figure 5. Some aspects of the diversity of wasp nest architecture. The following genera and species are representative of the nest types shown : a. *Mischocyttarus punctatus*, b. *Polistes*, c. stenogastrinae, *Parischnogaster mellyi*. Redrawn from HANSELL 1984 and JEANNE 1975.

stigmergic algorithm, only the local configuration met by the wasp determines its behavior, which is here reduced to fill or not the corresponding cells. The local configurations are the first four neighbors cells which surround the cell occupied by the wasp. From the 16 possible configurations, only three configurations stimulate the filling of the cell (see figure 6). Two correspond to the horizontal mode and one to a vertical mode. Each mode is characterized by a probability P_{H1}, P_{H2} and P_V of filling the corresponding cell met by the wasp. The second type of algorithms we tested is called sequential. In this case, the past activities of a wasp affect its building activity. The local configuration does not play a stimulating role, but only authorizes the wasp to fill, or not, the cell. It is the state of the wasp which controls its activity. At time t the wasps can be in the state "horizontal filling" or "vertical filling". The wasp in the state horizontal (vertical) can only fill a cell in the configuration "horizontal filling" ("vertical filling"). Having exhibited a vertical (horizontal) filling, the animal has a probability P_{VH} (P_{HV}) of becoming an horizontal (vertical) builder. The wasps move randomly on the nest and in its neighborhood.

3.3.2. *The colony' behavioral patterns*. Comparing the rate of building per insect for different colony sizes, the sequential algorithm shows a decrease of efficiency. The stigmergic mechanism generally shows an increase as in the digging model. Moreover, the form produced is much more

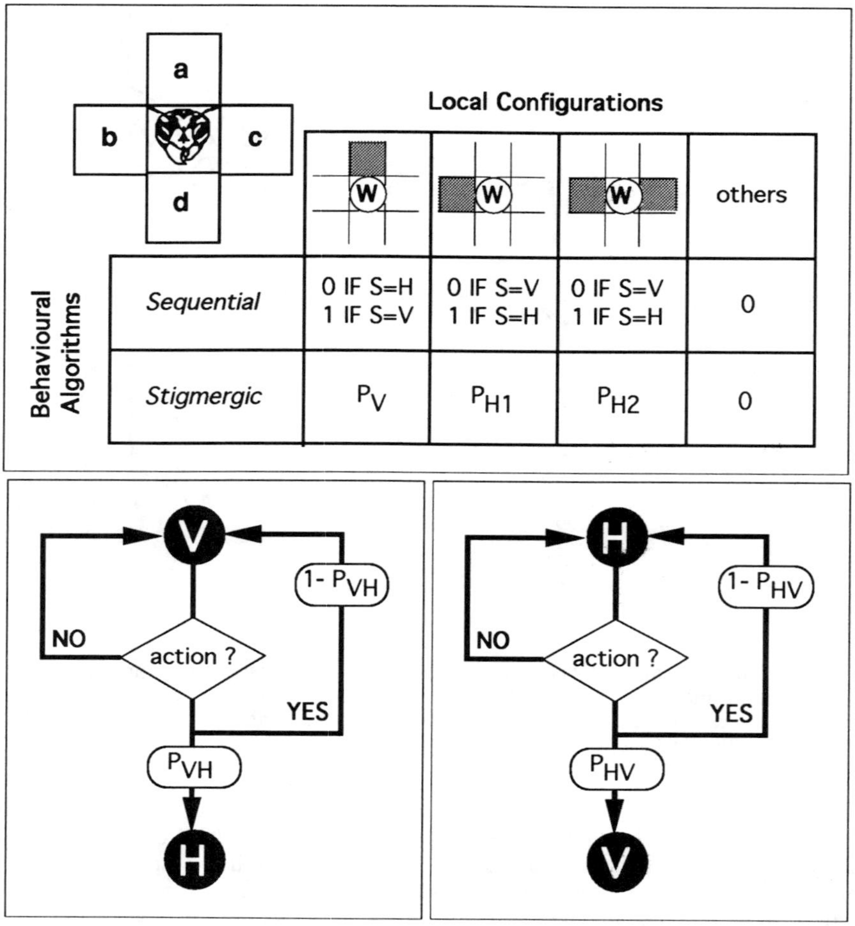

Figure 6. Definition of the behavioral algorithms used by artificial wasps. W: wasp; S: individual's state; H: "horizontal filling" state; V: "vertical filling" state; P_{H1}, P_{H2}, P_V: probabilities associated with the corresponding local configurations that a wasp fill the cell it occupies; P_{HV}, P_{VH}: probability that an horizontal (vertical) builder becomes a vertical (horizontal) builder.

stable (the same form is almost always produced) in the sequential case than in the stigmergic one (see figure 7). When the number of agents is increasing, the stigmergic algorithm produces forms that are more "complex", in terms of mutual information between the states of two neighboring cells, than the form produced by the sequential algorithm. This model shows that the sequential program seems to be more adapted to a solitary individual or a small group than to large numbers. The analysis of the architecture of wasps' nest shows that their complexity

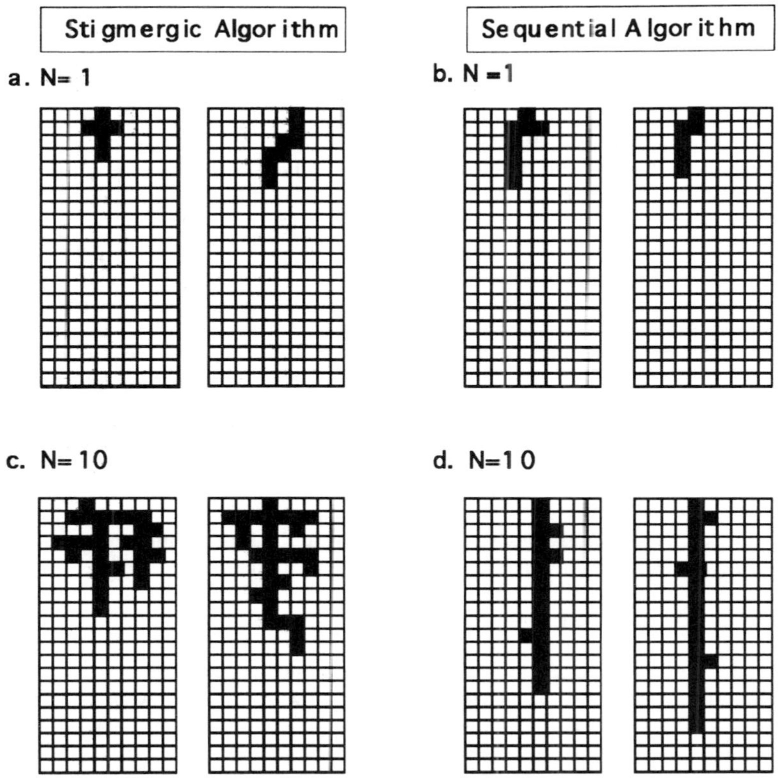

Figure 7. Architectural patterns obtained in a 2-dimension world with the stigmergic and the sequential algorithms. a. Solitary insect (N=1), stigmergic algorithm, $P_V = 0.8$; $P_{H1} = 0.2$; $P_{H2} = 0.4$. b. solitary insect (N=1), sequential algorithm, $P_{VH} = 0.2$; $P_{HV} = 0.8$. c. Idem as a), but the structure is obtained with N=10 individuals. d. Idem as b), but the structure is obtained with N=10 individuals.

increases with the transition from solitary to social life. We can see that one way to reach this complexity is to shift from a sequential to a stigmergic behavioral algorithm. One interesting question which still remains is : what could be the functionality of such an increase ?

4. Discussion.

4.1. THE ENVIRONMENT AS A MAP-MEMORY.

We have seen that two processes gave its identity to the swarm functioning : the random movements of the individuals and positive feedback mechanisms which strongly interact with the geometrical and topological constraints of the environment to give its own characteristics to the swarm patterns. The fundamental logic is then given by the interplay that exist between multiple random perceptions and actions and their consequences on both the individual behavior and the structure of the environment. Each action changes the landscape in a way that affects the resulting behavior of the colony. Small fluctuations affect some environmental variables as the result of random activities of the insects. These fluctuations (*e.g.* : the decrease of larval demand in the area occupied by the brood in the wasp nest, the creation of a seed in the building behavior, etc....) will then affect the behavior of the insects in the colony, which results in a decrease of the randomness. As a result of their past action the behavior of the insects is modified; sometimes the spatio-temporal activity is only affected through changes in the environment; sometimes it is affected by the probability to do a particular action on the environment through learning processes. In other words the swarm self-organizes through its own produced distributed activity. One could say that the environment is playing the role of a spatio-temporal memory which selects and directs the present action of the swarm.

It is worth noting that the swarm functioning does not imply any global representation or planning of the task to be doing, nor coded representation of the environment in a map form. Indeed, nor their processing abilities nor the properties of the signal they perceived enable the individuals to elaborate a global representation of their environment and to plan their actions. The indirect interactions that result from the actions of the individuals on the environment replace the plan of action. They produce a kind of "landmarks" in the environment which evolve in space and time. It is no more necessary that a "map" exists at the level of the swarm. The directional data are directly picked up in true-size at the level of the environment. One could say that the environment is playing the role of a full-sale model.

4.2. TRUE AND FAKE CULTURAL TRADITIONS IN SWARM DYNAMICS.

In social insects' societies an interesting phenomenon is occurring. Indeed it seems that something like a culture appears at the level of the swarm. Indeed, we saw that the swarm can be characterized by collective structures and nest architectures. So it seems to be the form (social organization and artifacts produced) which is perpetuated from one generation to another generation. But we cannot, so to speak, use the term "cultural tradition" since there is no culture at the individual level. Indeed if we transpose an individual from a colony A to a colony B it will not be able to transmit the features of its own social organization. This is not surprising since we saw that these specific patterns results from the interactions between the individuals and generally speaking from the group functioning. Thus some processes allow for the perpetuation of a structure without requiring any cultural transmission. In insects societies (and in other groups) we find numerous examples of permanent behavior at the group's level which are interpreted as a cultural effect (transmission and learning) or the identical response of individuals to environmental cues. However a number of examples can not be explained by this hypothesis and are much more like fake tradition randomly selected. When very strong mass effects exist inside a group, the new arrivals are oblige to adopt the same behavior as others. Chemical trails in ants are a good example . The new ants which "discover" a trail for the first time, follows this trail and reinforce it. The trail is like an external and collective memory. Its form depends on its history and may be perpetuated for a long time, and this without individual memory. However the trails can be also the touchstone of "cultural phenomena" in ants.

In order to get such a true cultural tradition, there must be a behavioral acquisition and the transmission of this one at the individual level. We can find such processes where the communication and the transmission of an information occur. One of the best example of such a true cultural tradition is that of "trunk-trails" in ants (the classical exemple is the genus *Formica*, ROSENGREN 1971). These trails remain stable from year to year, and the ants follow the same path before and after the winter which is a period of inactivity for the colony. When ants discovered a food source they recruited by laying a trail as we saw in section 3.2. This trail attracted naive ants which became specialists to the trail after several trips. When winter arrived the trails are erased. In spring, the old ants remembered the path (this process implies a visual memory) and re-traced the trail. This will guide the young ants and the system is perpetuated in this way. With this phenomenon one is faced with numerous questions. In particular one could be interested by the reliability of the individuals' memory and the extent of the amplification phenomena through the trail if we refer to the size and the life of the food source. Indeed if the learning process at the individual level is too strong or the memory too high, then the colony may be trapped into a non-optimal solution. This will appear when the landscape changes and the food sources disappear. Thus it is important to maintain a certain level of plasticity which can be randomness and forgetting at the individual level, in order to discover new food sources and allow different kind of organizations. In this way the individual behavioral strategy is better adapted to the evolving structure of the environment, what we have seen with studying swarm dynamics.

5. Acknowledgements

Many thanks to J. GERVET, S. GOSS, C. LANGTON and G. NICOLIS for their help and constant interest and stimulating discussions. Many thanks are due to the SANTA FE INSTITUTE, where G. THERAULAZ was supported by the C.N.R.S. (Programme Cognisciences), for the warm hospitality during the writing of the present review. J.L. DENEUBOURG. is a fellow of the Belgium F.N.R.S.

6. References

ARON S., DENEUBOURG J.L., GOSS S., and PASTEELS, J.M. (1990) 'Functional self-organisation illustrated by inter-nest traffic in ants : the case of the Argentine ant', in: W. ALT and G. HOFFMANN (eds.), Biological Motion, Lecture Notes in Biomathematics 89, Springer-Verlag, pp. 533-547.

BECKERS R., DENEUBOURG J.L., GOSS S., and PASTEELS, J.M. (1990) 'Collective decision making through food recruitment', Insectes Sociaux 37, 258-267.

BELIC M.R., SKARKA V., DENEUBOURG J.L., and LAX, M. (1986) 'Mathematical model of a honeycomb construction', Journal of Mathematical Biology 24, 437-449.

BONABEAU, E. and THERAULAZ, G. (1994) 'Collective Intelligence', Hermes, Paris.

CAMAZINE, S. (1991) 'Self-organizing pattern formation on the combs of honey bee colonies', Behavioral Ecology and Sociobiology 28, 61-76.

CAMAZINE, S. and SNEYD, J. (1991) 'A mathematical model of colony-level nectar source selection by honey bees : self-organization through simple individual rules', Journal of Theoretical Biology 149, 547-551.

CAMAZINE S., SNEYD J., JENKINS M.J., and MURRAY, J.D. (1990) 'A mathematical model of self-organized pattern formation on the combs of honey bee colonies', Journal of Theoretical Biology 147, 553-571.

CORBARA B., DENEUBOURG J.L., FRESNEAU D., GOSS S., LACHAUD J.P., and PHAM-NGOC, A. (1991) 'Simulation de la genèse d'une division du travail au sein d'une société de fourmis Ponerines : un modèle d'auto-organisation', Actes Colloques Insectes Sociaux 7, 205-206.

DENEUBOURG, J.L. (1977) 'Application de l'ordre par fluctuations à la description de certaines étapes de la construction du nid chez les termites', Insectes Sociaux 24, 117-130.

DENEUBOURG, J.L. and GOSS, S. (1989) 'Collective patterns and decision-making', Ethology Ecology & Evolution 1, 295-311.

DENEUBOURG J.L., GOSS S., FRANKS N., and PASTEELS, J.M. (1989) 'The blind leading the blind : modeling chemically mediated army ant raid patterns', Journal of Insect Behavior 2, 719-725.

DENEUBOURG J.L., ARON S., GOSS S., and PASTEELS, J.M. (1990a) 'The self-organizing exploratory pattern of the argentine ant *Iridomyrmex humilis*.' Journal of Insect Behavior 3, 159-168.

DENEUBOURG J.L., GOSS S., PASTEELS J.M., FRESNEAU D., and LACHAUD, J.P. (1987) 'Self-organization mechanisms in ant societies (II) : learning in foraging and division of labour', in: J.M. PASTEELS and J.L. DENEUBOURG (eds.), From individual to collective behaviour in social insects, (Experientia Supplementum 54), Birkhäuser Verlag, Basel, pp. 177-196.

DENEUBOURG J.L., GOSS S., SANDINI G., FERRARI F., and DARIO, P. (1990b) 'Self-organizing collection and transport of objects in unpredictable environments', in Proceedings of Japan-USA Symposium on Flexible Autiomation (in press).

DENEUBOURG J.L., GOSS S., FRANKS N., SENDOVA-FRANKS A., DETRAIN C., and CHRETIEN, L. (1991) 'The Dynamics of Collective Sorting : Robot-Like Ant and Ant-Like Robot', in J.A. MEYER and S.W. WILSON, (eds.), Simulation of Adaptive Behavior : From Animals to Animats, MIT Press/Bradford Books, pp. 356-365.

EMERSON, A.E. (1939). 'Social Coordination and the Superorganism', Amer. Midl. Nat. 21, 182-209.

EMERSON, A.E. (1952). 'The supraorganismic aspects of the society',.in Structure et Physiologie des Sociétés Animales, Colloques Internationaux du CNRS, Paris Mars 1950. Publications du CNRS. pp. 333-353

FRANKS, N.R. (1989) 'Army ants: a collective intelligence', American Scientist, March-April, 139-145.

FRANKS N.R., BRYANT S., GRIFFITHS R., and HEMERIK, L. (1990) 'Synchronization of the behaviour within nests of the ant *Letptothorax acervorum* (Fabricius) — I. Discovering the phenomenon and its relation to the level of starvation', Bulletin of Mathematical Biology 52, 597-612.

FRANKS N.R., GOMEZ N., GOSS S., and DENEUBOURG, J.L. (1991) 'The blind leading the blind : Testing a model of self-organization (Hymenoptera : Formicidae)', Journal of Insect Behavior 4, 583-607.

GOSS, S. and DENEUBOURG, J.L. 1988. Autocatalysis as a source of synchronised rhythmical activity. *Insectes Sociaux.*, 35: 310-315.

GOSS S., ARON S., DENEUBOURG J.L., and PASTEELS, J.M. (1989) 'Self-organized shortcuts in the argentine ant', Naturwissenschaften 76, 579-581.

GRASSE, P.-P. (1959) 'La reconstruction du nid et les coordinations interindividuelles. La théorie de le stigmergie', Insectes Sociaux 6, 41-84.

GRILLOT, P.J. (1975) Form, Function and Design, Dover, New-York.

HANSELL, M.H., (1984) Animal architecture and building behaviour, Longman, London.

JEANNE, R.L. (1975) 'The adaptiveness of social wasp nest architecture', Quaterly Review of Biology 50, 267-287.

KRAFFT, B. and PASQUET (1991) 'Synchronized and rhythmical activity during the prey capture in the social spider *Anelosimus eximius* (Arameae, Theridiidae)', Insectes Sociaux 38, 83-90.

PASTEELS J.M., DENEUBOURG J.L., and GOSS, S. (1987) 'Self-organization in ant societies (I): Trail recruitment to newly discovered food sources', in: J.M. PASTEELS. and J.L. DENEUBOURG (eds.), From individual to collective behaviour in social insects, (Experientia Supplementum 54), Birkhäuser Verlag, Basel, pp. 155-175.

ROSENGREN, R. (1971) Route fidelity, visual memory and recruitment behaviour in foraging wood ants of the genus *Formica* (Hymenoptera, Formicidae). Acta Zool. Form. 133, 1-150.

SEELEY T.D., CAMAZINE S., and SNEYD, J. (1991) 'Collective decision-making in honey bees : how colonies choose among nectar sources', Behavioural Ecology and Sociobiology 28, 277-290.

SKARKA V., DENEUBOURG J.L., and BELIC, M.R. (1990) 'Mathematical model of building behavior of *Apis mellifera*', Journal of Theoretical Biology 147, 1-16.

STEELS, L. (1990) 'Cooperation between distributed agents through self-organisation', in Y. DEMAZEAU and J.-P. MULLER (eds.), Decentralized AI, Elsevier Science Publishers B.V., (North-Holland), pp. 175-196.

STEELS, L. (1991) 'Toward a theory of emergent functionnality', in J.A. MEYER and S.W. WILSON, (eds.), Simulation of Adaptive Behavior : From Animals to Animats, MIT Press/Bradford Books, pp. 451-461

THERAULAZ, G. and GERVET, J. (1992) 'Les perfomance collectives des sociétés d'insectes', Psychologie Française 37, pp. 7-14.

THERAULAZ G., GOSS S., GERVET J., and DENEUBOURG, J.L. (1990a) 'Self-organization of behavioural profiles and task assignement based on local individual rules in eusocial wasp *Polistes dominulus* Christ', in G.K. VEERESH, B. MALLIK and C.A. VIRAKTAMATH (eds.), Social Insects and the Environment, Proceedings of the symposium Models and Theoretical Approaches to the Study of Social Insects, 11th International Congress of IUSSI, Bangalore, India, Oxford & IBH Publishing Co. Pvt. Ltd., New-Delhi, pp.535-537.

THERAULAZ G., GOSS S., GERVET J., and DENEUBOURG, J.L. (1990b) 'Swarm Intelligence in Wasps Colonies : a case for tasks assignment in multi-agents systems', presented at the 1990 IEEE International Symposium on Intelligent Control, Philadelphia, PA.

THERAULAZ G., GERVET J., and SEMENOFF, S. (1991a) 'Social regulation of foraging activities in *Polistes dominulus* Christ : a systemic approach to behavioural organization', Behaviour 116, 292-320.

THERAULAZ G., GOSS S., GERVET J., and DENEUBOURG, J.L. (1991b) 'Task differentiation in Polistes wasp colonies : a model for self-organizing groups of robots', in J.A. MEYER and S.W. WILSON, (eds.), Simulation of Adaptive Behavior : From Animals to Animats, MIT Press/Bradford Books, pp. 346-355.

VARELA, F.J. (1991) 'Organism, Cognitive Science and the Emergence of Selfless Selves', Revue Européenne des Sciences Sociales 29, 173-198.

WHEELER, W.M. (1911). 'The Ant colony as an organism'. Journ. Morph. 22, 307-325.

MOTHER-PUP TRANSMISSION OF A FEEDING TECHNIQUE IN THE GOLDEN HAMSTER (*MESOCRICETUS AURATUS*)

E. PRATO PREVIDE[1], M. D. POLI

Istituto di Psicologia
Facoltà di Medicina e Chirurgia
Università degli Studi di Milano
Via Francesco Sforza, 23
20122 Milano, Italy

ABSTRACT. Many laboratory and field studies have shown that the transmission of acquired patterns of behavior from an individual to another is a relatively common and important mode of adaptation in vertebrates. The aim of this study was to evaluate whether the golden hamster (*Mesocricetus auratus*) would acquire a new feeding technique through interaction with a knowledgeable conspecific. The behavior to be acquired consisted in retrieving a piece of food dangling from a small chain by using paws and teeth. Instrumental learning techniques were used to shape the skilled demonstrators.

We focused on mother-pup and pup-pup relationships. Three different groups of pups were examined and their behavior was videorecorded during experimental and test sessions. In one group the pups were allowed to observe, and interact with, a skilled mother; in a second group they could learn the task together, but without the skilled adult; a third group of pups was given the possibility of individual trial and error learning.

The results obtained suggest that the possibility of observing and interacting with a skilled mother had a positive effect on the acquisition of a new pattern of behavior by pups. Furthermore, a preliminary ethological analysis indicates the occurrence of both quantitative and qualitative differences in behavior among groups, and suggests that a number of social cues might be relevant in the acquisition of the new response by pups.

1. Introduction

As is the case for so many topics in the biological and behavioral sciences, Charles Darwin was well ahead of his times in realizing that the presence of imitative learning in animals different from man was a relevant question. In *The Descent of Man*, he stated that "animals possess some power of reasoning", and, while acknowledging that "the difference between the mind of the lowest man and that of the highest animal is immense", he was nevertheless convinced that the difference "... certainly is one of degree, and not of kind. We have seen that the senses and intuitions, the various emotions and faculties, such as love, memory, attention, curiosity, *imitation*, reason, etc., of which man boasts, may be found in an incipient, or even sometimes in a well-developed condition in the lower animals".

However, although he collected a large amount of material, Darwin never studied in depth animal behavior and its evolution; this was attempted by George Romanes, whose

---

[1] Currently at the Dipartimento di Psicologia, Facoltà di Magistero, Università di Torino, Torino, Italy.

R. A. Gardner et al. (eds.), The Ethological Roots of Culture, 125–142.

book *Animal Intelligence* (1882), partly based on the notes collected by Darwin himself, presented a large collection of data, mostly anecdotal, trying to show that many animals, even of fairly low phylogenetic level, were capable of "intelligent" behavior. Romanes was severely criticized for his uncritical anthropomorphic views; one of the most severe critics was C. L. Morgan, who argued that most of the apparently complex behaviors described by Romanes could be more economically explained in simpler terms.

Morgan's arguments influenced E. L. Thorndike's ideas on animal learning, convincing him that animals were capable of learning many behavioral responses in a predictable and lawful way, and that their learning abilities were related to the phylogenetic level of the species. Thorndike's attempts to experimentally demonstrate imitation learning in different species proved unsuccessful, convincing him that the only learning mechanism available to animals was a gradual process of trials and errors: neither complex cognitive abilities nor imitation learning had any place in his system.

He dismissed the possibility that imitation, latent learning and social transmission of acquired habits played a role in animal behavior; his attitude caused research on these topics to become unfashionable within animal psychology, whose prevailing ideology remained for a long time to explain all behavior in terms of S-R contingencies. However, a limited but significant amount of research on imitative phenomena continued to be performed in some psychological laboratories, often with controversial results.

More than a century of laboratory investigation on the learning processes supporting traditions and culture in animals (see the reviews by Davis, 1973; Galef, 1976, 1988, 1990a, 1992; Pallaud, 1977; Whiten and Ham, 1992) clearly indicates that the problem of imitation learning is of great theoretical interest for comparative psychology, and fully deserves a thorough study. However, the issue is still far from being resolved. Although it is generally agreed that at least higher primates are capable of observational learning (but see Galef's comment, reported by Roberts, 1983; and Galef, 1992), many authors consider the reports of socially learned habits in non-primates to be examples of local enhancement or mindless copying (Gould, 1982). Furthermore, there has often been some disagreement in the results obtained by different authors. Reports of observational learning in the literature are frequently open to alternative interpretations, either because they fail to separate the different mechanisms by which information can be transmitted, or because experimental groups are inadequate or insufficient, or because the tasks (e.g., key-peck or bar press) are too simple to allow for discrimination between alternative explanations. Moreover, the importance of the social and physical situation in which observation should take place, and of the choice of the response to be learned have frequently been underestimated: Oldfield-Box (1970) has correctly stressed the importance of using experimental settings that optimize the possibility of observation. Finally, it is conceivable that the members of any single species might not possess an unlimited capacity for observational learning; that is, even when they are able to perform or to learn a response through a process of trials and errors, there might be clear-cut boundaries to observational learning, set up by the adaptations reached in the course of evolution. For instance, although the role played by imitation in song pattern acquisition by some species of birds is well-known, this capability is in general limited to some particular song patterns among the infinite variety of possible ones, namely to the patterns matching certain species-specific templates. A similar selectivity could possibly exist also for other species and behavioral patterns.

On the other hand, a large body of ethological studies on topics such as observational learning and "cultural" transmission of adaptive habits is now available. Ethological studies, both in the laboratory and in the field, have revealed a wealth of phenomena which can be gathered under such headings in vertebrates as far apart as fishes (Bronstein, 1986), birds (Hinde and Fisher, 1951; van Lawick-Goodall and van Lawick, 1966; Sherry and Galef, 1984, 1990; Palameta and Lefebvre, 1985; Johnson, Hamm and Leahey, 1986), non-primate mammals (Shaller, 1972; Galef, 1976; Mainardi, 1980; Laland and Plotkin, 1990), and, of course, many primates (Kawai, 1965; van Lawick Goodall, 1968, 1973; Menzel, Davenport and Rodgers, 1972; Itani and Nishimura, 1973; Lethmate, 1976; McGrew *et al.*, 1979; Tomasello *et al.*, 1987; Hauser, 1988; Tomasello, 1990; McGrew, 1992).

As a whole, the ethological observations agree with the fact that there seem to be good biological reasons for the diffusion of such an economically advantageous and adaptive learning modality. For many socially living animals, the opportunity to learn adaptive behaviors not through a time-consuming and potentially dangerous trial and error process, but through the interaction with experienced conspecifics and the observation of their behavioral patterns, would apparently be very advantageous and well within the limits of their capabilities. Of the many possible advantages of a non-genetic transmission of information, two seem to be particularly relevant: first, the social spread of newly acquired information through generations is by several orders of magnitude faster than genetic transmission; second, it allows for a unique possibility of intra-generational diffusion of acquired habits. Possibly, most responses functionally essential for survival are inherited, but many aspects of predator avoidance and foraging behavior are learned or, at least, largely modified by experience during the animal's lifetime.

The adaptiveness of some behavior being learned rather than inherited is especially important for opportunistic species and for those exploiting a wide variety of ecological niches. For example, a given species might encounter a whole range of predators in different situations, and a flexible, learned response would allow for wider and faster modifications of its behavior according to a variety of specific situations. It would also be adaptive for such learning to occur through the observation of other individuals performing the relevant responses. It is therefore conceivable that for group living animals there could exist a selective pressure for imitative capabilities to evolve.

Although the most impressive examples of cultural transmission are probably found in primates, a number of studies suggest that the social transmission of learned habits is a rather common phenomenon among rodents as well. In particular, it has been shown that rats and mice can acquire both natural behaviors and arbitrary responses by interacting with other individuals or, in some cases, by observing them (Mainardi and Pasquali, 1968; Manusia and Pasquali, 1969; Benel, 1975; Destrade and Cardo, 1977; Lore and Flanelly, 1977; Pallaud, 1977; Collins, 1988; Valsecchi *et al.*, 1989; Galef, 1990b; Heyes and Dawson, 1990; Zohar and Terkel, 1991; Valsecchi *et al.*, 1993).

However, much of the available data are still controversial and, with a few notable exceptions, very little is known about social learning in other rodent species. Our research was suggested by two previous studies by Mainardi *et al.* (1972, 1976) on the interspecific transmission of behavior in rodents. Mainardi and colleagues showed that a food related behavior could be learned by golden hamsters (*Mesocricetus auratus*) through interaction with an individual of another species: the house mouse (*Mus musculus*). The behavior to be acquired consisted of retrieving a piece of food dangling from a small metal chain by using paws and teeth; the authors reported that hamsters rarely use this technique to solve the problem of obtaining the food, but spontaneously adopt a quite different strategy. The studies by Mainardi *et al.* were interesting both for theoretical and methodological reasons, and were, as far as we know, the only studies on social learning in rodents using hamsters as experimental subjects. Therefore, on the basis of the extant literature and of the above mentioned theoretical considerations, we decided to set up an experimental situation similar to the one described in their studies. However, we focused on mother-pup relationship, evaluating whether hamster pups acquired a new feeding technique interacting with their knowledgeable mother: moreover, to obtain efficient demonstrators performing a motor response closely resembling the one described in their experiments, we used instrumental training techniques.

Our intent was to simulate in the laboratory, at least to some extent, a natural situation in which the mother knew *where* and *how* to get the food, and pups could take advantage of this knowledge interacting with her. The purpose was to investigate, integrating psychological and ethological approaches, the role of social interaction with a skilled conspecific in the acquisition of the new feeding behavior by pups.

2. Method

2.1 SUBJECTS

Thirty-nine weanling hamster pups (*Mesocricetus auratus)*, belonging to eight different family groups, served as subjects in this study. All the litters were born in our laboratory from outbred parents obtained from a commercial supplier (Charles River, Italy). All mothers were experimentally naive at the start of the study and their age ranged from 4 to 9 months. Since birth, each litter was housed with the mother in the experimental cage that will be described below, with food and water ad libitum. Colony room lights were on from 8 p.m. to 10 a.m. (14L:10D). All the animals were fed on a standard diet for laboratory hamsters (GLP Basic 5SB17 by Mucedola, Italy); in addition they were given commercial food for hamsters (Vitobel Friandises), sunflower seeds and apples. Each family group was left undisturbed until the pups were 22 days old; then, the pups of each litter were randomly assigned to the different experimental conditions.

2.2 APPARATUS

Two different types of apparatus were employed, one during the training of the demonstrators and the other throughout the remaining phases of the study. Demonstrator mothers were trained in a plastic container (60 × 40 × 30 cm) with a 35 cm high platform in the middle. This platform, which closely resembled the one employed by Mainardi and colleagues (1972, 1976), was made of plexiglass with a small wire mesh ladder that allowed the animal to reach the top landing (6 × 10 cm). A small chain 13 cm long with a sunflower seed fixed at the end dangled from the short side of the top landing (see Figure 1).

The experimental apparatus consisted of square cages (50 × 50 × 50 cm) with clear plexiglass walls in order to allow for observation and videorecording of behavior; the floor

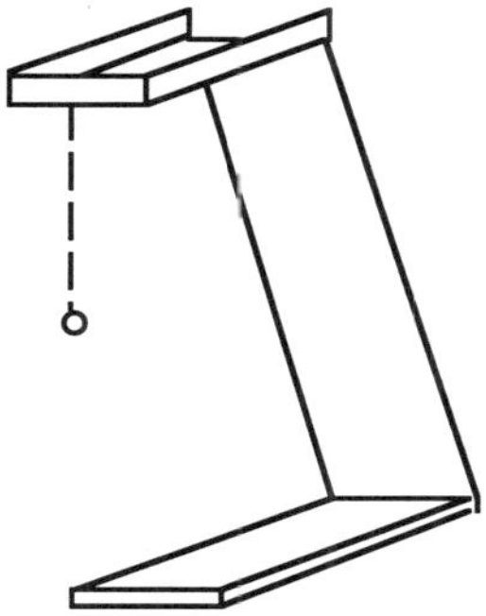

Figure 1. *Apparatus for training the demonstrators.*

of the cage was made of plastic and covered with sawdust. Each cage was arranged to house a family group and contained material for building the nest and a water reservoir.

In the cage there was an L-shaped wooden platform (1 cm thick; 14 cm wide) fixed on the walls at a height of 35 cm from the floor; the animals could get on the platform through a plexiglass ladder at one end of it. During the appropriate periods of the study, 10 small metal chains 12 cm long were hanging from the platform. A piece of food weighting 1–2 grams and a sunflower seed were fixed at the end of each chain (see Figure 2). An L-shaped plexiglass cover prevented mothers from getting out of the cage. Experimental and test sessions were videorecorded using a Panasonic F10 CCD video camera and a VHS Hitachi VT-8E video recorder.

2.3 PROCEDURE

The study consisted of four different and consecutive phases: (1) training of demonstrator mothers, (2) birth and weaning of pups, (3) transmission/acquisition of the feeding technique, and (4) testing.

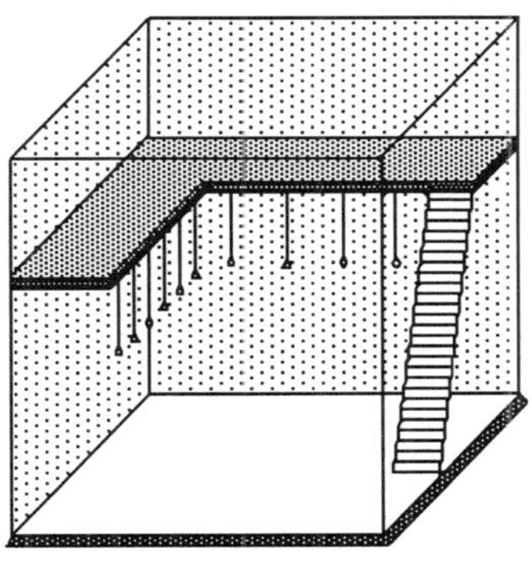

Figure 2. *The experimental apparatus.*

Phase 1: *Training of the demonstrators*. The female hamsters, randomly selected to become skilled demonstrators, were trained to use exclusively and efficiently a particular feeding technique similar to the one studied by Mainardi and colleagues (1972, 1976). This technique, that we called *fishing technique*, consisted in getting a piece of food dangling in space from a small chain by using forepaws and possibly teeth to retrieve the food, while maintaining the body on the platform (see Figure 3a). Such a behavior was remarkably different from the spontaneous technique described by Mainardi *et al.* (see Figure 3b). The demonstrators were shaped to perform the fishing technique through instrumental training.

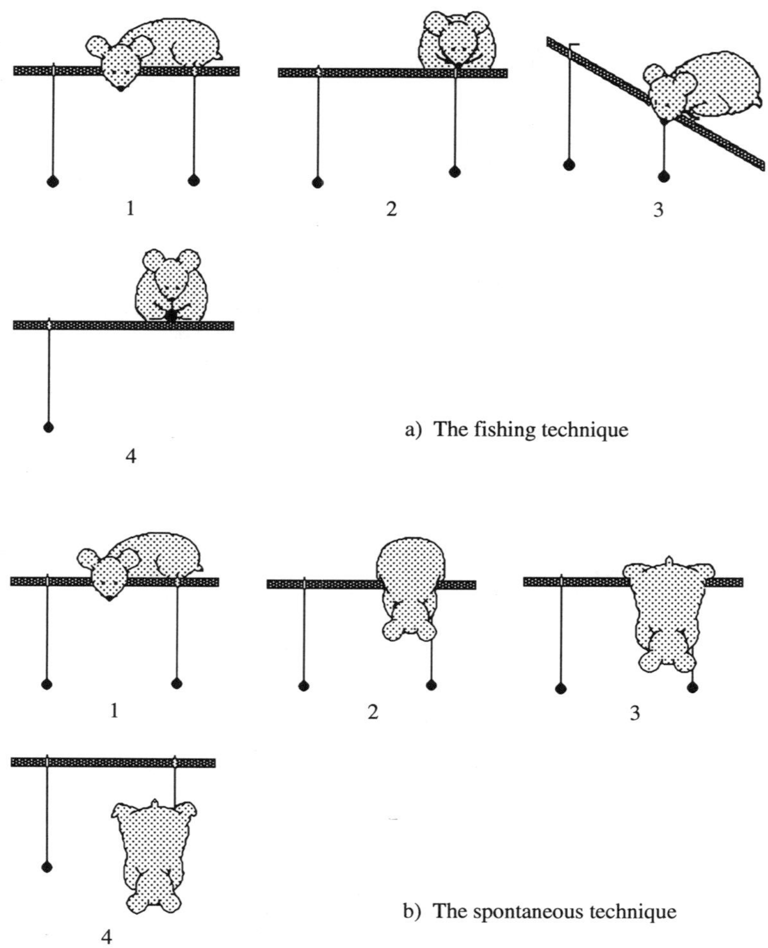

Figure 3. *The two different feeding techniques.*

Altogether this phase lasted two weeks; for about a week the hamsters were exposed to a daily 30 minutes training session in the training apparatus; as soon as the required behavior appeared with some consistence, the females were moved into the experimental apparatus to complete their training.

Phase 2: *Birth and weaning of the litters.* This phase started with mating and lasted 38 days, until the litter was three weeks old and therefore almost weaned. During pregnancy, the skilled females were allowed to use the fishing technique two times a day (at 12 noon and 4 p.m.). In these occasions they could retrieve the food pellets hanging from the 10 small metal chains; except for a piece of apple, this was the only food available to them (about 20 g a day). This procedure was suspended during the 2 days preceding parturition and was resumed on day 4 postpartum.

Phase 3: *Acquisition of the feeding behavior.* This phase had a duration of 10 days. On day 22 after birth the pups of each litter were weighed, marked on the ears to make individual recognition possible, and randomly assigned to three experimental groups that differed exclusively in the kind of social context pups were exposed to during acquisition sessions: Trained Mother group (TM-group), Social Pups group (SP-group), and Individual Pups group (IP-group).

In the TM group (N=15), 3 pups at a time were tested with their mother in the experimental apparatus for 10 consecutive days. During two daily experimental sessions (one at noon and one at 4 p.m.), these pups were allowed to observe and interact freely with their mother while she was recovering the food with the fishing technique. Each experimental session lasted 20 minutes and included two 10 minute trials; in each trial, 10 small metal chains (12 cm long) with a piece of food and a sunflower seed at the end were hanging from the platform. At the end of the session the chains were removed from the apparatus. Therefore, the pups could attend, at least in principle, to 20 performances of the fishing response during each experimental session, and to 40 performances each day.

In the SP group (N=12), 3 pups at a time were examined together during each experimental session. The procedure was exactly the same as in the previous group, except that the pups could not observe their mother, which was removed from the apparatus before each experimental session.

In the IP condition (N=12), each pup was alone in the apparatus during the experimental sessions; therefore, it could learn to get the food dangling from the small metal chains only through individual trial and error. Apart from this, the pups in this condition received the same treatment as the others.

Outside the experimental sessions each family group was housed together in the apparatus, to prevent possible effects due either to isolation from littermates or to the removal of the mother from the litter. Furthermore, pups were always tested in their own cages, in order to minimize possible effects of the temporary removal of the mother and/or littermates by preserving familiar visual and olfactory cues.

As the number of pups per litter ranged from 4 to 8, and the number of pups in two of the three experimental conditions (i.e., Trained Mother and Social Pups) was fixed at 3, a single litter could rarely be exposed to all three experimental conditions; however, at least two of the three experimental conditions were always represented. For instance, in a litter consisting of 5 pups, 3 were assigned either to the Trained Mother or to the Social Pups group, and two were assigned to the Individual Pups group.

Phase 4: Testing. At the end of the 10 days of the acquisition phase, on day 11, the pups of each litter were individually tested to evaluate their ability to solve the problem of retrieving the food. During the test 10 small metal chains, each with a sunflower seed at the end, were dangling from the platform and the pup could try to retrieve them. The test had a duration of 30 minutes, but was ended by the experimenter if the pup obtained all the sunflower seeds before this time had elapsed. The test was repeated on the following day.

3. Results

The results of this study can be summarized as follows. In the Trained Mother group (N=15), 11 pups acquired the mother's fishing technique (Figure 3a) and used it exclusively and efficiently on both tests days. Only two pups were unable to perform the mother's behavior and resorted exclusively to the spontaneous technique (Figure 3b) when trying to get the food. The remaining 2 pups used, at least to some extent, both feeding techniques during testing; in particular, one of them used the fishing technique on the first test day, and shifted to a mixed technique on the second day.

In the Social Pups group (N=12), only one pup learned the fishing technique, while all the others employed the spontaneous technique on both test days. Finally, in the Individual Pups group (N=12), 2 pups acquired the fishing technique by individual trial and error. It appears therefore that the presence of a skilled mother had a remarkable effect on the acquisition of the new motor pattern by pups.

The detailed analysis of the behavior of the subjects in the different groups during the acquisition period (*Phase 3*) has not been carried out yet; however, a preliminary ethological analysis indicates the occurrence of quantitative and qualitative differences in behavior among the three groups, suggesting that a number of social cues might be relevant for the transmission of the feeding technique.

The performance of the three groups of pups on test sessions (*Phase 4)* has been evaluated focusing on a number of behavioral variables, selected because two previous pilot studies carried out at our laboratory (Prato Previde *et al.*, 1991; Prato Previde, 1992) suggested that they were suitable for describing behavioral differences among pups.

The following behavioral variables have been recorded for each pup of each group during testing:

1) *Climb*: the number of times each pup climbed on the top of the platform.

2) *Spontaneous*: the number of times it adopted the spontaneous technique (Figure 3b, 1, 2, 3)

3) *Spontaneous+*: the number of times the spontaneous technique was successful and therefore rewarded. Given the number of sunflower seeds, a maximum of 10 rewards were possible.

4) *Fall*: the number of times the spontaneous technique ended with a fall from the platform (Figure 3b, 4).

5) *Fishing/Incomplete*: the number of times a pup performed the fishing technique to retrieve only part of the chain (Figure 3a, 1, 2, 3).

6) *Fishing:* the number of times it adopted the fishing technique and retrieved the chain completely (Figure 3a, 1, 2, 3, 4).

7) *Fishing+*: the number of times the fishing technique was successful and therefore rewarded. As for the spontaneous technique, only 10 rewards could be obtained.

Furthermore, as test duration could vary according to the efficiency of the pups in solving the problem, the duration of each test session was recorded.

Pup behavior was videorecorded during testing, and the total number of occurrences of each behavior was subsequently scored from videotapes. Furthermore, in order to assess interobserver agreement, a random subsample of the test sessions (30% of the sessions) was scored by a second independent observer. For each behavior, concordance between observers was calculated as the ratio:

$$\frac{A}{A+(D/2)}$$

where A is the total number of agreements, and D the total number of disagreements (see Caro *et al.* 1979). Interobserver agreeement was never less than .95, except for the category Fishing/Incomplete, for which it was .85.

Table 1 shows the behavior of the pups of the three different groups (Trained Mother, Social Pups and Individual Pups) on the first test day; the total number of occurrences of each behavior during the test is reported for each pup, together with the duration of the test session in minutes.

Table 2 shows the mean number of occurrences of each behavior for each group on both test days, and the mean duration of the two test sessions. It clearly appears from both tables that the behavior of the pups of the Trained Mother group during test sessions was remarkably different from that of the pups of the other two groups: on the one hand, most of them successfully adopted the mother's technique (*Fishing, Fishing+*); on the other hand, such a technique completely replaced the spontaneous one (*Spontaneous*). Finally, these pups were faster in solving the problem of obtaining the food (*Test duration*).

Figure 4 reports the percentage of pups of each group that during the first test session adopted the fishing technique, the spontaneous technique, or a mixture of the two (i.e., both techniques to a comparable extent).

Finally, the differences in performance among the three groups of pups on Test 1 are outlined in Figure 5. In this figure, the pups of each group are assigned to one of 5 different classes of performance according to the proportion of times they used the fishing technique to retrieve food on the first test day (see Table 1 – *Fishing* and *Spontaneous*). The pups that adopted the fishing behavior 0–20% of the times are included in Class 1; those that exhibited it 20–40% of the times are included in Class 2; Classes 3 and 4 include the pups that adopted it 40–60%, and 60–80% of the times respectively. Finally, Class 5 includes the pups that adopted the fishing technique more than 80% of the times. Most of the pups that were given the possibility to observe a skilled mother became skilled performers, while the majority of the pups of the other two groups remained unskilled.

Table 1. *Behavior of each pup of the three groups in Test 1.*

Subject	Climb	Spont.	Spont.+	Fall	Fish. inc.	Fishing	Fishing+	Test dur.
1	2	0	0	0	2	10	10	14:36
2	4	0	0	0	4	10	10	27:50
3	20	6	2	7	0	0	0	30:00
4	2	0	0	0	1	11	10	11:50
5	2	0	0	0	2	10	10	13:30
6	2	0	0	0	1	10	10	15:55
7	2	0	0	0	2	10	10	12:10
8	4	0	0	0	9	9	9	30:00
9	6	0	0	0	4	4	3	30:00
10	15	3	2	3	0	2	0	30:00
11	16	12	2	12	0	0	0	30:00
12	3	0	0	0	0	10	10	10:35
13	1	0	0	0	7	10	10	9:00
14	6	0	0	0	5	13	10	15:10
15	2	0	0	0	5	11	10	9:30

Trained Mother Group

Subject	Climb	Spont.	Spont.+	Fall	Fish. inc.	Fishing	Fishing+	Test dur.
1	15	10	5	11	0	0	0	30:00
2	9	18	8	6	0	0	0	30:00
3	11	10	6	10	0	0	0	30:00
4	15	15	10	15	0	0	0	26:02
5	13	9	8	9	0	0	0	30:00
6	1	0	0	0	1	10	10	11:14
7	18	6	4	6	1	0	0	30:00
8	10	4	2	4	0	0	0	30:00
9	7	5	2	5	0	0	0	30:00
10	17	16	7	15	0	0	0	30:00
11	5	14	7	1	0	0	0	30:00
12	13	12	7	11	4	0	0	30:00

Social Pups Group

Subject	Climb	Spont.	Spont.+	Fall	Fish. inc.	Fishing	Fishing+	Test dur.
1	20	5	2	4	0	0	0	30:00
2	6	1	0	1	0	10	10	20:10
3	20	1	0	0	0	0	0	30:00
4	5	3	0	3	1	0	0	30:00
5	9	8	2	1	2	0	0	30:00
6	15	10	3	10	1	0	0	30:00
7	16	7	3	7	3	0	0	30:00
8	6	1	0	1	4	10	10	15:20
9	5	13	6	1	0	0	0	30:00
10	7	7	3	4	0	0	0	30:00
11	6	10	5	2	1	0	0	30:00
12	33	11	6	11	1	0	0	30:00

Individual Pups Group

Table 2. *Mean number of occurrences of the different behaviors in each group on Tests 1 and 2.*

TM Gr. = Trained Mother Group – SP Gr. = Social Pups Group – IP Gr. = Individual Pups Group

Group	Climb	Spont.	Spont.+	Fall	Fish. inc.	Fishing	Fishing+	Test dur.
TM Gr.								
Mean	5.80	1.40	0.40	1.47	2.80	8.00	7.47	19:20
(SD)	(6.06)	(3.38)	(0.83)	(3.48)	(2.78)	(4.24)	(4.26)	(8:56)
SP Gr.								
Mean	11.17	9.92	5.50	7.75	0.50	0.83	0.83	28:06
(SD)	(5.06)	(5.42)	(2.97)	(4.92)	(1.17)	(2.89)	(2.89)	(5:26)
IP Gr.								
Mean	12.33	6.42	2.50	3.75	1.08	1.67	1.67	27:57
(SD)	(8.70)	(4.21)	(2.28)	(3.70)	(1.31)	(3..89)	(3.89)	(4:53)

Test 1

Group	Climb	Spont.	Spont.+	Fall	Fish. inc.	Fishing	Fishing+	Test dur.
TM Gr.								
Mean	5.33	2.40	0.73	2.27	2.00	9.07	7.93	17:13
(SD)	(6.02)	(5.07)	(1.49)	(4.82)	(1.93)	(4.54)	(3.86)	(9:25)
SP Gr.								
Mean	12.17	10.67	6.08	7.67	0.33	0.92	0.83	27:02
(SD)	(6.00)	(5.68)	(3.75)	(4.23)	(0.89)	(2.87)	(2.89)	(4:59)
IP Gr.								
Mean	12.00	7.42	3.42	3.92	1.75	2.08	1.92	27:11
(SD)	(7.95)	(5.85)	(3.03)	(3.20)	(3.44)	(3.80)	(3.82)	(6:34)

Test 2

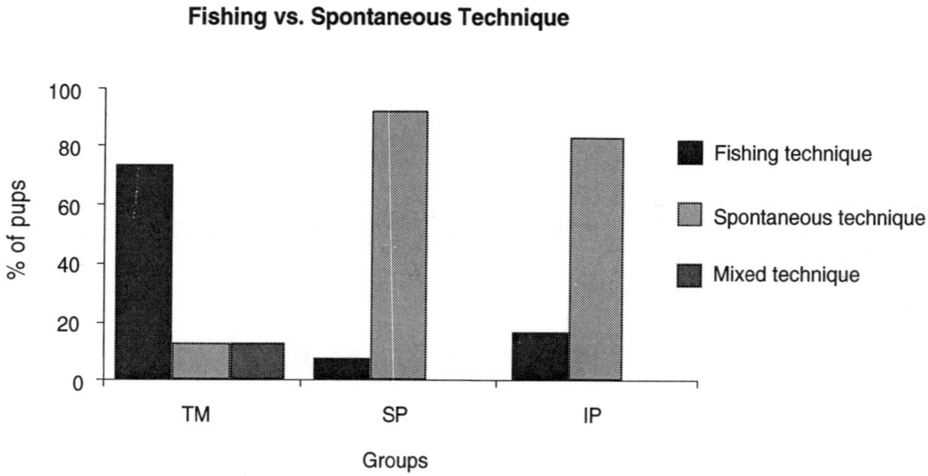

Figure 4. *Percentage of pups of each group that adopted each feeding technique during Test 1.*

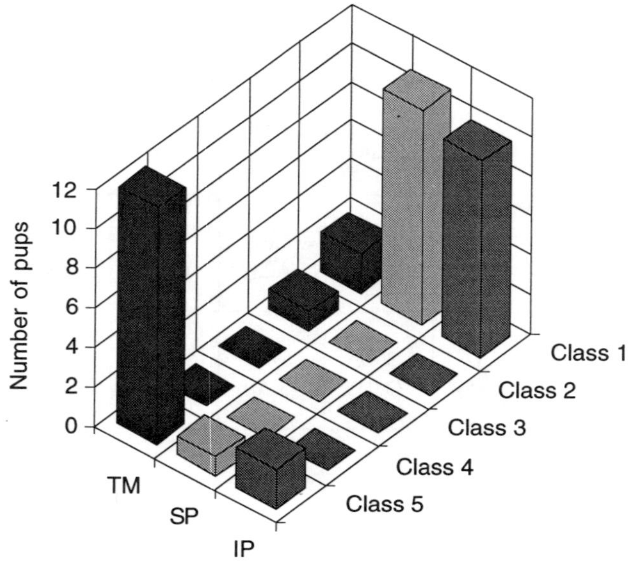

Figure 5. *Performance of the pups of the three groups during Test 1.*

The analysis of the difference in performance among the three groups of pups was carried out using a nonparametric one-way ANOVA (Kruskal-Wallis test) on the proportion of fishing responses on Test 1 (number of fishing responses/total number of responses). The test showed that differences in fishing behavior between groups were highly significant ($H = 19.836$, $p < .0001$). Comparisons between single groups, carried out using the Mann-Whitney U test, revealed that the pups in the Trained Mother group differed significantly from the pups in the other two groups in their ability to use the fishing technique (TM *vs.* SP: $z = -3.873$, $p = .0001$; TM *vs.* IP: $z = -3.975$, $p < .0001$). No difference was found between the other two groups (SP *vs.* IP: $z = -.503$, $p = .6153$).

4. Discussion

The aim of this study was twofold. Our first goal was to test whether the golden hamster (*Mesocricetus auratus*) could learn a new motor pattern observing and interacting with a knowledgeable conspecific. The general purpose, however, was to acquire a deeper knowledge of the mechanisms involved in the social transmission of behavior in animals.

This experiment was suggested and stimulated by two previous studies by Mainardi *et al.* (1972, 1976) on the interspecific transmission of behavior in rodents. Mainardi and his colleagues showed that hamsters could benefit from observing and interacting with a skilled mouse (*Mus musculus*) carrying out a particular method of food retrieval; however, they found no evidence of intraspecific transmission of this feeding behavior.

Our results confirm the findings by Mainardi *et al.*, as they show that hamsters can socially acquire a new and otherwise unlikely feeding technique, but in addition they indicate the occurrence of an intraspecific transmission of that technique, at least within the family group. When tested, 11 of the 15 pups in the Trained Mother group (73.33%) solved the problem using exclusively the mother's fishing technique (Figure 3a). This technique completely replaced the spontaneous one (Figure 3b) on both test sessions. No similar phenomenon was observed in the other two groups: in the Individual Pups group only 16.6% of the pups acquired the fishing behavior, while in the Social Pups group only 8.3% did (see figures and tables). Furthermore, the analysis of the behavior during testing showed that in the TM group the technique was rather homogeneous and closely resembling the one taught to the mother. Therefore, it seems that hamster pups can replace a spontaneous behavior pattern (Figure 3b) with a new one (Figure 3a) after interacting with a skilled mother.

According to Galef's definition (1976, 1988), the phenomenon observed in the present study can be considered as a case of social transmission of behavior. After interacting with their mother, hamster pups independently exhibited a specific behavior pattern, that initially was not part of their behavioral repertoire. In particular, it seems to us that the three criteria suggested by Galef (1976, pp. 79–80) for distinguishing transmitted behaviors from other observable changes in behavior resulting from social interaction have been completely fulfilled: first, social interaction was an alternative or optional route to direct transaction with the nonsocial environment in the development of the fishing behavior; second, social interaction promoted homogeneity of behavior between individuals; and third, this homogeneity of behavior extended temporally beyond the period of interaction.

What kind of process can account for the transmission of such a novel technique from mother to pups? As different explanations could account for our present results, at this stage of the study it is premature to draw conclusions on the mechanisms underlying this imitative phenomenon. On the one hand, the detailed analysis of the behavior during *Phase 3* has not been completed; on the other hand we do not know yet if and how the behavior of an unskilled mother would affect the performance of the pups. In particular, a number of studies indicate that among rats and mice the presence of an unskilled adult conspecific can affect, at least to a certain degree, the acquisition of new behaviors by pups (see Valsecchi *et al.*, 1989; Zohar and Terkel, 1991; Valsecchi *et al.*, 1993); therefore, the effect of an unskilled mother on the acquisition of the fishing technique by pups needs to be tested. The preliminary analysis of the recordings carried out during Phase 3 indicates the occurrence of both quantitative and qualitative differences in behavior among the three groups, suggesting that some social cues might be more relevant than others for the acquisition of the fishing technique by pups. On the basis of the available data, we believe that an explanation in terms of observational learning should not be excluded, although the possibility that the phenomenon observed in this study has been the outcome of a number of different processes acting in concert cannot be ruled out.

One of the main problems when dealing with social learning in animals is the difficulty of distinguishing among different processes leading to similar changes in the observed behavior; as several authors have pointed out (e.g., Davis, 1973; Galef, 1976, 1988, 1992; Hauser, 1988; Whiten and Ham, 1992) the "observer" may be influenced by various aspects of the demonstrator's performance, and it is extremely difficult to control for all alternative social learning processes. Last but not least, it is not an easy task to distinguish learning that is influenced socially from instances of individual learning in which behavior acquisition is not affected by interactions with others. Social and individual factors might be difficult to separate: for instance, social interaction may facilitate the introduction of a new response into an individual's repertoire, but the maintenance of such a response might be influenced by consequences resulting from its performance.

However, in our experimental setting we tried to overcome some of these difficulties and to control, as far as possible, for different processes. In our experimental situation, an individual knew *where* and *how* to obtain food, and other individuals could take advantage of this knowledge. As remarked by Galef (1988), focusing on the kind of response an "observer" performs after exposure to a demonstrator is a good step toward solving problems of control for different types of social learning.

Many laboratory studies have been unsuccessful in identifying the processes by which information is acquired during social learning because the tasks were too easy to allow for discrimination between alternative explanations. We selected a complex and well-defined motor pattern that was not part of the behavioral repertoire of the animal (see Mainardi *et al.*, 1972, 1976; Prato Previde *et al.,* 1991; Prato Previde, 1992). As noted by Hauser (1988), it is rather difficult to assess what "response already in the animal's repertoire" means. The feeding technique selected in our study was not part of the behavioral repertoire of the hamsters in the sense that they did not have this behavior in their repertoire *sensu stricto;* however they were capable of performing each of the behavioral components of the feeding technique. Therefore, the novelty of the behavior consisted in combining together the different components to obtain a particular motor pattern (Figure 3a).

A further problem with much experimental research on imitative phenomena has been the failure to control for social factors unrelated to imitation, like social facilitation and local or stimulus enhancement. For example, the comparison between an imitation group and a group that acquires the behavior on its own (by trial and error) fails to consider the potential effect on that behavior of the simple presence of another individual (social facilitation). In our experimental design, three different experimental conditions were arranged in order to control, at least to some degree, for both individual and social factors; the Individual Pup group controlled for individual trial and error learning, while the Social Pup group was the control for the potential effects of the simple presence of littermates on the acquisition of the feeding technique.

A final point worth mentioning is the introduction of systematic observation and videorecording of behavior during experimental sessions. In our opinion, this methodology is a powerful way for gathering quantitative and qualitative information on the nature of the cues responsible for the social transmission of acquired behavior. In this respect our research is an instance of "problem oriented approach" (Platt, 1964; Davis, 1973), where psychological and ethological methods are integrated in order to gain knowledge on a particular problem. Instrumental learning techniques have been employed to obtain reliable demonstrators capable of performing a particular motor pattern to obtain food, while observation and videorecording of behavior and interactions among individuals are a possible tool-kit for analyzing the relevant social cues and the underlying mechanisms. The use of an operant conditioning procedure is not new in the psychological approach to the study of imitative phenomena in vertebrates (see Davis, 1973; Pallaud, 1977; Zentall and Galef, 1988); however, this technique has been seldom employed in combination with an ethological approach.

Acknowledgements

This research has been supported by the C.N.R. (Italian National Research Council) and by a M.U.R.S.T. (Italian Ministry for University and Research) postdoctoral grant to E. Prato Previde. We would like to thank Allen and Trixie Gardner for their useful comments and help in improving the quality of our paper. We are also grateful to L. Baldasarre for help in data collection and scoring. No animal has been suppressed for this research; we are grateful to the "Animal World" Pet Shop, Milan, for valuable help in outplacing many experimental subjects.

References

Benel, R.A., 1975. Intra- and interspecific observational learning in rats, *Psychological Reports*, 37, 241–242.

Bronstein, P.M., 1986. Socially mediated learning in male Betta splendens. *Journal of Comparative Psychology*, 100, 3, 279–284.

Caro, T.M., Roper, R., Young, M. and Dank, G.R., 1979. Inter-observer reliability. *Behaviour*, 69, 303-315.

Collins, R.L., 1988. Observational learning of a left-right behavioral asymmetry in mice (*Mus musculus*). *Journal of Comparative Psychology*, 102, 222–224.

Davis, J.M., 1973. Imitation: A review and a critique. In P. P. G. Bateson and P. H. Klopfer, eds., *Perspectives in ethology*, Plenum Press, New York, London, 43–72.

Destrade, C., and B. Cardo, 1977. Apprentissage par observation chez la souris: Effects a long terme de l'observation et influence de la privation alimentaire. *Biology of Behavior*, 2, 83–91.

Galef, B.G., Jr., 1976. Social transmission of acquired behavior: A discussion of tradition and social learning in vertebrates. In J.S. Rosenblatt, R.A. Hinde, E. Shaw and C. Beer, eds., *Advances in the Study of Behavior,* 6, Academic Press, New York, 77–100.

Galef, B.G., Jr., 1988. Imitation in animals: History, definition, and interpretation of data from the psychological laboratory. In T.R. Zentall and B.G. Galef Jr., eds., *Social learning: Psychological and biological perspectives*, Lawrence Erlbaum, Hillsdale, NJ, 3–28.

Galef, B.G., Jr., 1990a. Tradition in animals: Field observations and laboratory analyses. In M. Bekoff and D. Jamieson, eds., *Interpretations and Explanations in the Study of Behavior: Comparative perspective*, Westview Press, Boulder, Colorado, 74–95.

Galef, B.G., Jr., 1990b. An adaptationist perspective on social learning, social feeding, and social foraging in Norway rats. In D.A. Dewsbury, ed., *Contemporary issues in comparative psychology*, Sinauer Associates, Sunderland, MA, 55–79.

Galef, B.G., Jr., 1992. The question of animal culture. *Human Nature*, 3, 2, 157–179.

Gould, J., 1982. *Ethology*. W.W. Norton, New York.

Hauser, M.D., 1988. Invention and social transmission: New data from wild vervet monkeys. In R. Byrne and A. Whiten, eds., *Machiavellian intelligence: Social expertise and the evolution of the intellect in monkeys, apes, and humans*, Oxford Science Publications, Oxford, 327–343.

Heyes, C.M., and G.R. Dawson, 1990. A demonstration of observational learning in rats using a bidirectional control. *Quarterly Journal of Experimental Psychology*, 42B, 59–71.

Hinde, R.A., and J. Fisher , 1951. Further observations of the opening of milk bottles by tits. *British Birds*, 44, 393–396.

Itani, J., and A. Nishimura, 1973. The study of infrahuman culture in Japan. A Review. In E.W. Menzel Jr., ed., *Precultural behavior*, Karger, Basel, 26–50.

Johnson, S.B., R.J. Hamm and T.H. Leahey, 1986. Observational learning in *Gallus gallus domesticus* with and without a conspecific model. *Bulletin of the Psychonomic Society*, 24, 3, 237–240.

Kawai, M., 1965. Newly-acquired precultural behavior of the natural troop of Japanese monkeys on Koshima islet. *Primates*, 6, 1–30.

Laland, K.N., and H.C. Plotkin, 1990. Social learning and social transmission of foraging information in Norway rats (*Rattus norvegicus*). *Animal Learning and Behavior*, 18, 3, 246–251.

Lethmate, J., 1976. Versuche zur Doppelstockhandlung mit einen jungen Orang-Utan. *Zoologische Anzeigen*, 197, 262–271.

Lore, R., and K. Flanelly, 1977. Rat societies. *Scientific American*, 236, 106–116.

Mainardi, D., 1980. Tradition and the social transmission of behavior in animals. In G.W. Barlow and J. Silverberg, eds., *Sociobiology: Beyond nature/nurture?,* AAAS Selected Symposium, 35, 227–255.

Mainardi, D., and A. Pasquali, 1968. Cultural transmission in the house mouse. *Atti della Società Italiana di Scienze Naturali*, 107, 147–152.

Mainardi, D., M. Mainardi and A. Pasquali, 1972. Interspecific observational learning: From house mice to golden hamsters. *Bollettino Zoologico*, 39, 634–635.

Mainardi M., M.P. Favoriti e A. Pasquali, 1976. Influenza del rapporto sociale sul fenomeno della trasmissione culturale interspecifica. *Rendiconti dell'Istituto Lombardo Accademia di Scienze e Lettere*, Classe di scienze, 110, 151–160.

Manusia, M., and A. Pasquali, 1969. A new case of imitation in the house mouse. *Atti della Società Italiana di Scienze Naturali*, 109, 457–462.

McGrew., W.C., 1992. *Chimpanzee material culture: Implications for human evolution.* Cambridge University Press, Cambridge, U.K.

McGrew., W.C., C.E.G. Tutin. and P.J. Baldwin, 1979. Chimpanzees, tools, and termites: Cross-cultural comparison of Senegal, Tanzania, and Rio Muni. *Man*, 14, 185–214.

Menzel, E.W., R.K. Davenport and C.M. Rodgers, 1972. Protocultural aspects of chimpanzee's responsiveness to novel objects. *Folia Primatologica*, 17, 161–170.

Oldfield-Box, H., 1970. Comments on two preliminary studies of "observation" learning in the rat. *Journal of Genetical Psychology*, 116, 45–51

Palameta, B., and L. Lefebvre, 1985. The social transmission of a food finding technique in pigeons: What is learned? *Animal Behaviour*, 33, 892–896.

Pallaud, B., 1977. L'apprentissage par observation chez la souris et le rat. Thèse presentée a l'Université Pasteur de Strasbourg, 1–286.

Platt, J.R., 1964. Strong inference. *Science*, 146, 347–353.

Prato Previde, E., 1992. Trasmissione sociale di un comportamento appreso nel criceto dorato (*Mesocricetus auratus*): uno studio pilota. Rapporto di Ricerca, Istituto di Psicologia della Facoltà di Medicina e Chirurgia, Università degli Studi di Milano.

Prato Previde, E., M. Oldani, C. Modonesi e M.D. Poli, 1991. Trasmissione sociale di un comportamento acquisito nel criceto dorato (*Mesocricetus auratus*): uno studio preliminare. *Riassunti del X Congresso della Divisione ricerca di base in psicologia*, 189–190.

Roberts, L., 1983. Insight into the animal mind. *Bio-Science*, 33, 362–364.

Romanes, G.J., 1882. *Animal Intelligence*. Appleton Century Crofts, New York.

Shaller, G.B., 1972. *The Serengeti Lion*, The University of Chicago Press, Chicago.

Sherry, D.F., and B.G. Galef Jr., 1984. Cultural transmission without imitation: Milk bottle opening by birds. *Animal Behaviour*, 32, 3, 937–938.

Sherry, D.F., and B.G. Galef Jr., 1990. Social learning without imitation: more about milk bottle opening by birds. *Animal Behaviour*, 40, 5, 987–989.

Tomasello, M., 1990. Cultural transmission in the tool use and communicatory signalling of chimpanzees? In S.Parker and K. Gibson, eds., *Language and Intelligence in Monkeys and Apes: Comparative Developmental Perspectives*, Cambridge University Press, Cambridge, 274–311.

Tomasello, M., M. Davis-Dasilva, L. Camak and K. Bard, 1987. Observational learning of tool use by young chimpanzees. *Primates*, 30, 35–50

Valsecchi, P., M. Mainardi, D. Mainardi and I. Bosellini, 1989. On the role of the demonstrator for the solution of a problem in the house mouse. *Ethology, Ecology and Evolution*, 1, 213–216.

Valsecchi, P., I. Bosellini, M. Mainardi and D. Mainardi, 1993. Further analysis of a case of social learning in the house mouse (*Mus domesticus*). *Ethology, Ecology & Evolution*, 5, 121–125.

van Lawick-Goodall, J., 1968. The behavior of free-living chimpanzees on the Gombe Stream Reserve (Tanzania). *Animal Behavior Monographs*, 1-3, 165–329.

van Lawick-Goodall, J., 1973. Cultural elements in a chimpanzee community. In A. Menzel, ed., *Precultural primate behavior*, Karger, Basel, 144–184.

van Lawick-Goodall, J., and H. van Lawick, 1966. Use of tools by the Egyptian vulture (*Neophron percnopterus*). *Nature*, 212, 1468–1469.

Whiten, A. and R. Ham, 1992. On the Nature and Evolution of Imitation in the Animal Kingdom: Reappraisal of a Century of Research. *Advances in the Study of Behavior,* 21, Academic Press, New York, 239–283.

Zentall, T.R., and B.G. Galef Jr., eds., 1988. *Social learning: Psychological and biological perspectives*, Lawrence Erlbaum, Hillsdale, NJ.

Zohar, O., and J. Terkel, 1991. Acquisition of pine cone stripping behaviour in black rats (*Rattus rattus*). *International Journal of Comparative Psychology*, 5, 1, 1–6.

A study of social, genetic, and environmental determinants of cultural transmission in the House Mouse (*Mus domesticus*)

Paola Valsecchi, Ilaria Bosellini, Danilo Mainardi and Marisa Mainardi
Dipartimento di Biologia e Fisiologia Generali
Università di Parma Viale delle Scienze
43100 Parma Italy

Abstract

In the following experiments we analyzed the ability of house mouse (*Mus domesticus*) to open an apparatus to get food. The first two studies investigated the effects of genetic and social factors, on the transmission of this ability from one individual to another. The results showed that two factors are likely to enhance the probability that the offspring will be able to solve the problem: being offspring of parents excellent at solving the problem and seeing the parents solving the problem. Two inbred strains C57 and DBA proved to be different in acquiring this capacity testifying for a genetic control over their social learning abilities. Experiments 3 and 4 showed that mice that lived with a demonstrator opening the apparatus and mice that lived with an adult interacting but not opening the apparatus improved their ability in solving the problem; whereas mice that had different non social experiences with the apparatus did not.

1. Introduction

Recent years have seen the growth of knowledge about the ability, present in all major vertebrate classes and in the Insecta, to transmit, from one individual to another, information or new patterns of behavior acquired through experience. This phenomenon, generically called "social learning", actually collects several different behavioral processes characterized by a common trait: social interactions may facilitate introduction of a pattern of behavior into an individual's repertoire. An extensive terminology grew in the attempt to explain and classify the increasing body of instances of social learning studied in the laboratory and in the field, but the proliferation of explanatory labels led to a greater confusion rather than to a clarification of this issue (see Galef 1976 and 1988 for a review). As Galef (1988) argued: " (...) in the absence of clearly defined, mutually exclusive categories, each reflecting an unique process underlying social learning (...) there is little to be gained by explaining instances of social learning as produced by one type of social learning rather than another." Nevertheless, the difficulties associated with the creation a framework to categorize social learning phenomena stimulated the use of more and more proper experimental designs and aroused a theoretical debate, not unraveled so

R. A. Gardner et al. (eds.), The Ethological Roots of Culture, 143–154.

far, on the definition of imitative and pseudo-imitative behavior (Galef 1992; Whiten and Ham 1992).

Both individual and social learning allow organisms to cope with different environments and they act in different ways. Behavioral modifications acquired by individual learning are not transmitted from one generation to the next, and the information gained through the direct transaction with the external environment are lost with the individual's death. In contrast, behaviors acquired by some sort of social learning may be transmitted from one individual to another and from one generation to the next. The simple dichotomy between individual and social learning is only apparent and it should be kept in mind that it is always the individual that learns. It is therefore worth understanding " (...) how individual and social learning interact to determine the evolutionary dynamics of the behavioral variants themselves as well as the genes that underlie learning processes" (Boyd and Richerson 1988).

The aim of this paper is to review a series of laboratory experiments performed in order to investigate the role of social and individual learning on the capacity to solve a problem and to analyze the genetic basis of this ability in domesticated house mouse *Mus domesticus*. As reference, it is important now to provide the operational simplified classification of social learning processes followed in evaluating the results of this study. Social interaction may produce the transmission of information at three different levels:

1) A behavior pattern already present in an animal's repertoire may increase the frequency or intensity of responses in a novel context by the presence of a conspecific doing that behavior, a phenomenon termed "social facilitation" (Thorpe 1963, Clayton 1978).

2) Animals may be attracted to a novel environmental stimulus by the presence of a conspecific and then acquire the new behavior by individual learning; this process has been called local and/or stimulus enhancement.

3) A relatively improbable behavioral response can be acquired directly through observation of another animal, a process defined as true imitation or inferential imitation (Hogan 1988).

2. Materials and Methods

In general, laboratory studies have been unsuccessful in convincingly identifying the process by which information is acquired during social interactions due to inadequate procedure and controls or use of tasks that are too easy to allow discrimination between alternative explanations (Lefebvre and Palameta, 1988).

We set a series of experiments to investigate the ability of domesticated mice to socially acquire a new behavioral skill. For the purpose of these studies we used an apparatus, consisting of a suspended cylindrical container, which could be opened by rotating a round door counterclockwise (Fig.1). The difficulty in opening the door was adjusted in such a way that in a population of mice a small number of individuals would be capable of learning the technique spontaneously. Approximately 50 turns were needed to open the door wide enough for the mouse to get into the cylinder, at the end of which there were a few sunflower seeds. If the animal rotated the door clockwise it closed the door more tightly. At the start the door was set to make it absolutely impossible for the mouse to get inside although it could smell the sunflower seeds.

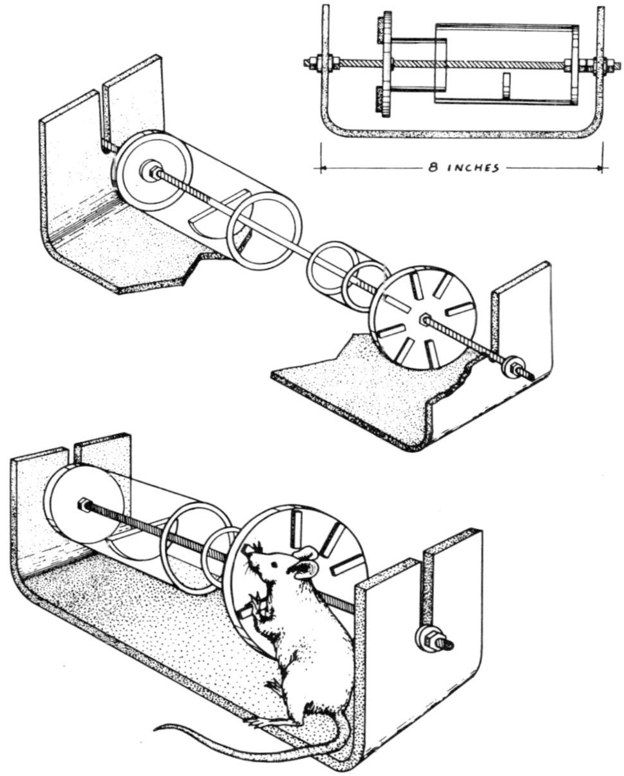

FIGURE 1. Experimental apparatus, a detailed description is given in the text (from Mainardi and Mainardi 1988).

The subject's capacity to open the door was evaluated on the basis of a trial in which the mouse spent 24 h, with food and water ad libitum, in a plexiglass box (42.5 x 26.6 x 15cm) containing the apparatus. The experimenter set the apparatus to the start position at the beginning of the test and checked the position of the rotating door 24 h later recording:
a "+" for the opened door;
a "-" for the door still closed;
a "i" for an insufficient rotation;

An individual test consisted of three of these trials performed on 3 consecutive days.

In order to analyze the information provided by the different combinations of results on the three-day test, every animal was assigned a score according to the following scheme.

+++	++i	++-	+ii	+i-	+--	iii	ii-	i--	---
9	8	7	6	5	4	3	2	1	0

Five hundred 2-3 month old mice (Swiss outbred CD1) underwent a preliminary test with the apparatus (Mainardi et al. 1986). Of these only 24 mice (4.8%) managed to open the cylinder on all three trials (+++); 176 animals (35.2%) failed completely (---), and 300 mice opened the cylinder either once or twice, or rotated the small door an insufficient number of times to open it.

Data of Experiments 1 and 2 were published (Valsecchi et al. 1989; Valsecchi et al. 1993) using a different statistical analysis. For the sake of simplicity they were made comparable to those of experiments 3 and 4 and so reanalyzed using the score scheme reported above. Possible differences that should turn out using this method of analysis will be considered in the discussion.

3. The Role of Social and Individual Experience

These two experiments were designed to show the existence of social transmission of the ability to solve this problem and to understand what sort of social learning mechanism is involved.

3.1. EXPERIMENT 1

Experimentally naive house mice were individually exposed to the apparatus in the presence of an adult female demonstrator that was able to open it. We compared the effects of this condition with the effects of the presence of an adult female unable to open the apparatus and with the effects of trial-and-error learning.
The subjects, 210, 20-day-old Swiss mice, were assigned to three experimental groups:
1. ABLE group: each pup spent seven days in a cage with an adult female demonstrator and the apparatus. The demonstrator opened the door allowing access to seeds every day.
2. UNABLE group: each pup spent the same amount of time with an adult female unable to solve the task.
3. APPARATUS group: each pup spent 7 days in a cage with the apparatus at his disposal.

After this training all animals underwent the test described below. In addition a fourth group (NAIVE) of 70 Swiss mice of the same age, with no previous experience with the apparatus, were tested in the same way to establish the proportion of naive animals able to perform the task spontaneously.

3.2. RESULTS

The general analysis of data reported in Figure 2 was carried out comparing the mean scores by Kruskall-Wallis ANOVA for non parametric measures. Data clearly showed that the four groups were significantly different and suggested that social interactions were crucial for the acquisition of this skill ($H=18.38$, $p<0.001$).

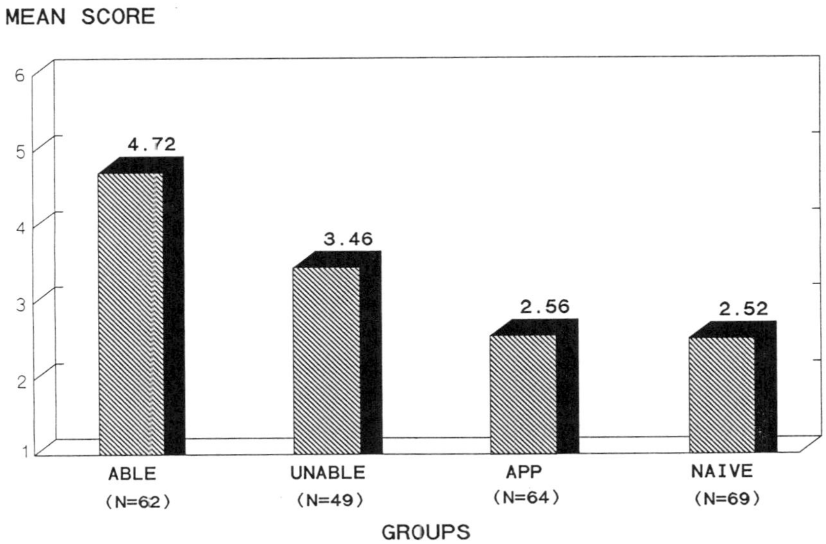

FIGURE 2. The mean scores obtained by the animals of ABLE, UNABLE, APPARATUS and NAIVE groups during the three-day test and the number of subjects for each group. See text for details.

When the pups were tested for the opening of the apparatus, mice that had been given a full demonstration had learned, but those that were shown only where to look or had the possibility to learn by themselves had not. In fact, single comparisons between groups carried out using Mann-Whitney U test, indicated that: pups of the ABLE group got a significantly higher score than the pups of all the other groups, (ABLE vs UNABLE: $Z=-2.14$, $p=0.03$; ABLE vs APP.: $Z=-3.65$, $p<0.001$; ABLE vs CONTROL: $Z=-3.79$ $p<0.001$). The experience for 7 days with the apparatus, both in the presence and absence of an adult female produced pups as skilled as the naive ones, showing that in this case trial-and-error learning was not sufficient to improve their ability. However pups that shared the experience with the apparatus together with unable mice got higher score than pups of the other two groups, even if this difference was not significant.

3.3. EXPERIMENT 2

Results of the first experiment raised two different hypotheses:

1) the inability of pups experiencing the apparatus by themselves might be due to the depression of their activity, resulting from being isolated and socially deprived across the 7 days' training.

2) the change produced in the environment by an adult (i.e. finding the apparatus already opened and having the possibility to eat the seeds) could be a sufficient

stimulus to influence the subject's behavior, as discussed by Sherry and Galef (1984, 1990).

In order to clarify these two points, we replicated the previous experiment adding two new training conditions. In the first condition (FELLOW group) the female was confined to a portion of the cage and her access to the apparatus was prevented; the pup could visit her, sleep and eat with her but could interact with the apparatus by itself. In the second condition (OPENED APPARATUS group) each pup spent the 7 days in the same conditions as in the FELLOW group but the apparatus was fixed in the opened position so that the individual could explore it and get the seeds placed inside. NAIVE mice served as a control group: each animal spent 7 days in a cage with an adult female, but without the apparatus. After training, all animals of each group underwent the three-day test.

3.4. RESULTS

A Kruskall-Wallis non parametric ANOVA was performed to analyze the data reported in Figure 3, showing a significant difference between groups ($H=39.98$, $p<.001$).

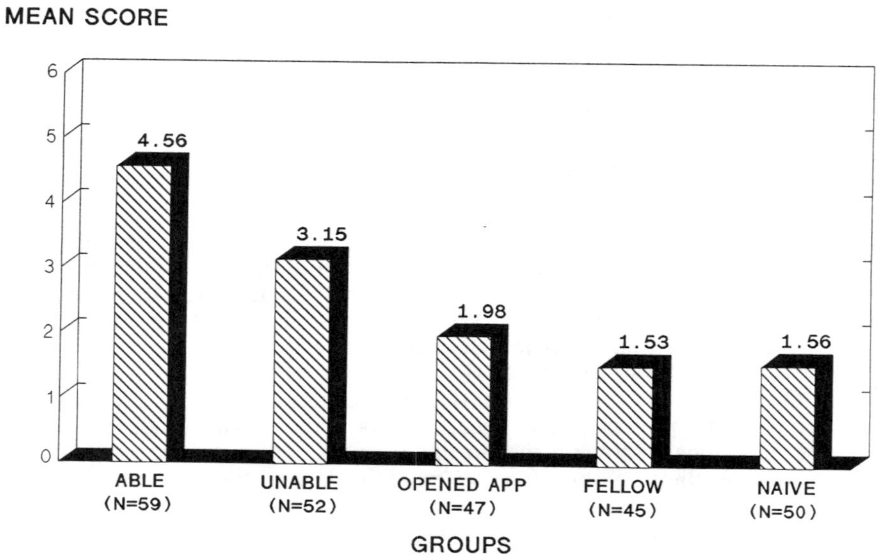

FIGURE 3. The mean scores obtained by the animals of ABLE, UNABLE, OPENED APPARATUS, FELLOW and NAIVE groups during the three-day test and the number of subjects for each group. See text for details.

Further comparisons (Mann-Whitney U test) between single groups indicated that the pups with the demonstrator acquired the food-finding technique better than did those of UNABLE, FELLOW, OPENED APPARATUS, and NAIVE subjects

(ABLE vs UNABLE: Z=-2.42, p=0.01; ABLE vs FELLOW: Z=-5.14, p<0.001; ABLE vs OP APP: Z=-4.31, p<0.001; ABLE vs NAIVE: Z=-5.29, p<0.001). These data revealed that the opportunity to practice with the apparatus already opened and to get the food reward is not sufficient in itself to improve this ability in young mice.

Pups of the UNABLE group got a significantly higher score than the pups of FELLOW group (Z=-2.31, p=0.02): the presence of an adult that could interact with the apparatus focusing the pups' attention on it, was more effective in increasing the pups' ability than trial-and-error learning even when the subjects were active and not socially deprived.

3.5. DISCUSSION

The improved ability showed by mice of ABLE and UNABLE groups, due to the interaction with an adult opening or just interacting with the apparatus, might suggest a case of stimulus (or local) enhancement (Spence 1937). As pointed out by Galef (1988), "the notion of stimulus enhancement extends the concept of local enhancement to include the entire class of objects sharing stimulus characteristics with an object a demonstrator manipulates, contacts, or marks". Concerning the analysis previously performed (Valsecchi et al. 1989; Valsecchi et al. 1992 in press) a discrepancy emerges between the results obtained using different scoring methods.

The position of animals of UNABLE group is in some way puzzling since it wavers between those of ABLE and the other groups according to the scoring system utilized. The most important feature of the score system 0 to 9 performed throughout this paper, is that it considers insufficient rotations in opening the door, measuring a general activity and interest in the apparatus, as well as positive and negative outcome.

In this case pups of UNABLE group manipulated the apparatus less than pups of ABLE group, but not than pups of other groups. This could be partially due to a reward effect of the seeds kept inside the apparatus that were for 7 days at disposal of pups of ABLE group. However it is worth emphasizing that this reward effect in itself is insufficient to stimulate learning as demonstrated by mice of OPENED APPARATUS group. Considering only the number of successful trials (as we did in the previous publication) pups of UNABLE group are more similar to those of ABLE group and particularly interesting, are significantly better performers than pups of other groups.

4. The Role of Genetics.

Social transfer of acquired behavior is normally said to be "non-genetic" and this is correct in that it takes place "via experience". Nonetheless the capacity for cultural transmission depends in various ways on the genetic characteristics of the species and it is, even if indirectly, genetically controlled (Mainardi 1980; Mundinger 1980; Cavalli Sforza & Feldman 1981).

A second series of experiments (Mainardi et al. 1986; Mainardi et al. 1988) was carried out in order to evaluate how genetic makeup affects the aptitude to solve or to socially learn to solve the problem. Since mothers may transmit behavioral propensities to their young either genetically or as result of behavioral interaction with them, the transfer of information between mother and young was analyzed.

4.1. EXPERIMENT 3

Positive animals (P) able to solve the problem consistently, and negative animals (N) which never succeeded in solving it, were selected from Swiss outbred population. Pairs were then formed within the two categories in order to obtain offspring from positive and negative parents.

In each category the litters were raised by their dam with the apparatus inside the breeding cage (YES) or without it (NO). Thus four groups of animals were obtained.

1. P-YES: offspring of spontaneously able parents that gave a full demonstration of the task every day.
2. P-NO: offspring of spontaneously able parents reared without the apparatus.
3. N-YES: offspring of unable parents but having the opportunity to familiarize with the apparatus.
4. N-NO: offspring of unable parents and reared without the apparatus.

At the same time offspring of parents taken at random from the naive population were reared to form a group of CONTROLS of the same age as the experimental subjects. All mice underwent the individual test when 50 days old.

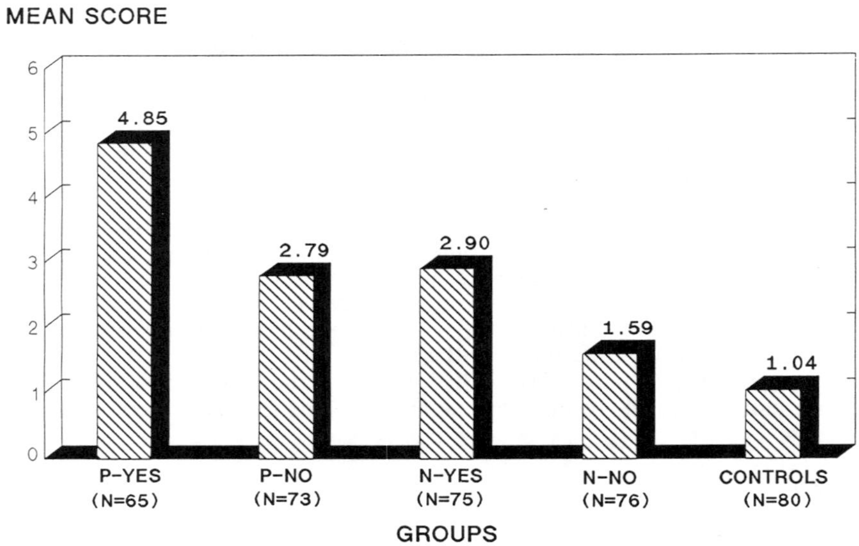

FIGURE 4. The mean scores obtained by the animals of P-YES, P-NO, N-YES, N-NO and CONTROL groups during the three-day test and the number of subjects for each group. See text for details.

4.2. RESULTS

The results, presented in Figure 4, showed significant differences between the offspring of P and N individuals, which testifies to a genetic effect on the acquisition of the food-finding technique. We also found differences between individuals that during growth had experience with the task and those that had none. Genetic and experiential effects in this experimental situation were of comparable magnitude. The population is distributed along a continuum of individuals characterized by a quantitatively different ability. In fact, 32.3% of P-YES mice, 15.1% of P-NO mice, 13.3% of N-YES mice, 3.9% of N-NO mice and 1.3% of Control animals were able to open the apparatus for three consecutive days.

Analysis of variance yielded a highly significant value ($F=18.17$, $p<0.001$). Single comparisons, carried out using Student t test showed that the score of P-YES mice is about four times that of CONTROLS mice ($t=8.37$, $p<0.001$) and three times that of N-NO animals ($t=6.41$, $p<0.001$). Animals in the other two categories obtained intermediate values and are very similar (P-NO vs N-YES $t=0.18$, n.s).

4.3. EXPERIMENT 4

To look more precisely into the genetic control of the ability shown by mice to socially acquire information about the apparatus, the animals used for the next experiment were mice belonging to two inbred strains, C57BL/6NCr1BR and DBA/2NCr1BR that have been proved to be different in learning abilities (Bovet et al., 1969).

Experimental animals were cross-fostered immediately after birth into mixed litters composed of half C57 and half DBA pups and adopted by a recently delivered Swiss Webster dam. The experiment was conducted under two conditions:

CONDITION 1: EARLY EXPERIENCE

The foster dam gave a full demonstration opening the apparatus for eight consecutive days when pups were 17 to 25 days of age.

CONDITION 2: LATE EXPERIENCE

In this condition, the training with the foster dam opening the apparatus began at a later stage of the pups' development, when they were 30 to 38 days old. In both cases pups were then submitted to the opening test.

4.4. RESULTS

Average scores and relative comparisons carried out using the Mann-Witney U test are reported in Figure 5 and show that there is a significant difference between the DBA and C57 strains about the capability to open the cylinder. Considering the results from Condition 1, DBA trained mice appeared able to solve the problem much more often than their controls ($Z=-2.98$, $p=0.003$) and C57 trained mice ($Z=-2.83$, $p=0.005$). However, in Condition 2, C57 trained pups were more likely to open the apparatus than DBA trained pups ($Z=-2.85$, $p=0.004$) and DBA trained pups did not differ from their controls.

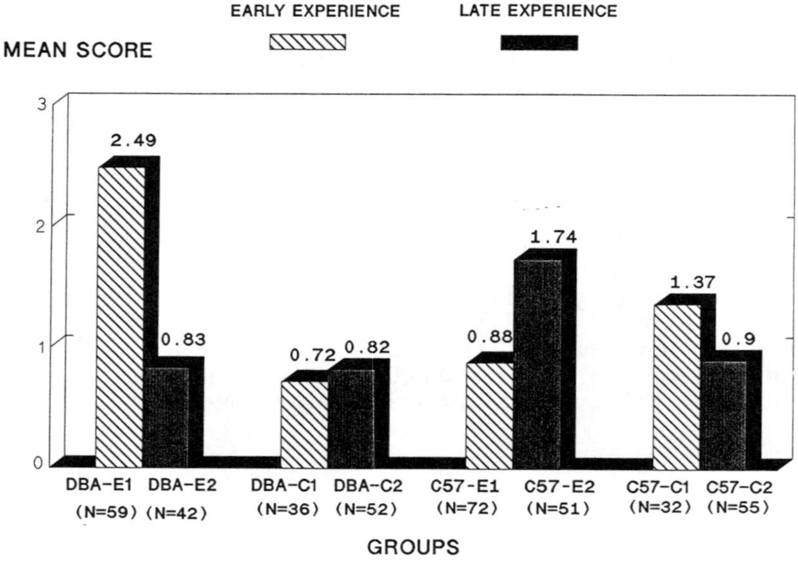

FIGURE 5. The mean score obtained by DBA and C57 mice during the three-day test and the number of animals in each group. E= Experienced animals; C= Controls; 1= Early experience; 2= Late experience.

An attempt to clarify the results of Condition 1 should consider Oliverio's (1983) demonstration of differences in brain maturity in inbred strains at birth. C57, at birth and for the first seven days of life, show a more advanced ontogenetic development than DBAs. Oliverio suggested that a more mature species at birth may be less sensitive to environmental influences later on. This refers to a more detailed and general statement that the behavior of species with early developing offspring is more directly under genetic control, whereas in altricial species learning or environmental influences in general have a greater weight (Rosenblatt 1979). There is another factor that should be considered in interpreting the results obtained by C57-E2 mice compared with DBA-E2 subjects: C57 are characterized by a greater locomotor activity, more sniffing, rearing and climbing than DBAs by a higher exploratory tendency (Van Abeelen 1974). This could have facilitated the direct experience with the apparatus during Condition 2 by C57. On the contrary the experience gained during Condition 1 occurred when the pups were very young and had a very little chance of manipulating the apparatus. In this condition C57 may have added to their experience of performing mothers (local enhancement), the effect of greater direct experience, such as a chance rotation of the door or exploration into the cylinder already opened by the mother gained as a result of a more lively exploratory activity.

5. Conclusions and Perspectives

These studies confirm the considerable role of social experience in the acquisition of this food-finding technique. In fact, mice exposed to varying quality of social context increased, at different extent, their ability to solve the problem. In contrast inexperienced animals that are not given demonstrations by a conspecific model are not able to learn the innovation by themselves. Furthermore the change in environment represented by the apparatus already opened does not increase the probability that naive mice will learn to open it.

The data also show that there is variability concerning the predisposition to respond positively to the test and a similar magnitude of the influence of the genetic and social factors. Results agree with the hypothesis that in a species like the mouse, which is programmed to adapt rapidly to changing environments, it is advantageous if a certain number of individuals are capable of solving an unfamiliar problem, the solution of which can be learned rapidly at the same time by a sufficient number of animals genetically inclined towards social learning.

Our first concern was to create a task that would present a potential problem for the mouse, and that would in some way resemble a situation a mouse might face in the wild, considering the well known tendency of this species to explore the minutest details of their environment. Future development of this study should consider both the nature of the information transmitted from a demonstrator to an observer, and the amount of the information transmitted.

References

Bovet, D., Bovet-Nitti F. and Oliverio A. (1969). Genetic aspects of learning and memory in mice. Science, N.Y. 163: 139-149.

Boyd, R. and Richerson, P.J. (1988). An evolutionary model of social learning: the effects of spatial and temporal variation. In: T.R. Zentall and B.G. Galef (eds.), Social Learning, Lawrence Erlbaum, Hillsdale, New Jersey, pp.29-48.

Cavalli-Sforza, L.L. and Feldman, M.W. (1981). Cultural transmission and evolution: a quantitative approach. Princeton University Press, Princeton, N.J.

Clayton, D.A. (1978). Socially facilitated behavior. Quart. Rev. Biol., 53: 373-392.

Galef, B.G. Jr. (1976). Social transmission of acquired behavior: a discussion of tradition and social learning in vertebrates. Adv. St. Behav. 6: 77-100.

Galef, B.G. Jr. (1988). Imitation in animals: history, definition, and interpretation of data from the psychological laboratory. In: T.R. Zentall and B.G. Galef (eds.), Social Learning, Lawrence Erlbaum, Hillsdale, New Jersey, pp.3-28.

Galef, B.G. Jr. (1992). The question of animal culture. Hum. Nat. 3: 157-178.

Hogan, D.E. (1988). Learned imitation by pigeons. In: T.R. Zentall and B.G. Galef (eds.), Social Learning, Lawrence Erlbaum, Hillsdale, New Jersey, pp.225-238.

Lefebvre, L. and Palameta, B. (1988). Mechanisms, ecology, and population diffusion of socially-learned, food-finding behavior in feral pigeons. In: T.R. Zentall and B.G. Galef (eds.), Social Learning, Lawrence Erlbaum, Hillsdale, New Jersey, pp.141-164.

Mainardi, D. (1980). Tradition and the social transmission of behavior in animals. In: Sociobiology: Beyond Nature/Nurture? G.W. Barlow & J. Silverberg (eds.), AAAS Selected Symposium 35, pp.227-255.

Mainardi, D., Mainardi, M. and Pasquali, A. (1986). Genetic and experiential features in a case of cultural transmission in the house mouse (*Mus musculus*). Ethology 72: 191-198.

Mainardi, D. and Mainardi, M. (1988). Culture and genetics in the house mouse. In: T.R. Zentall and B.G. Galef (eds.), Social Learning, Lawrence Erlbaum, Hillsdale, New Jersey, pp.239-252.

Mainardi, M., Mainardi, D. and Valsecchi, P. (1988). Social learning differences between C57/6NCr1BR and DBA/2NCrBR mice. Monitore zool. ital. 22: 183-192.

Mundinger, P.C. (1980). Animal cultures and a general theory of cultural evolution. Ethol. Sociobiol. 1: 183-223.

Oliverio, A. (1983). Genes and behavior, an evolutionary perspective. Adv. St. Behav., 13: 191-217, N.Y. Academic Press.

Rosenblatt, J.S. (1979). The sensory motion and motivational bases of early behavior development of selected altricial mammals. In: N.E. Spear & B.A Campbell (Eds.), Ontogeny of learning and memory, Lawrence Erlbaum Associates, Hillsdale, pp.1-38

Sherry, D.F. and Galef, B.G. Jr. (1984). Cultural transmission without imitation: milk bottle opening by birds. Anim. Behav. 32: 937-938.

Sherry, D.F. and Galef, B.G. Jr. (1990). Social learning without imitation: more about milk bottle opening by birds. Anim. Behav. 40: 987-989.

Spence, K.W. (1937). Experimental studies of learning and the higher mental processes in infra-human primates. Psychol. Bull. 34: 806-850.

Thorpe, W.H. (1963). Learning and istinct in animals (3rd edition), Methuen, London.

Valsecchi, P., Mainardi, M., Mainardi, D. and Bosellini, I. (1989). On the role on the demonstrator for the solution of a problem in the house mouse. Ethol. Ecol. Evol. 1: 213-216.

Valsecchi, P., Bosellini, I., Mainardi, D. and Mainardi, M. (1993). Further analysis on a case of social learning in the house mouse (*Mus domesticus*). Ethol. Ecol. Evol. 5:121-125.

Van Abeelen, J.H.F. (1974). Genotype and the colinergic control of explorative behavior in mice. In: J.H.F. Abeelen (ed). The genetic of behavior, North-Holland Publishing Co., Amsterdam, pp.347-372.

Whiten, A. and Ham, R. (1992). On the nature and evolution of imitation in the animal kingdom: reappraisal of a century of research. Adv. Stud. Behav. 21: 239-283, N.Y. Academic Press.

CAN CHIMPANZEES USE TOOLS BY OBSERVATIONAL LEARNING?

DANIEL PAQUETTE
Jardin zoologique du Québec
8191 Avenue du Zoo
Charlesbourg, Québec
Canada, G1G 4G4

ABSTRACT. Wild chimpanzees commonly use sticks to fish for termites, ants or honey. This ability seems to be socially transmitted to juveniles by their mothers across generations. In a natural environment, the limited visibility of this behavior with regards to the extent of stick's insertion and about the success or failure in fishing hinders the study of the underlying learning processes. This study explores the discovery and learning of tool use for fishing honey in an artificial hive by a group of four captive chimpanzees. The discovery of tool use was accidental and coactive. The speed with which the group of experimentally naive chimpanzees discovered and learned tool use suggests that wild chimpanzees of different populations independently discovered the fishing behavior. The alpha male and his ally learned before the subordinates. Here, trial-and-error learning was, as in monkeys, the main process allowing the acquisition of the tool-use technique. However, the observation of conspecifics allowed the orientation of the experimentation by the selection of clues. As suggested by Tomasello *et al.*, it is the understanding of the function of the tool, *i.e.* the cause-effect relations between the action of the demonstrator, the type of tool and the task to accomplish which confers to chimpanzees an advantage over monkeys.

1. Introduction

Learning allows quick adjustment of individuals to the environment's variations. Therefore, the learning mechanisms, although probably genetically determined, constitute an advantage over the genetical transmission of behaviours. Innovative behaviour may be strongly adaptive when the environment is changing during the individual's life. In some species of

R. A. Gardner et al. (eds.), The Ethological Roots of Culture, 155–172.

primates, a new food can become incorporated into the diet of a troop as a result of a whole series of independent individual discoveries; other novel behaviour can be discovered by one or a few individuals and acquired by other troop members through social transmission (Kummer & Goodall, 1985).

Social learning refers to a class of behavioral mechanisms that may result in the production of a tradition in a population (Galef, 1986). Social transmission mechanisms are not well known. Galef (1988) distinguishes imitation as a descriptive term and imitation as a term referring to a process. According to Thorpe (1963), there are several processes by which imitative behavior may be socially transmitted among conspecifics. Social facilitation or contagion is a process in which the behavior of a conspecific acts as a social releaser of identical instinctive behavior in an observing conspecific (Hogan, 1988); the increase in production of target behaviors is due to the mere presence of the conspecific rather than observation of the behavior itself, which may facilitate trial-and-error learning because the trials are produced more rapidly (Meltzoff, 1988). Local (or stimulus) enhancement is the increase in salience of important stimuli produced by observing a demonstrator make contact with those stimuli (Thorpe, 1963); it plays, for example, a critical role in the acquisition of knowledge about what to eat and where to find it (Whiten, 1989). True imitation or inferential imitation (or observational learning — Galef, 1988) is the copying of a novel or otherwise improbable act through observation (Thorpe, 1963) and without trial-and-error (Beck, 1980).

Experimental and observational studies on captive and wild monkeys using tools concluded that only stimulus enhancement is implicated; they also converged on a sceptical verdict about a capacity for imitation in monkeys (Whiten, 1989). Tomasello, Davis-Dasilva and Camak (1987) also deny true imitation in chimpanzees.

Wild and captive chimpanzees use tools more often and in more different ways than do monkeys (Beck, 1980; Tomasello *et al.*, 1987). According to Beck (1974), the chimpanzees' competence in using tools is mainly due to their great capacity to learn through observation. In fact, experiments and anecdotes show that learning through observation and without practice is possible in order to solve problems requiring the use of tools if chimpanzees have the chance to imitate a model shortly after observation (Hall, 1963). However, it is noteworthy that the tasks were relatively simple to remember and accomplish (e.g., tunnel-stick). But what about complex tasks? According to McGrew (1977), chimpanzees are capable of acquiring a complex tool-use technique with limited exposure and practice.

Wild chimpanzees are known to fish for termites, ants or honey with stalks or sticks which are selected according to their length, diameter, shape and texture. Termite fishing has been recorded in many sites in Africa (McGrew & Rogers, 1983). Inter-populational differences among chimpanzess have been reported (McGrew, Tutin & Baldwin, 1979). Ecological factors explain most of these differences, but some could be linked to differing «social customs» or «cultural traditions» (McGrew *et al.*, 1979; Kummer & Goodall, 1985; Galef, 1986; Boesch & Boesch, 1990), which persist across generations (McGrew, 1977). According to Nishida (1987), fishing for termites and ants would be examples of true imitation. This ability seems to be gradually acquired by young chimpanzees by observing and imitating their mothers. Trial-and-error learning would also be involved in the selection and preparation of the tool, in perfecting the techniques, and in order to deal with the ants' antipredator responses (van Lawick-Goodall, 1973; McGrew, 1977).

I wish to present here some data on the respective roles of observational learning and trial-and-error learning in the acquisition of fishing techniques by captive chimpanzees. Generally, experiments on observational learning have consisted in allowing a naive subject to observe an experienced individual, and verifying whether observation accelerated learning. According to Hall (1963), two drawbacks are linked to this type of assessment: 1) the prior experience of the animals was most often unknown, and 2) experiments were conducted in conditions hardly comparable to the natural surroundings, which limited their external validity. The orientation used in this study was thus inspired by Hall (1963), who suggested that more attention be payed to group situations or group dynamics during the learning process. Therefore, a context of discovery of a food resource was reproduced in captivity with chimpanzees whose previous experience was well known, and required the use of a tool similar to the one used by wild chimpanzees to fish insects or honey.

2. Methods

2.1. STUDY GROUP

The group consisted of two males (Spock and Merlin) and two females (Sophie and Maya). Spock and Sophie were 8 years 8 months old, Merlin was 7 years 8 months old, and Maya, 7 years 5 months old at the beginning of the experiments. The dominance hierarchy was linear: Spock was alpha male, followed by Merlin, Sophie and finally Maya. Sophie challenged Merlin's rank since Spock regularly interfered to help her against Merlin.

Although these chimpanzees were separated from their respective mothers very soon after birth, and were never again in the presence of adults of their species, they may be considered as good subjects for observation because they never experienced restrictive captivity conditions such as social isolation (see Menzel, Davenport & Rogers, 1970). On the contrary, they were reared together in an environment rich in stimulation, benefiting from the care and attention of human surrogate mothers fulfilling their need for security. In fact, these chimpanzees have never expressed reactions of fear to new objects (Paquette, 1980) as is the case with chimpanzees who have lived in restricted environments (Menzel, 1963), and the frequency of their abnormal behavior is relatively low (Paquette & Prescott, 1988).

Finally, it is noteworthy that before this study these chimpanzees had never tasted honey, nor had they used sticks in tasks similar to fishing.

2.2. THE PUZZLE-BOX AND TOOLS

The puzzle-box, made of steel plate 0.6 cm thick, was pyramid-shaped with a truncated top (Figure 1a). The base was 91 x 66 cm, the height 79 cm, with a roof 76 x 10 cm. The box was set on four 13 cm-high feet, and fixed to the ground by a central rod cast in the cement floor.

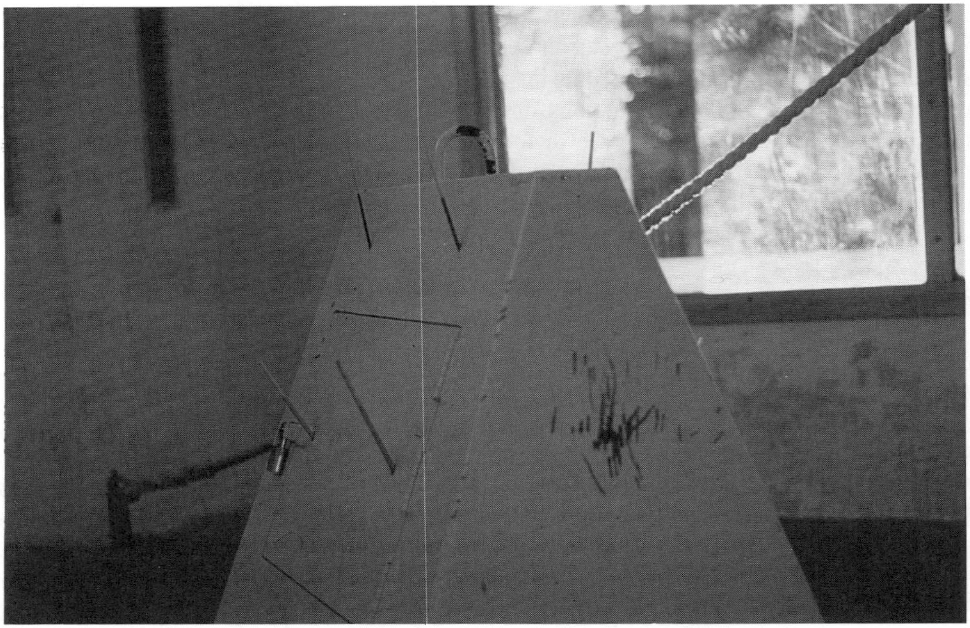

Figure 1a. Photo of the puzzle-box.

Four holes were pierced on each of the two larger sides of the puzzle-box. The two higher holes (R1 and R2) on the right side of the box (relative to the video camera) were 0.3 cm in diameter, whereas the two lower holes (R3 and R4) were 0.6 cm in diameter. This configuration was reversed on the left side of the puzzle-box, having the larger holes (L1 and L2) above the smaller holes (L3 and L4). The holes were vertically spaced at 30 cm intervals, and horizontally at 38 cm intervals. Inside the holes, 5 cm stakes served to orient the sticks in the right direction (Figure 1b).

Two bowls filled with honey were placed inside, in line with the holes (Figure 1b). A door (56 x 41 cm) secured with a large lock allowed the occasional addition of honey and the removal of sticks fallen inside.

Sticks were made of wood, 25, 50 or 75 cm long, and 0.6- or 0.3-cm wide. The larger ones were red (except at the tips) and the narrower ones were blue in order to be readily recognized on the video recordings. The 25-cm sticks were too short to fish honey in the puzzle-box. The 50 cm ones were long enough to fish from the lower holes (R3,R4, L3 and L4), but only the 75 cm sticks allowed fishing from the highest holes (R1,R2, L1 and L2).

Figure 1b. Closeup photo of the opened puzzle-box, showing arrangement of holes and honey bowls.

2.3. EXPERIMENTAL PROCEDURE AND ANALYSIS

In order to allow the animals to habituate to the box, it was placed in the chimpanzees' exhibit five days before the honey was added.

Twenty-four sticks, four of each length for the two series of diameter, were randomly placed on the floor of the exhibit before the chimpanzees entered.

The chimpanzees were filmed on video camera for three consecutives hours. Two sessions per day (morning and afternoon) took place between December 9 and 13, 1985. Having monopolized the sticks and especially the access to the puzzle-box, only Spock and Sophie (alpha male and his ally) learned the tool use during this period. However, considering that Merlin and Maya both had the opportunity to observe the successes of Spock and Sophie, it was interesting to verify their knowledge in the absence of the two others. A supplementary session thus took place on December 14 (the day 6) for Merlin and Maya. Another fishing session for the four chimpanzees was done on January 29 (1986) in order to verify whether Spock and Sophie remembered what they had learned. Finally, a period of two sessions per day took place for the whole group from May 7 to 10 inclusively, to allow Merlin and Maya to learn the technique.

I scored myself the videotapes. During each session, the frequency of insertion of sticks in the holes was recorded according to the type of stick, the technique used, and the final result (success or failure). Here the chimpanzee had a success when he collected honey after a complete insertion of a stick of correct length in the puzzle-box, and did an error when he did not collect honey after any manipulation of a stick on the puzzle-box. Three levels of insertion without gathering honey were coded: a) on the surface of holes, b) until hand grip, and c) complete insertion of the stick.

My analysis was essentially centered on the various types of errors made by the subjects according to the number of days of exposure, in order to verify their understanding of the task to be accomplished, as was done by Visalberghi and Trinca (1989) with capuchin monkeys.

Also, for each trial (success or error), the presence and the name of an observing chimpanzee was recorded. Finally, I recorded the frequency of times that a chimpanzee took away a stick from a peer, as well as any agonistic behavior observed within the group.

3. Results

3.1. DISCOVERY OF TOOL USE

3.1.1. *First day.* During the morning session, after 2 h 54 min, Sophie was the first to intensively explore the puzzle-box: she smelled it, kissed the holes, tried to peek inside, shook the box, tried to open the door and pulled at the lock. She undoubtebly smelled the honey, since the apparatus had been in the exhibit for five days and seemed no longer of interest by itself. After eleven minutes of intensive investigation, Sophie shook the rope which was beside the box, Spock and Sophie pant-hooted, then Spock displayed and in turn came to investigate the box for a few minutes. It is noteworthy that Sophie scratched herself very often during her investigations.

During the afternoon session, on the other hand, the box remained unexplored. The chimpanzees, however, investigated the sticks during both sessions that day: they manipulated them, chewed at them and poked them into ventilation holes and other apertures around their exhibit. By the end of the day, more than 76% of the sticks were no longer than twelve centimetres.

3.1.2. *Second day.* In order to compensate for the low level of interest in the box exhibited by the chimpanzees during the first day, a little honey was smeared around the edge of each hole.

As soon as he entered, Spock pulled at the lock and licked a wide hole on the left side of the box. Two minutes later, Merlin licked several holes, then five minutes later, it was Sophie's turn.

Eleven minutes after the beginning of this day's first session (6 hours 11 min. of total exposure), Sophie introduced a long narrow stick in hole R3 (a lower/wider hole), plunged it several times to the bottom, scratched herself and moved away without pulling it out. From this moment, chimpanzees always have fished from 2 of the 8 holes, R3 and R4 (on the right side), which were lower/wider holes. In these holes, both width of sticks, and two of the three lengths were effective tools.

Forty-two seconds later, Spock approached and pulled out the stick covered with honey, smelled its tip, took it, placed it in a ventilation hole, pulled it out, touched the tip and finally licked his fingers.

Twenty-two minutes later, Maya started experimenting on the puzzle-box with sticks. She inserted and pulled out pieces of narrow sticks in R4. Maya then inserted a wide stick in cracks in the door of the box, which Sophie also did with a narrow stick a few minutes after having observed Maya. During this time, Spock touched R4 with the tip of a narrow stick. Then, Maya tried to insert a narrow stick into the lock, and Spock did the same less than one minute later, followed by Merlin nine minutes later, both with narrow sticks. Spock then inserted a wide stick into R4, pulled it out, and later left it in a ventilation hole. Sophie, a few minutes later, inserted a wide stick into R4, and licked the tip. After having seen her, Merlin also introduced a wide stick into R4. Maya then explored other holes on the opposite side of the box. It is noteworthy that they all frequently licked the holes.

During the second session, Sophie introduced a wide stick into a ventilation hole, then she approached the box with a bunch of long sticks that she tried to put into R3, moved away and five minutes later inserted a short wide stick into R4, pushed it as far as she could and looked at the tip before putting it on the ground. She then put a long wide stick into R3 and moved away leaving it there. Spock pulled it out more quickly than the first time and licked the tip. Seeing that, Sophie pant-hooted while trying to insert a wide stick into the two small holes on the left side, she shook the box, then looked at both holes and scratched herself. Finally, Sophie came back, picked up and licked the stick that she had previously left on the ground. A few minutes later, Spock came back and slowly dipped the stick that he had picked up into the same hole (R3), and for the first time, he completed a successful sequence of fishing honey (9 h. 13 min. of exposure). Sophie tried experimenting with a wide stick that was too short. But, after having observed Spock's success, she plunged a long stick of the same width and in the same hole (R3) as that used by Spock and had her first success (9 h. 16 min of exposure).

3.2. LEARNING OF TOOL USE

3.2.1. *Spock and Sophie*. Figure 2 shows that Spock and Sophie had a success rate of more than 75% from the fourth day on.

What were the main types of errors made by these two chimpanzees? Sophie most often attempted to introduce a wide stick in a narrow hole, but only on days 2 and 3 (18/498). Figure 3 also shows that Sophie and Spock essentially experimented on the surface of holes during days 2 and 3. From the beginning, Spock had a tendency to use sticks of the appropriate length, whereas Sophie mostly used sticks that were too short (Figure 4). However, from day 4,

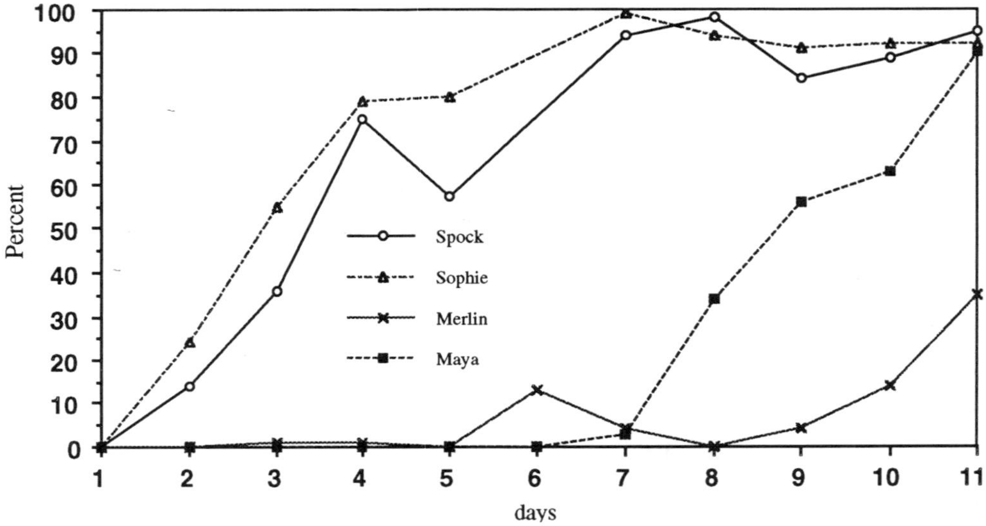

Figure 2. Percent of stick insertions that led to obtaining honey in the course of 11 days of exposure. (Note: Days 1-6 are successive, 12/9 to 12/14/85; Day 6, only Merlin and Maya present; Day 7 is 1/29/86; Days 8-11 are 5/7 to 5/10/86.)

both mostly used the technique of complete insertion with sticks of the correct length, which considerably increased their rate of success. The occurrence of complete insertions either down to the tip of the sticks or to the tips of the fingers with sticks that were too short reveals that the two chimpanzees did not yet understand the operation. The persistence of surface manipulations during days 4 and 5 may be interpreted as errors or as means of recovering the honey accumulated around the holes. In fact, such manipulations were generally made at the end of sessions, when few sticks of appropriate length were available. These rates of recovery were 14% and 5% for Spock and Sophie, respectively. It is noteworthy that these decreased to less than 2% of manipulations between days 7 and 11.

The technique of inserting a stick several times before licking the tip was used by Sophie as soon as day 2. During days 2 and 3, 67% (24/36) of such manipulations were successful, and 22% (8/36) were done at the surface with short sticks.

Spock used the technique of successive dipping only from day 3 on: 45% of such manipulations were successful, and 32% were made on the surface, mostly with sticks of the correct length. From day 4 on, this technique was exclusively used by Spock and Sophie with insertions down to the tip of the sticks. However, respectively 6% (n =97 for Spock) and 21% (n =86 for Sophie) of these insertions were still made with short sticks during days 4 and 5, as opposed to 0% (n =38) and 2% (n =303) during days 7 and 11. This technique was gradually abandoned by Spock, but not by Sophie: during days 10 and 11, it was respectively used in 2% and 28% of their successful trials.

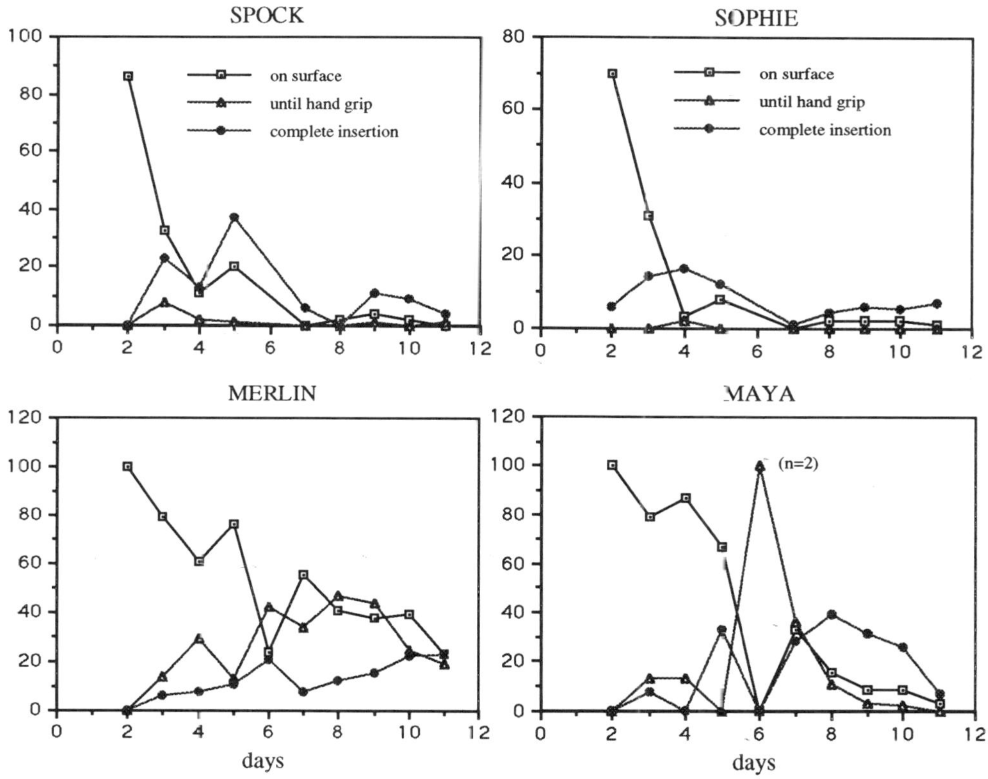

Figure 3. Percent of three levels of stick insertions without gathering honey in the course of 11 days of exposure. (Note: Days 1-6 are successive, 12/9 to 12/14/85; Day 6, only Merlin and Maya present; Day 7 is 1/29/86; Days 8-11 are 5/7 to 5/10/86.)

During days 3 and 4, 81% (*n* =94) of Spock's trials were done with wide sticks in hole R3. Sophie preferred using wide sticks also, but equally so in holes R4 and R3 (two lower holes on the right side). On the other hand, during day 4, Spock and Sophie both prefered hole R4 (77% for Spock and 67% for Sophie), mainly with wide sticks (62% and 46%), but also with narrow sticks (15% and 21%). During day 5, 79% of Spock's trials were made in hole R3 with wide sticks, whereas 66% of Sophie's trials were made in the same hole but with narrow sticks. This return to hole R3 was due to the fact that hole R4 was blocked by brocken sticks. From days 7 to 11, Spock fished in holes R3 and R4, but almost always with wide sticks (88% of times). Sophie, on the other hand, continued to show a preference (50%) for narrow sticks in R3 (n =1676).

3.2.2. *Merlin and Maya.* Figure 2 shows that Merlin and Maya learned the fishing skills only during the second intensive period of exposure to the puzzle-box.

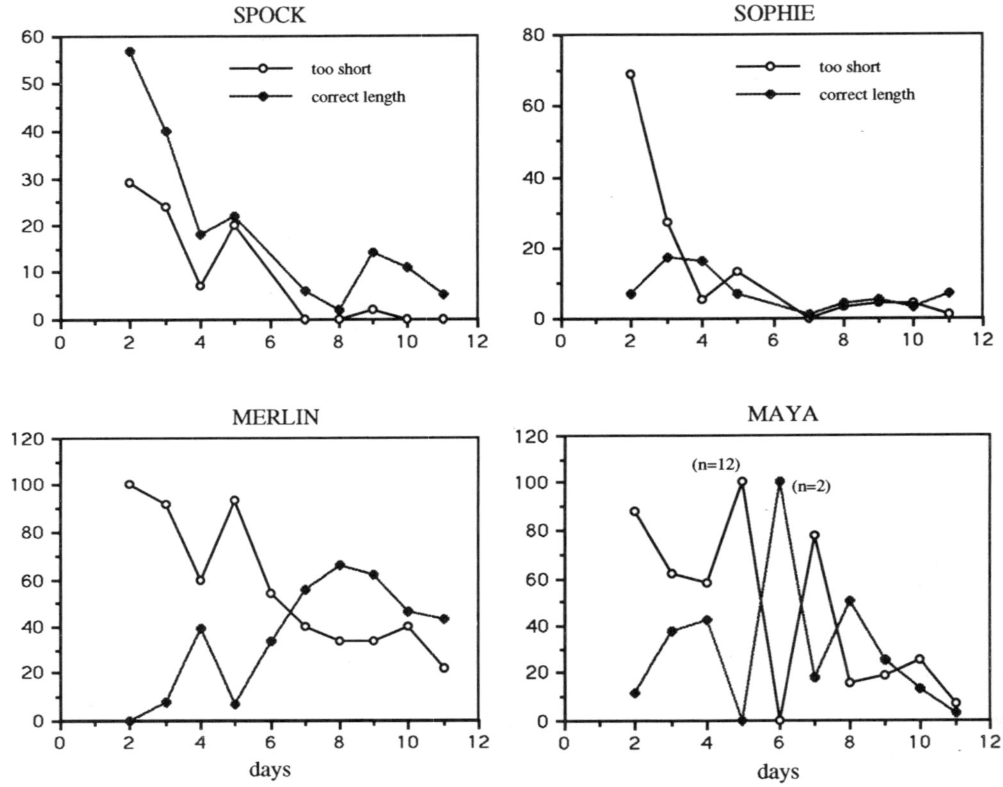

Figure 4. Percent of insertions of sticks too short and sticks of correct length, both without gathering honey in the course of 11 days of exposure. (Note: Days 1-6 are successive, 12/9 to 12/14/85; Day 6, only Merlin and Maya present; Day 7 is 1/29/86; Days 8-11 are 5/7 to 5/10/86.)

From day 2 to day 5, Merlin and Maya essentially used short sticks (Figure 4) at the surface of the holes (Figure 3). During days 3 and 4, they used sticks of the correct length more often than before (Figure 4), but generally dipped them only as far as the position of the hand grip allowed it (hand level dipping) or used them to scrape the edge of the holes (Figure 3).

The data reveal that Maya had fewer and decreasing chances to manipulate the sticks on the box. Merlin, however, had access to the device, but seemed to specialize in retrieving the honey at the edge of the holes right after Spock or Sophie had fished. In fact, he frequently licked the edge of the holes.

Merlin shook the sticks laterally in the holes with a relatively high frequency between days 2 and 5, as opposed to Spock and Sophie who only used this technique occasionally to retrieve honey around the holes. In fact, 100% (n =35) of shakings were done at the edge of the holes during day 2, 37% (n =98) were done with the hand level dipping on day 4, and 57% (n =14) with complete insertion but without collecting honey on day 5. It must be noted that the frequency of this technique decreased with time.

The technique consisting in retrieving the stick slowly was also widely used by Merlin. Whereas this technique was almost always associated with successfully dipping the stick to the bottom for Spock (16/17) and Sophie (11/13) between days 2 and 4, Merlin used this technique with sticks too short 73% of the time, and essentially at the edge of holes (Figure 5). The slow retrieval was mainly associated with sticks of the correct length on days 8 and 9, but more often with the hand level dipping. Only from day 10 were these retrievals associated with complete and successful insertions (Figure 5).

Maya used slow retrievals only from day 7, essentially with short sticks dipped until the hand grip (Figure 5). But from day 8, Maya used this technique with complete insertions of sticks of the correct length (Figure 5).

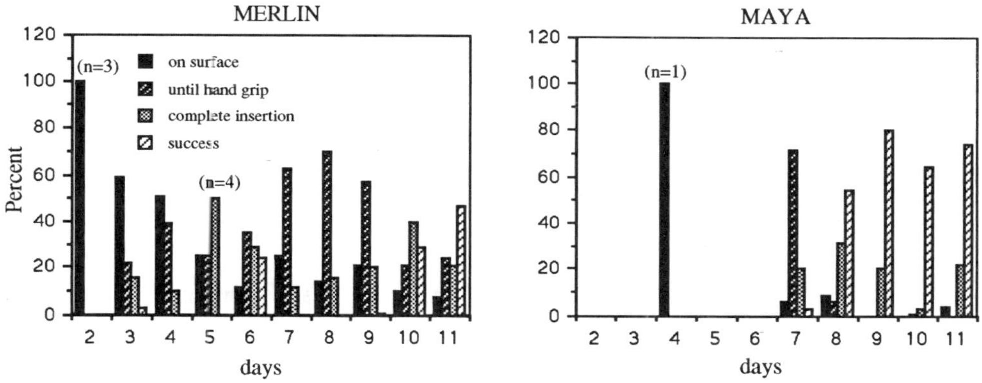

Figure 5. Percent of slow retrievals of sticks according to three levels of insertion without gathering honey and according to insertions with success in the course of 11 days of exposure. (Note: Days 1-6 are successive, 12/9 to 12/14/85; Day 6, only Merlin and Maya present; Day 7 is 1/29/86; Days 8-11 are 5/7 to 5/10/86.)

At the beginning (days 2 to 4), Merlin and Maya mainly associated successive insertions using short sticks with surface fishing (61% and 50%, respectively). On day 10, Merlin associated this technique with the appropriate sticks but generally inserted them until the hand grip. Only on day 11 were 51% of his successive insertions complete and therefore successful.

On day 7, Maya mostly used this technique of successive dipping with the hand level (49% of times) and completely (33%), but almost exclusively with short sticks. However, from day 8, she used this technique with the suitable sticks (89%) and in complete insertions.

On day 2, 45% of Merlin's manipulations were done at hole R3 with narrow sticks. But from day 3 and until day 11, he showed a clear preference (84%) for wide sticks, in hole R3 (69%). Maya also showed a slight tendency for fishing in R3 with wide sticks until day 7. However on days 8 and 9, she clearly prefered wide sticks in holes R4 as well as R3. On day 10, this preference changed for narrow sticks: in R3 during day 10, and in R4 on day 11.

3.3. SOCIAL CONTEXT

Why did Merlin and Maya learn the fishing techniques later, although they both had the opportunity to observe Spock' and Sophie's successes? Could the fact that they did not acquire the fishing skills during the first intensive period of exposure be associated with their low dominance ranks in relation to Spock and Sophie?

An attention index was computed by dividing the frequency of successes and failures observed (respectively by Merlin and Maya) by the total frequency of trials made by other chimpanzees in presence of Merlin and Maya. Figure 6 shows that in December 1985 (days 2 to 5), Merlin was not very attentive. In fact, he generally looked at a hole and rushed to lick it as soon as another chimpanzee removed his or her stick, instead of paying attention to the actions of Spock or Sophie. Similar behaviors were observed by Sumita, Kitahara-Frisch and Norikoshi (1985) in unsuccessful chimpanzees using a stone-tool to break open nuts. Being inattentive, Merlin selected the wrong information and thus experimented on the basis of wrong clues leading to inappropriate techniques, such as the too frequent use of short sticks, shaking the stick laterally in the hole, slowly removing the stick, or dipping it until the hand holding the stick. It is possible that his lack of attention was due to a particular socio-affective situation. Indeed, Merlin and Sophie were in competition for the dominant rank during this period, in a context where Spock often interfered to help Sophie against Merlin. In fact, when Spock and Sophie were absent (day 6), Merlin was much more successful: the hourly rate of success increased from 0.33 to 3.0 (Figure 2). This rate dropped when Spock and Sophie returned.

The case of Maya (the lowest ranking) is different but nonetheless dominance-related. She was very attentive at first, but her attention gradually decreased (Figure 6). Maya had the opportunity to approach the box and observe the others, but could only use the tools when no one was near, because Sophie and Merlin frequently took the sticks away from her or chased

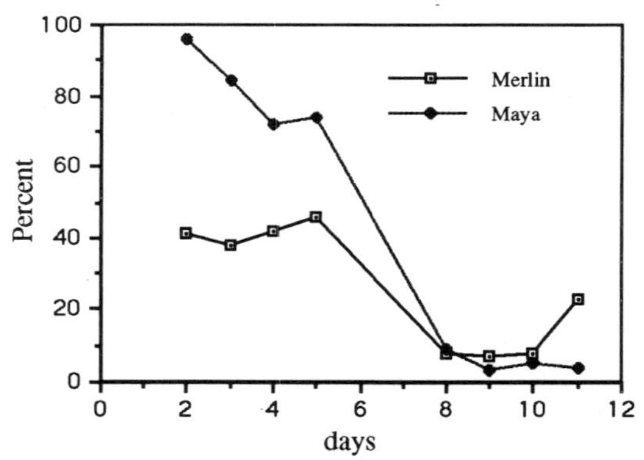

Figure 6. Percent of stick insertions by other chimpanzees that were watched by Merlin and Maya in the course of 11 days of exposure. (Note: Days 1-6 are successive, 12/9 to 12/14/85; Day 6, only Merlin and Maya present; Day 7 is 1/29/86; Days 8-11 are 5/7 to 5/10/86.)

her outright. This decreased her interest in the device although she could lick the honey remaining at the edge of the holes. Menzel (1972) observed the same situation in ladder-using by chimpanzees: the dominant male chased all animals other than his two closest companions who incidentally were the first, after him, to acquire the skill.

During the last intensive exposure to the device (day 8 to 11), more than three months after the preceding one, Merlin and Maya acquired the fishing skill, although they had not been very attentive to the others' manipulations (Figure 6). The social context had changed: Merlin was more self-confident and increasingly threatened Spock and Sophie. Although Merlin only charged at Sophie and Spock simultaneously once in 40 hours during December 1985, he chased them from the mound 12 times in 40 hours in May 1986. Taking advantage of her existing alliance with Merlin, Maya also acquired more self-confidence and was then able to manipulate sticks in the box more often. When an animal took a stick away from her, Maya would find another one and continue to fish instead of stopping as she did before. She also groomed the others in order to have access to the box. At the onset of the first day of this last period of exposure (day 8), Maya associated the «technique of complete dipping» with «sticks of the correct length», while also using successive insertions and slow retrievals. This drastic change in performance suggests that she had made some associations during her few previous successes from day 7.

In May, Maya had her sticks taken away from her 16 times more often than in December, especially by Merlin (69%) and Spock (27%). Merlin acted this away towards Maya especially on day 9, when she had more than a 50% rate of success. From that day on, I observed that Maya got rid of her long stick on the ground whenever Merlin got close to the puzzle-box, she then used a short stick in his presence, and resumed fishing with the long stick when he left. It is noteworthy that Merlin seldom took the short sticks away from her.

4. Discussion

4.1. DISCOVERY

The discovery of the use of sticks in fishing honey was here accidental, as was the case in the acquisition of a novel form of tool use by captive hamadryas baboons, guinea baboons, pigtailed macaques (Beck, 1978), and capuchin monkeys (Visalberghi & Trinca, 1989). Moreover, this discovery was coactive in the sense that many chimpanzees provided elements of the solution. First, Sophie detected by smelling the presence of something in the box and seems to have communicated it to the others, at least to Spock judging by his immediate exploration of the puzzle-box after Sophie's screams. The experiments of Menzel (1971) suggested that chimpanzees might be able to communicate the presence, direction, quality, and relative quantity or preference value of distant hidden objects. Sophie was also the first to insert a stick in the box. This action seems totally fortuitous since Sophie did not withdraw the stick, and also since all individuals in the group had previously inserted sticks in various apertures all around their exhibit during the exploratory manipulation of sticks.

Spock was the first to withdraw a stick from the puzzle-box. Maya was then the first to experiment with sticks in various holes of the box; Spock and Sophie observed and successively imitated her. But until then, all trials were done with short sticks. It is the fortuitous insertion of another long stick, abandoned by Sophie and retrieved again by Spock, that allowed the latter to associate «long sticks» with «access to honey» since, shortly after, Spock made the complete sequence allowing him to fish honey for the first time.

Two hypothesis have been put forward to explain the cross-cultural variations in termite fishing by wild chimpanzees (McGrew *et al.*, 1979): 1- long ago, only some individual chimpanzees discovered and then transmitted this technique to others, who afterwards modified it to local conditions; and 2- chimpanzees of different populations independently discovered tool-use techniques for obtaining termites. Considering that the complex task of tool-use was mastered within a few days by a group of naive captive chimpanzees, it is more likely that the fishing behavior of different populations came from separate inventions, and that traditions independently developed. The discovery could very well be reproduced at every generation, since similar dietary requirements in different environments could have stimulated chimpanzees to develop similar techniques of tool use (Yamagiwa, Yumoto, Ndunda & Maruhashi, 1988).

4.2. LEARNING

No doubt local enhancement was implicated in the discovery process by the three other chimpanzees of the group, as is the case in monkeys (Whiten, 1989). Then, the learning of a task as complex as tool use for fishing honey was here essentially made by trial-and-error. For example, all four chimpanzees gradually associated the technique of successive dipping (allowing to bend the stick and then to gather more honey) with the use of long sticks and with the technique of inserting the stick completely. Spock and Sophie's reciprocal observation of one another allowed them to orient their experimentation by the selection of one or several clues. Spock had the opportunity to readily associate «long stick» with «access to honey» since he licked the honey off a long stick inserted by Sophie. After observing Spock, Sophie dipped a long stick of the same color and in the same hole as that used by Spock and obtained her first success. It is already known that wild chimpanzees use fishing holes abandoned by others (McGrew & Collins, 1985), and the sight of a chimpanzee opening an ant nest immediately attracts others who will try to use the same hole (Boesch & Boesch, 1990). However, the association between useful sticks, the techniques and the access to honey were not yet complete, since Sophie often used short sticks during the first days.

Merlin chose appropriate sticks, not after watching Maya fish, but rather after taking away from her and licking the sticks with which she was successful. This nonetheless required some attention from Merlin to the results of Maya's trials.

Altogether, the observation of a peer and the selection of relevant clues could allow a chimpanzee to accelerate his learning by orienting his individual experimentation in a profitable direction. Maya's case suggests that, in order to be profitable, the information selected must be imitated or experimented a short time after observation of the demonstrator, which was impossible for her during the first period of exposure.

The chimpanzees were not able to imitate the demonstrator's precise use of sticks for fishing honey, as was the case in the study of Tomasello *et al.* (1987) with the use of a T-bar to have access to a food resource.These two tasks were probably too complex to be learned by true imitation. But we maybe must question the notion of imitation or observational learning. It could be normal that individuals periodically take information from the demonstrator to allow quick adjustment of the technique to the task, rightly to avoid the copying of wrong acts.

Tomasello *et al.* (1987) suggested that it is not just the tool as an object that is made more salient (local enhancement), but it is the object in its function as a tool that is made more salient. Accordingly, I believe that observation would allow the chimpanzee to understand the relation between the tool and the goal, and to remember a clue from which he may discover the technique by trial-and-error. The ability of chimpanzees to select particular objects as tools, to modify them appropriately and beforehand (see Boesch & Boesch, 1990), and finally the use of a few types of tools in sequence, each performing a different function (see Brewer & McGrew, 1990) give evidence that chimpanzees do have some understanding of causality (see Tomasello *et al.*, 1987). Young chimpanzees do not become skilled in fishing until the age of five or six (van Lawick Goodall, 1973), and ladder-use first occurs when they are six to seven years old (Beck, 1978), the age at which they complete the sixth sensorimotor intelligence stage in the Piagetian perspective (see Chevalier-Skolnikoff, 1989).

The recurrence of certain errors made at the beginning of the experiment shows that chimpanzees needed some time to fully master the technique. It is noteworthy that the chimpanzees always fished from the same two holes, without generalizing their experimentation to the other two holes of same diameter on the left side of the box, nor to the four narrow holes placed at two different heights. However, it must be stressed that it was indeed easier to fish from larger lower holes because two out of three lengths of sticks and sticks of both diameters could be used in them indiscriminantly. Only towards the end of their learning period did the chimpanzees begin using narrow sticks as well as wider ones, in both wide holes of the right side. This partial generalization shows that the chimpanzees then understood the technique: indeed, Spock and Sophie had a 90% success rate, even after a prolonged period of non-exposure.

I am certain that my laboratory task was far simple than the fishing of wild chimpanzees because, among other things, my chimpanzees didn't have to adapt the tool to the task (i.e., strip off leaves), or be distracted from the task by termite bites or bee stings. Moreover, they could choose the appropriate stick according of his length, width and suppleness. In spite of everything, my results show that chimpanzees learned tool use through trial-and-error, as monkeys do (Beck, 1975). Trial-and-error learning appears to be necessary for integration and coordination of the elements into efficient sequences (see Beck, 1978). The difference between monkeys and chimpanzees lies in the fact that the former failed to show evidence of imitation (Visalberghi & Fragaszy, 1990; Whiten, 1989). Whereas only local enhancement is involved in monkeys (Galef, 1988), chimpanzees appear to understand the function of the tool, which makes them pay more attention to the relations between the actions of the demonstrator, the type of tool and the task to be accomplished. Capuchin monkeys made a variety of errors which seemed inconsistent with an understanding of the task (Visalberghi & Trinca, 1989). Moreover, they often discarded the appropriate tools and tried to use inappropriate ones, and instead of modifying the tool beforehand, they mainly modified it when their action was unsuccessful.

Finally, the delayed acquisition of the tool use behavior by Merlin and Maya shows clearly the extent to which social factors determine the performance of individuals within a group, especially those at the lower end of the dominance hierarchy. On the other hand, Merlin's motivation to fish and his frustration in not being able to do so appear to have been at the origin of the rank-reversal of Spock by Merlin (see Paquette, 1990). Maki, Alford, Bloomsmith and Franklin (1989) had already underlined the fact that competition for tool use may induce aggression in some users or discourage others.

ACKNOWLEDGMENTS

I thank Louis Lefebvre for his helpful comments on the manuscript, Suzanne Lampron St-Germain for the text editing, Robert Morin and Jacques Prescott for their friendly support, and also the Jardin zoologique du Québec for constructing the artificial hive. This research was supported by grants from the Fonds F.C.A.R. (#86-AR-0245) de la Province du Québec, Canada. Correspondence should be addressed to Daniel Paquette, Ecole de Psycho-Education, Université de Montréal, 750 Gouin E., Montréal, Québec, H2C 1A6, Canada.

REFERENCES

Beck, B.B. (1974). Baboons, chimpanzees and tools. *Journal of human evolution 3*, 509-516, .

Beck, B.B. (1975). Primate tool behavior. In R.H. Tuttle (Ed.), *Socioecology of primates*, (pp. 413-447). Paris: Mouton Publishers, The Hague.

Beck, B.B. (1978). Ontogeny of tool use by nonhuman animals. In G.M. Burghart and M. Bekoff (Eds), *The development of behavior*, (chap. 20). New York and London: Garland STPM Press.

Beck, B.B. (1980). Animal tool behavior, New York: Gartland Press.

Boesch, C. and Boesch, H. (1990). Tool use and tool making in wild chimpanzees. *Folia Primatol. 54*, 86-99.

Brewer, S.M. and McGrew, W.C. (1990). Chimpanzee use of a tool-set to get honey. *Folia Primatol. 54*, 100-104.

Chèvalier-Skolnikoff, S. (1989). Spontaneous tool use and sensorimotor intelligence in Cebus compared with other monkeys and apes. *Behavioral and brain sciences 12*, 561-627.

Galef, B.G. Jr. (1986). Tradition and social learning in animals. In R.J. Hoage and L. Goldman (Eds), *Animal Intelligence: insights into the animal mind*, (pp.149-163). Washington and London: Smithsonian Institution Press.

Galef, B.G. Jr. (1988). Imitation in animals: history, definitions, and interpretation of data from the psychological laboratory. In T. Zentall & B.G. Galef (Eds.), *Psychological and biological perspectives*, (pp. 1-28). Hillsdale, NJ: Erlbaum.

Hall, K.R.L. (1963). Observational learning in monkeys and apes. *Brit. J. Psychol. 54*(3), 201-226.

Hogan, D.E. (1988). Learned imitation by pigeons. In T. Zentall & B.G. Galef (Eds.), *Psychological and biological perspectives*, (pp. 225-238). Hillsdale, NJ: Erlbaum.

Kummer H. and J. Goodall (1985). Conditions of innovative behaviour in primates. In L. Weiskrantz (Ed), *Animal intelligence, Oxford psychology series (7)*, (pp. 203-214). Oxford: Clarendon Press.

Maki, S.. Alford, P.L., Bloomsmith, M.A. and Franklin, J. (1989) Food puzzle device simulating termite fishing for captive Chimpanzees (Pan troglodytes). *American Journal of Primatology Supplement 1*, 71-78.

McGrew, W.C. (1977). Socialization and object manipulation of wild chimpanzees. In S. Chevalier-Skolnikoff and F.E. Poirier (Eds), *Primate bio-social development: biological, social and ecological determinants*, (pp. 261-288). New York and London: Garland Publishing Inc.

McGrew, W.C. and Collins, D.A. (1985). Tool use by wild chimpanzees (Pan troglodytes) to obtain termites (Macrotermes herus) in the Mahale Mountains, Tanzania. *American Journal of Primatology 9*, 47-62.

McGrew, W.C. & Rogers, M.E. (1983). Chimpanzees, tools, and termites: new record from Gabon. *American Journal of Primatology 5*, 171-174.

McGrew, W.C., Tutin, C.E.G. & Baldwin, P.J. (1979). Chimpanzees, tools, and termites: cross-cultural comparisons of Senegal, Tanzania and Rio Muni. *Man 14*, 185-214.

Meltzoff, A.N. (1988). The human infant as «homo imitans». In T. Zentall & B.G. Galef (Eds.), *Psychological and biological perspectives*, (pp. 319-341). Hillsdale, NJ: Erlbaum.

Menzel, E.W. Jr. (1963). The effects of cumulative experience in responses to novel objects in young isolation-reared chimpanzees. *Behaviour 21*, 1-12.

Menzel, E.W. Jr. (1971). Communication about the environment in a group of young chimpanzees. *Folia Primat. 15*, 220-232.

Menzel, E.W. Jr. (1972). Spontaneous invention of ladders in a group of young chimpanzees. *Folia primat. 17*, 87-106.

Menzel, E.W. Jr., Davenport, R.K. and Rogers, Ch. M. (1970). The development of tool using in wild-born and restriction-reared chimpnzees. *Folia primat. 12*, 273-283.

Nishida T. (1987). Local traditions and cultural transmission. In B.B. Smuts, D.L. Cheney, R.M. Seyfarth, R.W. Wrangham & T.T. Struhsaker (Eds), *Primates societies*, (pp. 462-474). Chicago and London: University of Chicago Press.

Paquette, D. (1980). *Quelques aspects de l'ontogénie du comportement d'exploration chez deux chimpanzés (Pan troglodytes)*. Mémoire de maîtrise, Université de Montréal.

Paquette, D. (1990). *La dynamique des renversements du rang de dominance dans un groupe de chimpanzés adolescents en captivité*. PhD thesis, Université de Montréal.

Paquette, D. and Prescott, J. (1988). Use of novel objects to enhance environments of captive chimpanzees. *Zoo Biology 7*, 15-23.

Sumita, K., Kitahara-Frisch, J. & Norikoshi, K. (1985). The acquisition of stone-tool use in captive chimpanzees. *Primates 26*(2), 168-181.

Thorpe, W.H. (1963). *Learning and instincts in animals,* Methuen, London.

Tomasello, M., Davis-Dasilva, M. and Camak, L. (1987). Observational learning of tool-use by young chimpanzees. *Human evolution* 2(2), 175-183.

van Lawick-Goodall, J. (1973) Cultural elements in a chimpanzee community. In E. Menzel (Ed), *Precultural primate behavior,* (pp. 144-184), New York: Karger.

Visalberghi, E. and Trinca, L. (1989). Tool use in captive monkeys: distinguishing between performing and understanding. *Primates 30*(4), 511-521.

Visalberghi, E. and Fragaszy, D.M. (1990). Do monkeys ape? In S.T. Parker and K.R. Gibson (Eds), *«Language» and intelligence in monkeys and apes,* (pp. 247-273), Cambridge University Press.

Whiten, A. (1989). Transmission mechanisms in primate cultural evolution. *Trends in ecology and evolution* 4(3), 61-62.

Yamagiwa, J., Yumoto, J. T., Ndunda, M. & Maruhashi, T. (1988). Evidence of tool-use by chimpanzees (Pan troglodytes schweinfurthii) for digging out a bee-nest in the Kahuzi-Biega National Park, Zaïre. *Primates 29* (3), 405-411.

SOCIAL TRANSMISSION OF STIMULUS RECOGNITION BY BIRDS, FISH, AND MOLLUSCS

MILTON D. SUBOSKI
Department of Psychology
Queen's University
Kingston, Ontario
Canada K7L 3N6

ABSTRACT: An ethologically oriented learning model, termed "releaser-induced recognition learning", was developed from research on remarkable pecking behavior of hatchling chicks. Chicks that watch an angular stimulus make "pecking" movements toward small food-like objects of a particular color will, when tested for choice, show an enhanced preference for pecking at objects that match the "pecked" object in color. Further research eliminated perceptual artifacts and other nonlearning interpretations for the enhanced preference. Instrumental and Pavlovian conditioning were each analyzed and found to be inconsistent with the phenomenon. The social learning paradigm was also found to provide only inadequate explanations. The model that does work starts with the extension-of-control mechanism of stimulus substitution theory combined with the releasing-stimulus -- released-response mechanism of classical ethology and assumptions of intrinsic taxic direction of released responses and nontemporal, particularly spatial, learning relationships. Recent research on cultural transmission of predator recognition by minnows and social transmission of food location by pond snails suggests that the model has the ability to generate interesting research and predict unlikely research findings.

KEYWORDS: Alarm pheromone/Cultural transmission/Imitation/Instrumental conditioning/Local enhancement/Pavlovian conditioning/Released responses/Releasing stimulus/Social facilitation/Social learning.

ACKNOWLEDGEMENTS: Support for preparation of this paper and for the author's research that is described here was provided by Grant A-OGP0007059 from the Natural Sciences and Engineering Research Council of Canada. I thank the Organizing Committee of the NATO/Advanced Study Institute on the Ethological Roots of Culture, particularly R.A. Gardner and B.T. Gardner. This paper was prepared while the author was on sabbatical leave from Queen's University at Oxford University. I thank Darwin Muir for carefully reading this manuscript.

1. Introduction

The purpose of this exposition is to introduce and trace the development of the *releaser-induced recognition learning model* and then show how the model works to produce cultural transmission of stimulus recognition in chickens, fish, and snails. Briefly, releaser-induced recognition learning is a simple and general learning model appropriate to stimulus recognition learning by infrahumans (Suboski, 1988; 1989b; 1990; 1992a; 1992b; Suboski, Muir, & Hall, 1993). The model has the extension-of-control version of the Pavlovian stimulus substitution principle as one unifying mechanism combined with the classical ethology concept of releasing-stimulus--released-response interactions.

R. A. Gardner et al. (eds.), The Ethological Roots of Culture, 173–198.

1.1. LEARNING

The releaser-induced learning model was developed in the context of the disarray in animal learning research and theory of the past two decades. Because of its importance, therefore, that context is briefly characterized before further discussion of the model. The traditional conceptualization of learning is in terms of a number of separate paradigms. Two of these, instrumental (operant) conditioning and Pavlovian (classical) conditioning, are discussed here. The social learning paradigm (observational learning/imitation) is presented later. Together with imprinting (perceptual learning), discussed later and elsewhere (Suboski, 1989b), these make up a set of basic learning paradigms. Each of these terms has two different referents. First, each refers to a procedure, the operations that comprise the necessary and sufficient antecedent conditions that produce behavior change. Second, each term designates a class of behavioral phenomena, the outcomes of imposing that procedure or operation. Together, each procedure and outcome combines to constitute a paradigm. An adequate descriptive paradigm specifies the operations necessary and sufficient to produce all of the set of predesignated outcome behavior and no behaviors other than the outcome behavior. Similarly, a paradigm is inadequate when the specified operations fail to produce paradigmatic phenomena, when they produce nonparadigmatic phenomena, and also when outcome behavior is produced by nonparadigmatic procedures. A completely successful paradigm specifies general procedures that uniquely generate a broad class of homogeneous behavioral events and provides a descriptive analysis to included behavioral phenomena (cf. Shettleworth, 1983, pp. 5-7).

1.2. LEARNING: EXPLANATORY MODELS

At the present time none of these four paradigms can be considered to be completely successful. Despite a long history of learning research and theory centered on analysis of paradigms, there seems to be widespread opinion among investigators of animal learning that the study of paradigms has not provided adequate explanatory models. The issue is whether specifying the "rules" or antecedent conditions that generate learned behavior provides an adequate explanation of that behavior. Guthrie (1935) to the affirmative stated that scientific "explanation of any event consists in stating the rule or generalization of which the event is an instance" (p. 12). This view, according to Bunge (1985), remains current: "The dominant view on scientific explanation among philosophers is that it is identical with subsumption under laws" (p. 155). From this position, the antecedent conditions of an adequate descriptive paradigm should be considered as subsuming and thus explaining the behavioral events generated within the paradigm. However, Dodwell (1985, p. 149) observed and Bunge (1985, p. 155) concurred that "explanation in psychology is not a well understood and agreed upon matter." Nevertheless, Bunge (1985) considered subsumption to be descriptive rather than explanatory, asserting that "an explanation proper involves some *mechanism* or other" (p. 155). Thus, satisfactory explanation seems to require explication of the mechanisms by which antecedent conditions produce learned outcome behavior. A distinction, then, is made in the present discussion between descriptive paradigm (specifying procedure and outcome behavior) and explanatory model (specifying procedure, mechanism, and outcome behavior). Before turning to a discussion of the releaser-induced recognition learning model, each of the conditioning paradigms will be examined for its descriptive and explanatory adequacy.

1.3. LEARNING: PARADIGMS

1.3.1. *Instrumental conditioning*. Instrumental conditioning is foremost a paradigm descriptive of response acquisition and modification rather than for stimulus recognition learning, the major concern of the present review, and thus is only briefly considered. The antecedent conditions for instrumental conditioning specify that elements of an arbitrarily chosen response class be selectively followed in time by a reinforcer. The behavioral outcome is usually well defined: Frequency of occurrence of responses in the class either increases or decreases depending on whether the reinforcer is an appetitive or aversive stimulus. Thus, a change in the probability of a response when that response is temporally juxtaposed with a reinforcing stimulus has been considered as virtually diagnostic of instrumental conditioning.

Nevertheless, instrumental conditioning is not a completely adequate descriptive paradigm. First, conditioning operations often fail to produce conditioning effects. Instances of behavior classes resistant or impervious to the effects of reinforcement are not rare (see, e.g., Shettleworth, 1983; Gilbertson, 1975, p. 737). Second, conditioning operations often produce nonparadigmatic outcomes, that is, changes in the probability of responses in classes other than those reinforced. Examples include the intrusion of unreinforced consummatory behaviors during instrumental conditioning (e.g., Boakes, Poli, Lockwood, & Goodale, 1978; Breland & Breland, 1961), the feature positive effect (see Hearst, 1978), and the emergence of temporally noncontingent, but reinforcer-specific consummatory responses directed to manipulanda (e.g., Davey, Phillips, & Cleland, 1981; Moore, 1971; 1973; Wolin, 1968). Third, paradigmatic behavior occurs to nonparadigmatic operations, as shown by autoshaping. Many of the responses commonly chosen to be operantly conditioned have been found to occur readily to stimulus-contingent rather than response-contingent presentation of reinforcers. Finally, when placed in direct opposition to each other, paradigmatic procedures of response-contingent reinforcement are sometimes inferior to nonparadigmatic stimulus-contingent procedures in controlling responses. This outcome is clearly shown in the appetitive omission effect (e.g., Peden, Browne, & Hearst, 1977).

In view of the preceding evaluation, the conclusion seems inevitable that instrumental conditioning is not yet an adequate exploratory model. The specifications for antecedent conditions and/or outcome behavior remain incomplete. One possibility is that identification of a unifying mechanism could clarify the paradigmatic requirements of instrumental conditioning. At present, the widely accepted mechanism is the traditional "law of effect", the tenet that response occurrence is governed by reinforcement resulting from the consequences of such responses. However, considering the problems already noted, it appears that response modification by reinforcement is not the unifying mechanism that is required to make instrumental conditioning an explanatory model. Another possibility is an alternative to the law of effect recently outlined by Gardner and Gardner (1988). This interpretation, based on the ethological concept of releasing stimuli rather than reinforcers, may integrate Pavlovian and instrumental conditioning.

1.3.2. *Pavlovian conditioning*. Pavlovian conditioning is the paradigm most frequently invoked for recognition learning phenomena. In terms of procedures, there are three elements in Pavlovian conditioning: 1) The requirement for a behaviorally-inert stimulus, the conditional stimulus (CS), 2) The requirement for an unconditional stimulus (US) that reliably produces some behavior change, the unconditional response (UR), and 3) The requirement for some temporal relationship between these stimuli. The behavioral phenomenon that results from presenting, to an animal, the CS and US in particular temporal juxtapositions, is that the CS acquires ability to evoke conditional responses (CRs) that usually have topographical resemblance to the URs.

The problems with invoking Pavlovian conditioning as an explanatory model begin with the fact that no element of the paradigm is completely adequate. Temporal pairing of a suitable CS and US under adequate parameters of relevant conditioning variables usually produces overt CRs -- but not always. Noteworthy exceptions were reported by Bruner (1965) and Colavita (1965). Other investigations have shown that the context in which CS and US are paired can completely determine whether or not CRs will occur. Relevant findings include interference with conditioning by the presence of another CS, as in overshadowing and, particularly, blocking (Kamin, 1968; 1969). Similarly, conditioning may be degraded by presentation of additional USs that are not paired with the CS (e.g., Rescorla, 1968). Thus, the same number of CS-US pairings may or may not yield conditioning depending entirely on other contextual factors. As well, the equipotentiality principle, the implication that Pavlovian procedures will yield CRs to pairing of any combination of effective CS and US, remains highly suspect. Despite prolonged controversy, it seems distinctly unlikely that the formation of conditioned taste aversions is based on simple temporal relations of stimuli rather than on stimulus novelty.

Temporal contiguity between CS and US, as noted above, clearly does not suffice to produce CRs in any and every case and, in fact, is often unnecessary. Many conditioning theorists currently accept Rescorla's (1967) conclusion that the critical temporal arrangement is contingency or correlation between CS and US such that CS provides predictive information about US occurrence (Dickinson & Mackintosh, 1978; Mackintosh, 1977; Pearce & Hall, 1980; Rescorla, 1978; 1984; Rescorla & Holland, 1976). As a further complication, altered responding to particular stimuli can be obtained by predictive spatial and dynamic stimulus relationships that are clearly different from the temporal relationships specified by the Pavlovian paradigm (e.g., Suboski & Bartashunas, 1984).

The Pavlovian paradigm is no less complicated at the outcome end. Gormezano (e.g., 1984) has steadfastly maintained that Pavlovian CRs are properly restricted to components of the UR. Rescorla (1978, see also Rescorla, 1988) presents another extreme -- that "any behaviors which are differential as a result of exposure to a relation are potentially of interest" (p. 26). In between, various temporal arrangements of putative CSs and USs produce a heterogeneous welter of behaviors with disparate topographical relationships to URs, ranging from no apparent relationship at one limit to CRs that are identical to URs at the other. No particular principle for arranging temporal relationships between CS and US yields consistent categories of CRs. All in all, there is little reason to dispute Jenkin's (1984) conclusion that "we still do not have a principle that explains the form of conditioned behavior" (p. 106).

The difficulties in establishing an adequate Pavlovian paradigm, as discussed earlier, may be conceptually solved by the identification of mechanisms that relate behavioral outcomes to Pavlovian procedures. Not surprisingly, elucidation of such mechanisms has occupied considerable effort of conditioning theorists beginning with Pavlov (1927). Nevertheless, no procedural arrangement or nuance of theoretical interpretation has yet yielded a Pavlovian model that has escaped contradiction by experimental results.

One recent approach to the development of a Pavlovian explanatory model merits closer examination. Hollis (1982) proposed that a functional analysis could provide an explanation of Pavlovian conditioning. The analysis begins from the premise that learned behavior ultimately functions to provide an evolutionary advantage in terms of increased inclusive fitness (enhanced reproductive success). Hollis (1982) then reviewed extensive evidence to document her contention that "the biological function of classically conditioned responding ... is to enable the animal to *optimize interaction with the forthcoming biologically important event (US)"* (Hollis, 1982, p. 3. Italics in original). Clearly, the ultimate function of CRs is seen as essentially anticipatory in nature. Moreover, a behavioral mechanism that could produce optimized CR-US interactions is readily available. Known as the preparatory response hypothesis, the mechanism is a version of the strong law of effect and holds that anticipatory CRs produce immediate reinforcement when they beneficially modify the sensory effects of the US. According to the preparatory response hypothesis, CRs occur because they optimize immediate CR-US interactions. Thus, the preparatory response hypothesis proposes a mechanism that is congruent to the ultimate function of optimizing CR-US interactions.

Unfortunately, close examination of the preparatory response hypothesis reveals that it does not provide a mechanism by which CR-US interactions are optimized. The critical evidence comes from the appetitive omission studies (e.g., Gormezano & Hiller, 1972; Patten & Rudy, 1967; Peden, Browne, & Hearst, 1977; Sheffield, 1965) where CS--appetitive-US (e.g., food) presentations to an animal at first occur as in ordinary Pavlovian conditioning. However, as soon as the animal makes a CR to the CS, the omission contingency takes effect and the US is not delivered on that trial. Thus, a CR and a US can never occur on the same trial and CRs cannot be acquired as a result of a CR-US interaction because such interactions are not permitted to occur. The CRs made in omission studies are maladaptive in that the only effect of the CRs is the loss of USs but CRs are still readily acquired and maintained. The outcome is that the preparatory response hypothesis, the obvious mechanism for producing the CR-US optimization function, does not do so.

The cognitive theories that currently dominate Pavlovian conditioning emphasize learning of stimulus events and predictive relationships. The weakness of cognitive theories in generating specific behavior has been acknowledged. Rescorla (1978) argued that although "cognitive accounts fail to anticipate the particular behaviors which will result from learning, ... this supposed failing is not without its virtues" (p. 16). It seems quite unlikely, however, that any such strategic advantage in the development of learning theory can be permanent.

The status of Pavlovian conditioning as an explanatory model was evaluated by Gormezano and Kehoe (1975) who perceived "a prevalent tendency to regard any phenomenon deemed to be associative in nature as 'classical conditioning' and hence, to blur the distinction between a specific phenomenon and a theoretical mechanism" (p. 145) and concluded that "the phenomenon of conditioning itself was in need of explanation" (p. 146). Jenkins (1984, p. 90) agreed: "classical and operant conditioning are not viewed as fundamental and distinct forms of learning but as phenomena which themselves require a theoretical analysis".

Very recent critics have also noted the need for adequate mechanisms to explain Pavlovian conditioning. Fields (1989) wrote "classical conditioning phenomena...can only be explained by showing how they result from the actions of mechanisms" (p. 142). In a similar vein, Locurto (1989) commented "we have yet to develop a *theory* of classical conditioning. The sort of theory that is needed...speaks...to underlying mechanisms" (p. 153). Rescorla (1988) also agreed: "We are badly in need of an adequate theory of performance in classical conditioning" (p. 158).

The state of disorder in learning theory of the past several decades can be safely attributed to the failure of available paradigms, models, and mechanisms to provide theoretical explanations for learning phenomena or guidance for learning research. The central feature of the releaser-induced recognition learning model is an adequate general mechanism that underlies and organizes a diversity of learning procedures and phenomena. Gaining an appreciation of the model may be facilitated by a review of the research findings and insights that inspired the development of the model.

2. The Origins of the Releaser-Induced Recognition Learning Model

My Ph.D. thesis research was in the area of Pavlovian conditioning with I. Gormezano and this continued to be my major area of research interest for some time -- continuing, in fact, after I thought that I had abandoned it. I was pursuing what seemed to be a good idea then -- the response shaping version of the preparatory response hypothesis (Suboski, Brace, Jarrold, Teller, & Dieter, 1972; Teller, Dieter, & Suboski, 1972). Murray and Carruthers (1974) (and I, unpublished research) had found clear evidence that eyeblink CRs attenuated the felt aversiveness of an electric shock US -- which I then regarded as powerful evidence that the strong law of effect worked in subtle and mysterious ways. Just about then I became aware that the omission studies, particularly those using appetitive USs, had eliminated the law of effect as a contender for a general mechanism to explain Pavlovian conditioning.

At that point, I left Pavlovian conditioning and decided to try to look at animal learning from as far away from classical conditioning as possible and that seemed to be from the standpoint of observational learning. Sifting through the sparse literature back then filtered out several studies that combined adequate controls with positive results. One was an elegant study by Helmut Adler published in 1955 in the *Journal of Genetic Psychology* (not, one suspects, Adler's first choice of publication outlet). Adler's (1955) results were particularly impressive because observer cats completed all observation of trained or untrained demonstrator cats before any testing was begun. Nevertheless, observer cats that watched trained demonstrators perform a learned response showed large and significant differences over observers of untrained cats on the very first test trial.

In search of research direction, I then made a series of videotapes, the first of an empty Skinner box with the key light signalling food hopper presentations, another of a magazine-trained pigeon receiving the same stimulus presentations, a third of the pigeon after key-peck training, and the last of the pigeon after training on a Variable Interval schedule, pecking more or less continuously without being reinforced. Each of the videotapes was shown individually to a separate group of magazine-trained pigeons. Only the pigeons that had watched key-peck trained pigeons showed significantly elevated levels of key pecking when then tested by themselves in the Skinner box. Furthermore, those pigeons also showed accelerated key pecking when subsequently shown the video-tape of the continuously key-pecking pigeon. Learning by observation seemed to be not at all difficult, ephemeral, or exotic.

2.1. Turner's preparation: "imitation" by hatchling chicks

Nevertheless, obtaining of research direction waited until I happened across a 1964 paper by E.R.A. Turner titled "Social feeding in birds". Buried in the middle of this long paper was what appeared to be an elegant technique for studying imitation in newly hatched chicks. The possibility that neonatal chicks *(Gallus gallus)* learn by imitation made Turner's (1964) preparation very appealing. First, a cognitively complex behavior appeared to occur in neonatal members of a bird species not otherwise noted as possessing high levels of cognitive processing ability. If chicks really do imitate, then clearly the extent to which cognitive abilities occur in simpler animals requires general reconsideration. A second simplifying implication of learning by imitation in hatchling chicks is that any life-history origin for such learning is immediately ruled out.

Turner's (1964) preparation is shown in Figure 1. A crude model of a hen was hand operated to "peck" at a grain of one color and the chick could peck at a series of matching grains alternated with another color. Turner rather summarily reported that chicks reliably and selectively pecked at the grains that matched the color of the one being "pecked" by the model hen and then went on to other things.

Fig. 1. Turner's Fig. 16. The mechanical hen pecked at a green coloured grain and the chicks had the choice of green or orange grains. All grains were firmly glued down. With permission of E.J. Brill, Leiden, Publisher.

2.2. ANALYSES OF THE PHENOMENON AS CONVENTIONAL LEARNING

Our subsequent research (Bartashunas & Suboski, 1984; Suboski, 1984; 1987; 1989a; Suboski & Bartashunas, 1984) fully confirmed Turner's (1964) findings. Chicks preference for pecking at a food-like stimulus of one color was increased if the chicks had observed a pointed object making "pecking" movements at another object of the same color. This phenomenon of one stimulus enhancing an animal's preference for another stimulus is remarkably, even deceptively simple, considering that all of the propositions that make up the releaser-induced recognition learning model were derived from it. Because of its importance to the model, this research is next evaluated in an effort to ensure that the possibilities of simpler interpretation were exhausted. Presuming an ultimately positive conclusion, determining whether the "pecking" arrow-induced stimulus preference can be accommodated as "food imprinting" or within one of the two traditional conditioning paradigms, instrumental or classical conditioning, is undertaken. If such accommodation is not possible, the social learning paradigm will be explicated and examined for its ability to accommodate stimulus-induced pecking. Failing all else, the possibility will be considered that a highly specialized idiosyncratic learning mechanism is responsible.

When I (later in collaboration with Carl Bartashunas) started to work on Turner's preparation some 16 years after its publication, advances in technology provided the preparation shown in Figure 2 -- a "pecking" arrow (mounted vertically, all else viewed from overhead) operated by an electric motor rather than by hand, and plastic pin heads rather than colored grains. Various combinations of colored pin-head stimuli can be displayed behind the Plexiglas barrier where the chicks can't reach them. The stimuli usually consist of a pin of one color located on or beneath the tip of the arrow, with pin heads of another color laterally displaced away from the arrow. We have also used little tricolor LED lamps for the stimuli. The chick may have access to one stimulus of each color, termed the target and nontarget.

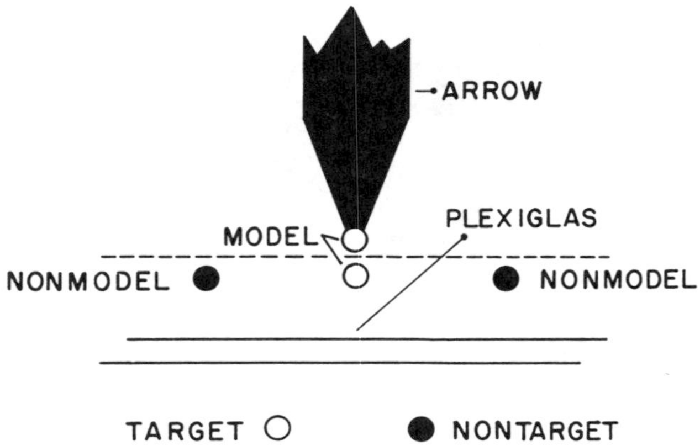

Fig. 2. General diagram of displays presented to hatchling chicks.

Chicks hatch overnight in a dark incubator with sound masked by ventilating fan noise and remain there until they are removed for experimentation without having been fed or exposed to a hen. Chicks are given 2 min habituation to the apparatus typically followed by a 20-trial training session. During each trial the arrow is motor driven to make 2.5-cm excursion vertical "pecking" movements in 10-s bursts at 1.5 - 3.0 "pecks" per second. The phenomenon of interest occurs when chicks are exposed to a display of the arrow "pecking" at a stimulus of one color (the model-target color) whereas stimuli of a second color (the nonmodel-nontarget color) are displaced laterally away from the arrow. When given an opportunity, the trained chicks peck significantly more to the target than they do to the nontarget.

2.2.1. *Counterbalancing for Color.* One of the most prominent behaviors of young chicks is directed pecking. Untrained chicks peck at virtually any small stimulus that contrasts with the background and are occasionally seen to peck at the air. One class of peck-inducing stimuli is small food-like objects that contrast with the background. From hatching, chicks show strong unlearned preferences among stimuli based on differences in color, size, shape, and movement of the peck-inducing stimuli. To eliminate effects of innate preferences as much as possible, we used either colored plastic pinheads or miniature LED lamps that differed in color only. Obviously, counterbalancing stimulus colors with model-target and nonmodel-nontarget locations as shown in Figure 2 is extremely important to control for the inevitable color preferences. Far more importantly, however, the effects of counterbalancing are critical because they reveal that it is the location in the display relative to the "pecking" arrow that determines which stimulus color becomes an enhanced elicitor of pecking. If chicks observe the arrow "pecking" at stimuli of one color placed in the model positions, while another color is in the nonmodel positions then those chicks will later tend to peck stimuli of the color that matches the model. If the positions of the colors in the display are reversed, then so is the color preference of the chicks. The effects of counterbalancing the stimulus colors, always conducted in our studies, revealed a strong preference for the arrow-"pecked" stimulus, a preference that was superimposed on color preferences.

2.2.2. *Controlling for Perceptual Artifacts* . We also always completely counterbalanced for position effects; for target right or left, and for nonmodel side if only one nonmodel was in the display. When position preferences were found (Suboski, 1989a), counterbalancing cancelled out their effect. A number of factors that could affect stimulus choice in the absence of learning were brought to light. We discovered early on that chicks are highly responsive to the relative density of food-like objects and preferentially peck at the stimulus with the lower relative density. With equal numbers of stimuli of each color, as is usual in our research, trained chicks selectively peck at the target. Another potential perceptual artifact was experimentally eliminated. In balanced displays such as is shown in Figure 2, the model can be identified either by its location with respect to the arrow or by its position at the center of the display. In principle, arrow operation is unnecessary for the chicks to determine which is the model color. If chicks tend to peck at objects in the middle of a group of peck-inducing objects unless they are physically prevented from doing so, then pecking may be redirected to stimuli that match the central stimulus in color. Quite unlikely, perhaps, and unsupported by the research findings, the hypothesis nevertheless generated some interesting data.

To test the "central matching" hypothesis, displays were arranged that contained one model, one nonmodel, a target and a nontarget (see Figure 4 in Suboski & Bartashunas, 1984). Unique to these displays, the target and nontarget were centered on the nonmodel stimulus rather than on the model. Now, when target side is counterbalanced, right or left, half of the resulting displays have model and target close together at one side of the display, while the other half of the displays have model and target far apart at opposite sides of the display. In neither type of display is the model stimulus at the center of the stimulus display and thus, by the "central matching" hypothesis, should not generate target-directed pecking by chicks. In fact, in several variations of these conditions, chicks showed a significant preference for pecking at the target. More interesting still, in those configurations in which the target and nonmodel were on the same side, chicks had to leave the model and move past the nontarget in order to peck at the target when it was at the opposite side of the display. The evidence indicates that the phenomenon is robust.

2.2.3. *Tests for Learning: Permanence and Acquisition Functions*. The obvious perceptual effects that might provide an explanation for arrow-directed pecking in the absence of learning mechanisms have been examined and eliminated. There are two other ways of evaluating whether or not experience based behavior change represents effects of learning. One way is to determine how persistent or permanent the change in behavior is. It is generally agreed that the effects of learning are relatively permanent. Unfortunately, there is no general agreement as to how long the effects of training must persist in order to be considered relatively permanent. Among vertebrates, learning is often undiminished after months or years. Chicks that had been given the standard 20-trial exposure to the display illustrated in Figure 2 were tested in a second session with just target and nontarget after all of the remaining stimuli had been removed from the display. Such chicks showed target-directed pecking in the test session that was equivalent to that elicited in the training session (Suboski & Bartashunas, 1984). Much longer persistence was reported by Bartashunas and Suboski (1984). Time between training and test did not reduce the acquired pecking preferences of chicks trained at 9 hr posthatch and tested at various times up to 72 hr posthatch. Whether or not persistence for three days of an experimentally-acquired behavior change is to be considered "permanent" is perhaps problematical. It certainly is not transient or ephemeral.

The second "litmus" test for learning is the presence of an acquisition function. Learning is expected to improve or increase in strength with additional training. Such effects were reported by Bartashunas and Suboski (1984). Also, Suboski and Bartashunas (1984) found that pecks to the target increased with training trials significantly more than did pecks to the nontarget. By this test as well, arrow-directed pecking qualifies as learning.

2.2.4. *Food Imprinting*. Since arrow-directed pecking appears to be an instance of learning, then learning paradigms and mechanisms should be where a proper explanation of arrow-directed pecking is to be found. One possible explanatory mechanism, the concept of food imprinting, can be summarily dismissed (see Bartashunas & Suboski, 1984). Food imprinting was offered (Hess, 1964) to account for primacy effects in the development of food preferences and an apparent "sensitive period" for the formation of food preferences in chicks. For the first few days after hatching, chicks peck indiscriminately at small objects but actually ingest very little and are unresponsive to the stimulus consequences of ingesting material (see, e.g., Hogan, 1977). The occurrence of food imprinting requires early and selective exposure to a particular foodstuff. In our research, chicks were nonselectively exposed to inedible objects, a procedure which cannot produce differential enhancement of responding to any particular stimulus. Our findings did share one general feature with imprinting: Only simple exposure to the stimuli was required to effect learned behavior change.

2.2.5. *Reinforcement*. The possibility that selective pecking by the chicks reflects reinforced learning by either instrumental or Pavlovian conditioning processes requires careful analysis. In fact, a "pecking" stimulus has many of the properties of a positive reinforcing stimulus. Chicks are strongly attracted to the "pecking" arrow. They approach the arrow, attempt to press up against it, and track the arrow movement with head and eyes.

The instrumental conditioning procedure is expected to produce changes in response probability as a result of selective temporal placement of a reinforcing stimulus with respect to that response. The frequency of a response is predictably altered by the response-contingent occurrence or removal of particular events, either positive and appetitive or negative and aversive. If movement of the arrow is a reinforcing stimulus, then the operation of instrumental conditioning in selective pecking by chicks should be easily detected by examining the contingencies between occurrence of responses and operation of the arrow to "peck". However, in the experiments under consideration, there were no such contingencies. No stimulus change of any kind was programmed to occur contingent on the chicks' behavior. Arrow operation began at arbitrarily preselected intervals, continued for a fixed time and then ceased regardless of what the chicks did.

An alternative possibility is that pecking responses are reinforced by the immediate consequences of pecking rather than by arrow operation. If so, it is difficult to see what such reinforcement might be. The pin heads available for the chicks to peck at were attached to the floor so that no external stimulus change occurred as a consequence of pecking. Moreover, Reymond and Rogers (1981) found that pecking by chicks toward immobile, inedible objects decreased rather than increased in frequency as a consequence of experience. Even if the act of pecking were intrinsically rewarding (Herrnstein, 1977), the basic fact of stimulus-selective pecking remains unexplained since pecking to target and nontarget are not differentially rewarded either intrinsically or extrinsically.

Finally, development of altered pecking preferences by chicks does not require that the chicks make any pecking responses at all. Chicks given visual exposure to the display of Figure 2 but without the target and nontarget pin heads, that is, with no stimuli available for the chicks to peck at, nonetheless come to prefer the arrow-"pecked" stimulus. Chicks given only visual exposure to display stimuli nevertheless preferentially peck at the target stimulus in a second session of arrow operations when all of the pin-head stimuli are removed from the display and the target and nontarget are made available to the chick (Suboski, 1989a; Suboski & Bartashunas, 1984). The development of pecking preferences by chicks thus resembles autoshaping in that only visual observation of stimulus relationships is required in order to alter behavior to those stimuli. Clearly, the paradigm involved is not instrumental conditioning.

2.2.6. *Autoshaping - Pavlovian Conditioning: Temporal Factors.* Autoshaping is widely recognized as a Pavlovian conditioning phenomenon. Pavlovian conditioning refers to behavior changes by an animal to a behaviorally-inert stimulus that occur as a result of a temporal relationship between that stimulus and a behaviorally-active stimulus. Rather than a temporally-contingent relationship between a response and a reinforcer as in instrumental conditioning, Pavlovian conditioning requires a temporally-contingent relationship between the two stimuli. Temporally-contingent relationships in Pavlovian conditioning may range from temporal contiguity where the stimuli occur either simultaneously or not at all, to prediction of the nonoccurrence of one stimulus by the occurrence of the other.

The structure of the development of arrow-directed pecking preferences of chicks parallels the structure of Pavlovian conditioning. In both cases, a stimulus, (relatively) behaviorally inert, acquires behavior-eliciting properties for an animal as a consequence of exposure of the animal to a particular relationship between that stimulus and a second, behaviorally active, stimulus. The major difference lies in the nature of the relationship between stimuli that enhances the behavior-eliciting properties of the neutral stimulus. As noted above, temporal contingencies are presumed to be the operative factors in Pavlovian conditioning whereas temporal contingencies are not necessary or even present during the development of pecking preferences in our research.

During training of the chicks, all of the stimuli were present at all times. The only stimulus change was in the arrow: Stationary versus activated to "peck". Neither occurrence nor nonoccurrence of "pecking" by the arrow temporally predicted or was predicted by any other stimulus. Neither was any stimulus selectively associated in time with arrow operation. Temporal contiguity is also irrelevant; all of the stimuli in the display were present both during and in the absence of arrow operation. Clearly, arrow-enhanced pecking is the result of spatial, not temporal, interaction.

2.2.7. *The Selective Attention Hypothesis.* A remaining possibility, however, is that arrow operation induces selective attention to the arrow and thus to any stimuli close to the arrow, resulting in selectively increased exposure of the chick to that stimulus during the "pecking" arrow reinforcer. Suboski and Bartashunas (1984) presented evidence that argues against this possibility. If selective attention to the arrow-proximate stimulus (i.e., the model) during arrow operation is the basis for subsequent enhanced stimulus preference, then increasing such differential exposure by removing nonmodel stimuli from the display should increase the preference. In fact, quite to the contrary, removal of nonmodel stimuli spatially removed from the arrow (Suboski & Bartashunas, 1984; Condition 1c) and increasing the number of spatially-proximate (model) stimuli (Condition 1g) reversed the chicks' preferences.

This conclusion was confirmed by Suboski (1989a) in a direct test of the selective attention hypothesis with the results shown in Figure 3. According to the selective attention hypothesis, chicks with no nonmodel stimuli should show the effects of undistracted attention to the model and have a strong preference for the target, whereas the group with two nonmodels should show a diminished preference for the target. In fact, however, the responses to the target as shown in Figure 3 are ordered exactly backwards from the order predicted by the selective attention hypothesis. The reason why the order is reversed is not obvious. Young chicks, unlike many other animals, are strongly neophagic. With a choice between familiar and novel objects, chicks often peck the novel (see e.g., Hogan, 1971, p. 159).

2.2.8. *Spatial Relationships*. The considerations discussed above vitiate against temporal relationships among stimuli as causal factors in the development of arrow-assigned pecking preferences by chicks. However, a particular *spatial* relationship between stimulus and "pecking" arrow is clearly a necessary condition. As noted earlier, this is shown simply by the effects of counterbalancing the color of the pin heads in the display. Reversing the color of the learned preference from one chick to another requires only exchanging the position of the pin heads of the two colors in the display. Thus, it is clear that exposure of a chick to a particular spatial relationship among stimuli in the display is a necessary condition for enhancing the peck-eliciting properties of a stimulus. What is so far uninvestigated is whether nonselective (i.e., continuous) temporal contiguity is also a necessary condition. Answering this question would require comparisons involving a group, for example, for which the pin-head stimuli were absent when the arrow was operated and present when the arrow was stationary. This would provide a spatially-paired, temporally-unpaired control condition that is also so far mostly absent from investigations of the role of spatial relationship in Pavlovian conditioning (cf. Arnaut & Shettleworth, 1981; Bronstein, 1985; Sullivan, 1979).

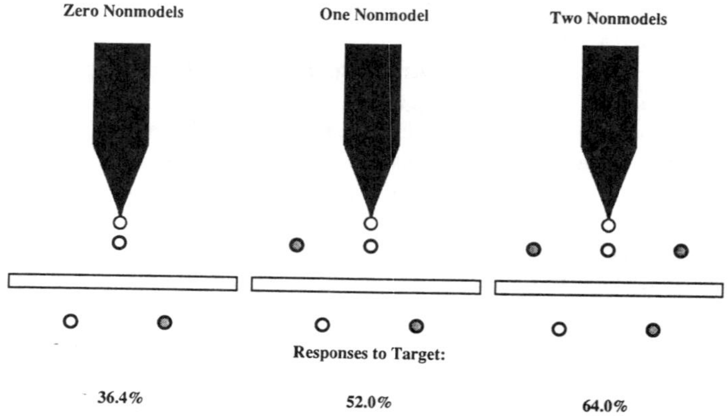

Fig. 3. Responses to target stimulus as a function of the number of nonmodel stimuli present at training.

2.3. ANALYSES OF THE PHENOMENON AS SOCIAL LEARNING

2.3.1. *Social Facilitation and Local Enhancement.* The acquisition of pecking preferences by chicks does not find easy explanation in terms of conventional conditioning processes. This leaves social learning as the major remaining paradigm to be considered. Three modes of social transmission of learned behavior among conspecifics form the social learning model as developed by ethologists (Thorpe, 1963). These are social facilitation, local enhancement, and imitation/observational learning. No consistent distinction between imitation and observational learning marks the relevant literature and the terms are hereinafter regarded as synonymous. Zajonc (1965) identified two types of social facilitation; "audience" and "coaction" effects. The audience effect refers to the potential for altered behavior in an animal as a result of the mere presence of a conspecific, presumably because of increased arousal.

Coaction refers to behavior that "occurs in the presence of other individuals also engaged in the same activity" (Zajonc, 1965, p. 269). In the latter sense of the term, social facilitation often describes an interaction between conspecifics wherein the behavior of one animal is a releasing stimulus for topographically identical behavior by an observing conspecific. There is no shortage of examples from a diversity of behaviors and species: Feeding releases feeding, flight releases flight, aggression releases aggression, etc. Topographic similarity between releasing stimulus and released behavior and immediacy of response production after releasing stimulus occurrence characterize social facilitation as "contagious" behavior.

Another noteworthy but usually unremarked feature of social facilitation is that released responses are typically taxic, or directed, with respect to the releasing stimulus. Many social facilitation releasers are autotaxic, that is, they recruit conspecifics to them as in feeding, mobbing, aggressive and sexual behavior, etc. Many other releasers are allotaxic, as in feeding (released feeding is directed at food, not feeding conspecifics), or as in mobbing (directed at predators, not mobbing companions), etc.

Local enhancement occurs when the behavior of a conspecific is a releaser that directs the behavior of an observer toward a particular area or aspect of the environment (Spence, 1937, p. 821). Thus, local enhancement may be considered as social facilitation with a directional component. Both the stimulus direction of the observer's behavior and topographic similarity between releasing and released behaviors are the distinguishing features of local enhancement.

Without question, social release, as in social facilitation and local enhancement, can be and is a powerful means of social transmission of acquired behavior (see, e.g., Valzelli, 1973). Thus, for example, social facilitation may result in the selection of novel food by an inexperienced observer as a result of feeding released in the presence of novel food located by an experienced conspecific (cf. Turner, 1964, p. 39; Hailman, 1961, 1967). Local enhancement effects can direct feeding behavior to initially unrecognized food identified by a knowledgeable conspecific. Nevertheless, while social facilitation and local enhancement may result in learning, they are not, by themselves, learning processes. That is, they contain no mechanism for effecting a permanent change in behavior. Only in conjunction with reinforcement effects can altered stimulus recognition be expected to emerge.

Klopfer (1961), in an analysis of the issues involved in social transmission of food recognition, emphasized the transitory nature of social facilitation but did not at that time explicate the potential for reinforcement processes to convert transitory response tendencies to permanent preferences (see Hailman, 1967). Klopfer (1961) correctly noted that a satiated hen placed with a hungry partner and food will promptly cease feeding when the hungry partner is removed. However, a hungry hen will not cease to feed if socially facilitated pecking brings the hen into contact with the reinforcing aspects of the food. Indeed, even the sated hen might show a permanent change in food selection if visually novel and palatable food is supplied with an experienced hungry partner.

2.3.2. *Observational/Imitation Learning.* In a still influential definition, Thorpe (1963) proposed imitation to be the active copying by an observer of a demonstrator performing a novel or otherwise improbable act for which there is clearly no instinctive tendency. Again, Klopfer (1961) noted that persistence of observer behavior in the absence of the demonstrator is a necessary condition to the inference of learning by imitation. Topographic similarity between demonstrator and observer behavior (copying) is the only inclusionary property of imitation in Thorpe's definition, with probable or instinctive behavior arbitrarily defined as not imitatable. This definition may work well in the search for an unequivocal example of imitation since any behavior that could possibly be an instance of socially-released responding is so classified. However, social release and imitation are certainly not mutually exclusive processes so distinction between them is crucial. In each case, a demonstration involving a behaving conspecific is provided and the result is similar behavior by the observer. The characteristics distinguishing the two processes are theoretical rather than directly observable. Social facilitation and probably local enhancement as well are members of a class of releasing-stimulus -- released-behavior relationships in which releasing stimulus and released response happen to be topographically identical. In contrast, imitation implies "cognitive" transformation of the demonstrated response into similar behavior by the observing conspecific.

Nevertheless, the possibility is worth examining that topographic similarity between demonstrator and observer behavior will provide the basis for a distinction between imitation and social release. Typically, only limited features of a natural releasing stimulus are necessary or effective in releasing a particular behavior (e.g., Goodwin & Hess, 1969; Hailman, 1967; Tinbergen, 1951). Such features may be extracted from the naturally occurring stimuli and may even prove to be more effective than the natural releaser ("supernormal" stimuli, Goodwin & Hess, 1969; Tinbergen, 1951). This suggests that released behavior will be unresponsive to any variation in the demonstration that does not alter the effective features of the releasing stimulus. Imitation, on the other hand, as active copying of the demonstrated behavior, should show considerable correspondence in observer behavior as demonstrator behavior is varied. (See, e.g., Miller & Dollard, 1941). The potential and problems of this approach are well illustrated in the controversy that followed the report by Meltzoff and Moore (1977) of apparent imitation in human neonates.

Meltzoff and Moore (1977) claimed that neonates specifically imitated four oral and one manual gesture subsequent to demonstration by an adult model. (The fact that their claim was based on an invalid statistical assumption of independent observations on 36 babies when there were only six may be laid aside for the present discussion.) Meltzoff and Moore (1977) argued that it was unlikely that five separate responses would each be a released response. In essence, their argument was that the neonatal response varied in form to match the demonstrated behavior. Jacobson and Kagan (1979) disputed this claim by showing that certain of the oral gestures occurred to a number of stimuli, such as a ball or pen, moved toward the baby's face. This behavioral constancy in response to stimulus variability is suggestive of released responses rather than imitative behavior (see Anisfeld, 1991, for a further discussion of these issues).

Another version of the same argument for distinguishing between social release and imitation could hold that imitation of complex acts is unlikely to be a simple product of social release by demonstrator behavior and is therefore likely to be the result of an active response-copying process. This frequently unstated assumption is presumably the basis for frequent claims of experimental demonstration of imitative behavior despite the absence of appropriate control conditions. Also in need of consideration is the possibility that observational experience may result in novel stimulus recognition rather than novel response acquisition. Imitation may be evidenced by the occurrence, not of an unlikely response, but rather the unlikely redirection of a probable response toward a particular stimulus. Thus, assuming that unreinforced observer behavior meets the qualifying criterion of persistence after removal of the demonstrator, imitation may be manifested in three ways: 1) Plasticity of observer behavior to match topographically variable demonstrator behavior; 2) Replication by an observer of complex demonstrator behavior; 3) Replication by an observer of the stimulus direction of demonstrator behavior.

A number of instances of observational acquisition of specific responses modeled by conspecifics have been found in experiments containing controls of greater or lesser adequacy. Some of these studies have found observers to match the topography of demonstrated responses to a degree unlikely to occur except by response copying. A few examples include birds: budgerigars, *Melopsittacus undulatus*, Dawson & Foss, 1965; parrot, *Psittacus Erithacus*, Moore, 1992; pigeons, *Columba livia*, Epstein, 1984; cats: *felis domesticus*, Adler, 1955; rats: Huang, Koski, & DeQuardo, 1983; Strupp & Levitsky, 1984; Will, Pallaud, Soczka, & Manikowski, 1974; and squirrels, *Tamiasciurus hudsonicus*, Weigl & Hanson, 1980). Despite these, other recent reports, and many more less controlled or less convincing reports suggestive of response copying, the research on observational learning to date has been almost exclusively directed toward demonstration of the phenomenon and elimination of confounding factors. Whether social learning processes can adequately account for the observation-based acquisition of selective pecking preferences remains to be examined.

2.3.3. *Social Learning and Arrow-Directed Pecking by Chicks*. Consideration of chick behavior suggests that pecking responses are initially under the control of nonsocial releasing stimuli. Chicks have a strong tendency to peck at virtually any stimulus of an appropriate size and contrast (e.g., Goodwin & Hess, 1969; see also Suboski & Bartashunas, 1984). In addition, chicks also peck at social stimuli such as other chick's beaks and at the tip of the hen's beak. They also peck at nonsocial stimuli: The apex of angular stimuli (Turner, 1964) and vertically moving stimuli such as a finger (Collias, 1952), a pencil (Tolman, 1964), or an arrow (Strobel & Macdonald, 1974; Suboski, 1984). These stimuli are releasers of pecking that is directed either to appropriate food-like objects (allotaxic) or toward the tip of the moving object (autotaxic). One immediate implication of pecking released by various moving objects is that "social facilitation" of pecking behavior by chicks does not necessarily depend on the presence of a conspecific companion. Rather, released pecking occurs to the essential features of the releasing stimulus, which may be any "pecking" angular object. The "social facilitation" phenomenon lacks a directional component and thus can only be expected to increase the general frequency of pecking to food-like objects (Suboski, 1984). "Social facilitation" predicts selective pecking only to stimuli nearer to the "pecking" arrow, whereas chicks selectively peck at stimuli that match the arrow-proximate stimuli in color, even if such a stimulus is farther away from the arrow than is an alternative stimulus (Suboski & Bartashunas, 1984). Clearly, social facilitation has little potential as an explanation of selectively enhanced pecking by chicks.

Stimulus control over the direction of released taxic pecking is suggestive of local enhancement except that, as noted, stimulus control of pecking extends well beyond cessation of the demonstration condition. There is another feature that distinguishes between local enhancement and the current phenomenon that is slightly subtle but nonetheless important. In local enhancement, the observer responds to the same stimulus as does the demonstrator. Chicks, however, direct pecks not only to the same stimulus as does the "demonstrator", but they also direct pecks toward *matching* stimulus objects, even well after the original display stimuli have been removed. Such stimulus selectivity cannot be accounted for in terms of demonstrator direction of observer behavior (or attention) toward a particular stimulus. Stimulus generalization based on local enhancement also fails to provide a mechanism to account for responding to color-matching stimuli after the removal of the demonstration display. The problem with stimulus generalization as an explanation is that it begs the question. Stimulus generalization accounts for responding to similar stimuli only after a stimulus preference has already been learned. Thus, the explanation presupposes the occurrence of the phenomenon that is to be explained. Local enhancement also fails to account for the phenomenon of interest.

Observation-based selective pecking clearly meets a basic qualifying criterion for imitation, that is, persistence in the absence of demonstrator behavior, and displays otherwise unlikely stimulus control over the direction of pecking by observer chicks. The evidence that most strongly suggests releasing stimulus and released response involvement in much apparent imitation is demonstrations of "imitation" in the absence of any demonstrator at all. Our previously described work on chicks (Bartashunas & Suboski, 1984; Suboski, 1984; 1987; 1989a; Suboski & Bartashunas, 1984) is obviously relevant here. Groesbeck and Duerfeldt (1971) found two demonstration conditions that significantly enhanced learning of a discriminated Y-maze by observer rats. One condition was the performance of a discriminating demonstrator in the maze. The other successful demonstration was the experimenter wielding a pointer to knock down the correct discriminative cue panel and tap the spout of a water bottle (See also Kohn, 1976; Kohn & Dennis, 1972).

Most fascinating are the observations of Neuringer and Neuringer (1974) and Bullock and Neuringer (1977) on pigeons that were first trained to eat from the experimenter's hand and then given various demonstrations in which the experimenter's hand moved about the experimental chamber, "pecked" at a keylight and food hopper, etc. Subsequent movement and pecking behavior by observer pigeons was distinctly nonrandom with respect to the demonstration.

Occasional reports of cross-species "imitation" may certainly reflect active copying of responses, but if so, they point toward an intelligence sufficient to effect a cognitive transformation of demonstrator behavior into replicated responses and also suggest the insight that beneficial consequences can be gained from imitation of a member of another species. On the other hand, explanation based on releasing stimulus interactions requires no more than noting that critical features of social releasing stimuli such as eating behavior are quite similar among many species.

2.3.4. *Imitation vs. Extension of Stimulus Control Over Preformed Responses*. Although some demonstrations of observational learning can be interpreted in terms of active copying of responses, many studies are equally consistent with the hypothesis that demonstrator behavior functions essentially as a releasing stimulus for topographically-similar behavior from an observer. Furthermore, the outcome of such studies is usually not a change in the form of behavior but rather a change in stimulus control over pre-existing responses. In studies of observational learning of food recognition, for example, a demonstrator makes feeding responses to a novel or distinctive stimulus and, after removal of the demonstrator, an observer is found to have an increased tendency to direct consummatory responses toward similar stimuli. The altered stimulus control over feeding behavior occurs well after the conclusion of the demonstration, thereby eliminating social facilitation and local enhancement as possible explanations.

Response copying accounts for such behavior but so does the concept of releasing ability or "valence" (Russell, 1935) acquired from an interaction between a releasing stimulus and a neutral stimulus. The visual features of demonstrator feeding behavior socially facilitate (release) feeding by an observer. If the feeding behavior of the demonstrator is directed toward novel food, the stimulus features of the food may thereby acquire ability to release subsequent feeding behavior from an observer. Numerous reports of observational learning of food and noxious stimulus recognition are consistent with both views.

2.3.5. *"Semantic" vs. "Imitative" Forms of Cultural Transmission*. Finally, there is one set of phenomena quite tractable to interpretation in terms of releasing stimulus interactions that is equally inexplicable in terms of imitation by active copying. First, we note that social facilitation and local enhancement refer to intraspecific interactions where the behavior of one animal is the releasing stimulus for released responses by observing conspecifics. More exactly, these terms refer to interactions where the releasing stimulus and released responses are topographically identical behavior. However, such interactions constitute only part of the conspecific interactions that involve social releasers. There are numerous social releasers where the released behavior has a more or less arbitrary (but not functionally irrelevant) topographical relationship to the behavior that provides the releasing stimulus. These releasers are usually signals of one sort or another given in the presence of food or predators. Socially-transmitted stimulus recognition resulting from such interactions obviously would not be susceptible to explanation by behavioral copying and thus might be termed "semantic" as opposed to "imitative" communication of stimulus recognition.

Food calls, displays, and pheromones are rampant in vertebrates (see, e.g., Galef & Beck, 1985). These species-specific behaviors attract conspecifics and release feeding but most have not yet been found to yield acquired releasing valence for feeding to novel foodstuffs. Our research on pecking preferences of neonatal chicks (Bartashunas & Suboski, 1984; Suboski, 1984; 1987; 1989a; Suboski & Bartashunas, 1984) provides one clear example (see also Galef, Kennett, & Stein, 1985). Alarm calls, displays, and pheromones are also widely distributed among vertebrate species and several reports show that acquired predator recognition does result (Magurran, 1989; Suboski et al, 1990).

Species-specific predator calls differing as to predator types (e.g., ground, aerial, etc.) are probably universal to social birds and mammals that are subject to predation. Herzog and Hopf (1984) have shown that such predator calls function in the social transmission of predator recognition. Two squirrel monkeys, *saimiri sciureus*, were kept isolated with a cloth surrogate mother equipped with an internal speaker. During their second and third months of age, squirrel-monkey predator calls were presented to the infants through the speaker. No consistent behavior occurred to recordings of ground-predator alarm calls until one of two stuffed toy animals was simultaneously placed in the cage, whereupon the monkeys avoided the toy and went to the mother. The other toy evoked approach and investigatory behavior when presented with a tone. A few days later, each of the monkeys was found to avoid the doll that had been presented with the predator call and approach the other toy. This example illustrates how "semantic communication" of predator recognition can result from innately-organized species-specific releasing stimuli.

3. Current Status of Releaser-Induced Recognition Learning

This review has revealed a decided intractability of observational modification of pecking preferences of chicks to conventional theoretical interpretation; an intractability that is to be found more in the phenomenon than in the reviewer. Perhaps the answer to the interpretative problem is that an *idiosyncratic* learning process is involved. A general description of the acquisition of pecking preferences by chicks notes that initially chicks peck indiscriminately at various peck-eliciting stimuli. After the demonstration, chicks tend to peck selectively at particular stimuli. Observer chicks appear to acquire enhanced valence to release pecking to such stimuli. Thus, learned acquisition of pecking preferences by chicks involves observation of an interaction between a releasing stimulus normally provided by a conspecific and another stimulus that is initially behaviorally indifferent. Observation of an angular stimulus "pecking" at a particular object enhances the peck-eliciting properties of that object for observer chicks. The enhancement requires only unreinforced observation of a dynamic and spatial relationship between stimuli in which temporal contingencies are absent and temporal contiguity does not differentially predict stimulus change. Furthermore, this description does not refer to an empirically-isolated, specialized, species- and situation-specific learning mechanism. The description fits many interactions among conspecifics and provides a simple and parsimonious interpretation for many otherwise quite remarkable instances of apparent social transmission of stimulus recognition.

The group on the "Biology of Learning in Nonhuman Mammals" at the 1984 Dahlem Conference set as their number one topic of discussion the question of whether learning is a general process or whether it consists of a set of particular specialized adaptations. The releaser-induced recognition learning model offers a definite answer: Learning consists of a set of particular specialized adaptations of a general process. In a series of review articles (Suboski, 1988; 1989b; 1990; 1992a; 1992b; Suboski, Muir, & Hall, 1993), I have extensively documented many examples of learning procedures and phenomena that are either consistent with the propositions of releaser-induced recognition learning or are more generally and parsimoniously explained by the model than by alternatives.

A detailed description of the basic features of the four propositions that comprise the model may be found in the general review article (Suboski, 1990) and will be only briefly presented here. Proposition 1, the most important tenet of the model, is a variant of the transfer-of-control version of the stimulus substitution mechanism. In at least simple forms of learning, animals learn what stimuli to respond to; they do not learn how to respond. In instance after instance, presentation of two stimuli to an animal results in an extension, from one stimulus to another, of the control over the release of responses from the animal (as suggested by Gary T. Walters, we now use the more precise term "extension" in preference to "transfer"). New responses do not occur in simple learning, rather, control over pre-existing responses is redirected to new stimuli. One other point is worth making here. Extension-of-control also provides a performance mechanism to the model. Essentially, it says that animals learn stimulus recognition because they are phylogenetically prepared to do so and not because of reinforcement.

Proposition 2 asserts that Pavlovian USs and ethological releasing stimuli do not differ in principle and are functionally identical. The behaviorally-active stimuli that enter into learning relationships are releasing stimuli that elicit endogenously pre-organized response sequences of varying degrees of complexity.

Proposition 3 notes that responses induced by releasing stimuli are usually taxically directed with regard to their releasing stimuli. Taxic direction is intrinsic to released responses and is transferred to a new stimulus when control over release is transferred. In the terms of Pavlovian conditioning, the taxic direction of the URs to the US determines the taxic direction of CRs to the CS. This is quite straightforward for autotaxic and antitaxic releasers, but becomes a little complicated for allotaxic releasers. Consider a simple example. An observer animal watches a demonstrator eat a novel foodstuff. Feeding by a conspecific is a powerful releaser of feeding by the observer. However, feeding behavior by the demonstrator is an allotaxic releasing stimulus that directs feeding behavior of the observer to the food and not at the demonstrator. The "pecking" arrow is an allotaxic releaser; it directs chicks to peck, not just at the arrow itself, but rather to stimuli specifically designated by the "pecking" arrow.

Proposition 4 deals with the relationships between Pavlovian conditioning and releaser-induced recognition learning. There is, in fact, no critical difference. Pavlovian conditioning has historically and technically been concerned with the effects on learning of temporal relationships between stimuli whereas releaser-induced recognition is specifically concerned with the broader set of potential learning relationships, particularly spatial and dynamic relationships, and the effects of novelty.

Support for Proposition 4 was waiting in Curio, Ernst and Vieth's excellent 1978 paper on cultural transmission of predator recognition by the mobbing behavior of blackbirds. Curio et al provided an early example of a spatial-dynamic learning relationship in predator recognition. In mobbing, the predator can only be identified to inexperienced blackbirds by its location at the focal point of the mobbing display (flying around and darting at the predator while giving species specific mobbing calls). We can assume that any novel stimulus object at the focus of the mobbing display acquires valence to release mobbing from previously naive observer birds. Spatial location, not temporal relations, determines which stimulus acquires control. Note also that mobbing behavior is an allotaxic releaser. Mobbing behavior, including mobbing of learned stimuli, is directed at the focal object, not at the mobbing birds. The importance of the spatial relationship could be shown by having two different objects as potential stimuli to be mobbed. If one object is mobbed and a laterally displaced object is ignored then observers will afterward mob the valence-enhanced and ignore the previously displaced objects, or so the releaser-induced recognition model predicts.

3.1. PREDICTED RESEARCH: LEARNED ALARM REACTION IN FISH

Recent research in my laboratory has been directed at evaluation of the ability of the releaser-induced recognition learning model to generate interesting research and predict new research findings. Two such theory-driven research areas have been opened in my laboratory in recent years; one on alarm behavior in fish (Suboski et al, 1990; recently with collaboration from David Hall and Robert Smith), the other on the role of feeding pheromone in food selection by fresh-water pulmonate snails (Suboski et al., 1993; with collaboration from David Hall, John Donohue, and Darwin Muir).

The alarm reaction was selected for investigation because of indications that learning of predator recognition by fish was involved. If so, then cultural traditions of predator recognition may also be found. Alarm pheromone ("alarm substance"), secreted from the skin of injured Ostariophysi (social, nonpredaceous) fish, yields an "alarm reaction" (apparent antipredator behavior) when placed in water containing an injured conspecific fish. According to Pfeiffer's (1974) description of research by Göz (1941), blinded minnows, *Phoxinus phoxinus*, placed in a tank with a pike, *Esox lucius*, first displayed an alarm reaction only after several of the minnows had been attacked by the pike. In later tests after removal of the pike, the minnows displayed an alarm reaction to the odor of the pike.

Learned alarm reactions by fish based on extension-of-control from alarm substance to visual and/or olfactory predator stimuli are likely to have small effect since they depend on exposure of individual fish to slow and limited diffusion of chemosensory stimuli through water and also require periodic successful visits from the predator. However, the alarm reaction in some fish (e.g., *Rasbora heteromorpha*) is socially facilitated and visually transmitted (Verheijen, 1956).

If visual or direct exposure to alarm behavior of conspecifics suffices to yield learned alarm to a concurrently presented novel chemosensory and/or visual stimulus, then it may be possible to establish cultural traditions of predator recognition among Ostariophysi fish. Outcomes consistent with the predictions of the model were obtained by Suboski et al, (1990; Hall & Suboski, unpublished) with zebra danio (*Brachydanio rerio*) fish. Thus, the findings of learning and social transmission of "predator" recognition were those predicted by the releaser-induced recognition model.

First, exposing fish to alarm substance from skin abrasions of conspecifics along with an innocuous stimulus (morpholine) yields alarm behavior to the alarm substance and quite identical learned alarm to morpholine when later tested by itself. Second, social transmission of predator recognition was demonstrated. Groups of fish that had been exposed to alarm substance combined with morpholine were mixed together with naive fish and later tested with morpholine alone. The mixed group displayed an alarm reaction -- learned by trained fish, socially facilitated by naive fish. Now the trained fish were sorted out and removed and a new naive group replaced them. Another test with morpholine alone revealed an alarm reaction by the mixed group, learned by the group with prior experience of alarm reacting conspecifics in the presence of morpholine, socially facilitated by the naive group. One further replacement of experienced fish with naive yielded a third socially transmitted learned alarm reaction. More recent work has replicated cultural transmission quite satisfactorily using a visual CS as well as the olfactory CS (Hall & Suboski, unpublished).

Several other features of this research are quite interesting. The response to alarm substance is allotaxic - the fish school and then move to the substrate and circle for prolonged periods of time. Their behavior is not directed toward the alarm substance. As predicted, learned alarm is also allotaxic. Learned alarm reactions have exactly the same form as do unlearned alarm reactions, whether precipitated by alarm substance or by an alarm reaction of conspecifics. It is interesting that social facilitation produced permanent learning. This finding suggests that further distinction may be necessary in the taxonomy of social learning (see Moore, 1992). As a last point here, note that the pheromones that release alarm responses are species-specific and not general interchangeable stimuli, whereas neutral stimuli do appear to be general; arbitrarily selected chemosensory and visual stimuli both work very well.

3.2. PREDICTED RESEARCH: FEEDING PHEROMONE IN POND SNAILS

Our research on snails was initiated by reports that some species of predatory marine snails are selectively attracted to feeding conspecifics in preference to the same stimuli when presented separately (see Suboski, 1992b). These findings suggest that social facilitation and recruitment occur in these species, presumably in response to a feeding pheromone. Research (in collaboration with Darwin Muir) on a North Atlantic snail, *Nucella lapillus*, failed to disclose recruitment in that species. However, an intriguing study by Bovbjerg (1965) strongly indicated that a herbivorous fresh-water pulmonate snail, *Stagnicola reflexa*, rapidly recruits conspecifics immediately upon finding food, again, presumably by a feeding pheromone. Quite fortuitously, a large colony of *Stagnicola reflexa* were located in an urban creek in Kingston, Ontario, sympatric with a coextensive colony of another fresh-water snail, *Physella gyrina gyrina*, both species feeding on *Cladophera* species of alga.

A series of two-choice tests showed that test snails approached conspecifics feeding on alga in preference to any of the other choices which included four conditions: "no stimulus", "alga alone", "snails alone", and "snails with alga but unable to feed". These results supported the prediction of recruitment based on a feeding pheromone. However, informal observations suggested that the response to feeding pheromone was more complex than simple recruitment. Choice tests between alga alone and alga with feeding snails were conducted in which the number of feeding snails was systematically varied. In contrast to the monotonically increasing function predicted by simple recruitment, inverted U-shaped functions were obtained for both snail species. The feeding pheromone appears to direct snails in a way that will optimize the distribution of foraging snails on a food patch. The pheromone directs snails toward areas that have an intermediate number of feeding snails and away from areas with either few feeding snails or a large number of feeding snails.

The most important prediction from the releaser-induced recognition learning model is that exposing snails to feeding pheromone along with a gustatory stimulus will alter the snails' response to that gustatory stimulus. According to the transfer-of-control mechanism, the gustatory stimulus will release from the snails responses that were previously released only by the feeding pheromone. We would then expect the gustatory stimulus to recruit conspecifics. However, in the light of the initial findings, recruitment to the gustatory stimulus should occur only when the solution of exposure had an intermediate concentration of feeding pheromone. Low and high concentrations of feeding pheromone should yield snails that are repelled by the gustatory stimulus. This is exactly the outcome of the relevant experiment. Test snails exposed overnight to effluent from varying numbers of snails feeding on alga were afterward repelled by that alga when large or small numbers of snails produced the effluent of exposure. Snails exposed to intermediate concentrations were strongly attracted to the alga.

4. summary and conclusions

No theory of animal learning yet devised has received general approbation from learning theorists. The lack of adequate underlying explanatory mechanisms for learning has resulted in limited, idiosyncratic, and contradictory theories that often encounter data inconsistent with predictions of the theory. One such outcome resulted from my research on apparent social transmission of learned food recognition by hatchling chicks. These findings are incompatible with conventional learning mechanisms but do seem amenable to interpretation in terms of an ethologically oriented version of the stimulus substitution theory of Pavlovian conditioning.

The resultant learning model provides a general mechanism that appears to underlie many forms of learning. This theory, the releaser-induced recognition learning model, has successfully predicted self-perpetuating cultural transmission of predator recognition in minnows and at least simple social transmission of food recognition and location in pond snails. Few other learning theories could do as well.

5. references

Adler, H. E. (1955). Some factors of observational learning in cats. Journal of Genetic Psychology, 86, 159-177.

Anisfeld, M. (1991). Neonatal imitation. Developmental Review, 11, 60-97.

Arnaut, L. & Shettleworth, S. J. (1981). The role of spatial and temporal contiguity in defensive burying in rats. Animal Learning & Behavior, 9, 275-280.

Bartashunas, C. & Suboski, M. D. (1984). Effects of age of chicks on social transmission of pecking preferences from hen to chicks. Developmental Psychobiology, 17, 121-127.

Boakes, R. A., Poli, M., Lockwood, M. J., & Goodall, G. (1978). A study of misbehavior: Token reinforcement in the rat. Journal of the Experimental Analysis of Behavior, 29, 115-134.

Bovbjerg, R. V. (1965). Feeding and dispersal in the snail *Stagnicola reflexa* (Basommatophora: Lymnaeidae). Malacologia, 2, 199-207.

Bronstein, P. M. (1985). Toxiphobia, "social reinforcement", comparative psychology, and Patrick J. Capretta. Annals of the New York Academy of Sciences, 443, 158-170.

Breland, K., & Breland, M. (1961). The misbehavior of organisms. American Psychologist, 16, 681-684.

Bruner, A. (1965). UCS properties in classical conditioning of the albino rabbit's nictitating membrane response. Journal of Experimental Psychology, 69, 186-192.

Bullock, D. & Neuringer, A. (1977). Social learning by following: An analysis. Journal of the Experimental Analysis of Behavior, 25, 127- 135.

Bunge, M. (1985). On research strategies in psychology: Reply to commentators. Annals of Theoretical Psychology, 3, 151-156.

Collias, N. E. (1952). The development of social behavior in birds. Auk, 69, 127-159.

Colavita, F. B. (1965). Dual function of the US in classical salivary conditioning. Journal of Comparative & Physiological Psychology, 60, 218-222.

Curio, E., Ernst, U., & Vieth, W. (1978). The adaptive significance of avian mobbing. II. Cultural transmission of enemy recognition in blackbirds: Effectiveness and some constraints. Zeitschift für Tierpsychology, 48, 184-202

Davey, G. C. L., Phillips, S., & Cleland, G. G. (1981). The topography of signal-centered behaviour in the rat: The effects of solid and liquid food reinforcers. Behaviour Analysis Letters, 1, 331-337.

Dawson, B. V. & Foss, B. M. (1965). Observational learning in budgerigars. Animal Behaviour, 13, 470-474.

Dickinson, A., & Mackintosh, N. J. (1978). Classical conditioning in animals. Annual Review of Psychology, 29, 587-612.

Dodwell, P. C. (1985). Is neuropsychology something new? Annals of Theoretical Psychology, 3, 143-150.

Epstein, R. (1984). Spontaneous and deferred imitation in the pigeon. Behavioural Processes, 9, 347-354.

Fields, C. (1989). Explaining classical conditioning: Phenomenological unity conceals mechanistic diversity. Behavioral & Brain Sciences, 12, 141-142.

Galef, B. G., Jr. & Beck, M. (1985). Aversive and attractive marking of toxic and safe foods by Norway rats. Behavioral & Neural Biology, 43, 298-310.

Galef, B. G., Jr., Kennett, D. J. & Stein, M. (1985). Demonstrator influence on observer diet preference: Effects of simple exposure and the presence of a demonstrator. Animal Learning & Behavior, 13, 25-30.

Gardner, R.A., & Gardner, B.T. (1988). Feedforward versus feedbackward: An ethological alternative to the law of effect. Behavioral & Brain Sciences, 11, 429-493.

Gilbertson, D. W. (1975). Courtship as a reinforcement for key pecking in the pigeon, Columba livia. Animal Behaviour, 23, 735-744.

Goodwin, E. B. & Hess, E. H. (1969). Stimulus generalization and responses to supernormal stimuli in the unrewarded pecking behaviour of young chicks. Behaviour, 34, 255-266.

Gormezano, I. (1984). The study of associative learning with CS-CR paradigms. In D. L. Alkon & J. Farley (Eds.), Primary Neural Substrates of Learning and Behavioral Change. Cambridge: Cambridge University Press. pp. 5-24.

Gormezano, I., & Hiller, G. W. (1972). Omission training of the jaw-movement response of the rabbit to a water US. Psychonomic Science, 29, 276-278.

Gormezano, I., & Kehoe, E. J. (1975). Classical conditioning: Some methodological-conceptual issues. In W. K. Estes (Ed.), Handbook of Learning and Cognitive Processes. Vol. 2. Hillsdale, N.J.: Erlbaum Associates. pp. 143-179.

Göz, H. (1941). Uber den Art- und Individualgeruch bei Fischen. Zeitschrift für Vergleichende Physiologie, 29, 1-45.

Groesbeck, R. W. & Duerfeldt, P. H. (1971). Some relevant variables in observational learning of the rat. Psychonomic Science, 22, 41-43.

Guthrie, E. R. (1935). The Psychology of Learning. New York: Harper.

Hailman, J. P. (1961). Why do gull chicks peck at visually contrasting spots? A suggestion concerning social learning of food-discrimination. American Naturalist, 95, 245-247.

Hailman, J. P. (1967). The ontogeny of an instinct. The pecking response of chicks of the laughing gull (Larus atricilla L.) and related species. Behaviour, Suppl. 15.

Hearst, E. (1978). Stimulus relationships and feature selection in learning and behavior. In S. H. Hulse, H. Fowler, & W. K. Honig (Eds.), Cognitive Processes in Animal Behavior. Hillsdale, N.J: Lawrence Erlbaum. pp. 51-88.

Herrnstein, R.J. (1977). The evolution of behaviorism. American Psychologist, 32, 593-603

Herzog, M. & Hopf, S. (1984). Behavioral responses to species-specific warning calls in infant squirrel monkeys reared in social isolation. American Journal of Primatology, 7, 99-106.

Hess, E. H. (1964). Imprinting in birds. Science, 146, 1128-1139.

Hogan, J. A. (1971). The development of a hunger system in young chicks, Behaviour, 39, 128-201.

Hogan, J. A. (1977). The ontogeny of food preferences in chicks and other animals. In L. M. Barker, M. Best, and M. Domjan (Eds.), Learning mechanisms in food selection. Waco, Texas: Baylor University Press. pp. 71-97.

Hollis, K. L. (1982). Pavlovian conditioning of signal-centered action patterns and autonomic behavior: A biological analysis of function. Advances in the Study of Behavior, 12, 1-64.

Huang, I.-N., Koski, C. A. & DeQuardo, J. R. (1983). Observational learning of a bar-press by rats. Journal of General Psychology, 108, 103-111.

Jacobson, S. W. & Kagan, J. (1979). Interpreting "imitative" responses in early infancy. Science, 205, 215-217.

Jenkins, H. M. (1984). The study of animal learning in the tradition of Pavlov and Thorndike. In P. Marler & H. S. Terrace (Eds.), The Biology of Learning. New York: Springer Verlag. pp. 89-114.

Kamin, L. J. (1968). "Attention-like" processes in classical conditioning. In M. R. Jones (Ed.), Miami Symposium on the Prediction of Behavior, 1967: Aversive Stimulation. Coral Gables: University of Miami press. pp. 9-31.

Kamin, L. J. (1969). Predictability, surprise, attention and conditioning. In B. A. Campbell & R. M. Church (Eds.), Punishment and Aversive Behavior, New York: Appleton-Century-Crofts. pp. 279-296.

Klopfer, P. H. (1961). Observational learning in birds: The establishment of behavioral modes. Behaviour, 17, 71-79.

Kohn, B. (1976). Observation and discrimination learning in the rat: Effects of stimulus substitution. Learning & Motivation, 7, 303-312.

Kohn, B. & Dennis, M. (1972). Observation and discrimination learning in the rat: Specific and nonspecific effects. Journal of Comparative & Physiological Psychology, 78, 292-296.

Locurto, C. (1989). The dark side of hegemony. Behavioral & Brain Sciences, 12, 153-154

Mackintosh, N. J. (1977). Conditioning as the perception of causal relations. In R. E. Butts & J. Hintikka (Eds.), Foundational Problems in the Special Sciences. Dordrecht, Holland: D. Reidel. pp. 241-250.

Magurran, A. E. (1989). Acquired recognition of predator odour in the European minnow (*Phoxinus phoxinus*), Ethology, 82, 216-223

Meltzoff, A. N. & Moore, M. K. (1977). Imitation of facial and manual gestures by human neonates. Science, 198, 75-78.

Miller, N. E. & Dollard, J. (1941). Social Learning and Imitation. New Haven: Yale University Press.

Moore, B. R. (1971). On directed respondents. Unpublished Ph.D. Thesis, Stanford University.

Moore, B. R. (1973). The role of directed Pavlovian reactions in simple instrumental learning in the pigeon. In R. A. Hinde & J. Stevenson-Hinde (Eds.), Constraints on Learning. London: Academic Press. pp. 159-186.

Moore, B. R. (1992). Avian movement imitation and a new form of mimicry: Tracing the evolution of a complex form of learning. Behaviour, 122, 231-263.

Murray, H. G., & Carruthers, B. C. (1974). Human eyelid conditioning with airpuff vs infraorbital shock as the UCS. Canadian Journal of Psychology, 28, 269-287.

Neuringer, A. & Neuringer, M. (1974). Learning by following a food source. Science, 184, 1005-1008.

Patten. R. L., & Rudy, J. W. (1967). The Sheffield omission of training procedure applied to the conditioning of the licking response in rats. Psychonomic Science, 8, 463-464.

Pavlov, I. P. (1927). Conditioned Reflexes, London: Oxford University Press.

Pearce, J. M., & Hall, G. (1980). A model for Pavlovian learning: Variations in the effectiveness of conditioned but not of unconditioned stimuli. Psychological Review, 87, 532-552.

Peden, B. F., Browne, M. P., & Hearst, E. (1977). Persistent approaches to a signal for food despite omission for approaching. Journal of Experimental Psychology: Animal Behavior Processes, 3, 377-399.

Pfeiffer, W. (1974). Pheromones in fish and amphibia. In M. C. Birch (Ed.), Pheromones. Amsterdam: North-Holland. pp. 269-296.

Rescorla, R. A. (1967). Pavlovian conditioning and its proper control procedures. Psychological Review, 74, 71-80.

Rescorla, R. A. (1968). Probability of shock in the presence and absence of CS in fear conditioning. Journal of Comparative & Physiological Psychology, 66, 1-5.

Rescorla, R. A. (1978). Some implications of a cognitive perspective on Pavlovian conditiong. In S. H. Hulse, H. Fowler, & W. K. Honig (Eds.), Cognitive Processes in Animal Behavior. Hillsdale, N. J.: Lawrence Erlbaum. pp. 15-50.

Rescorla, R. A. (1984). Comments on three Pavlovian paradigms. In D. L. Alkon & J. Farley (Eds.), Primary Neural Substrates of Learning and Behavioral Change. Cambridge: Cambridge University Press. pp. 25-45.

Rescorla, R. A. (1988). Pavlovian conditioning. American Psychologist, 43, 151-160.

Rescorla, R. A., & Holland, P. C. (1976). Some behavioral approaches to the study of learning. In M. R. Rosenzweig & E. L. Bennett (Eds.), Neural Mechanisms of Learning and Memory. Cambridge, Mass.: MIT Press. pp. 165-192.

Reymond, E. & Rogers, L. J. (1981). Deprivation of the visual and tactile aspects of food important to learning performance of an appetitive task by chicks. Behavioral & Neural Biology, 31, 425-434.

Russell, E. S. (1935). Valence and attention in animal behaviour. Acta Biotheoretica, 1, 91-99.

Sheffield, F. D. (1965). Relation between classical conditioning and instrumental learning. In W. F. Prokasy (Ed.), Classical Conditioning: A Symposium. New York: Appleton-Century-Crofts. pp. 302-322.

Shettleworth, S. J. (1983). Function and mechanism in learning. In M. D. Zeiler & P. Harzen (Eds.), Advances in Analysis of Behaviour. Vol 3. New York: John Wiley & Sons. pp. 1-39.

Spence, K. W. (1937). Experimental studies of learning and higher mental processes in infrahuman primates. Psychological Bulletin, 34, 806-850.

Strobel, M. G. & Macdonald, G. E. (1974). Induction of eating in newly hatched chicks. Journal of Comparative & Physiological Psychology, 86, 493-502.

Strupp, B. J. & Levitsky, D. A. (1984). Social transmission of food preferences in adult hooded rats (Rattus norvegicus). Journal of Comparative Psychology, 98, 257-266.

Suboski, M. D. (1984). Stimulus configuration and valence-enhanced pecking by neonatal chicks. Learning & Motivation, 15, 118-126.

Suboski, M. D. (1987). Environmental variables and releasing-valence transfer in stimulus-directed pecking of chicks. Behavioral & Neural Biology, 47, 262-274.

Suboski, M. D. (1988). Acquisition and social communication of stimulus recognition by fish. Behavioural Processes, 16, 213-244.

Suboski, M. D. (1989a). The acquisition of stimulus control over released pecking by hatchling chicks. Canadian Journal of Psychology, 43, 431-443.

Suboski, M. D. (1989b). Recognition learning in birds. Perspectives in Ethology, 8, 137-171.

Suboski, M. D. (1990). Releaser-induced recognition learning. Psychological Review, 97, 271-284.

Suboski, M. D. (1992a). Releaser-induced recognition learning by amphibians and reptiles. Animal Learning & Behavior, 20, 63-82.

Suboski, M. D. (1992b). Releaser-induced recognition learning by gastropod molluscs. Behavioural Processes, 27 1-26.

Suboski, M. D., Bain, S., Carty, A. E., McQuoid, L. M., Seelen, M. I., & Seifert, M. (1990). Alarm reaction in acquisition and social transmission of simulated-predator recognition by fish (Brachydanio rerio). Journal of Comparative Psychology, 104, 101-112.

Suboski, M. D. & Bartashunas, C. (1984). Mechanisms for social transmission of pecking preferences to neonatal chicks. Journal of Experimental Psychology: Animal Behavior Processes, 10, 182-194.

Suboski, M. D., Brace, T. G., Jarrold, L. A., Teller, K. G., & Dieter, R. (1972). Interstimulus interval and time estimation in ratings of signaled shock aversiveness. Journal of Experimental Psychology, 96, 407-415.

Suboski, M. D., Muir, D., & Hall, D. (1993). Social learning in invertebrates. Science, 259, 1628-1629.

Sullivan, L. (1979). Long-delay learning with extroceptive reinforcement in rats. Australian Journal of Psychology, 31, 21-32.

Teller, K. J., Dieter, R., & Suboski, M. D. (1972). Time estimation and the interstimulus interval function in classical conditioning. Journal of Experimental Psychology, 95, 445-448.

Thorpe, W. H. (1963). Learning and Instinct in Animals. Cambridge: Harvard University Press. (2nd ed.).

Tinbergen, N. (1951). The Study of Instinct. London: Oxford University Press.

Tolman, C. W. (1964). Social facilitation of feeding behaviour in the domestic chick. Animal Behaviour, 12, 245-251.

Turner, E. R. A. (1964). Social feeding in birds. Behaviour, 24, 1-45.

Valzelli, L. (1973). The "isolation syndrome" in mice. Psychopharmacologia, 31, 305-320.

Verheijen, F. J. (1956). Transmission of a flight reaction amongst a school of fish and the underlying sensory mechanisms. Experientia, 12, 202-204.

Weigl, P. D. & Hanson, E. V. (1980). Observational learning and the feeding behavior of the red squirrel Tamiasciurus hudsonicus: The ontogeny of optimization. Ecology, 61, 213-218.

Will, B., Pallaud, B., Soczka, M. & Manikowski, S. (1974). Imitation of lever-pressing 'strategies' during the operant conditioning of albino rats. Animal Behaviour, 22, 664-671.

Wolin, B. R. (1968). Differences in manner of pecking a key between pigeons reinforced with food and with water. In A. C. Catania (Ed.), Contemporary Research in Operant Behavior. Glenview, Ill.: Scott, Foresman. p. 286.

Zajonc, R. B. (1965). Social facilitation. Science, 149, 269-274.

ETHOLOGICAL ROOTS OF LANGUAGE

R. ALLEN GARDNER and BEATRIX T. GARDNER
Center for Advanced Studies
University of Nevada/296
Reno NV 89557

ABSTRACT. Sign language studies of cross-fostered chimpanzees measure the amount of human culture transmitted to very young chimpanzees under special conditions. Cross-fostered chimpanzees, like human children, only acquire culture and language gradually, in a process that takes many years. Here we discuss details of the procedure, the overlap between human and chimpanzee infants in the contents of the first 50-item vocabularies, and the ways in which the signs of the chimpanzees exhibit the fuzziness of natural language categories. We also compare the cross-fostering approach with more traditional modular approaches to the study of language-like behavior in nonhuman animals.

Mostly, this NATO ASI has looked for the ethological roots of culture in the transmission of behavior from individual to individual within the same species. Sign language studies of cross-fostered chimpanzees approach the problem from a somewhat different angle. They study the amount of human culture that can be transmitted to chimpanzees under special conditions. This is very different from attempting to find out whether or not chimpanzees can acquire human culture, where culture is defined as an indivisible, either-or, all-or-none entity of Aristotelian logic. As the problem becomes one of degree rather than of kind, it becomes scientifically practicable.

The question of whether or not a cross-fostered chimpanzee can acquire human culture to some degree was answered by the Hayes and Hayes (1951) in their study of the chimpanzee Viki. In spite of her many, appropriately childish, cultural accomplishments, Viki learned to speak only four English words, and those only with great difficulty. She died of encephalitis when she was barely seven years old. It takes human beings much longer than seven years to reach any adult level of human culture and chimpanzees should take more rather than less time to reach any given level of acculturation. Viki's premature death terminated the project before anything approaching a limit of social development could be expected. The lack of two-way communication in a natural language placed an addition limit on the level of human culture

R. A. Gardner et al. (eds.), The Ethological Roots of Culture, 199–222.

that Viki could have reached at any age.

In Project Washoe, and its sequel with Moja, Pili, Tatu, and Dar, infant chimpanzees also grew up in a human-style household, with human foster families. This laboratory improved on the cross-fostering of Project Viki by giving the infant chimpanzees total immersion in a naturally occurring human language, American Sign Language (ASL), the primary language of the deaf community in North America. Whenever the cross-fosterlings were present, all business and all casual conversation was in ASL.

1. A Cross-Fostering Laboratory

Many people keep pets in their homes. They may treat their pets very well, and they may love them dearly, but they do not treat them like children. True cross-fostering -- treating the fosterling like a human child in all respects, in all living arrangements, 24 hours a day every day of the year -- requires a rigorous experimental regime that has rarely been attempted (Kellogg, 1968).

For young chimpanzees no less than for human children, familiarity with simple tools such as keys, devices such as lights, articles of clothing such as shoes, are intimately involved in learning signs or words for keys, lights, shoes, opening, entering, lighting, and lacing. We kept the laboratory well stocked with such objects and activities, and the subjects had free access to them, or at least as much access as young human children usually have. While no more free than human children to go outdoors without permission, they were free of mechanical restraints both indoors and out. They not only learned to eat human style food, they learned to use cups and spoons and to clear the table and help wash the dishes after a meal. They not only learned to use human toilets (in their own quarters and elsewhere) but they learned to wipe themselves and flush the toilet, and even to ask to go to the potty during lessons and after bedtime. R. Gardner and Gardner (1989) is a detailed description of their daily indoor and outdoor life in human style surroundings.

1.1. SIBLING SPECIES

Chimpanzees are an obvious first choice for cross-fostering. They look and act remarkably like human beings and recent research reveals closer and deeper biological similarities of all kinds (Goodall, 1986). In blood chemistry, for example, the chimpanzee is not only the closest species to the human, but the chimpanzee is closer to the human than the chimpanzee is to the gorilla or to the orangutan (Stanyon, Chiarelli, Gottlieb, & Patton, 1986; Ruvolo, 1994).

For cross-fostering, the most important resemblance is that chimpanzees and humans mature very slowly. Infant chimpanzees are quite helpless:

warmth, bodily care, and food must be provided (Plooij, 1984). In the cross-fostering laboratory, Moja, Pili, Tatu, and Dar only began to roll over by themselves when they were four to seven weeks old, to sit at ten to fifteen weeks, and to creep at twelve to fifteen weeks. The change from milk teeth to adult dentition begins at about five years. Under natural conditions in Africa, infant chimpanzees depend on their mothers' care almost completely until they are two or three. They cannot survive if their mother dies before they are three, even when older siblings attempt to care for them. Weaning only begins when they are between four and five. In Africa, young chimpanzees usually live with their mothers until they are seven; and females often stay with their mothers until they are ten or eleven. Menarche occurs when wild females are ten or eleven, and their first infant is born when they are between twelve and fifteen (Goodall, 1986 pp. 84-85, 443). Captive chimpanzees have remained vigorously alive, taking tests and solving experimental problems when they were more than 50 verified years old (Maple & Cone, 1981).

Under favorable conditions, their behavioral repertoire continues to expand and develop throughout their long childhood (Goodall, 1967; Hayes & Nissen, 1971; Plooij, 1984). This gives us a long scale of comparative development for measuring the progress of cross-fostered chimpanzees.

1.2. SUBJECTS

Washoe was captured wild in Africa. She arrived in Reno on June 21, 1966, when she was about ten months old and lived as a cross-fosterling until October 1, 1970. In the next project, we started newborn chimpanzees at intervals to capitalize on the relationships between older and younger foster siblings. Moja, Pili, Tatu, and Dar, were born in American laboratories and each arrived within a few days of birth. Moja, a female, was born at the Laboratory for Experimental Medicine and Surgery in Primates, New York, on November 18, 1972, and arrived in our laboratory in Reno on the following day. Cross-fostering continued for Moja until late Spring 1979. Pili, a male, was born at the Yerkes Regional Primate Research Center, Georgia, on October 30, 1973, and arrived in our laboratory on November 1, 1973. (Pili died of leukemia on October 20, 1975, so that his records cover only the first two years.) Tatu, a female, was born at the Institute for Primate Studies, Oklahoma, on December 30, 1975, and arrived in our laboratory on January 2, 1976. Dar, a male, was born at Albany Medical College, Holloman AFB, New Mexico, on August 2, 1976, and arrived in our laboratory on August 6, 1976. Cross-fostering continued for Tatu and Dar until May, 1981.

1.3. OBSERVERS

Reveille was at 7:00 AM and lights out at 8:00 P.M., and through all of those hours the cross-fosterlings were never left alone. Infant chimpanzees are much more dependent than infant humans (Goodall, 1986 pp. 101-104, 203, 379). One or more adults had to be with each infant every day, seven days a week, every week of the year. Frequent periods of overlap were the time for games and other activities that required more than one player, and the time when adults modeled conversation in sign language.

Each cross-fosterling had an individual family, consisting of a small group of primary members, much like the nuclear family of a human child. Each observer was a primary member of one cross-fostering family, and most observers participated to some extent in at least one other family. In this way, each family consisted of a small nuclear group and several extended family members. As in a human family, the adults played with the young chimpanzees, carried them about, cared for them, kept house, and so on, teaching sign language and other skills in the course of these daily activities.

The human members of the foster families were trained observers. Most had completed their undergraduate degree in Psychology, Anthropology, or Biology and many were concurrently graduate students in Experimental Psychology. We recruited many for their expertise in sign language: persons who were deaf themselves, hearing persons who were native signers, and persons who had worked as teachers of deaf children. Some were deaf students enrolled as graduate students in Psychology, and thus combined both areas of expertise.

Each foster family was a stable group. During Washoe's last year in Reno for example, 100% of her waking hours were divided among just six individuals. Three of these six, had been members of her family continuously from the first day that Washoe arrived in Reno. Two others had been members of the family continuously for more than three of Washoe's four years in Reno. In the second project, the foster families were somewhat larger, and most individuals belonged to more than one family. Thus, during his final year in Reno, when Dar was between four and five years old, his waking hours were divided among eight individuals. Three individuals left the project and were replaced during the course of that year, so there were, altogether, 11 members in Dar's family that year. Four had participated from the day that newborn Dar arrived in Reno, and three others had been members continuously for more than two years. All but two also participated in the families of Moja or Tatu and five had been members of Washoe's original foster family (see R. Gardner & Gardner, 1989, for more details of cross-fostering).

2. Fostering Development

The special conditions in this laboratory fostered all aspects of development.

2.1. MOTOR DEVELOPMENT

Authoritative psychologists used to teach that only the most basic postural and locomotor functions develop during the first year of human infancy, and these develop almost independently of the nurturing environment.

> Maturational factors (with the spontaneous exercise and practice derived from them) are sufficient to ensure the early development of typical postural [and locomotor] behavior as long as nutrition, shelter, space for practice, and perhaps a minimum of kindness from mother substitutes for infants not left with their mothers are provided (Riesen & Kinder, 1952, p. 173).

Mounting evidence of the contrast between individual and institutional child rearing led to a different view. In institutions, Dennis (1960) and Hunt, Mohandess, Ghodessi, and Akiyama (1976), for example, reported profound retardation, especially in institutions with a high infant to caretaker ratio. Some infants could neither sit alone nor creep at age 1, and many, could not walk at age 2. Interventions that lowered the infant to caretaker ratio produced dramatic effects. The mean age for sitting alone dropped from 39 weeks to 27 weeks, and for standing and cruising while holding onto the crib edge, from 69 weeks down to 41 (Hunt et al., 1976, pp. 207-208)

B. Gardner and Gardner (1989a) compares the age of onset of 50 early postural, locomotor, manipulative, perceptual, and social behaviors in the cross-fostered chimpanzees, Moja, Pili, Tatu, and Dar with norms established for human infants by Dennis and Dennis (1937), Shirley (1931/1933), and Cohen and Gross (1979). The cross-fosterlings were ahead of human infants on many items such as fixating objects and playing with own hands, but the sequence of development was highly correlated (rho = .77). Fixating objects and following moving objects came before bringing objects to mouth and playing with own hands, while sitting unsupported and patting, beating, and striking objects came later still, both for human infants and for cross-fostered chimpanzees (B. Gardner & Gardner, 1989a, pp. 439-441, Tables 1 and 2).

Riesen and Kinder (1952) used the items of Gesell and Thompson's (1929) test for human infants to measure the age of onset of motor milestones for 14 infant chimpanzees reared at Yerkes Laboratories. They found a parallel pattern of development with laboratory infants somewhat ahead of human infants. B. Gardner and Gardner (1989a) compared the cage-reared chimpanzees of Riesen and Kinder (1952) with wild chimpanzees reared by their own mothers and observed by Goodall (1967) and Plooij (1984). Just as human infants reared by their own mothers are ahead of human infants reared in institutions, so chimpanzee infants reared in the wild are well ahead of chimpanzee infants reared in cages: e.g., sit when propped at 10 weeks for wild, 19 weeks for caged infants; stand alone at 12 weeks for wild, 39 weeks for caged infants.

Table 3 of B. Gardner and Gardner (1989a) compares cage-reared, wild-reared, and cross-fostered chimpanzees on the overlapping items that provide observations for all three groups. On these milestones of development the cross-fostered chimpanzees are quite close to wild chimpanzees reared by their own mothers, while both wild and cross-fostered infants are distinctly advanced compared with cage-reared infants. With respect to these milestones of the first year, then, cross-fostering successfully brought Moja, Pili, Tatu, and Dar close to the full potential of their species.

2.2. OBJECT PERMANENCE

Object permanence is a major achievement of the sensorimotor period of human infancy. There is now a body of evidence that, in a variety of species, at least some individuals can achieve some form of object permanence (Doré & Dumas, 1987). So far however, these studies have missed the developmental dimension of Piaget's object permanence; human infants go through an orderly sequence of stages from continued visual pursuit of a disappearing object (Stages I and II) to a pattern of search that follows a consecutive series of hidden placements (Stage VI).

Pursuing the notion of orderly sequential development, Uzgiris and Hunt (1975) constructed a scale consisting of 15 tasks appropriate to each of Piaget's six stages of object permanence. These tasks use objects and play routines appropriate to infants growing up in a human household. The Uzgiris and Hunt scale showed that human infants proceed through the Piagetian stages of object permanence in the predicted orderly sequence. The tasks are equally appropriate for cross-fostered chimpanzees and Wood, Moriarty, Gardner, and Gardner (1980) took advantage of this to administer the scale to Pili and Moja starting when the cross-fosterlings were 18 and 30 months old. Concurrently, the same experimenters administered the identical scale to three human infants, starting when they were 8, 18, and 24 months old. They tested both groups of infants over a period of 3 - 4 months so that the results traced a developmental pattern in which chimpanzees and children progressed through the graded stages scaled by Uzgiris and Hunt (1975). The cross-fostered chimpanzees closely resembled the human children in the number of steps achieved at comparable ages, the sequence and rate of achievement, and even in detailed characteristics of searching behavior.

3. Fostering Language

3.1. TEACHING

In teaching sign language to Washoe, Moja, Pili, Tatu, and Dar we imitated human parents in a human home. We called attention to everyday events and objects that might interest the young chimpanzees, e.g., THAT CHAIR. SEE PRETTY BIRD. MY HAT. We probed and questioned and we tried hard to answer questions and to comply with requests. We expanded on fragmentary utterances using the fragments to teach and to probe. We also followed the parents of deaf children by using an especially simple and repetitious register of ASL and by making signs on the youngsters' bodies to capture their attention (Maestas y Moores, 1980; Marschark, 1993; Schlesinger & Meadow, 1972).

The procedures of operant conditioning were plainly inappropriate. Withholding food or water or any other necessity, administering formal trial by trial training, paying them off with treats for each sign or each sentence, all would have defeated the primary purpose of the cross-fostering laboratory. Many operant behaviorists have insisted that the results must have been "established in chimps through rigorous application of conditioning principles" (Schwartz, 1978, p. 374). In fact, attempts at drill or bribery only interfered with the task at hand (see for example, B. Gardner & Gardner, 1971, pp. 123-140). There was a pattern to the failure of operant methods with Washoe, Moja, Pili, Tatu, and Dar that, together with the failures of operant conditioning in many other laboratories, calls for a thorough reanalysis of the traditional view that learning depends on contingent reinforcement (see R. Gardner & Gardner, 1988, for an extensive analysis of this question).

Washoe, Moja, Pili, Tatu, and Dar signed to friends and to strangers. They signed to themselves and to each other, to dogs, cats, toys, tools, even to trees. We did not have to tempt them with treats or ply them with questions to get them to sign to us. Most of the signing was initiated by the young chimpanzees rather than by the human adults. They commonly named objects and pictures of objects in situations in which we were unlikely to reward them.

When Washoe signed to herself in play, she was usually in a private place -- high in a tree or alone in her bedroom before going to sleep. All of the cross-fosterlings signed to themselves when leafing through magazines and picture books, but Washoe resented our attempts to join in this activity. If we persisted in joining her or if we watched her too closely, she often abandoned the magazine or picked it up and moved away. Washoe not only named pictures to herself in this situation, she also corrected herself. On one occasion, she indicated a certain advertisement, signed THAT FOOD, then looked at her hand closely and changed the phrase to THAT DRINK, which was correct.

The cross-fosterlings also signed to themselves about their own ongoing or impending actions. We often saw Washoe moving stealthily to a forbidden part of the garden signing QUIET to herself, or running pell-mell for the potty chair while signing HURRY (B. Gardner & Gardner, 1974).

3.2. FIELD RECORDS

Project Washoe continued for 51 months. In the second project, Moja, Pili, Tatu, and Dar altogether spent a total of 229 months in the cross-fostering laboratory. Throughout this period the observers recorded spontaneous utterances along with other behavioral developments in a daily journal. The field records for each of the five cross-fostered subjects, occupy 100 to 300 handwritten pages per month, and contain detailed descriptions of sensorimotor, social, and communicative behavior, as well as reports on food intake, wakefulness and sleep, etc. (see B. Gardner, Gardner, & Nichols, 1989, for details).

3.2.1. *Text and Context.* The human observers recorded the events that preceded and followed chimpanzee utterances. They described routine events quite briefly and special or unusual events in more detail. They also recorded adult utterances or other chimpanzee utterances that preceded and followed a chimpanzee utterance. When, as often happened, there were several rounds of utterance and rejoinder, the observers recorded these in the form of a dramatic script with appropriate notes on nonverbal context.

3.2.2. *Interobserver Agreement.* Even film or taped samples must be transcribed and transcriptions must be checked for accuracy by the agreement between a pair of independent observers. Transcribing rich samples is expensive in time and effort, so investigators usually check a subsample only for agreement between a pair of independent transcribers. Rimpau, Gardner and Gardner (1989) presents an analysis of eight videotaped sessions of Dar interacting with a familiar adult when he was 40 to 49 months old. Two independent observers transcribed one of the sessions or 12% of the sample to measure agreement in identifying individual signs and agreement in identifying utterance boundaries or segmentation. They agreed in their identification of 84% of the individual signs and 81% of the utterance boundaries (Rimpau et al., 1989, pp. 245-250). For a videotaped sample of Tatu interacting with a familiar adult, two independent observers transcribed 14% of the sample and agreed on 81% of the individual signs and 85% of the utterance boundaries.

We also measured agreement between independent observers making written transcriptions of live signing. In Drumm, Gardner, and Gardner (1986) Tatu and Dar responded to a systematic experimental set of announcements and events. A second observer witnessed 10% of the trials, and the agreement was 80% for individual signs and 95% for utterance boundaries. In their vocabulary tests Washoe, Moja, Tatu, and Dar named photographs. Two independent observers, who were prevented from seeing either the photographs or each other, transcribed live all of the responses on every trial. There were 760 trials in all and median agreement was 90% for correct and incorrect responses combined (B. Gardner and Gardner, 1989b, p. 191). Vocabulary tests represent a particularly stringent measure of the reliability of live transcription because the responses consisted of single signs and the tests consisted of randomly distributed vocabulary items. In tests of the intelligibility of human speech, random examples of single lexical items represent the most difficult case.

3.3. FIRST FIFTY SIGNS

Washoe, Moja, Pili, Tatu, and Dar grew up in the world of a human child and learned signs from their human foster parents the way children learn signs and words in such an environment. Consequently, their vocabularies resembled the vocabularies of human children. Nelson's (1973) detailed report of the first 50 words in the vocabularies of 18 English-speaking children is particularly suitable for comparatison.

Nelson carefully instructed the mothers to act as observers for her sample of 18 toddlers and checked their reports in detail in monthly interviews. On the date that each new word appeared the mothers recorded pronunciation, context and "the adult word that it means (p. 128)." In important respects then, Nelson's records are comparable to the records for the first appearance of signs in the vocabularies of cross-fostered chimpanzees.

Nelson listed the items of vocabulary according to the context notes and English glosses provided by the mothers rather than the actual lexical terms. For example, Nelson listed the child Leslie's "pa" for apples as "apple" and her "ba" for bottles as "bottle". Similarly, Nelson listed different child forms such as DOG, DOGGIE, PUPPIE, and WOOF-WOOF as the same term, DOG; forms such as CAT, KITTY, and MEOW, as CAT; and so on. The result is an inventory of the conceptual contents of the vocabularies. Since agreement depends on identity of referents rather than identity of morphemes, Nelson's method shows agreement in the conceptual contents of these early vocabularies in spite of widely different pronunciation and even widely different terms in the vocabularies of different children. The conceptual contents of different vocabularies can be compared even for children acquiring languages as different as English and ASL.

Even though the cross-fosterlings used signs of ASL, Nelson's technique of listing referents rather than actual terms shows that the contents of their 50-sign vocabularies were very similar to the contents of the 50-word vocabularies in her sample of English-speaking children. Both had terms for Nelson's Specific Nominals such as proper names for themselves and their parents or foster family members, terms for Nelson's Actions, such as GO and HUG, terms for Nelson's Modifiers, such as HOT and MINE, terms for

Nelson's Social Expressives such as PLEASE and GOOD-BYE, and so on. Nelson's General Nominals accounted for about half of all of the terms in both sets of vocabularies and both included terms for common household objects such as HAT, SHOE, DOG, and TOOTHBRUSH.

Nelson published the complete 50 item vocabularies for each of eight children as Appendix A of her monograph. As a result of the overlap in conceptual contents, there were 188 rather than 400 different items in the eight sets of 50. Furthermore, there were 356 rather than 900 different items in the 18 sets of 50 that comprised the entire sample. This decline in the proportion of new items when new vocabularies are added to the pool of items suggests that individual samples converge on a common child vocabulary. The overlap measures the amount of agreement between individual children and within the group as a whole.

For example, we can measure how well the vocabulary of one child, Paul, agrees with the vocabularies of the other seven children in Nelson's completely reported subsample. When pooled, the other seven vocabularies contain not 350 but only 162 different items and 24 of the 50 items in Paul's vocabulary are already included in this pool. We repeated this process eight times for each of the eight children to obtain the data that appear in Table 1. Table 1 shows that the amount of overlap between each child and the remaining seven children ranges from a low of 24 vocabulary items to a high of 33 with a median of 29.5. This is an index of how much each child vocabulary agrees with the rest in Nelson's subsample of eight speaking children. Table 1 as a whole estimates the amount of agreement in a representative sample of eight children.

Table 1. Overlap in the first 50 vocabulary items of eight hearing children.

Child	Paul	Beth	Rob't	Mark	Jane	Ellen	Lisa	Leslie	Median
Overlap	24	27	28	29	30	31	31	33	29.5

Overlap = number of items matched by items in the pooled vocabularies of the remaining seven children.

We can use this statistic to measure the amount of overlap of any child with this standard group, including a deaf child acquiring sign language. Collins-Ahlgren (1975) describes the early signed vocabulary of a deaf child of deaf parents (Subject #2) who was exposed to "total communication" (the simultaneous use of sign and speech), but whose own early expressive vocabulary consisted of signs, only. Like the first words of speaking children,

> first signs were usually simplifications of the adult model and had the charm of 'baby sign' (p. 525). . . . DOG meant all animals, MOTHER signified an adult caretaking person of either sex (p. 533).

The parents and Collins-Ahlgren kept chronological records of these early signs which "frequently included the contribution of the other participant in the dialogue" and "a brief description of the context, when the meaning of the recorded expression was not obvious". On the basis of these chronological records, Collins-Ahlgren compiled a table of expressive vocabulary for Subject #2 in which she listed, month by month, an English gloss for the items that had appeared in the records for this child. With the information published in Collins-Ahlgren (1975) the first 50 signs in the vocabulary of her Subject #2 can be identified and matched to terms in the vocabularies of Nelson's subjects.

We measured the degree of agreement by counting the number of items in the vocabulary of the signing child that were predicted by each of the eight sets of seven Nelson vocabularies. When this is done, the number of matching items ranges from 27 to 33 with a median of 32. Thus, the contents of the first 50 signs in the vocabulary of this deaf child of deaf parents are as predictable from the vocabularies in Nelson's sample as the contents of the first 50 words in the vocabulary of a hearing child of hearing parents. With Nelson's method of classifying items on the basis of reference, the measure of agreement is independent of the ease or difficulty of articulation of particular words or signs.

For Washoe, Moja, Pili, Tatu, and Dar the records that are most comparable to Nelson's are the first three independent reports of each new sign in their vocabularies. From these records we took the first 50 items that met the 15-day criterion of reliability for each subject. At the time that he died, only 35 of Pili's signs had met the 15-day criterion. In order to include his vocabulary in the comparisons, we included the first 50 of his signs that were confirmed by three independent observers.

We compared each chimpanzee vocabulary with each of the eight sets of seven vocabularies in the Nelson sample in the same way that we compared the vocabulary of Collins-Ahlgren's signing child. The results in Table 2 show that the agreement between the chimpanzees and the children is about the same as the agreement of the children with each other; any of the 50-sign chimpanzee vocabularies is as easy to predict from the 50-word vocabularies published by Nelson as any child vocabulary.

Table 2. Overlap in the first 50 vocbulary items of children and chimpanzees.

	Children	Washoe	Moja	Pili	Tatu	Dar
Median	29.5	31.5	28.5	30.5	29.0	29.0
Range	24-33	29-33	27-30	29-32	27-30	28-31

Overlap = number of items matched by items in the pooled vocabularies of eight sets of seven speaking children.

The 18 children in Nelson's sample took an average of 20 months to acquire a vocabulary of 50 words with a range of 15 to 24 months and Collins-Ahlgren's Subject #2 reached 50 signs at 21 months. Moja reached 50 signs at 25 months, Pili at 21 months, Tatu at 22 months, and Dar at 25 months. At this stage of their development the chimpanzees are within the range of the English-speaking children, but slightly behind them.

3.4. NATURAL LANGUAGE CATEGORIES

The signs in the vocabularies of the cross-fostered chimpanzees represented natural language categories.

3.4.1. *Semantic Range.* The first objective of vocabulary tests (B. Gardner & Gardner, 1989; R. Gardner & Gardner, 1984) was to demonstrate that chimpanzees can communicate information under conditions in which the only source of information available to a human observer was the signing of the chimpanzees. Washoe, Moja, Tatu, and Dar demonstrated this by naming pictures that were out of sight of their human interlocutors under rigorous testing conditions. An equally important objective of these tests was to demonstrate that their signs referred to natural language categories -- that DOG referred to any dog, FLOWER to any flower, and so forth. The chimpanzees accomplished this by naming a varied set of exemplars selected from a large library of photographs. In the tests, each slide appeared once and once only so that each trial was a first trial (B. Gardner & Gardner, 1989b; R. Gardner & Gardner, 1984). That is to say, on each trial the chimpanzees named a picture of an object that they had never seen before.

The cross-fosterlings did well on these tests, but they also made errors. In forced-choice tests of understanding, as for example, when subjects must choose between a few plastic tokens (Premack, 1971) or a few pictures on a testing board (Savage-Rumbaugh, McDonald, Sevcik, Hopkins, & Rubert, 1986) the tester provides the errors. In productive tests, errors contain

Figure 1a. Exemplars of APPLE (A), BERRY (B), FLOWER (C), and SHOE (D) used in tests of vocabulary. Test slides projected in color.

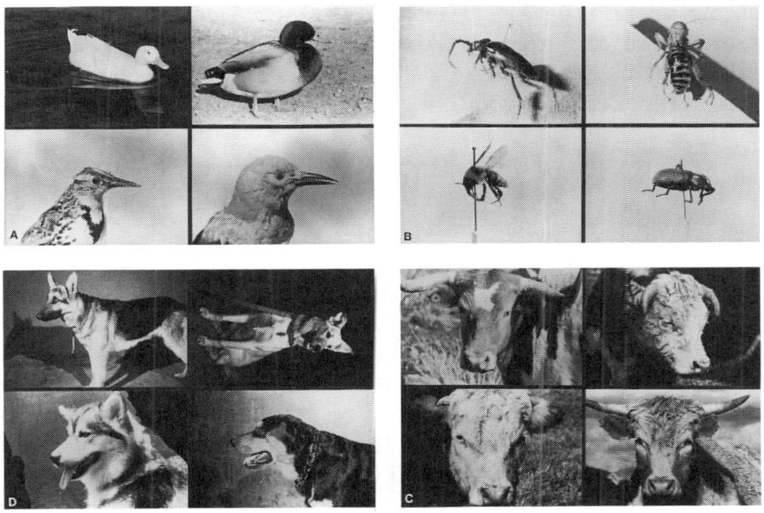

Figure 1b. Exemplars of BIRD (A), BUG (B), COW (C), and DOG (D) used in tests of vocabulary. Test slides projected in color.

information because the subjects are free to choose their own errors. On each trial of a test, signing chimpanzees can produce with their own hands any sign or combination of signs in their vocabularies. Most of the errors on the vocabulary tests depended on semantic relationships among the objects in the pictures or cheremic relationships among the signs. Thus, DOG was a common error for a picture of a cat, SODAPOP was a common error for a picture of ice cream, and so on. Meanwhile, signs placed on the nose such as BUG and FLOWER were common errors for each other, as were signs made by touching one hand with the other such as SHOE and SODAPOP (R. Gardner & Gardner, 1984, pp. 393-398).

Even the dithering between signs showed this effect. When the chimpanzees replied with more than one sign on one trial, the answer was often appropriate, as in WHITE DOG for a picture of a white dog. When on 14% of the trials, the chimpanzees responded with terms for different objects the observers reported only one of these as the response on that trial. Statistical analyses showed that the response was more likely to be wrong on such trials suggesting that the chimpanzees were more confused when they dithered in their choices. The signs that the chimpanzees made in these indecisive responses were also conceptually and cheremicaly related. Conceptual pairs, such as CAT and DOG, and cheremic pairs such as BUG and FLOWER were the most common. Sometimes, the dithering consisted of a string of related signs such as when Washoe signed CAT, BIRD, DOG, MAN for a picture of a cat, FLOWER, TREE, LEAF, FLOWER for a picture of daisies, or when she signed OIL, BERRY, MEAT -- all signs made by grasping different places on the passive hand -- for a picture of frankfurters (R. Gardner & Gardner, 1984, p. 398).

3.4.2. *Broader Categories*. Daisies are flowers, flowers belong to a broader category of botanicals such as trees and leaves, and botanicals are objects as distinguished from actions or traits. Appropriate answers to Wh-questions depend on membership in these broader semantic categories. In a published film (R. Gardner & Gardner, 1973; 1974), Greg asks Washoe a series of questions about her red boot. Her reply to WHAT THAT? is SHOE, to WHAT COLOR THAT? is RED, and to WHOSE THAT? is MINE. If, for example, she had replied GREEN when asked WHAT COLOR THAT of her red boot, she would have been incorrect, but her reply would still be appropriate to the question in a way that replies such as GREG or HAT or MINE would be inappropriate. Brown (1968) and Veneziano (1985) used the replies of human children to Wh-questions to show that children use different functional categories of words as sentence constituents.

B. Gardner and Gardner (1975) and Van Cantfort et al. (1989) embedded a systematic series of Wh-questions in the daily conversational interactions between adults and cross-fostered chimpanzees. The experimental questions restricted appropriate replies to one of a predefined set of semantic categories. At the same time, all possible replies were assigned unambiguously to exactly one semantic category.

Van Cantfort et al. (1989) analyzed a longitudinal series of Wh-questions that started when Moja, Tatu, and Dar were between 18 and 20 months old and continued through their first five years. As children grow older, the percent of replies to questions increases together with the percent of appropriate replies. Moja, Tatu, and Dar progressed in the same way (Van Cantfort et al., 1989, pp. 229-234, Figures 5.1 and 5.2). The cross-fosterlings also mastered particular kinds of Wh-questions in a sequence like the sequence reported for children. Both children and chimpanzees, initially provide nominals for What questions and locatives for Where questions. Later they provide verbs for What-do/predicate questions and proper nouns and pronouns for Who questions, and still later appropriate replies to Whose questions (for children see Ervin-Tripp, 1970, p. 105; for chimpanzees see Van Cantfort et al., 1989, pp. 234-236, Tables 5.16 and 5.17). Finer distinctions between the major interrogatives yield still finer parallels. For example, both children and chimpanzees provide appropriate replies to Who subject questions earlier than Who object questions (for children see Ervin-Tripp, 1970, p. 89; for chimpanzees see Van Cantfort et al., 1989, Tables 5.7 and 5.9).

The functional categories specified by Wh-questions also determined the errors that the chimpanzees made. As in the vocabulary tests, the Wh-question tests were productive tests in which the subjects could respond with any item in their vocabulary. They could create their own errors and these errors could be factually incorrect, yet functionally appropriate to the Wh-question. For example, the reply STRING to the question WHAT NAME THAT of a white leather belt, was factually incorrect with respect to the object, but functionally appropriate with respect to the question. Meanwhile, the reply WOOD to the question WHAT NAME THAT of a metal bell was both factually incorrect with respect to the object and functionally inappropriate with respect to the question. Gardner, Van Cantfort, and Gardner (1992) showed that 92% of Washoe's factually incorrect replies, 72% of Moja's, 82% of Tatu's, and 69% of Dar's were, nevertheless, functionally appropriate with respect to the Wh-question asked, and all of these results yielded highly significant chi-squares.

4. Degrees of Language

Toddlers are less than fully developed human beings, particularly with respect to intellectual functions, and they have only partially acquired their native languages. Their continuous development yields a relative scale of human development and a comparative scale of nonhuman development. The cross-fostering laboratory simulated as closely as possible the conditions in which human children grow and develop in all intellectual functions. From the beginning of Project Washoe, the comparisons were comparisons with what human children actually do and say rather than rich, theoretical interpretations or philosophical analyses of what children should do and say.

In a cross-fostering laboratory, the sign language of chimpanzees grows and develops. As shown in this article and in (B. Gardner & Gardner, this volume), size of vocabulary, appropriate use of sentence constituents, number of utterances, proportion of phrases, variety of phrases, length of phrases, complexity of phrases, inflection, all grew robustly throughout five years of cross-fostering. The growth was patterned growth and the patterns were consistent across chimpanzees. Wherever there are comparable measurements, the patterns of growth for cross-fostered chimpanzees parallel the characteristic patterns reported for human infants. Clearly, cross-fostering chimpanzees with ASL as a medium of communication transmitted some degree of language to the cross-fosterlings.

5. Modular Approaches

A much more popular approach has been to analyze language into theoretically defined components, such as concept formation or rules of order, and study each component separately. Investigators working in this modular tradition train, or attempt to train, nonhuman animals to perform arbitrary tasks that are related by theory to inferred cognitive functions. Success at an experimentally isolated, concrete task is evidence that this or that nonhuman being possesses this or that abstract, cognitive function. The conclusions depend entirely on the theoretical reasoning that ties the task to the cognitive target.

In this modular spirit, Harlow trained caged rhesus monkeys to solve oddity problems arguing that:

> We would be very much surprised if there is any fundamental difference in the learning of the oddity problem and the learning of differential equations -- other than that of complexity. (1958, p. 288)

The success of Project Washoe stimulated modular studies of a variety of nonhuman animals (e.g., Herman, Richards, & Wolz, 1984; Matsuzawa, 1985; Pepperberg, 1990; Premack, 1971; Rumbaugh, 1977; Schusterman & Gisiner, 1988; Terrace[1], Petitto, Sanders, & Bever, 1980; Savage-Rumbaugh, 1984) each study with its own theoretically defined components, highly specific tasks, and relentless application of operant conditioning. In modular studies, it would be pointless to avoid operant conditioning or any other training techniques, regardless of how foreign these techniques would be in a human nursery. The entire performance usually consists of requests for goods and services or answers to questions with goods or services as reward for correct answers or correct obedience to commands.

> With regard to the intensity of training, it was decided that Lana would live in the language environment 24 hours a day. There, her linguistic expressions would provide repeated, reinforcing engagement with the system, since she would have to obtain all of her necessities and social interactions by making appropriate requests of it. (Gill & Rumbaugh, 1977, p. 158)

Appropriately, the "symbol systems" of modular studies are usually synthetic and deliberately designed to be different from any naturally-occurring human language. The more arbitrary the task, the stronger the argument that mastery depends on some abstract, cognitive function. In the typical "language-training" study subjects only have access to their apparatus-based symbols during formal testing sessions and formal tests usually represent the entire population of relevant responses. Conversation and communication are irrelevant. Comparisons with what human beings actually do or say are equally irrelevant. The question of growth and development is also irrelevant because investigators only test one module, or only one module at a time in an arbitrary sequence. The modular approach uses very different procedures from the developmental approach of cross-fostering.

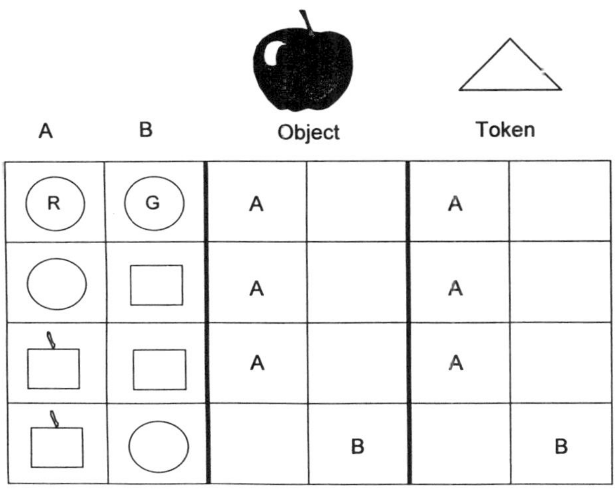

Figure 2. Premack's "feature analysis" of *apple*

5.1. MODULAR SEMANTICS

Figure 2 illustrates Premack's typically modular "feature analysis" of *apple* (Premack, 1971, p. 224-226). At the top of the figure the object on the left represents an actual example of an apple and the object on the right represents the flat plastic token, colored blue and shaped like a triangle that represented the "apple" in Premack's token language. The first pair of columns in the figure represent a series of tests presented to Premack's chimpanzee subject, Sarah, who had to choose between the following alternatives; 1) between a red-colored patch and a green-colored patch, 2) between a round shape and a square shape, 3) between a square shape with a stem-like protuberance at the top and a plain square shape, and finally 4) between square shape with a stem-like protuberance at the top and a round shape. The second pair of columns represents the choices that Sarah made when the apple object was present, and the third pair of columns represents the choices that Sarah made when the apple object was removed and replaced by the apple token.

According to Premack, when Sarah repeated with the apple token the same choices she had made with the apple object she demonstrated that the object and the token were equivalent to her. Premack's testing procedure, as reported, cannot demonstrate such equivalence as we have pointed out elsewhere (R. Gardner & Gardner, 1978). What concerns us here, however, is the relation of Premack's "feature analysis" to the way human beings use terms for *apple* in the various natural languages of the world.

Consider an attempt to present the same series of choices to a two- or three-year old human child. In the experience of most toddlers only some apples are red. Apples also come in green, yellow, and many shades in between. Some apples are more than one color at the same time. Peeled or sliced apples are white until they sit for a while and become brownish. Apples also come in many shapes, particularly when cut up for eating. Only some apples have stems, and so on. A human child growing up in human world would probably be baffled by Premack's "feature analysis." A cross-fostered chimpanzee who had equally broad experience with apples would be equally baffled.

Outside of certain special laboratory conditions, the natural world resists Aristotelian attempts to divide it into reds and nonreds, apples and nonapples, fruits and nonfruits, and so on. For certain natural philosophers this only means that the laboratory world is more faithful to the underlying truths of nature. Yet, the terms of natural languages mostly represent fuzzy categories that are faithful to the variegated, overlapping categories of the natural world (Zadeh & Kacprzyk, 1992). Moreover, human speakers and signers of natural languages use these fuzzy terms with great ease, beginning in infancy. For those who are interested in what actually happens in the natural world, this prominent characteristic of the human use of natural language is the clearest evidence that traditional Aristotelian logic is irrelevant to human language.

5.2. MODULAR CONVERSATION

Modular studies make sweeping and definitive theoretical claims (like Harlow's claim about oddity problems and differential equations), but they are always claims about aptitudes and capacities. Both investigators and commentators find it easy to label them as studies of "ape language" or "animal language" hence set apart from language proper.

Savage-Rumbaugh and her associates (Greenfield & Savage-Rumbaugh, 1990; Savage-Rumbaugh et al., 1986) have made particularly strong claims for their modular studies of the bonobo, Kanzi. Bonobo is the common name of an animal that seemed at one time to be a pygmy variety of chimpanzee. Modern taxonomists distinguish two separate species, the bonobo (*Pan paniscus*) and the true chimpanzee (*Pan troglodytes*). Genetically and physiologically, bonobos are more specialized and further removed than chimpanzees from the main hominid line, though not so far removed as gorillas and orangutans (Stanyon et al., 1986).

Kanzi communicated with human beings by pointing at abstract forms displayed on a large response panel. Savage-Rumbaugh and her associates call these forms "lexigrams," "words," and "elements," interchangeably. When Kanzi pointed at someone or touched someone present, this was transcribed as if he had pointed at a lexigram for "person." Nearly all Kanzi's lexigrams stand for different kinds of foods, drinks, destinations or services (Savage-Rumbaugh et al., 1986, Table 1).

Earlier work in the Rumbaugh laboratory with true chimpanzees used a computer-assisted device that recorded all responses, automatically (Rumbaugh, Gill, & Von Glasersfeld, 1973). Possibly because the computer-assisted device became too cumbersome for use outside of a cage, it was mostly replaced by a folding panel with painted patches representing the array of lexigrams. When Kanzi moved from place to place, his trainers folded the response panel for portability and set it up again when they were ready to receive Kanzi's next utterance.

Without automatic recording, records of Kanzi's lexigram utterances depended on observer reports like the records of the signed utterances of Washoe, Moja, Pili, Tatu, and Dar. Like other serious investigators Savage-Rumbaugh et al. (1986) saw that this procedure requires some evaluation of observer reliability. Accordingly, they video taped a sample of Kanzi's sessions with the response panel when he was in his fourth year. Independent transcriptions of the video tape agreed with the written reports of observers on the scene for 80% of individual lexigrams (Savage-Rumbaugh et al., 1986, p. 217).

Thus, there are three comparable video tape records, two for signing chimpanzees (described in 3.2.2., this article), and one for the lexigram utterances of a bonobo. All three were made under comparable conditions with comparable interobserver reliability when the subjects were at comparable ages. In their taped records, Tatu averaged 441 signed utterances per hour and Dar averaged 479 signed utterances per hour. In his tapes Kanzi averaged 10.2 lexigram utterances per hour. In Tatu's tapes 35% of her utterances consisted of combinations of two or more signs, and in Dar's tapes 49% of his utterances consisted of combinations of two or more signs. Savage-Rumbaugh has not yet published the proportion of phrases in Kanzi's video taped sample, but Greenfield and Savage-Rumbaugh (1990) describe a sample of Kanzi's utterances written down by observers on the scene when he was 60 - 69 months old. In the larger and later sample Kanzi averaged 10.1 utterances per hour and 10.4% of these contained two or more elements.

There are several differences that could explain why signing chimpanzees are so much more loquacious than Kanzi. First, bonobos are farther removed from the hominid line than chimpanzees (Stanyon et al., 1986). Second, the lexigram panel impedes the communication of any species. Greenfield and Savage-Rumbaugh (1990) report that Kanzi's human trainers, themselves, "most frequently inserted only one or two lexigrams per sentence, reflecting the mechanical difficulty of the lexigram mode" (p. 568). Third, the laboratory regimen of modular training and testing, itself, may depress any tendencies for spontaneous communication.

6. A Robust Phenomenon

Cuts in the research budget ended cross-fostering when Moja, Tatu, and Dar were just beginning to change from milk teeth to adult teeth, and far from intellectual maturity. Conditions in their next laboratory ended comparisons with human children, but signing went on. Cross-fostered chimpanzees converse among themselves without any human presence, when conversations must be recorded with remotely controlled cameras (D. Fouts, this volume). The infant, Loulis, adopted by Washoe when he was about a year old learned over 50 signs of ASL that he could only have learned from other chimpanzees (R. Fouts, this volume).

In 1994, twenty-four years after she left Reno, Washoe was still signing, not only to humans but to other chimpanzees whether or not there were any human beings in sight. This is more remarkable when we consider the procedure of Project Loulis. To show that Washoe could teach signs to an infant without human intervention, the Foutses prohibited all human signing when Loulis was present and Washoe lost almost all her input from human signers. Later, Moja joined the group in Oklahoma, and still later Tatu and Dar joined the group in Ellensburg, Washington. The signing chimpanzees continued to sign to each other and to the humans. Indeed, there was no way to stop them. They became part of Loulis's input.

Once introduced, sign language is robust and self-supporting, unlike the systems that depend on special apparatuses -- keyboards, plastic tokens, and so on. The regimen that the Foutses enforced to demonstrate that the infant Loulis could learn signs from Washoe, Moja, Tatu, and Dar, deprived them of human input and slowed the growth of their sign language. It also demonstrated that the sign language acquired by cross-fostered chimpanzees becomes a permanent and robust aspect of their behavior.

Note

1. Terrace and his associates introduced some confusion into the literature by blaming their failure to replicate Project Washoe on the incompetence of chimpanzees. Apart from teaching signs to a chimpanzee, however, these investigators failed to replicate the cross-fostering, the immersion in sign language, the human style of teaching, the stringent procedures of recording and verification, and the thorough analyses in the studies of Washoe, Moja, Pili, Tatu, and Dar (Gardner, 1981; Gardner & Gardner, 1986; Gaustad, 1981; O'Sullivan & Yeager, 1989; Van Cantfort & Rimpau, 1982). They have also failed to respond to extensive published criticism on these points.

References

Brown, R. (1968). The development of wh questions in child speech. *Journal of Verbal Learning and Verbal Behavior*, *7*, 277-290.

Cohen, M.A., & Gross, P.J. (1979). *The developmental resource: Behavioral sequences for assessment and program planning*. New York: Grune & Stratton.

Collins-Ahlgren, M. (1975). Language development of two deaf children. *American Annals of the Deaf*, *120*, 524-539.

Dennis, W. (1960). Causes of retardation among institutional children: Iran. *Journal of Genetic Psychology*. *96*, 47-59.

Dennis, W., & Dennis, M.G. (1937). Behavioral development in the first year as shown by forty biographies. *Psychological Record*, *1*, 349-361.

Doré, F.Y. & Dumas, C. (1987). Psychology of animal cognition: Piagetian studies. *Psychological Bulletin*, *102*, 219-233.

Drumm, P., Gardner, B.T., & Gardner, R.A. (1986). Vocal and gestural responses of cross-fostered chimpanzees. *American Journal of Psychology*, *99*, 1-29.

Ervin-Tripp, S. (1970). Discourse agreement: How children answer questions. In J.R. Hayes (Ed.), *Cognition and the development of language*, (pp. 78-106). New York: Wiley .

Gardner, B.T., & Gardner, R.A. (1971). Two-way communication with an infant chimpanzee. In A. Schrier & F. Stollnitz (Eds.), *Behavior of nonhuman primates*, (Vol. 4, pp. 117-184). New York: Academic Press.

Gardner, B.T., & Gardner, R.A. (1974). Comparing the early utterances of child and chimpanzee. In A. Pick (Ed.), *Minnesota symposium on child psychology*, (Vol. 8, pp. 3-23). Minneapolis: University of Minnesota Press.

Gardner, B.T., & Gardner, R.A. (1975). Evidence for sentence constituents in the early utterances of child and chimpanzee. *Journal of Experimental Psychology: General*, *104*, 244-267.

Gardner, B.T., & Gardner, R.A. (1989a). Prelinguistic development of children and chimpanzees. *Journal of Human Evolution*, *4*, 433-460.

Gardner, B.T., & Gardner, R.A. (1989b). A test of communication. In R.A. Gardner, B.T. Gardner, & T.E. Van Cantfort (Eds.), *Teaching sign language to chimpanzees*, Albany: SUNY Press.

Gardner, B.T., Gardner, R.A., & Nichols, S.G. (1989). The shapes and uses of signs in a cross-fostering laboratory. In R.A. Gardner, B.T. Gardner, & T.E. Van Cantfort (Eds.), *Teaching sign language to chimpanzees*, Albany: SUNY Press.

Gardner, R.A., & Gardner, B.T. (1973). Teaching Sign Language to the Chimpanzee, Washoe. (16 mm sound film). State College, PA: Psychological Cinema Register.

Gardner, R.A., & Gardner, B.T. (1974). Teaching sign language to the chimpanzee, Washoe. *Bulletin D'Audio Phonologie*, *4*(5), 145-173.

Gardner, R.A., & Gardner, B.T. (1978). Comparative psychology and language acquisition. *Annals of the New York Academy of Sciences*, *309*, 37-76.

Gardner, R.A., & Gardner, B.T. (1984). A vocabulary test for chimpanzees. *Journal of Comparative Psychology*, *98*, 381-404.

Gardner, R.A., & Gardner, B.T. (1988). Feedforward versus feedbackward: An ethological alternative to the law of effect. *Behavioral and Brain Sciences*, *11*, 429-447.

Gardner, R.A., & Gardner, B.T. (1989). A cross-fostering laboratory. In R.A. Gardner, B.T. Gardner, & T.E. Van Cantfort (Eds.), *Teaching sign language to chimpanzees*, Albany: SUNY Press.

Gardner, R.A., Van Cantfort, T.E., & Gardner, B.T. (1992). Categorical replies to categorical questions by cross-fostered chimpanzees. *American Journal of Psychology*, *105*, 25-57.

Gesell, A., & Thompson, H. (1929). Learning and growth in identical infant twins: An experimental study by the method of co-twin control. *Genetic Psychology Monographs*, *6*, 3-32.

Gill, T.V., & Rumbaugh, D.M. (1977). Training strategy and tactics. In D.M. Rumbaugh (Ed.), *Language learning by a chimpanzee*, (pp. 157-162). New York: Academic Press.

Goodall, J. (1986). *The chimpanzees of Gombe*. Cambridge, MA: Harvard University Press.

Goodall, J. van Lawick (1967). Mother-offspring relationships in free-ranging chimpanzees. In D. Morris (Ed.), *Primate Ethology*, (pp. 287-346). London: Weidenfeld and Nicolson.

Greenfield, P.M, & Savage-Rumbaugh, E.S. (1990). Grammatical combination in *Pan paniscus:* Processes of learning and invention. In S.T. Parker & K.R. Gibson (Eds.), *"Language" and intelligence in monkeys and apes: Comparative developmental perspectives*, New York: Cambridge University Press.

Harlow, H. (1958). The evolution of learning. In A. Roe & G.G. Simpson (Eds.), *Behavior and evolution*, (pp. 269-290). New Haven: Yale University Press.

Hayes, K.J., & Hayes, C. (1951). The intellectual development of a home-raised chimpanzee. *Proceedings of the American Philosophical Society*, *95*, 105-109.

Hayes, K.J., & Nissen, C.H. (1971). Higher mental functions of a home-raised chimpanzee. In A.M. Schrier & F. Stollnitz (Eds.), *Behavior of nonhuman primates*, (Vol. 4, pp. 59-115). New York: Academic Press.

Herman, L.M., Richards, D. and Wolz, J. (1984). Comprehension of sentences by bottlenosed dolphins. *Cognition*, *16*, 129-219.

Hunt, J. McV., Mohandessi, K., Ghodssi, M., & Akiyama, M. (1976). The psychological development of orphanage-reared infants: Interventions with outcomes (Tehran). *Genetic Psychology Monographs*, *94*, 177-226.

Kellogg, W.N. (1968). Communication and language in the home-raised chimpanzee. *Science*, *162*, 423-427.

Maestas y Moores, J. (1980). Early linguistic environment: Interactions of deaf parents with their infants. *Sign Language Studies*, *26*, 1-13.

Maple, T.L., & Cone, S.G. (1981). Aged apes at the Yerkes Regional Primate Research Center. *Laboratory Primate Newsletter*, *20*, 10-12.

Marschark, M. (1993). *Psychological development of deaf children*. New York: Oxford University Press.

Matsuzawa, T. (1985). Use of numbers by a chimpanzee. *Nature*, *315*, 57-59.

Nelson, K. (1973). Structure and strategy in learning to talk. *Monograph of the Society for Research in Child Development*, *38*, (1-2, Serial No. 149), 1-137.

Pepperberg, I. (1990). Cognition in an African gray parrot (*Psittacus erithacus*): Further evidence for comprehension of categories and labels. *Journal of Comparative Psychology*, *104*, 41-52.

Plooij, F.X. (1984). *The behavioural development of free-living chimpanzee babies and infants*. Norwood, NJ: ABLEX Publishing Corporation.

Premack, D. (1971). Language in chimpanzee? *Science*, *172*, 808-822.

Riesen, A.H., & Kinder, E.F. (1952). *Postural development of infant chimpanzees: A comparative and normative study based on the Gesell behavioral examination*. New Haven: Yale University Press.

Rimpau, J.B., Gardner, R.A., & Gardner, B.T. (1989). Expression of person, place and instrument in ASL utterances of children and chimpanzees. In R.A. Gardner, B.T. Gardner & T.E. Van Cantfort (Eds.), *Teaching sign language to chimpanzees*. Albany, NY: SUNY Press.

Rumbaugh, D.M. (Ed.) (1977). *Language learning by a chimpanzee*. New York: Academic Press.

Rumbaugh, D.M., Gill, T.V., & Von Glasersfeld, E. (1973). Reading and sentence completion by a chimpanzee. *Science*, *182,* 731-733.

Ruvolo, M. (1994). Molecular evolutionary processes and conflicting gene trees: The hominoid case. *American Journal of Physical Anthropology*, *94*, 89-113.

Savage-Rumbaugh, E.S. (1984). Verbal behavior at a procedural level in the chimpanzee. *Journal of the Experimental Analysis of Behavior*, *41*, 223-250.

Savage-Rumbaugh, E.S., McDonald, K., Sevcik, R.A., Hopkins, W.D., & Rubert, E. (1986). Spontaneous symbol acquisition and communicative use by pigmy chimpanzees. *Journal of Experimental Psychology: General*, *115*, 211-235.

Schlesinger, H.S., & Meadow, K.P. (1972). *Deafness and mental health: A developmental approach*. Berkeley: University of California Press.

Schusterman, R.J., & Gisiner, R. (1988). Artificial language comprehension in dolphins and sea lions: The essential cognitive skills. *Psychological Record*, *38*, 311-48.

Schwartz, B. (1978). *Psychology of learning and behavior*. New York: W.W. Norton.

Shirley, M.M. (1931/1933). *The first two years: A study of twenty-five babies*. Vol. I & Vol. II, Minneapolis: University of Minnesota Press.

Stanyon, R., Chiarelli, B., Gottlieb, K., & Patton, W.H. (1986). The phylogenetic and taxonomic status of *Pan paniscus:* A chromosomal perspective. *American Journal of Physical Anthropology*, *69*, 489-498.

Terrace, H.S., Petitto, L., Sanders, R.J., & Bever, T.G. (1980). On the grammatical capacity of apes. In K.E. Nelson (Ed.), *Children's language*, (Vol. 2, pp. 371-495). New York: Gardner Press.

Uzgiris, I.C., & Hunt, J. McV. (1975). *Assessment in infancy: Ordinal scales of psychological development*. Urbana, IL: University of Illinois Press.

Van Cantfort, T.E., Gardner, B.T., & Gardner, R.A. (1989). Developmental trends in replies to Wh-questions by children and chimpanzees. In R.A. Gardner, B.T. Gardner & T.E. Van Cantfort (Eds.), *Teaching sign language to chimpanzees*, Albany, NY: SUNY Press.

Veneziano, E. (1985). Replying to mothers' questions. *Journal of Pragmatics*, *9*, 433-452.

Wood, S., Moriarty, K.M., Gardner, B.T. & Gardner, R.A. (1980). Object permanence in child and chimpanzee. *Animal Learning & Behavior*, *8*, 3-9.

Zadeh, L.A., & Kacprzyk, J. (Eds.) (1992). *Fuzzy logic for the management of uncertainty*. New York: John Wiley & Sons, Inc.

DEVELOPMENT OF PHRASES IN THE UTTERANCES OF CHILDREN AND CROSS-FOSTERED CHIMPANZEES

BEATRIX T. GARDNER and R. ALLEN GARDNER
Center for Advanced Studies
University of Nevada/296
Reno NV 89557

ABSTRACT. Project Washoe and its sequel with Moja, Pili, Tatu, and Dar, simulated with infant chimpanzees the conditions in which the language of human children develops gradually and piecemeal into the language of human parents. This article traces patterns of growth and development in the early utterances of children and chimpanzees. The evidence for continuous processes and variables contradicts the yes-no, either-or Aristotelian logic of philosophical linguistics that has prevailed for so long. What the children and chimpanzees actually say supports, instead, a position that is much more compatible with modern natural science.

Exciting as they are, the first steps only mark the beginning of walking and the first words only mark the beginning of talking. Studies of the development of language in human children show a pattern of development extending over a period of years. If the earliest utterances represent language, then they are best described as primitive, childish language. Gradually and piecemeal, but in an orderly sequence, the language of human children develops into the language of their parents.

Sign language studies of cross-fostered chimpanzees are a tool for studying development within an ecologically appropriate context. So much commentary has concentrated on this or that traditional barrier between human and nonhuman or between language and nonlanguage that many have failed to appreciate the developmental perspective of the research. This article concentrates on growth and development in terms of patterns and degrees rather than discrete junctures.

R. A. Gardner et al. (eds.), The Ethological Roots of Culture, 223–255.

1. Longitudinal Sample

Further details of the cross-fostering procedure, including descriptions of the chimpanzee subjects appear in R. Gardner and Gardner (1989; and this volume).

1.1. TESTING AND SAMPLING

Cross-fostering studies of chimpanzees properly include a great deal of formal testing. Indeed, test results make up the bulk of the published scientific record (R. Gardner, Gardner, & Van Cantfort, 1989). As in studies of human children, however, formal tests of cross-fostered chimpanzees are embedded in a rich learning environment and test results represent only a small fraction of the continuous stream of behavior that appears throughout the waking day.

Records of human linguistic development also consist of samples that represent only a fraction of daily output because complete records are impractical, except in the earliest months when a child begins to speak. In the same way, the signed output of cross-fostered chimpanzees throughout the day soon grew to the point where complete records were impractical. Consequently, the output of the cross-fosterlings, like the output of human children is only represented by samples.

1.2. FOCAL MONTHS

Washoe's experience with sign language began when she was about ten months old so that some of her usage was relatively advanced, considering the size of her vocabulary, and at the same time relatively retarded, considering her age. Because Moja, Pili, Tatu, and Dar arrived in the cross-fostering laboratory within a few days of their birth, they provide a much clearer developmental record. We selected as *focal month samples* the field records of the 6th, 12th, 18th, 24th, 36th, 48th, and 60th month for Moja and Tatu. Dar left Reno during his 56th month so we substituted his 54th month for his 60th month here. Pili died of leukemia during his 24th month and was very ill for more than eight weeks before that so we analyzed only his 6th, 12th, and 18th months here.

2. Vocabulary

In Project Washoe we had to pioneer new procedures for establishing the reliability of individual signs and for verifying field notes with blind testing (R. Gardner & Gardner, 1969; B. Gardner & Gardner, 1971; 1974). We continued and extended these procedures in the second project with Moja, Pili, Tatu, and Dar.

2.1. CRITERIA

Early in Project Washoe, we developed procedures to determine when a sign first entered the vocabulary and when it became a reliable item (B. Gardner & Gardner, 1971, pp. 140-141). After three separate and independent reports of well-formed, unprompted, and appropriate, observations of a new sign by three different observers, we placed the new sign on the list of candidates for reliability. A sign remained on this list until there was at least one report of a well-formed, unprompted, and appropriate observation on each of fifteen consecutive days. We restarted the count after any day without a qualified report. After meeting the fifteen day criterion, the sign was added to the list of reliable vocabulary items. The record of the fifteen day criterial period contained independent reports by most if not all of the members of the foster family.

At least three human members of the foster family of each chimpanzee subject judged each of the three initial reports and the fifteen reports in the criterion series for acceptable diction, deciding difficult cases on the basis of the likelihood of confusion with other signs and the acceptability of similar variants in human children. During the second project with Moja, Pili, Tatu, and Dar there was always at least one and usually several members of the staff whose experience with young deaf children guided these decisions.

Each report also had to be judged as spontaneous and appropriate by at least three familiar human companions. It was spontaneous if it appeared without informative prompting -- such as direct modeling or guidance that could have induced any portion of the target sign. It was appropriate if it suited the verbal and situational context. The staff decided, for example, that SODAPOP was inappropriate for juice and CANDY was inappropriate for gum.

Some signs became reliable in just fifteen days, most required several weeks, and some required months. In all cases the date that the sign entered the vocabulary was the date of the third independent report, the report that met the criterion for putting the sign on the list of candidates for reliability. After a sign met the fifteen day criterion, the form and usage was reassessed regularly to be sure that old signs did not drop out as new signs entered the vocabulary, and also to chart developments in usage and diction (see B. Gardner et al., 1989, for more details, including Table 3.2 which lists and describes all of the signs of Washoe, Moja, Pili, Tatu, and Dar).

2.2. BASE VOCABULARIES

Figure 1 shows the number of signs that met the initial three observer criterion by the end of each focal month and that also met the fifteen day criterion of reliability before the chimpanzee left Reno. The solid line plots the median values, and the symbols indicate the scatter of individual chimpanzee values and the parallel patterns of development over time.

These, then, were the base vocabularies of individual signs in the phrases of each focal month. Vocabulary grew robustly throughout the 60-month period of cross-fostering. The rate of growth was limited, of course, by the stringent, time-consuming criterion which competed with a constantly growing number of other verification procedures. The last points for each chimpanzee underestimate the vocabulary during the last focal month because they exclude signs that had met the starting three observer criterion, but had not yet met the fifteen day criterion of reliability.

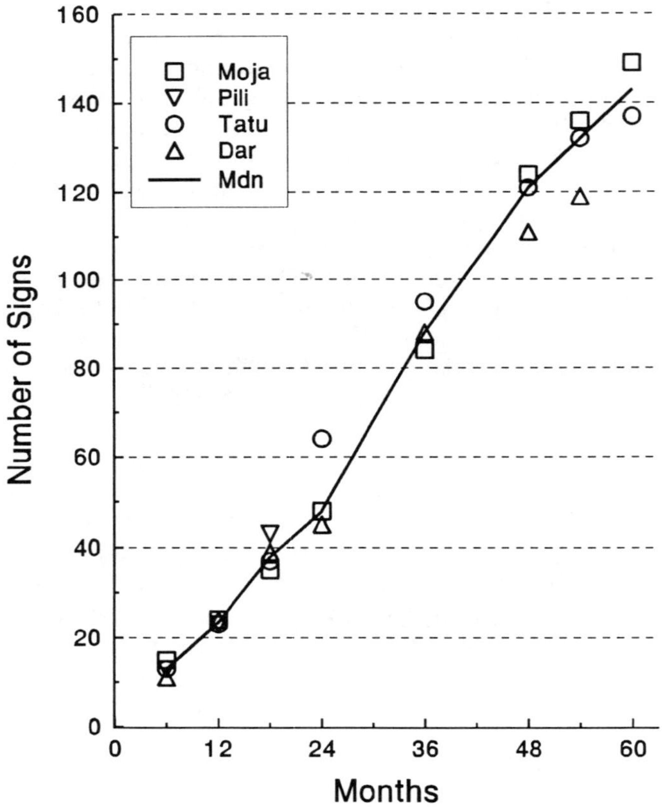

Figure 1. Growth of Vocabulary

Table 1. Semantic Categories

Category	Description	Examples
person	names of chimpanzees	DAR, TATU
	names of adult familiars	SUSAN, TOM
	generics for persons	BOY, GIRL
	pronouns	ME, YOU
object	animates	BUG, DOG
	inanimates	BALL, TREE
	edibles	APPLE, CEREAL
verb	actions	CHASE, KISS
noun/verb	either noun or verb	BRUSH, DRINK
locative	directions or places	OUT, UP
demonstrative	indicatives	THAT
attribute	colors	BLACK, RED
	possessives	MINE, YOURS
	materials	METAL, WOOD
	numbers	ONE, TWO
	comparatives	BIG, SAME
	sensory quality	HOT, SWEET
trait	personal attribute	GOOD, QUIET
negative	negative markers	CAN'T, FINISH
request	request/emphasis markers	HURRY, MORE

For detailed description see Gardner, Gardner, & Nichols (1989).

2.3. SEMANTIC CATEGORIES

For cross-fostered chimpanzees, as for human children, Golden Delicious and Rome Beauties are apples, apples are foods, and foods are objects as distinguished from actions, or attributes (see R. Gardner and Gardner, this volume, 3.3, for more details).

Like children, the cross-fostered chimpanzees also used broader, functional categories in forming their early phrases. Table 1 groups the signs in their vocabularies into these functional semantic categories. B. Gardner et al. (1989) discusses the evidence for appropriate usage in the field records and describes the sign categories in detail in Table 3.1. B. Gardner and Gardner (1975) and Van Cantfort et al. (1989) provide experimental evidence from Wh-question tests (see also R. Gardner & Gardner, this volume, 3.4.2). In addition, R. Gardner, Van Cantfort, and Gardner (1992) showed that even when replies to Wh-questions were incorrect Tatu and Dar usually replied with a sign from the semantic category specified by the question showing that the semantic categories controlled errors as well as correct replies. The classification of Table 1 rests firmly on a body of evidence that the cross-fostered chimpanzees used their signs in semantic categories.

Table 1 lists all signs for animate, inanimate, and edible objects as *objects* and all signs for actions as *verbs*. As in English, nouns in ASL can be used as verbs and *vice versa*, but some dual usages are more acceptable than others. Thus, in English, DRINK is freely used either as a verb or as a noun, but FOOD is less common as a verb and EAT is less acceptable as a noun. We classified as *objects* signs such as NUT, that the cross-fosterlings virtually always used for objects, and as *verbs* signs such as OPEN, that they virtually always used for actions. Signs such as BRUSH, that they used both for objects and for actions present a classification problem that studies of human children usually solve on the basis of context notes. To avoid this dependence on subjective interpretation and to keep the system as close as possible to the observations (see this article, 4.1.3), we classified separately as *noun/verbs* all signs that the chimpanzees used freely both for objects and for actions. The criteria used to distinguish between the *demonstrative* indexical, THAT, and the *locative* indexical, THERE, should become more clear in the discussion of sign order later in this article (see 4.3.2).

Ten signs appeared in phrases within the focal month samples but failed to appear in Table 3.1 of B. Gardner et al. (1989) because that table only includes signs that met the criterion of reliability by the last month of cross-fostering. Of these ten signs, LONG was classified as a *comparative attribute* in Table 1; WIFE, as a *generic person*; ASK, FIRST, and QUICK, as *request markers*; REFUSAL, as a *negative marker*; STUPID as a *trait*; DARK as a *noun/verb*; and WHOSE, as a *possessive attribute*. The tenth not-yet-reliable sign, THINK, appeared in three phrase types but was omitted from further analysis because its usage failed to place it unambiguously into one of the categories of Table 1. Altogether, not-yet-reliable signs appeared in only 21 of the 4601 phrases in the focal month samples analyzed in this article.

3. Phrases

From the time that Washoe had eight reliable signs in her vocabulary, she began to combine them in meaningful phrases such as, GIMME SWEET (when shown a jar of baby food dessert), COME OPEN (after the bathroom door was locked) and MORE TICKLE (after Naomi had tickled Washoe). Like human children, Washoe formed combinations without direct tuition, other than the modeling of simple phrases and sentences in normal conversation. Most of her phrases were spontaneous in the sense that they usually initiated an interchange and, even when they followed in response to an adult utterance, all or at least part of her phrases introduced new signs into the conversation. Many of Washoe's combinations were her own inventions, quite unlike any that adults had modeled for her. Washoe, for example, called her refrigerator the OPEN FOOD DRINK, and her toilet chair the DIRTY GOOD even though her human companions referred to these as the COLD BOX and the POTTY CHAIR.

Shortly after the publication of R. Gardner and Gardner (1969), which only covered the first 22 months of Project Washoe, D. O. Hebb sent us a reprint of Hebb and Thompson's chapter on comparative psychology which appeared in the (1968) *Handbook of social psychology*. Hebb put a bracket around the following passage.

> We propose therefore that the minimal criterion of language, as distinct from other purposive communication, is twofold. First, language combines *two or more* representative gestures or noises purposefully, for a single effect; and second, it uses the *same* gestures in different combinations for different effects, changing readily with circumstances. (p. 739)

In the margin, Hebb wrote "This criterion means Washoe has achieved language."

As youngish investigators and great admirers of Hebb, we were enormously flattered. Nevertheless, we resisted then and we continue to resist now all such either-or, yes-no criteria. Just as in the speech of human children, the first combinations only mark the beginning of a long process of growth and development in the sign language of cross-fostered chimpanzees.

3.1. CRITERIA

The definition of a phrase in this analysis is an utterance with two or more different signs within two utterance boundaries.

3.1.1. *Utterance boundaries.* In the last third of the film, *Teaching Sign Language to the Chimpanzee Washoe* (R. Gardner & Gardner, 1973; 1974), there are several examples of extended utterances embedded in bouts of play between Washoe and a human companion. In the following typical example Washoe is signing to Susan Nichols.

Washoe: YOU [hold] ME/ YOU ME WASHOE ME [hold] GO/

There are two hesitation pauses in this example, first when Washoe holds the first sign, YOU, and later when she holds the next to the last sign, ME. In both cases, she holds her hand in the configuration and the place of the sign throughout the pause. Most viewers, whether or not they are familiar with sign language, can see that these pauses are typical of the halting signing and speech of young children (Stokoe, 1978, pp. 83-85). It is even clearer that the final GO ends with a boundary marker because Washoe abruptly drops her hands out of the normal signing space and begins to "go". The boundary between the first phrase, YOU ME/ and the second phrase, YOU ME WASHOE ME GO/ is marked with a "/" here and in our field records. This type of boundary is more subtle because Washoe continues to hold her hands in the normal signing space. She relaxes them, however, and pauses briefly before beginning the second phrase. Utterance boundaries of this type are quite clear, and judgments yield high interobserver agreement (see this article, 1.7.3).

3.1.2. *Inflection.* Dictionaries of ASL, like dictionaries of spoken languages, show signs in citation form - the form that is seen when an informant responds to the question 'What is the sign for X?' In normal conversation, fluent signers inflect their signs in a variety of ways. The sign, GIVE, for example, may start near the signer's body and move out toward the addressee to indicate, "I give you". The same sign with the direction of the movement reversed, indicates, "You give me". Inflection makes signs more versatile and more expressive; a single lexical item can become several different signs. Typical inflections have parallel effects on many different signs. At least 20 different types of inflection appear in the field records. Each type can be characterized by an aspect of sign form - e.g., Place or Movement - and by the way in which this aspect differs from citation form (see B. Gardner & Gardner, 1989; Rimpau et al., 1989, and this article, 4.5 for more detailed discussion).

Even though inflection is a highly significant development, we have omitted reports of inflection from this analysis of phrase development so that we can concentrate on the combination of signs into phrases. Thus, this analysis of phrases treats all records of a given sign as if the sign appeared in citation form, whether or not the field notes indicate inflection.

3.1.3. *Incorporation and Reiteration*. Like human children (Bohannon & Stanowicz, 1989; Brinton & Fujiki, 1984; Keenan, 1977; Uzgiris, Broome, & Kruper, 1989), the cross-fostered chimpanzees often incorporated into their utterances a sign that had appeared in the immediately preceding utterance of their human companion. For example, in response to the question, YOU WANT ICECREAM? Dar signed, ICECREAM HURRY GIMME. Also, like human children (e.g., Keenan, 1977; Nelson, 1980; and Scollon, 1979, for speaking children; and Hoffmeister, Moores, & Ellenberger, 1975, p. 123, for signing children), the chimpanzees often reiterated a sign within an utterance. For example, in response to the statement YOU PLAY THERE (indicating a large cupboard where Dar liked to play) Dar signed OPEN OPEN. And, again like human children Tatu and Dar often reiterated an incorporated sign. In response to the statement TIME ICECREAM NOW, Tatu signed ICECREAM ICECREAM ICECREAM ICECREAM ICECREAM ICECREAM.

In child speech reiteration and incorporation are pragmatic devices indicating assent or emphasis (Keenan, 1977). Drumm et al. (1986) systematically varied the affective content of adult utterances that announced positive and negative future events and experimentally presented these announcements to Tatu and Dar in the normal course of daily life. Tatu and Dar used incorporation, reiteration, and reiteration of incorporation significantly more often in responses to announcements of positive events than in responses to negative or neutral announcements.

Clearly, ICECREAM ICECREAM ICECREAM ICECREAM ICECREAM ICECREAM is more than one sign and expresses much more than a single ICECREAM. Although reiteration is a significant phenomenon, it is beside the point of the present analysis of the use of signs in combinations. As in other cases of inflection, the analyses in this article treat all reiterations of the same sign within the same utterance as a single sign.

3.1.4. *Expansion*. When Dar responded to the question, YOU WANT ICECREAM? with the reply, ICECREAM HURRY GIMME, he took the sign ICECREAM from the question and incorporated it into an expanded utterance. In human conversation incorporation with expansion is an integral feature of topic maintenance, and the development of this device shows developing conversational skill in human children (Bloom, Rocissano, & Hood, 1976; Keenan, 1977).

In another exchange, Tatu responded THAT HURT to Cathy's comment, THAT MY HURT. In this case her entire response consisted of signs incorporated from the preceding adult utterance without any additional signs. Important as incorporation may be as a pragmatic device, this is beside the point of the present analyses. Therefore, these analyses only include replies and rejoinders that incorporated signs from an adult utterance if the chimpanzee also expanded the utterance with additional signs. They omit all chimpanzee utterances that consisted entirely of signs found in the immediately preceding adult utterance, even when the chimpanzee utterance rearranged the sequence of incorporated signs as when Dar responded to ME HUG YOU MORE with YOU MORE ME. This is the procedure that we have used in all analyses of phrases since B. Gardner and Gardner (1971).

3.2. PHRASE TOKENS

Many phrases were likely to be particularly common in certain situations, e.g., CHASE DAR and TOM CHASE in a bout of chase play between Dar and Tom. Often in such situations, the observer only reported that Dar signed CHASE DAR "many times" without reporting a specific number. For the purposes of the present analysis, whenever a report in the field records indicated more than one token of a particular phrase without specifying the number, we counted the number of *phrase tokens* of that phrase in that report as exactly two. Figure 2 plots the number of phrase tokens recorded during each focal month. After a slow start up to month 18, the volume of signing in phrase-length utterances grew steadily from the 24th to the 60th month of cross-fostering.

3.3. PHRASE TYPES

To measure the variety of phrases, we grouped all the tokens into types according to the signs that they contained. For this purpose all phrases that contained the same signs, as in CAN'T POTTY, POTTY CAN'T, POTTY CAN'T CAN'T, and CAN'T POTTY CAN'T POTTY CAN'T (recorded for Dar in his 36th month) counted as different tokens of a single *phrase type* containing the same two signs, CAN'T and POTTY. Thus, in this article the number of phrase types for each chimpanzee during each focal month is the number of phrases that contained at least two signs and differed from all other phrases by at least one sign, regardless of sign order within the phrase.

Figure 3 plots the number of different phrase types recorded during each focal month and shows that the variety of phrase types also grew steadily throughout the period of cross-fostering, again after a slow start up to month 18. Figure 4 plots the number of different phrase types containing three or more signs and shows that the number of longer phrase types also grew steadily throughout the period of cross-fostering. The curve of Figure 4 rises more sharply than the curve of Figure 3 because variety of longer phrase types grew both absolutely and proportionately faster than the base rate of phrases.

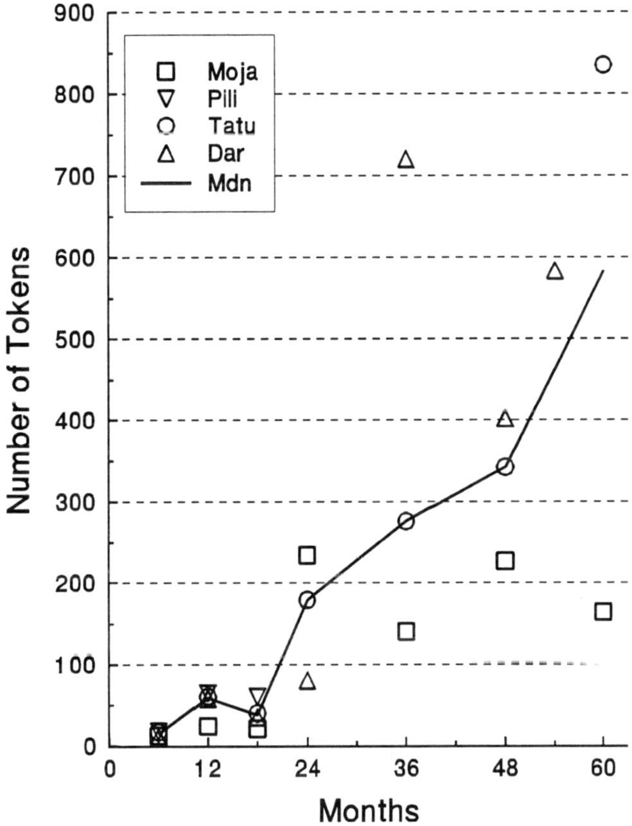

Figure 2. Growth of Phrase Tokens

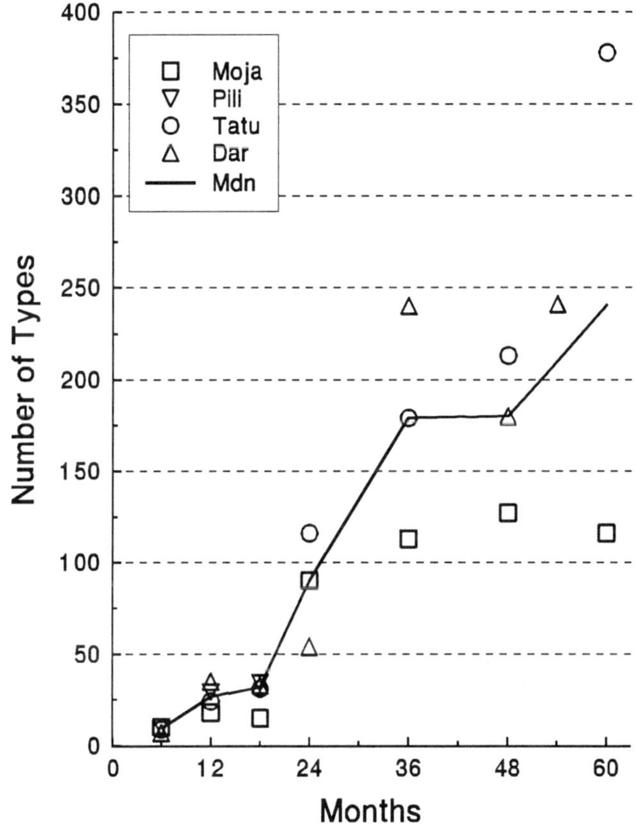

Figure 3. Growth of Phrase Types

3.4. PHRASE PATTERNS

In this article, phrase types are sets of phrase tokens that all contain exactly the same signs. A *phrase pattern* is a set of phrase types that are structurally related because all contain signs that belong to the same semantic categories. Thus, GROOM DAR, YOU TICKLE, and SUSAN CHASE are three distinct phrase types, but each contains one sign that belongs to the category, *person*, and a second sign that belongs to the category, *verb*, so all three phrase types belong to the structurally related set called here *person* + *verb*. Similarly, BLACK HAT, GLASS MIRROR, and YOUR SHOE, are three distinct phrase types, but each contains one sign that belongs to the category, *object*, and a second sign that belongs to the category, *attribute*, so all three of these belong to the structurally related set

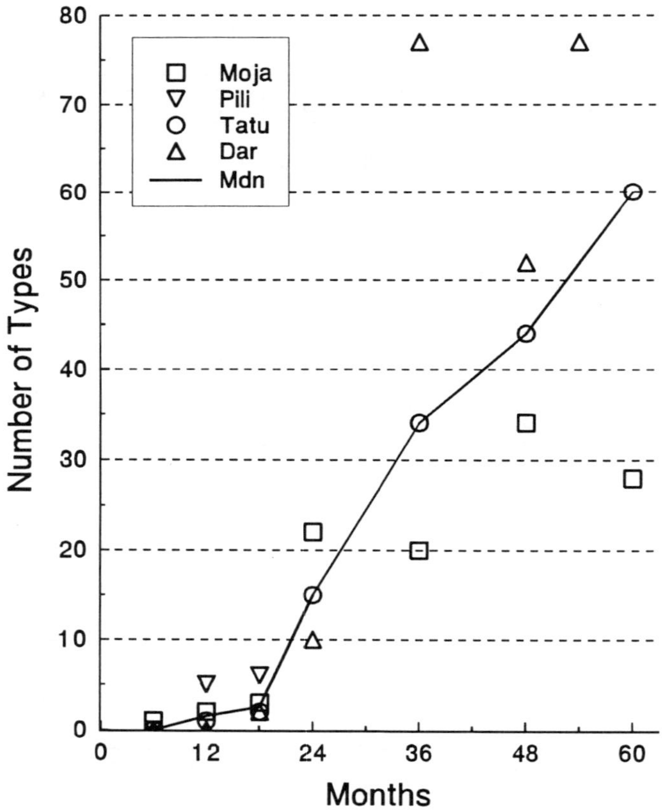

Figure 4. Growth of Phrase Types: Three or More Signs/Phrase

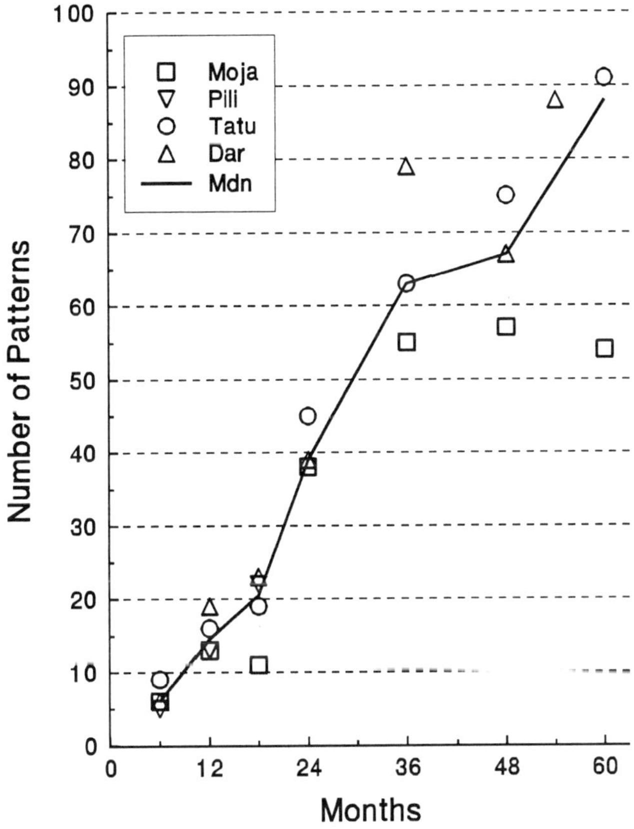

Figure 5. Growth of Phrase Patterns

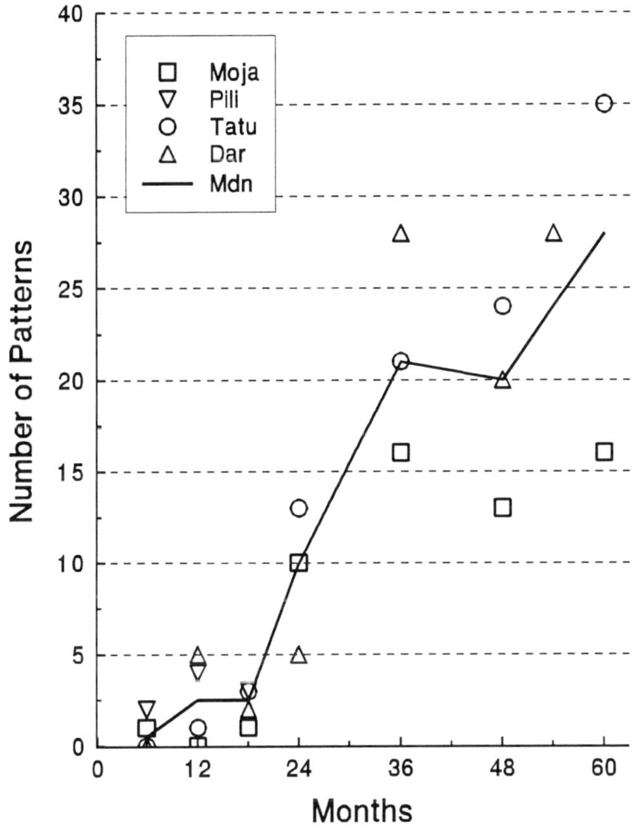

Figure 6. Growth of Productive Phrase Patterns

called here *attribute* + *object.* And, GOOD BOY, QUIET ME, and DAR SORRY each contain one *person* sign and one *trait* sign so all belong to the phrase pattern *person* + *trait.* Throughout this paper the terms for phrase patterns list the component semantic categories in alphabetical order regardless of the order used by the chimpanzees. We preassigned all of the items in each chimpanzee vocabulary to one of the ten semantic categories in Table 1. Consequently, each phrase type belongs unambiguously to one and only one of the possible phrase patterns. Indeed, our wordprocessor assigned phrase types to phrase patterns, automatically.

Figure 5 plots the number of different phrase patterns in each focal month and shows that the variety of phrase patterns grew steadily throughout the period of cross-fostering. The phrase patterns in the early utterances of human children are said to be *productive* when they generate three to five different phrase types within the same sample (Bloom, 1991, pp. 8-9). Figure 6 plots the number of different phrase patterns that generated three or more different phrase types during each focal month. After a slow start up to month 18, this measure of variety of productive phrase patterns also grew steadily throughout the period of cross-fostering.

3.4.1. *Productivity.* Certain phrase patterns, because they are much more productive than others, account for a disproportionately large share of the phrases in any sample of child or chimpanzee utterances. The labels on the vertical axis of Figure 7 are a list of all of the two-sign phrase patterns that appeared in the focal month samples. The lengths of the horizontal bars represent *productivity*, the number of different phrase types generated by each phrase pattern for each chimpanzee, after pooling the phrases in all seven focal months. Phrase patterns varied widely in productivity, and Moja, Tatu, and Dar agreed closely about the relative usefulness of the different patterns for generating phrases. The Pearson product moment correlation between Moja and Tatu is .65, between Moja and Dar is .78 and between Tatu and Dar is .87. All three values are statistically reliable beyond the .0001 level of significance.

As phrases of children and chimpanzees grow to three or more units they begin to look more like sentences. Like children (Reich, 1986, p. 86), Moja, Tatu, and Dar tended to combine elements of their most productive two unit phrases in Figure 7. They added person signs to request patterns as in PLEASE ME EAT GUM for Moja, YOU TICKLE HURRY for Tatu and GIMME TOOTHBRUSH PATRICK for Dar. They added person signs to action + object patterns as in YOU SMELL COAT for Moja (handing a coat to Ken), TATU EAT BANANA for Tatu, and BALL DAR CHASE for Dar. They also added a second person sign to person + action phrases as in YOU ME BATH for Moja, YOU ME CHASE for Tatu, and SUSAN CHASE DAR for Dar.

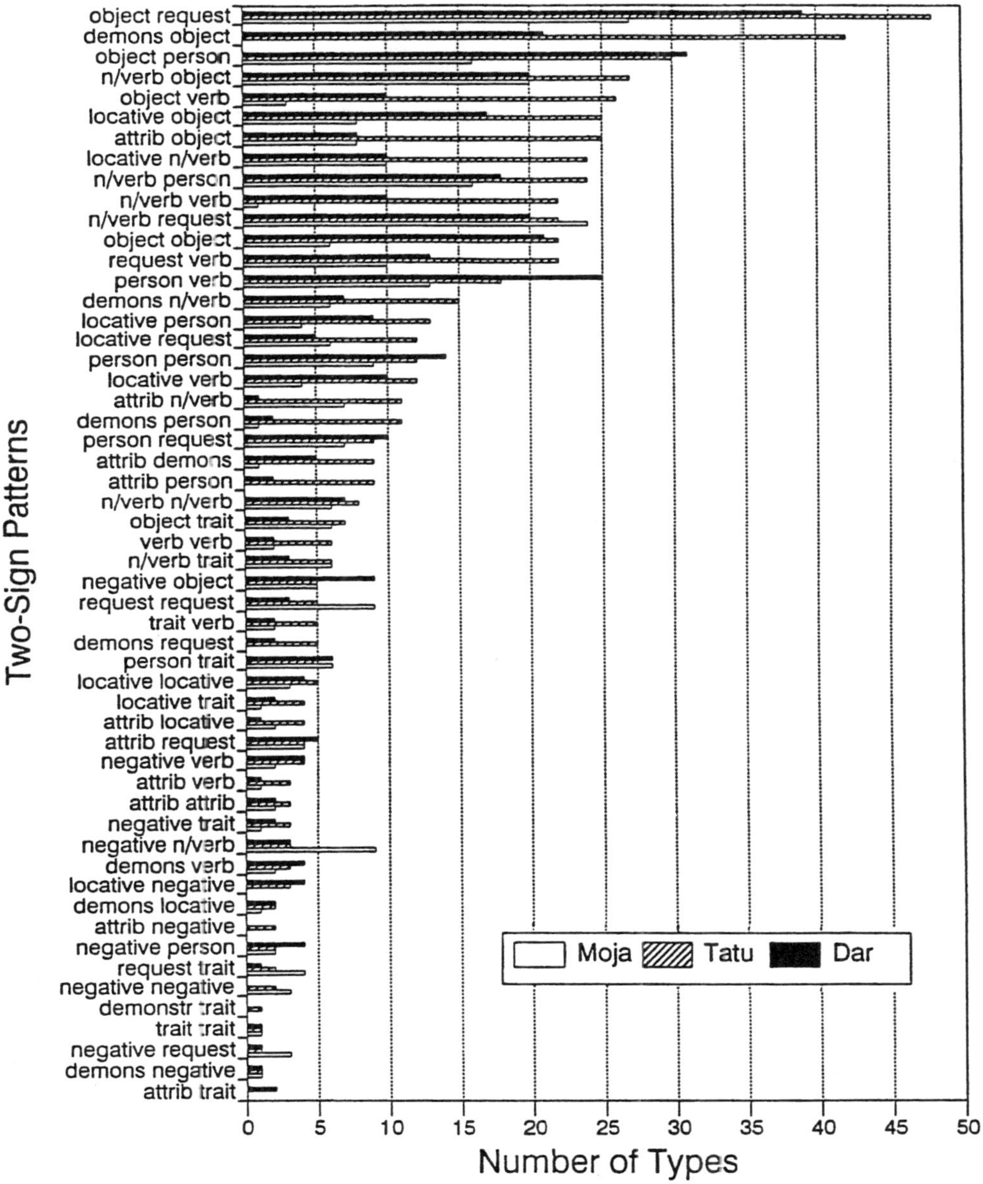
Two-Sign Patterns
object request
demons object
object person
n/verb object
object verb
locative object
attrib object
locative n/verb
n/verb person
n/verb verb
n/verb request
object object
request verb
person verb
demons n/verb
locative person
locative request
person person
locative verb
attrib n/verb
demons person
person request
attrib demons
attrib person
n/verb n/verb
object trait
verb verb
n/verb trait
negative object
request request
trait verb
demons request
person trait
locative locative
locative trait
attrib locative
attrib request
negative verb
attrib verb
attrib attrib
negative trait
negative n/verb
demons verb
locative negative
demons locative
attrib negative
negative person
request trait
negative negative
demonstr trait
trait trait
negative request
demons negative
attrib trait
0 5 10 15 20 25 30 35 40 45 50
Number of Types
Moja
Tatu
Dar

Table 2. Agreement on the Sequence of Appearance of Productive Phrase Patterns

Note. The letters M, T, and D represent Moja, Tatu, and Dar. Thus, the letter T in column 5 of row 3 indicates that Tatu's second different phrase type of the ***person verb*** pattern appeared in her 18th month sample.

		Focal Month					
Pattern	**Example**	**6**	**12**	**18**	**24**	**36**	**48**
request verb	HUG MORE	M	T D				
n/verb request	MORE VACUUM	D	M T				
person verb	YOU TICKLE	M	D	T			
n/verb person	BRUSH ME		M D	T			
demons n/verb	THAT TOOTHBRUSH		T D	M			
person request	MORE SUSAN		M D		T		
locative n/verb	POTTY THERE		D	T	M		
n/verb n/verb	EAT DRINK		D	T	M		
locative verb	GROOM THERE			T D			
demons object	THAT DOG			T D			
locative person	ME OUT			D	T		
n/verb object	SEE APPLE			T	M D		
object request	SHOE GIMME			D	M T		
person person	YOU ME	M			T	D	
n/verb trait	GOOD EAT		T			M	
locative request	PLEASE UP		D		M	T	
object person	SHOE DAR				M T	D	
request request	MORE GIMME				M	T	
object verb	NUT CATCH				T	D	
locative object	BALL THERE				T	M D	
attrib object	RED GUM				D	M T	
object object	BIB MILK				T	D	M
n/verb verb	DRINK SWALLOW					T D	
person trait	QUIET ME					M D	T
negative object	BALL CAN'T				D	T	
object trait	GOOD FLOWER					T	M
attrib n/verb	BLACK EAT					M	T
attrib demons	THAT METAL						T D

Spearman rank order correlations:

Moja/Tatu $r = .58$ $p < .005$, Moja/Dar $r = .56$ $p < .02$, Tatu/Dar $r = .68$ $p < .0005$

3.4.2. *Developmental Sequence.* Figure 5 shows that the number of different phrase patterns in the samples grew steadily throughout the first five years. The longitudinal samples for Moja, Tatu, and Dar also agreed closely as to which patterns became productive early in development and which became productive later. Considering only the two sign patterns for the moment, there are 55 possible patterns that can be formed from the ten semantic categories -- (10 x 9)/2 possible combinations of two signs belonging to different semantic categories plus 10 possible combinations of two signs belonging to the same semantic category. For the pooled sample of all seven focal months we called a phrase pattern productive if it generated at least five different phrase types within the pooled sample. By this criterion, 28 of the possible 55 patterns were productive for at least two cross-fosterlings. We dated the age of the chimpanzee at the time that a phrase pattern first became productive as the month of the sample in which the second different phrase type of that pattern appeared.

Table 2 shows how well Moja, Tatu, and Dar agreed as to the developmental sequence in which the patterns became productive. The Spearman rank order correlation between Moja and Tatu is .58 ($p < .005$), between Moja and Dar is .56 ($p < .02$), and between Tatu and Dar is .68 ($p < .0005$).

4. Comparisons with Children

The phrases of human children also grow in variety, length, and productivity, of course. Here we discuss more refined parallels in the early development of children and chimpanzees.

4.1. SEMANTIC RELATIONS

For human children, the first indications of phrase structure appear in the semantic relations expressed by two-word and two-sign phrase patterns that Brown (1970) refers to as "the first sentences" (for recent reviews see Barrett, 1989; Bates, O'Connell, & Shore, 1987; De Villiers & De Villiers, 1986; Reich, 1986).

4.1.1. *Parallel Schemes.* Studies of human children generally agree about the major semantic relations (De Villiers & De Villiers, 1986; Reich, 1986). Table 3 shows the agreement of four representative studies. Table 4 shows how well Moja, Tatu, and Dar fit the same general scheme. Examples of each semantic relation in Tables 3 and 4 are very similar because the contents of daily life are very similar for toddlers and for cross-fostered chimpanzees, with the same sorts of objects, events, and activities and a similar cast of characters. The vocabularies of children and chimpanzees serve the same semantic functions.

Table 3. Schemes of Semantic Relations Used in Studies of Children

Pattern	Bloom et al. (1975)	Bowerman (1973)	Braine (1976)	Brown (1973)
Nominative	Existence THIS BOOK, THAT DADDY	Predicate Nominative THAT CANDY, THAT HOLE	Indicating/Identifying HERE APPLE, THAT LIGHT	Demonstrative and entity THAT BOOK, THAT CAT
Agent + action	Action MOMMY PUSH, I FINISH	Subject-verb HUG MOMMY, KENDALL BITE	Actor-Action MOMMY SIT, DADDY WALK	Agent and action ADAM PUT, EVE READ
Object + action	Action TOUCH MILK, OPEN IT	Verb-object READ BOOK, OPEN LOTION	Verb-object HOLD HAND, DO BABY	Action and object PUT BOOK, HIT BALL
Locative action	Locative action, Place of action UP KATHRYN, GO OUTSIDE	Verb-locative PLAY BED, SIT HERE	Action-locative GO THERE, PUT DOWN	Action and locative WALK STREET, GO STORE
Object + locative[a]	Locative state LIGHT THERE, BABY IN	Noun-locative THERE COW, LOTION TUMMY	Locative THERE TOWER, LADY THERE	Entity and location SWEATER CHAIR, BOOK TABLE
Attributive	Attribution FUNNY DUCK, LITTLE THAT	Adjective-noun BIG BED, POOR DOGGIE	Property Indicating HOT TEA, RED BALLOON	Entity and attributive BIG TRAIN, RED BOOK
Possessive[a]	Possession MOMMY APPLE, MY PEN	Genitive-noun BALL MINE, KIMMY BIKE	Possessive DADDY COOKIE, MY PENNY	Possessor and possessed ADAM CHECKER, MOMMY LUNCH
Experience/Notice	Notice SEE JUICE, I SEE	with HEAR, LOOK AT, SEE LOOK KENDALL, SEE CANDY	with HEAR, LOOK AT, SEE SEE CAR, SEE PLANE	Experience ADAM SEE, HEAR HORN
Negative	Negation NO BIB, CAN'T SEE	Negative NO PAGE, NO RABBIT	Negative NO PACIFIER, NO WANNA	Negative NO WASH, NO WET
Requests	(not used)[b] –	(not used)[b] –	Requesting WANT COOKIE, GIMME BALL	(not used)[b] –

[a] noun + noun richly interpreted as object + location; name + noun richly interpreted as possessor + possessed, see text.

[b] request category only in Braine (1976) others distribute phrases with request terms in other categories, see text.

Table 4. Child Scheme of Semantic Relations Applied to Chimpanzees

Child Pattern	Chimpanzee Pattern				
Nominative	demonstrative object THAT HAT THAT KEY	demonstrative n/verb THAT DRINK THAT LIGHT	demonstrative person THAT NAOMI THAT PAT		
Agent + Action	person verb ME OPEN NAOMI HUG	n/verb person TATU DRINK YOU WRITE			
Object + Action	object verb BALL CATCH GROOM HURT	n/verb object DRINK SODAPOP EAT APPLE	demonstrative verb THAT OPEN THAT CATCH		
Locative action	locative verb GO OUT TICKLE THERE	locative n/verb BRUSH THERE DRINK HOME	locative person ME UP TATU DOWN		
Object + locative	locative object IN KEY NUT THERE				
Attributive	attribute object BLACK DOG RED ICECREAM	attribute n/verb RED TOOTHBRUSH HOT DRINK	attribute demonstrative THAT WHITE THAT RED	person trait GOOD ME SORRY DAR	
Possessive	object + MINE, YOURS MINE HAT YOUR SHOE	n/verb + MINE, YOURS DIRTY MINE MINE DRINK	demonst'ive + MINE, YOURS THAT MINE		
Experience/Notice	object + SEE, SMELL, HEAR FLOWER SMELL SEE FLOWER	person + SEE, SMELL, HEAR SEE SUSAN TATU HEAR			
Negative	negative object NO HAT ROCK NO ROCK	n/verb negative BRUSH NO POTTY CAN'T	negative verb HUG NO NO CHASE	locative negative CAN'T OUT FINISH OUT	
Requests	object request GIMME MILK MORE BALL	n/verb request BRUSH MORE GIMME DRINK	request verb HUG PLEASE TICKLE MORE	locative request MORE UP OUT PLEASE	demonst'ive request GIMME THAT THAT MORE

Even though all of the children in these studies also used request terms of Table 1 like GIMME, WANT, and MORE in their early phrases, only one of the child schemes in Table 3 includes a *request* pattern corresponding to the chimpanzee request patterns in Table 4. The other three studies assigned request patterns to other categories such as *recurrence* for phrases with MORE (Brown, 1973, p. 177), *dative* for phrases with GIMME and *state* for phrases with WANT (Bloom, Lightbown, & Hood, 1975, pp. 12, 13, 43; Brown, 1973, pp. 177, 198). This partial lack of correspondence, then, is a matter of schematic style rather than any substantive difference between children and cross-fostered chimpanzees.

In general, the scheme of phrase patterns for chimpanzees makes more distinctions than the child schemes. Table 4 indicates this by listing the chimpanzee phrase patterns that have to be grouped into broader categories to fit the more inclusive divisions of the child schemes. In two cases, however, we had to make finer divisions in order to show child-chimpanzee parallels.

First, all four of the systems for children include a separate *possessive* relation based on the use of possessive attributes combined with other terms. In the chimpanzee scheme we classified possessive attributes together with all other attributives (see Table 1). Separating phrases that included possessives from others in attributive phrases and placing them in the appropriate row of Table 4 shows that the chimpanzees expressed the same semantic relation.

Second, Brown (1973, pp. 177, 198) and Bloom et al. (1975, p. 12) found that phrases with terms such as SEE, LOOK, and HEAR combined with terms for persons or objects appeared relatively late. Brown labeled these as phrases of *experience* and Bloom et al. called them phrases of *notice*. Similar phrases appeared in the Bowerman (1973) and Braine (1976) as in Table 3. In the chimpanzee scheme we classified SEE, HEAR, and LOOK as noun/verbs because the cross-fosterlings not only used them to refer to the act of seeing or hearing, but they also often called optical devices SEE's, devices that made characteristic sounds such as ticking watches HEAR's, and so on. Separating these phrases from other action phrases and placing them in the appropriate row of Table 4 shows that the chimpanzees expressed the same semantic relation. The context notes that accompanied phrases like SEE FLOWER (Tatu and Naomi are walking through park area) and HEAR WRISTWATCH (Moja playing with Jeff's wristwatch) are very similar to context notes for *experience/notice* phrases in the child samples.

Both the child vocabularies and the chimpanzee vocabularies contain clearly possessive terms such as MINE and YOURS which form a solid basis for including a category of possessive phrases in all schemes. Most of the child schemes also include phrases consisting of a person term combined with an object term, as in MOMMY APPLE. Child studies often infer that the child used the person term as a possessive adjective by *rich interpretation* of context notes. The chimpanzee samples also contain the object + person pattern (Figure 7), but we have avoided interpreting these as possessive phrases (see this article, 4.1.3).

Both the child vocabularies and the chimpanzee vocabularies contain clearly locative terms such as THERE, IN, and OUT which form a solid basis for including a category of locative phrases in all schemes. Most of the child schemes also include phrases consisting of two object terms as in SWEATER CHAIR. Child studies often infer that the child used one of the object terms as a locative by *rich interpretation* of context notes. The chimpanzee samples also contain the object + object pattern (Figure 7), but we have avoided interpreting these as locative phrases (see this article, 4.1.3).

4.1.2. *Developmental Patterns*. Studies of human children generally agree that the major semantic relations appear in a characteristic developmental sequence (Bloom et al., 1975; Braine, 1976; Leonard, 1976; and Wells, 1974). De Villiers and De Villiers (1986, pp. 50-51) and Reich (1986, p. 83) summarize this agreement. Nominative phrases, the first group in Tables 3 and 4, together with action phrases, the second group, are the earliest to appear. Next, come attributive phrases expressing the properties of objects (attribution, possession, location) the third group in the tables. Experience/notice phrases are relatively late in child development and these are the fourth group in the tables. With respect to negatives and requests at the bottom of Tables 3 and 4, studies of children have so far either failed to report developmental order or reported inconsistent orders.

Studies of child speech have used a variety of criteria to say when a particular child acquired a particular phrase pattern. Usually, they have avoided using the appearance of the first phrase type of a particular pattern, on the argument that it takes more than one type to establish a pattern. The number of different phrase types required before crediting a child with a pattern has varied from study to study, but most have required at least three. To compare Moja, Tatu, and Dar with children, we first grouped their phrases into the four major relations as shown in Table 3. We then dated the acquisition of each of the four major relations as the first focal month sample in which at least three different phrase types of the same major pattern appeared for each chimpanzee. By this measure, nominatives and actions appeared first, attributives second, and experience/notice appeared latest in the developmental samples of each chimpanzee -- the same sequence that appears in studies of child development.

4.1.3. *Rich Interpretation.* Brown (1973) explains "the method of rich interpretation" with two instances in which the same utterance, MOMMY SOCK, appears in Bloom's (1970) sample of Kathryn.

> On the two occasions Bloom's transcriptions describe the context in these terms:
>
> Kathryn picked up her mother's sock
> Mother was putting Kathryn's sock on Kathryn
>
> In the first of these contexts, an 'expanding' adult would be likely to produce a genitive or possessive gloss: perhaps 'That's right, it's Mommy's sock.' In the second context, such an adult would be likely to produce a very different gloss: 'That's right, Mommy is putting on your sock.' In the second case, the adult takes MOMMY SOCK to be a subject-object relation rather than a possessor-possessed relation. Of course, the investigator does not need the expanding adult to instruct him in the probable meanings of such utterances. He can be his own 'expander,' operating as the parent does from context and assigning the structural meanings genitive (or possessive) and agent-object. (p. 107)

Bloom (1970, p.8) and Brown (1973, p. 106) claim that the fact that Kathryn said the same thing in two different situations proves that there is deep structure in her infant grammar. The two meanings that the investigator found in the two different situations represent the different deep structures that produced the same surface structure, MOMMY SOCK. Of course, if Kathryn really did mean two different things in these two different situations, then she failed to express them. The linguistic ingenuity and intelligence in rich interpretation belong to the investigators rather than to the children.

Agreement between parents and investigators is said to validate the method of rich interpretation. Fortunately, in the case of the second MOMMY SOCK Bloom published complete notes (1970, p. 48). Rather than agreeing with Brown's "That's right, Mommy is putting on your sock," what the mother actually said was, "That's not mommy's sock." Bloom (p. 55) dismisses the mother's interpretation on the grounds that a mistake on Kathryn's part was unlikely. But, 21-month old children must be at least as likely as their mothers to make mistakes and even to misspeak themselves. For that matter, the mother had a much larger sample of Kathryn's speech and Kathryn's behavior and must have been at least as likely as the visiting investigator to interpret Kathryn correctly, if not richly.

4.2. CHILD WORD ORDER

The early utterances of human children are highly telegraphic; functors such as "of" and "to" and inflectional morphemes such as "ing" are virtually absent. The lack of other signs of early grammar led early psycholinguists to make strong claims about word order. Brown (1973), for example, seemed to claim that in their informal speech toddlers are much more precise than college professors.

> In all the samples of Stage I English listed in Table 9 the violations of normal order are triflingly few [six examples] and perhaps 100 or so others. Of utterances in normal order there are many thousands. (p. 156)

The following examples from Brown (1973) illustrate how infant speech can achieve nearly errorless word order when it is richly interpreted by fluent adult investigators. RED BALL is an example of normal English order for noun + adjective, but PILLOW DIRTY is an example of a "predicate adjective" because it is a telegraphic version of [the] PILLOW [is] DIRTY. ADAM PUT is in normal English order for agent + action, but GO MOMMY is also correct because it is an example of a "commitative". PUT TRUCK MOMMY is correct because it is a "vocative construction". Both ADAM SIT CHAIR and SIT ADAM CHAIR are correct because ADAM SIT CHAIR is in normal subject-verb-object order while SIT ADAM CHAIR is an example of an "action-object relation" in which the object becomes a "noun-phrase" by the expansion ADAM['s] CHAIR as in MOMMY['s] SOCK. EVE LUNCH is an "agent-action construction" by the expansion EVE [is eating] LUNCH as in MOMMY [is putting Kathryn's] SOCK [on Kathryn]. Meanwhile, EAT EVE LUNCH is also correct by the possessive expansion EVE['s] LUNCH.

Bloom (1970, pp. 85-87) is explicit about the circularity of rich interpretation in her discussion of Gia's phrase WHEEL TRUCK uttered when Gia was playing with the wheel of a toy truck. Bloom classified this as a *genitive* relation by expanding it to TRUCK['s] WHEEL, but she goes on to say:

> if (23) had occurred with reversed order, that is, "wheel truck," it could have been classed with (25) through (28) as a *locative* relational phrase. (p. 86)

4.3. CHIMPANZEE SIGN ORDER

Obviously, it is easy to find a rich interpretation to agree with any word order found in almost any example of child speech or any sign order in chimpanzee signing. What is needed is independent evidence of order as a structural device.

4.3.1. *Modifiers.* The focal samples yielded 186 cases in which a phrase contained a term for a modifier (trait or attributive) adjacent to a term for a person or an object, e.g., GOOD DAR, QUIET ME for traits and persons and BLACK DOG, MINE HAT for attributes and objects. There were 12 cases of dual adjacency, such as MINE SHIRT WHITE and ME GOOD GIRL, which we eliminated from the analysis in order to consider only clear cut cases.

Table 5. Sign Order of Modifiers in Phrases

	Before X	After X	Chi-square	*df*	*p*
Moja	20	4	10.67	1	<.005
Tatu	51	24	9.72	1	<.005
Dar	66	9	43.32	1	<.0001

Table 5 shows how Moja, Tatu, and Dar were significantly more likely to place terms for traits and attributes before terms for persons and objects. The argument for structure here is that the dominant order, modifier before modified, is consistent across subjects and appears in many different phrases. A variety of attributes, including colors, possessives and materials, came before a variety of objects, including animates, edibles, and inanimate objects, and the traits GOOD, SORRY, and QUIET came before a variety of person signs, including pronouns, proper names, and generic names such as BOY and GIRL.

4.3.2. *THAT and THERE.* Like signing children and signing adults, the cross-fosterlings, used the indexical sign THAT/THERE both in nominative phrases, as in THAT (indicating hairbrush) BRUSH, in reply to the question WHAT THAT?, and in locative phrases, as in BRUSH THERE (indicating own shoulder), in reply to the question WHERE BRUSH? Observers wrote either THAT or THERE, depending on which sounded more appropriate, but the vocabulary records only credited the subjects with one sign, THAT/THERE (B. Gardner et al., 1989, p. 157). We have two sources of evidence that the chimpanzees used order to distinguish the demonstrative indexical from the locative indexical.

First, *Wh-question tests* included both WHAT Demonstrative questions such as, WHAT THAT?, WHAT NAME THAT? and WHERE Action/Nominal questions such as, WHERE TOOTHBRUSH? TICKLE WHERE? (B. Gardner & Gardner, 1975; Van Cantfort et al., 1989). Table 6 shows the distribution of all cases in which THAT/THERE appeared in a phrase reply with single adjacency - that is, excluding cases of dual adjacency such as, THAT/THERE MILK THAT/THERE and OIL THAT/THERE MOJA. The statistical analysis in Table 6 shows how Moja, Tatu, and Dar were significantly more likely to place THAT/THERE before the object term in their replies to WHAT Demonstrative questions and after the object or location in reply to WHERE Action/Nominal questions.

Table 6. Sign Order of THAT/THERE in Replies to Wh-questions

		Before X	After X	Chi-square	*df*	*p*
Moja	What X?	1	0	2.18	1	< .20
	Where X?	3	8			
Tatu	What X?	7	2	2.70	1	< .10
	Where X?	10	12			
Dar	What X?	2	0	7.20	1	< .01
	Where X?	1	9			
		Sum of chi-squares		12.08	3	< .01
MTD Pooled	What X?	10	2	9.83	1	< .01
	Where X?	14	29			
			Heterogeneity	2.25	2	< .50

Note insignificant Heterogeneity indicates lack of difference among subjects.

Second, in the original *field records*, the observers wrote either THAT or THERE and the other signs in the phrase, together with notes on the verbal and nonverbal context such as "naming pictures in a book" or "in reply to WHERE TICKLE?" To evaluate demonstrative vs. locative uses of THAT/THERE in the focal month samples, we first selected all the records of phrases containing THAT/THERE in single adjacency. Then, we prepared special records for coding, by removing any indication as to whether the original observer had written THAT or THERE and whether the indexical sign had appeared before or after the other signs in the phrase. We printed these amended records on special forms that preserved all of the context notes together with the signs in the phrase. The coders classified the phrases into four categories: demonstrative, locative, could-be-either, or insufficient contextual information.

Table 7. Sign Order of THAT/THERE in Demonstrative vs Locative Contexts

		Before X	After X	Chi-square	*df*	*p*
Moja	Demonstrative	17	7	20.00	1	< .0001
	Locative	1	23			
Pili	Demonstrative	18	9	7.42	1	< .01
	Locative	0	7			
Tatu	Demonstrative	319	69	56.99	1	< .0001
	Locative	16	33			
Dar	Demonstrative	104	19	46.04	1	< .0001
	Locative	7	25			

The coders agreed on 80% of the 908 judgments. Table 7 shows the distribution of all cases in which both coders agreed that the context was either demonstrative or locative. The statistical analysis in Table 7 shows that independently classified context notes significantly predict the order of the indexical sign for Moja, Pili, Tatu and Dar.

Wherever there is independent evidence to distinguish between the two cases, the cross-fosterlings mostly used THAT/THERE, before other terms in demonstrative phrases and after other terms in locative phrases. Locative phrases were numerous and varied, e.g., WRITE THERE, GROOM THERE, FRIEND THERE, and IN THERE. Demonstrative phrases were even more numerous and varied, e.g., THAT BIRD, THAT RED, THAT MORE, THAT NAOMI. The variety of signs, and the consistency across subjects, shows that they used order as a structural device to distinguish demonstrative THAT/THERE from locative THAT/THERE.

4.4. INFLECTION

Among the languages of the world, English is unusual in its heavy reliance on word order. Most human languages rely more on inflection and ASL is one of the heavily inflecting languages of the world (Klima & Bellugi, 1979, p. 299; p. 314).

> The key to understanding ASL syntax, particularly word order, is the recognition that locations in space are used for inflectional purposes. Within

> the "signing space" (the allowable area in which signs may be made), signs may be moved from one location to another to indicate differences in subject and object. (Wilbur, 1980, p. 19)

Adult skill with the inflectional devices of ASL is the product of a long developmental history (Cokely & Gawlik, 1974). A deaf child of deaf parents studied by Ellenberger and Steyaert (1978, p. 265) was still indicating participants by placing action signs on the body of a person or on an object (a childish, immature version of this inflection) at 54 months. He was only beginning to move action signs towards the person or object as fluent adult signers do.

> While one might expect spatial modifications to appear earlier because of their pictorial nature, they are, in fact, relatively late acquisitions, perhaps because such representations may require a fairly advanced mastery of cognitive skills involving spatial relationships. (p. 268)

In ASL, QUIET is signed with an upward orientation of the extended index finger placed near the lips of the signer in its citation form, but the sign is directed outward, towards the addressee to tell the addressee to be quiet. Like young human children, the cross-fostered chimpanzees exaggerated this inflection by placing the index finger on the lips of the addressee to sign HUSH or BE-QUIET. Similarly, the throat is the citation place for SWALLOW, but the cross-fosterlings often signed SWALLOW on the throat of their companion to indicate that the addressee had swallowed or should swallow something. The back of the hand is the citation place for TICKLE, but the cross-fosterlings often placed TICKLE on the back of a companion's hand as a way of requesting tickling. Through month 60, Moja, Tatu, and Dar still used the childish, touching form of spatial inflection. Observers noted that Moja, Tatu, and Dar often placed GROOM, QUIET, SWALLOW, and TICKLE on the addressee, in phrases such as TICKLE MORE and GROOM THERE.

Videotape records of Dar taken when he was between 40 and 49 months old show how he indicated participants in action with the childish form, touching the person, place, or object (Rimpau et al., 1989). For example, Dar signed TICKLE (on the side of his head) TICKLE (on his dinosaur toy) ME indicating that Tony was to tickle Dar's head with the toy. Statistically, when Dar's utterance contained a name or a pronoun to indicate person he usually omitted the inflection, but when he omitted names and pronouns he usually inflected the sign (Rimpau et al., 1989, pp. 250-262).

Hoffmeister et al. (1975) describes another inflection, conjunction or the "and" inflection, which consists of a series of THAT signs indicating a set of two or more objects, such as a dog and a cat, or several pictures. Hoffmeister and Wilbur (1980) reported that their subject, a deaf daughter of deaf parents, only started to use conjunction at 38 months. Moja, Tatu, and Dar all inflected THAT in a very similar way. Dar, for example, signed THAT THAT THAT THAT THAT EAT indicating five different tomatoes in a picture of a basket of cherry tomatoes. Tatu signed THAT THAT THAT BLACK indicating three black things on a magazine page containing illustrations of shoes and gloves. Within the focal month samples, reports of this inflection only appeared in the last or 60th month for Moja and Tatu, and in 48th and 54th months for Dar. Inflected signs show the acquisition of morphology in ASL (Newport & Meier, 1986, pp. 896-903). Inflections that appear late in the signing of human children also appeared late in the development of cross-fostered chimpanzees. According to Brown (1973) and De Villiers and De Villiers (1986), the appearance of grammatical morphemes and inflections in the utterances of speaking children marks the beginning of Stage II.

5. Functional Variables and Modular Categories

Language is probably the most significant product of human evolution -- so significant and so awesome that it is difficult to study on the same objective terms as other natural phenomena. If language is a product of biology, however, then it is a function of variables; a highly complex function, yes, many variables, doubtless, but a function of variables, nevertheless. Where Plato urged his followers to carve nature at the joints, modern natural scientists advance, instead, by searching for fundamental laws that unify rather than divide nature. For centuries the winning strategy has been a strategy of discovering and relating general continuities, functions, and laws, rather than erecting logical partitions.

Since Darwin, the principle that all biological functions vary in degree rather than in kind has dominated biology. It is hard to imagine modern human biology and medicine without this principle. Darwinism of blood and bone prevails in our time, but Darwinism of mind remains embattled.

> Possession of human language is associated with a specific type of mental organization, not simply a higher degree of intelligence. (Chomsky, 1972, p. 70)

A logical division between human and nonhuman and between language and nonlanguage seems plausible only because divisions of this kind agree with Aristotle's law of the excluded middle -- as still taught reverently in modern universities. We would argue instead, that whenever natural scientists succeed in discovering continuous, functional variables they quickly abandon categorical, either-or divisions (Kosko, 1993; Zadeh & Kacprzyk, 1992).

5.1. HUMAN DEVELOPMENT

If language and humanity are either-or, all-or-none categories then partial language is impossible in human beings or in any other species. Philosophical linguists must therefore deny human linguistic development.

> We are presenting an 'instantaneous model' of language acquisition which is surely false in detail, but can very well be accepted as a reasonable first approximation. . . . Given an instantaneous model that is empirically well-supported, as a first approximation, there are many questions that can immediately be raised . . . (Chomsky, 1967, pp. 441-442)

Chomsky's early followers in child psycholinguistics stretched their observations as far as they could to support a virtually 'instantaneous model'.

> Children all over the world learn to speak their native language at approximately the same time -- 3 to 4 years of age. Within a relatively brief period, the child appears to learn a complicated and abstract system of rules . . . without teaching or training, [they] acquire their native language at about the same time -- regardless of just about any variable one cares to look at, short of deafness or severe retardation. (Moore, 1973, p. 4)

Thus, loyalty to the logical separation between language and nonlanguage flattens the landscape of development from toddler to college student. Loyal followers must deny both the negative effect of impoverished environments and the positive effect of enriched environments. They also debase the concept of syntax by claiming that it is virtually complete in the speech of the same four-year-olds who cannot yet use a knife to spread jam on a cracker or tie their own shoelaces (Cohen & Gross, 1979). They place syntax on a par with the most primitive, vegetative functions that appear soon after birth, making syntax totally unlike the advanced intellectual functions that continue to develop until maturity.

Perhaps the worst damage done in the name of the virtually 'instantaneous model' demanded by Aristotelian logic is the damage to normal rules of evidence. In the method of rich interpretation (see this article, 4.1.3), the syntactical imagination of the investigator replaces straightforward reports of what children actually say. In most cases, the original observations are so layered over with rich interpretation that they can never be recovered.

5.2. TRENDS AND PREDICTIONS

By measuring patterns of growth and development rather than static modules we can make predictions about more extended sign language studies of cross-fostered chimpanzees. More advanced developments appeared with each succeeding year of cross-fostering. The proof that Moja, Tatu, and Dar had not yet reached any limit at three years is their growth during the fourth year, and the proof that they had not yet reached a limit at four years is the growth during the fifth year. Nevertheless, after three years of cross-fostering, they had clearly fallen behind human three-year-olds, and they fell farther behind after four years, and still farther behind after five years. From this we can predict that chimpanzees should be even farther behind human children after six years of cross-fostering, but by the same token, we can predict that at six they should achieve more than they achieved at five.

Sign language studies of cross-fostered chimpanzees reveal robust growth and development. Further intellectual development should continue until sexual maturity which takes more than twelve years. Much more can be accomplished in future studies with long-term support.

References

Barrett, M. (1989). Early language development. In A. Slater & G. Bremmer (Eds.), *Infant development*, (pp. 211-241). Hillsdale, NJ: Lawrence Erlbaum.

Bates, E., O'Connell, B., & Shore, C. (1987). Language and communication in infancy. In J.D. Osofsky (Ed.), *Handbook of infant development*. (2nd ed.), (pp. 149-203). New York: John Wiley & Sons, Inc.

Bloom, L. (1970). *Language development: Form and function in emerging grammars*. Cambridge MA: M.I.T. Press.

Bloom, L. (1991). *Language development from two to three*. New York: Cambridge University Press.

Bloom, L., Lightbown, P., & Hood, L. (1975). Structure and variation in child language. *Monographs of the Society for Research in Child Development*, *40* (2) Ser. No. 160.

Bloom, L., Rocissano, L., & Hood, L. (1976). Adult-child discourse: Developmental interaction between information processing and linguistic knowledge. *Cognitive Psychology*, *8*, 521-552.

Bohannon, J.N., & Stanowicz, L. (1989). Bidirectional effects of imitation and repetition in conversation: A synthesis within a cognitive model. In G.E. Speidel & K.E. Nelson (Eds.), *The many faces of imitation in language learning*, (pp. 121-150). New York: Springer-Verlag.

Bowerman, M. (1973). *Early syntactic development*. London: Cambridge University Press.

Braine, M.D. (1976). Children's first word combinations. *Monographs of the Society for Research in Child Development*, *11* (1, Serial No. 164), 1-96.

Brinton, B. & Fujiki, M. (1984). Development of topic manipulation skills in discourse. *Journal of Speech and Hearing Research*, *27*, 350-358.

Brown, R. (1970). The first sentences of child and chimpanzee. In R. Brown (Ed.), *Selected psycholinguistic papers*, (pp. 208-281). New York: Macmillan.

Brown, R. (1973). *A first language: The early stages*. Cambridge, Mass.: Harvard University Press.

Chomsky, N. (1967). The formal nature of language. In E.H. Lenneberg, *Biological foundations of language*. New York: John Wiley & Sons.

Chomsky, N. (1972). *Language and mind*. (Extended ed.). New York: Harcourt Brace Jovanovich.

Cohen, M.A., & Gross, P.J. (1979). *The developmental resource: Behavioral sequences for assessment and program planning*. New York: Grune & Stratton.

Cokely, D.R., & Gawlik, R. (1974). Childrenese as Pidgin. *Sign Language Studies*, *5*, 72-81.

De Villiers, J.G., & De Villiers, P.A. (1986). The acquisition of English. In D.I. Slobin (Ed.), *The crosslinguistic study of language acquisition*, (Vol. 1, pp. 27-139). Hillsdale, NJ: Lawrence Erlbaum.

Drumm, P., Gardner, B.T., & Gardner, R.A. (1986). Vocal and gestural responses of cross-fostered chimpanzees. *American Journal of Psychology*, *99*, 1-29.

Ellenberger, R., & Steyaert, M. (1978). A child's representation of action in American Sign Language. In P. Siple (Ed.), *Understanding language through sign language research*, (pp. 261-269). New York: Academic Press.

Gardner, B.T., & Gardner, R.A. (1971). Two-way communication with an infant chimpanzee. In A. Schrier & F. Stollnitz (Eds.), *Behavior of nonhuman primates*, (Vol. 4, pp. 117-184). New York: Academic Press.

Gardner, B.T., & Gardner, R.A. (1974). Comparing the early utterances of child and chimpanzee. In A. Pick (Ed.), *Minnesota symposium on child psychology*, (Vol. 8, pp. 3-23). Minneapolis: University of Minnesota Press.

Gardner, B.T., & Gardner, R.A. (1975). Evidence for sentence constituents in the early utterances of child and chimpanzee. *Journal of Experimental Psychology: General*, *104*, 244-267.

Gardner, B.T., & Gardner, R.A. (1989). Cross-fostered chimpanzees: II. Modulation of meaning. In P.G. Heltne & L.A. Marquardt (Eds.), *Understanding chimpanzees*, (pp. 234-241). Cambridge, MA: Harvard University Press.

Gardner, B.T., Gardner, R.A., & Nichols, S.G. (1989). The shapes and uses of signs in a cross-fostering laboratory. In R.A. Gardner, B.T. Gardner, & T.E. Van Cantfort (Eds.), *Teaching sign language to chimpanzees*, Albany: SUNY Press.

Gardner, R.A., & Gardner, B.T. (1969). Teaching sign language to a chimpanzee. *Science*, *165*, 664-672.

Gardner, R.A., & Gardner, B.T. (1973). Teaching Sign Language to the Chimpanzee, Washoe. (16 mm sound film). State College, PA: Psychological Cinema Register.

Gardner, R.A., & Gardner, B.T. (1974). Teaching sign language to the chimpanzee, Washoe. *Bulletin D'Audio Phonologie*, *4*(5), 145-173.

Gardner, R.A., & Gardner, B.T. (1989). A cross-fostering laboratory. In R.A. Gardner, B.T. Gardner, & T.E. Van Cantfort (Eds.), *Teaching sign language to chimpanzees*, Albany: SUNY Press.

Gardner, R.A., Gardner, B.T., & Van Cantfort, T.E. (Eds.) (1989). *Teaching sign language to chimpanzees*, Albany: SUNY Press.

Gardner, R.A., Van Cantfort, T.E., & Gardner, B.T. (1992). Categorical replies to categorical questions by cross-fostered chimpanzees. *American Journal of Psychology*, *105*, 25-57.

Hebb, D.O., & Thompson, W.R. (1968). The social significance of animal studies. In G. Lindzey & E. Aronson (Eds.), *The handbook of social psychology*. (2nd ed.) Vol II. *Research methods*, (pp. 729-774). Reading, MA: Addison-Wesley.

Hoffmeister, R.J., Moores, D.F., & Ellenberger, R.L. (1975). Some procedural guidelines for the study of the acquisition of sign languages. *Sign Language Studies*, *7*, 121-137.

Hoffmeister, R., & Wilbur, R. (1980). Developmental: The acquisition of sign language . In H. Lane & F. Grosjean (Eds.), *Recent prespectives on American Sign Language*, (pp. 61-78). Hillsdale, NJ: Lawrence Erlbaum.

Keenan, E. Ochs (1977). Making it last: Repetition in children's discourse. In S. Ervin-Tripp and C. Mitchell-Kernan (Eds.), *Child discourse*, (pp. 125-138). New York: Academic Press.

Klima, E.S., & Bellugi, U. (1979). *The signs of language*. Cambridge MA: Harvard University Press.

Kosko, B. (1993). *Fuzzy thinking: The new science of fuzzy logic*. New York: Hyperion.

Leonard, L.B. (1976). *Meaning in child language*. New York: Grune & Stratton.

Moore, T.E. (Ed.) (1973). *Cognitive development and the acquisition of language*. New York: Academic Press.

Nelson, K. (March, 1980). First words of the chimp and child. Paper presented at the Southeastern Psychological Association Symposium on Apes and Language.

Newport, E.L., & Meier, R.P. (1986). The acquisition of American Sign Language. In D.I. Slobin (Ed.), *The crosslinguistic study of language acquisition*, (Vol. 1, pp. 881-938). Hillsdale, NJ: Lawrence Erlbaum.

Reich, P.A. (1986). *Language development*. Englewood Cliffs, NJ: Prentice Hall.

Rimpau, J.B., Gardner, R.A., & Gardner, B.T. (1989). Expression of person, place and instrument in ASL utterances of children and chimpanzees. In R.A. Gardner, B.T. Gardner & T.E. Van Cantfort (Eds.), *Teaching sign language to chimpanzees*, Albany, NY: SUNY Press.

Scollon, R. (1979). A real early stage: An unzippered condensation of a dissertation on child language. In E. Ochs & B. Schieffelin (Eds.), *Developmental pragmatics*, (pp. 215-227). New York: Academic Press.

Stokoe, W.C. (1978). Sign language versus spoken language. *Sign Language Studies*, *18*, 69-90.

Uzgiris, I.C., Broome, S., & Kruper, J.C. (1989). Imitiation in mother-child conversations: A focus on the mother. In G.E. Speidel & K.E. Nelson (Eds.), *The many faces of imitation in language learning*, (pp. 91-120). New York: Springer-Verlag.

Van Cantfort, T.E., Gardner, B.T., & Gardner, R.A. (1989). Developmental trends in replies to Wh-questions by children and chimpanzees. In R.A. Gardner, B.T. Gardner & T.E. Van Cantfort (Eds.), *Teaching sign language to chimpanzees*, Albany, NY: SUNY Press.

Wells, G. (1974). Learning to code experience through language. *Journal of Child Language*, *1*, 243-269.

Wilbur, R. (1980). The linguistic description of American Sign Language. In H. Lane & F. Grosjean (Eds.), *Recent perspectives on American Sign Language*, (pp. 7-31). Hillsdale, NJ: Lawrence Erlbaum.

Zadeh, L.A., & Kacprzyk, J. (Eds.) (1992). *Fuzzy logic for the management of uncertainty*. New York: John Wiley & Sons, Inc.

TRANSMISSION OF A HUMAN GESTURAL LANGUAGE IN A CHIMPANZEE MOTHER-INFANT RELATIONSHIP

ROGER S. FOUTS
Chimpanzee and Human Communication Institute
Central Washington University
Ellensburg, WA 98926
U.S.A.

ABSTRACT. Ten-months old Loulis was adopted by Washoe and acquired signing and other skills from her and the other signing chimpanzees in his community. To insure that Loulis would acquire his signs from the chimpanzees, the humans, in his presence, restricted their signing to seven specific signs or used vocal English to communicate with him. Washoe was observed to teach Loulis through modeling, molding and signing on his body. Most of his signs appeared to be delayed imitations of signs he had seen Washoe and the other chimpanzees use in similar contexts. We were able to examine the development of social behavior, communication, and other skills in Loulis without disrupting them. In this way, we obtained a comprehensive record of cultural transmission.

1. Introduction

When I was an undergraduate student, more than 30 years ago, I remember from my anthropology classes that culture was considered to be something quite unique to our human species. It was something we acquired and passed on from generation to generation. The idea that it might have biological basis was never entertained. If any of our fellow animals displayed any behavior that looked at all as if it might be cultural then the preface "proto" was added which robbed it of any consideration as an established behavior with some historical position.

I also remember that some of my anthropology professors maintained that what made culture so unique to our species was that a great deal of it was passed onto the next generation via our languages. Language had a very special position because it was not only the vehicle by which our culture was transmitted but it was itself also transmitted across generations this way. A popular notion at that time was that birds were born with their songs, and dogs with their barks, but only we humans had to acquire our language culturally from our parents and fellow humans. Much of this supposition was based on armchair speculation and ignorance where the absence of evidence was considered evidence of absence. However, much has changed since we started asking our fellow animals to tell us about themselves by studying them in the field and in the laboratory. It is no longer acceptable to base our assumptions about the capacities of our fellow animals on our armchair preconceptions of human nature.

In regard to the chimpanzee the last 30 years has been a veritable explosion of information that has required our academic community to continually readjust and rethink previous ideas that were considered to be inviolable. The chimpanzees that live in Africa have shown us some amazing cultural behaviors, that I am sure they have been doing all along. The following statement made by Jane Goodall (1986) was considered heresy or at least evidence of insanity 30 years ago: "The difference in cultural tradition between the

R. A. Gardner et al. (eds.), The Ethological Roots of Culture, 257–270.

chimpanzees of Gombe and those of the Mahale Mountains, only 100 miles away, has already proved to be very great" (p. xii). Others at this conference provide us with recent tangible evidence of the African chimpanzee's cultural prowess.

Research from the laboratory has proven to be just as profound in its implication for established thought. In Project Washoe, my professors Allen and Beatrix Gardner demonstrated that a young chimpanzee was receptive to the cultural transmission of gestural signs from humans (B. Gardner & Gardner, 1971). The Gardners (1989) went on to demonstrate that this was a robust phenomenon, because every chimpanzee that was raised in Washoe's fashion readily acquired the signs of American Sign Language (ASL). Their early research with Washoe is the necessary precursor to the research I will report. After Washoe had acquired the signs of ASL the question was still there: even if chimpanzees could acquire signs from humans, would they pass it on to the next generation themselves?

The Gardners began Project Washoe in 1966 at the University of Nevada at Reno and I came onto the project in 1967 as one of their graduate students. In 1970, at the Gardners request, I brought Washoe with me to the Institute for Primate Studies (IPS) at the University of Oklahoma. Because the IPS was a colony it often maintained about 20 or more chimpanzees of varying ages and backgrounds. We saw the IPS as an opportunity to expand the horizons in regard to signing chimpanzees. We could now address questions that were impossible to answer in Reno. Could chimpanzee other than Washoe acquire signs? Would Washoe use her signs to communicate with other chimpanzees? Would the chimpanzees in Oklahoma begin to pick up the signs from Washoe or human signers? One of the most interesting questions would have to wait until Washoe was old enough to have her own offspring. That question was: Would Washoe pass her signing skills onto her baby, or in other words, is the cultural transmission of ASL possible in chimpanzees?

2. Expanding Horizons

The horizons really did expand. We found that signing was not something unique to Washoe, but that other chimpanzees could acquire the signs. With the help of other human caregivers the young chimpanzees Booee, Bruno, Cindy, and Thelma acquired signs (Fouts, 1973). Other chimpanzees associated with the IPS who were being home-reared by humans, such as Lucy and Ally (aka Ali) were also taught the signs of ASL. Ally was able to comprehend and produce novel prepositional phrases demonstrating his ability to use sign order (Fouts, Shapiro & O'Neil, 1978). In a study with Lucy it was found that she was able to categorize and conceptualize fruits as different from vegetables and that she would describe novel objects by combining signs already existing in her vocabulary. For example, Lucy referred to a watermelon as a CANDY DRINK and a DRINK FRUIT, and she called an old bitter radish a CRY HURT FOOD after she took a bite of it and spits it out (Fouts, 1975). Ally also demonstrated that he could be taught ASL signs using vocal English words as exemplars and then in a blind condition transfer these signs to their physical referents (Fouts, Chown & Goodin, 1976). In addition to these new chimpanzees, Washoe continued to learn new signs. For example, we did a follow up on my dissertation (Fouts, 1972) examining the relative efficacy of the procedures of modeling and molding for teaching new signs. In my dissertation, with a 3 to 4-year-old Washoe, molding was the superior technique. In the follow-up study with a nine-year-old Washoe and an eight-year-old Lucy the two procedures were found to be equally effective (Fouts & Goodin, 1974). Like Lucy, Washoe also combined signs to produce novel phrases to describe objects for which she did not have signs in her vocabulary such as WATER BIRD for a swan on lake, and ROCK BERRY for Brazil nuts (Fouts & Rigby, 1977).

3. Washoe Converses with Chimpanzees

Before Washoe went to Oklahoma she had not been around chimpanzee since she was an infant. At Oklahoma she initially did not seem to like the chimpanzees, in fact during her first introduction to them she referred to them as BLACK CATS, and BLACK BUGS neither of which animals she cared for at all. In other words, she did not apparently put them in her own class of being. However, we humans insisted that she live with them and she was housed in the colony inside a laboratory building that contained a complex of interconnected enclosures.

In addition to the laboratory there was also a 1/3 acre artificially moated island where she spent considerable time with other juvenile chimpanzees. My students, Debbi and I continued to sign with Washoe as well as with some of the other young chimpanzees that were familiar with human contact. Because Debbi and I were the only humans fluent in ASL and my students were learning and all of the other personnel at the IPS did not sign, Washoe was exposed to spoken English as well as ASL. When she first arrived in Oklahoma I had to act as interpreter for her with the IPS personnel.

One indication that Washoe began to accept the IPS chimpanzees as peers was that it was not long after her arrival that she was observed to sign to the other chimpanzees. None of the chimpanzees at that time had acquired or even been taught any ASL signs and they did not appear to understand her or attempt to respond to her signed remarks. Washoe would often sign COME HUG when some of the other juvenile chimpanzees became upset. One young chimpanzee, Manny, was not used to human contact because he was being raised by his mother in the adult colony. When Washoe was in the adult colony enclosures she would play with him through the chain-link fence that separated their enclosures. When Washoe would attempt to end the interaction and move away Manny would throw himself on the floor in a tantrum, screaming. Washoe would respond to his tantrum by signing COME HUG to him and then approach and comfort him through the wire once more. Manny began to use her COME HUG sign to replace his tantrums. Manny also began to use it in appropriate contexts as when greeting or comforting another chimpanzee (Fouts & Rigby, 1977).

Another similar incident was observed when Washoe was on the island with the other juveniles shortly after her arrival at IPS. On this occasion she was with three juveniles on the island and was grooming the young colony-born home-reared male in the middle of the island. The two juvenile female chimpanzees apparently discovered something at the east end of the island (most probably a water snake) and began giving alarm calls. Washoe stood up bipedally and watched the two females move toward the west end of the island, while they continued to give occasional alarm calls. Washoe started in that direction as well, but her young friend remained lounging on the ground. When Washoe was approximately four meters away, she turned and signed to him, COME HUG COME HUG, but he remained where he was. Washoe then went back to him, took his hand, and led him to the safe west end of the island.

On another occasion, when she was in a group of chimpanzees who were being given fruit, Washoe indicated a water spigot (aka licksip) in a corner and signed GO DRINK to one of her competitors. She often signed HUG to the young chimpanzees, but after Washoe was introduced to the adult population, the sign she used most was TICKLE. If, as was often the case, the other chimpanzees did not respond to her sign, Washoe pursued them until a tickle game was started.

The fully adult alpha male in the colony even began to use the FOOD sign. I presume that he learned it from watching Washoe. He would sign FOOD repeatedly to humans, but it was interesting when we would go and get some food for him and return with it he would stop signing, and cling to the chain-link fence with his mouth open ready to receive it. I would even hold the food out and ask in ASL "WHAT THAT", but he would only continue to stare at the food with his mouth ready to receive it.

Another incident of transmission was reported to me by Lucy's human parents. Her parents were planning to send Lucy back to Africa and in preparation they obtained a young female chimpanzee by the name of Mary Ann in the hopes that Mary Ann would supply some comfort to Lucy when she was sent to Africa. Because Lucy was being sent away I was not giving her any signing lessons, but Lucy was still a prolific and spontaneous signer with her human friends. Her human parents rarely used any sign, because they had always used vocal English to communicate with Lucy which she understood very well. Her human parents told me that Mary Ann was beginning to pick up signs from Lucy, but that Lucy apparently did not approve of this because every time Mary Ann would sign Lucy would threaten or discipline the smaller chimpanzee. Her parents were amused by the fact that Mary Ann would try to sign to them when Lucy was not looking at her, and if Lucy did turn when Mary Ann was making a sign, Mary Ann would try to disguise what she was doing by turning the sign into a scratch or a rub.

The fact that chimpanzees could so easily acquire signs from humans as well as other chimpanzees made the prospect of seeing if Washoe's infant would acquire signs from her even more promising.

4. Birth, Death and Adoption

Washoe's first infant was born in August of 1976, but the baby only lived four hours because of a congenital heart defect. Her second infant was born in January 1979 and was named Sequoyah. Unfortunately Sequoyah was also a sickly infant and died of pneumonia when he was two months old. Washoe's short time with Sequoyah showed that she had excellent maternal skills, and she had become so attached to him that she became very depressed when he died (Fouts, Hirsch & Fouts, 1982). For several days after his death Washoe would ask me "BABY?" using the questioning modulation of raising her eyebrows and holding the cradled "baby" sign in its final position. I would answer her "BABY DEAD, BABY GONE, BABY FINISHED". In response to my answer Washoe would then drop her cradled arms to her lap and break eye contact and slowly move away to a corner of the enclosure. For the next three days after I had announced his death, Washoe would greet me each morning with the question "BABY?" and I would sign to her again as I had before. In the following days she continued to isolate herself from any interactions with her human caregivers and her signing dropped off to almost nothing. Her eyes appeared vacant or distant. We put Ally, a slightly younger adult male, in with her in hopes that his active nature would serve to distract her from her obvious depression.

In order to relieve Washoe's depression and continue our study to see if Washoe would pass her language skills to her offspring we contacted several primate facilities around the country in hopes of finding an infant that could replace Washoe's dead baby. The Yerkes Regional Primate Research Center at Emory University in Atlanta Georgia kindly agreed to provide the project with a ten-month-old male. On March 24, 1979, 15 days after the death of Sequoyah, Loulis arrived at IPS. I left Loulis in the car with a student and went in to see Washoe first, and I signed to her "I HAVE BABY FOR YOU". This simple announcement seemed to snap Washoe out of her severe depression. For the first time since Sequoyah's death she became very excited. Every hair on her body stood on end. She began to pant hoot and swagger bipedally while signing "BABY, MY BABY" repeatedly. I went out to the car to fetch Loulis and returned carrying him in my arms. When I entered Washoe's enclosure and got close enough to her for her to get a good look at Loulis her high excitement disappeared immediately. I had expected her to reach for Loulis, but instead she sat down, looked at him and very calmly signed "BABY". I had expected Loulis to want to go to her because he had been away from his chimpanzee aunt only 27 hours, but instead he clung even more tightly to me as he peaked around at Washoe. I managed to pull him off me by his arms and turn him around and offer him to Washoe. The minute she took him I let go and beat a hasty retreat. No sooner had I left when Loulis struggled out of Washoe's arms and approached the enclosure fence to get near to me. Both he and Washoe set a short space from each other, looking out, and in Washoe's case, interacting with the humans outside her enclosure.

After the first hour together, Washoe began to approach Loulis playfully, touching him gently and then moving away, in an attempt to start a tickle game or play chase. Loulis did not respond, but would turn away from her advances. He continued to sit alone on the floor and look at his new human friends just outside his enclosure. When night came, Washoe tried to get Loulis to sleep in her arms, as her own infant had done. But Loulis refused her attempts to pick him up, and chose instead to sleep alone, on the only bench available in the enclosure, the same one Washoe was on. At 4:00 A.M. the next morning, Washoe woke up, stood bipedally, swaggered, and signed COME BABY to Loulis, slapping her arms loudly as she made the "baby" sign. This commotion woke Loulis, and no doubt frightened him. He jumped into Washoe's arms and they both went back to sleep. From then on they have slept together at night. Thus within 24 hours, Washoe had begun to care for Loulis as if she were his mother, and he too, began to depend on her mothering. According to plan, the human caregivers restricted their signing in Loulis's presence and the cultural transmission study was begun.

5. Project Loulis

5.1. SIGNING CONTROLS

To make sure that Loulis would learn his signs only from chimpanzees, the human beings did not use ASL signs in his presence (with the exception of seven question signs, WHO, WHAT, WHERE, WHICH, WANT, SIGN and NAME). Instead of signing to Washoe in Loulis' presence we used vocal English which Washoe had been exposed to for nine years at IPS, and which Loulis had been exposed to since his birth at Yerkes. In addition to vocal English we also communicated using a rich repertoire of human and chimpanzee nonverbal gestures, postures, and vocalizations to interact with Washoe and Loulis. When Washoe would sign to us we would respond appropriately: for example, if she signed DRINK, we would give her a drink, or tell her in English that she would get a drink in a little while. If anyone erred and signed when Loulis was present, the instance was recorded. In the more than five years of the experiment, there were fewer than 40 such instances.

While we humans refrained from signing to Loulis, his fellow chimpanzees were not bound by this rule. In the five years and three months of this project, in addition to his adoptive mother Washoe, three other signing chimpanzees interacted with Loulis. In December 1979, seven-year-old Moja, who had been reared in the Gardners' cross-fostering laboratory (Gardner & Gardner, 1989, p.1-28) joined Washoe and Loulis at IPS. The following year, Washoe, Loulis, and Moja, along with Debbi, three students and me moved to Central Washington University. The three chimpanzees were housed in a complex of interconnected enclosures. In May 1981, Washoe, Loulis, and Moja were joined by five-and-a-half-year-old Tatu and five-year-old Dar, who had also been cross-fostered in the Gardners' laboratory.

5.2. CRITERIA FOR FORM AND USAGE

In order to qualify as a sign, a gesture that Loulis made had to resemble the form of an actual ASL sign in regard to the place where it was made (e.g., where on the body or in relation to the body of signer), the configuration of the hand, and the movement of the sign. In addition, before it was counted as a sign in Loulis' vocabulary, the gesture had to be reported by three different observers as occurring in an appropriate context. The observers recorded the gesture, the context, and the accompanying nonverbal behavior.

They also noted whether Loulis was oriented toward someone (e.g., toward Washoe, or Dar, or a human caregiver), and whether he was looking someone in the face, when he made the sign. Either or both of these behaviors were used to determine the addressee. In addition, level of arousal was scored as high, medium, or low by noting the degree to which his hair was piloerected. Loulis' body postures and facial expressions were recorded as well, as were any vocalizations (e.g., laughs, screams, or pant-hoots). If the sign and the nonverbal behaviors were consistent with each other and with the referent, then the sign was judged to have been used appropriately by Loulis.

Gestures developed into signs gradually, as in the case of COME/GIMME, which Loulis started to make soon after he joined Washoe. In early COME/GIMMEs, Loulis extended his arm in front of himself, palm up, for one to three seconds. In the following months, the form of the sign improved, as Loulis began to flex the extended hand. He also began to babble or play with this sign, making it in a variety of unrelated contexts. By August 1980, he was using COME/GIMME almost entirely in appropriate contexts, and at this time, his nonverbal behavior started to complement the COME/GIMME signs. For example, when offered a highly preferred food or drink, Loulis now looked directly at the object while he food-grunted and signed GIMME. He began to orient the sign toward humans or chimpanzees, and he gazed at the face of the addressee. During the second meeting of Loulis and Moja, when the connecting cage door was opened, Loulis oriented toward Moja and signed COME directly to her. Moja approached the signing infant, and initiated a play interaction. COME/GIMME became a distinct sign for Loulis, quite separate from reaching or the natural chimpanzee begging gesture. In approximately 25% of the COME/GIMME signs that Loulis made, he used one hand to sign and the other hand to reach or to make the begging gesture.

In a number of cases, such as the signs TICKLE, DRINK, and HUG, there was an initial period during which Loulis used the correct form in a variety of contexts before he restricted their use to appropriate contexts. Loulis was not credited with these signs until there were three observations, by different observers, of his use of the sign in appropriate contexts. In the case of TICKLE, for example, Loulis had many opportunities to observe Washoe soliciting tickling from the human caregivers. Washoe would first sign TICKLE, then press her side against the wire of her enclosure so that the person could tickle her. Initially, Loulis imitated Washoe by pressing his body to the wire of the enclosure, without making the sign, and almost always the caregiver would tickle him when he did this. Later, Loulis began to make TICKLE signs, but did not press his body to the wire of the enclosure. Then he went through a period of signing TICKLE repeatedly to himself, without approaching the human caregiver. Finally, he began to incorporate the nonverbal elements that Washoe used in signing TICKLE. Thus, he would approach persons with a play-face, slightly bipedal, and sign TICKLE either on himself or on the person, and then press his side or bottom against the wire of the enclosure for tickling. Thirteen of Loulis' first 22 signs went through a transition of this sort, before he began to use them in their appropriate contexts (Fouts, Hirsch & Fouts, 1982; Fouts, Fouts & Van Cantfort, 1989).

5.3. LOULIS' SIGN VOCABULARY

The first sign to appear in Loulis' vocabulary was the name sign that Washoe used for George Kimball. Loulis was first observed to use this sign eight days after his introduction to Washoe. George's name sign was made by moving the open hand down the back of the head, indicating the long hair, fashionable for men in the 1970's. However, Washoe changed the form of the sign in a very effective way. George was in charge of serving Washoe and Loulis breakfast, so his attention was much in demand. If George had his back turned, he could not tell that Washoe was signing. Washoe's response to this was to make his name sign into a noisy sign, by slapping the top of her head. The sign proved to be so effective in getting George's attention, that Washoe began to use it to refer to persons who did not have name signs, and finally, to anyone who was not looking at her. We glossed this noisy attention-getting sign as the PERSON (or "Hey you!") sign. In addition to Loulis, Moja, Tatu, and Dar also acquired this unusual sign from Washoe, and all used it in the same way as Washoe did.

At 15 months of age, Loulis started to use combinations of two signs, such as HURRY GIMME and PERSON COME. By 29 months, Loulis was using at least 17 different signs (Fouts, Hirsch & Fouts, 1982). By age 63 months, his vocabulary had grown to 47 signs (Fouts, Fouts & Schoenfeld, 1984). On June 24, 1984, after five years and three months of the experimental procedure, we ended the restriction on human signing in Loulis' presence, because, in essence, it was a form of deprivation for Loulis as well as for the other chimpanzees. At that time Loulis was 73 months old, and his vocabulary consisted of the 51 signs shown in Table 1 (Fouts, Fouts, & Van Cantfort, 1989). A few of his signs, such as FLOWER/DEB and HAT/PERSON, appeared to be homonyms, and the form described in the table was assigned two different English glosses, for the two distinct contexts in which it was used. If the signs that are homonyms in form are counted as separate signs, then his vocabulary was 55 signs at that time.

Table 1 shows that the vocabulary of Loulis included signs from several different categories: names (e.g., ALAN, DAR), pronouns (ME, YOU), nouns (e.g., BALL, BIRD), verbs (e.g., CHASE, HUG), locatives (e.g., IN, OUT) and such markers and traits as NO, SORRY, and WANT. This last sign has two distinct forms in ASL--a one-handed version, placed on the signer's chest, and a two-handed version, made in the space before the signer's body (cf. DESIRE and WANT in Stokoe, Casterline, & Croneberg, 1976, Dictionary of American Sign Language). Washoe used the first of these forms, while one of the few signs used by human signers in Loulis' presence was the other form of WANT. Even though Loulis was exposed to both forms of the sign, the WANT that he acquired was the one used by his mother.

In June 1986, the policy of speaking and signing in the presence of the chimpanzees was changed. From that time on, only signing was used in the presence of the chimpanzees with few exceptions (e.g., experiments that demonstrated that the chimpanzees continued to understand spoken English and could actually translate vocal English words into their ASL glosses, See H. Shaw, 1989, "Comprehension of the spoken word and ASL translation by chimpanzees (Pan troglodytes)", Friends of Washoe Newsletter, 9, No.1/2, 8-19.) and vocal speech was used only outside the chimpanzee living area. Since that time Loulis' observed vocabulary has increased by another 18 signs. Eight of the signs have met the three observation criteria (BABY, CRY, GROOM, ORANGE, QUIET, SODAPOP, STUPID, and YES) and other have not (BERRY, BITE, BOY, HAIR, HOOD, KATHLEEN, KISS, LOOK, RED, and SMELL). Of course we cannot be sure if he acquired these signs from humans or chimpanzees.

5.4. WASHOE TEACHES SIGNS

When Washoe was an infant in Reno, her human foster family taught her signs by modeling, molding, and signing on her body the way human parents teach deaf infants (Gardner & Gardner, 1989; Maestas & Moores, 1980; Schlesinger & Meadow, 1972). She used all of these methods with her own infant, Loulis. Teaching has also been observed between mothers and their infants among wild chimpanzees in regard to tool use (Boesch, 1991).

During the first three days that they were together, Washoe often turned toward Loulis, signed COME, approached, and then grasped his arm and retrieved him. During the next five days, she signed COME, and only approached Loulis. Then after eight days, Washoe no longer approached but only signed COME while orienting and looking at Loulis until he responded by coming to her. COME was among the first signs that Loulis came to use.

Parents of deaf children often mold the infant's hand into the configuration and then put the hand through the movement of the sign (Gardner & Gardner, 1989, pp. 18-19; Maestas y Moore, 1980, pp. 5-6). Washoe also molded Loulis' hands. For example, while waiting for a candy bar that a human friend was bringing her, Washoe signed FOOD repeatedly with much excitement and food-grunts. Loulis was sitting next to her, watching. Washoe stopped signing, took Loulis' hand, molded it into the FOOD configuration, and put it through the FOOD movement several times. In a similar context, Washoe formed the sign GUM, but with her hand on Loulis' cheek. During the first few months after his arrival, Washoe was also observed to place DRINK on Loulis' lips and HAT on his head, the way parents of deaf children place signs on their infants (Gardner & Gardner, 1989, p. 19; Maestras y Moores, 1980, p. 5; Rimpau, Gardner & Gardner, 1989, pp. 249-250; Schlesinger & Meadow, 1972, pp. 67-68).

Sometimes the first observation of a new sign involved direct imitation. For example, Loulis first used DRINK during a meal after Washoe used this sign in answer to a human caregiver who had asked WHAT about a drink. As Washoe was signing DRINK, Loulis watched her and signed DRINK, himself. Washoe also modeled directly for Loulis. For example, she signed BRUSH and then brushed Loulis with a hairbrush. On another occasion, Washoe placed a small plastic chair in front of Loulis, and then signed CHAIR/SIT to him several times in succession, watching him closely throughout.

It is important to remember that Washoe and Loulis were not under constant observation. Funds available in those critical early days permitted only four hours per day of scheduled observation by trained observers. In only a handful of cases could we be sure that we had observed the events surrounding the first use of a sign by Loulis. In most cases Loulis' signs appeared to be delayed imitations of signs that he had seen Washoe or another signing chimpanzee using in similar context (Fouts, Hirsch & Fouts, 1982).

5.5. CULTURAL TRANSMISSION OF NON-SIGNING SKILLS

In addition to signs, Loulis acquired other skills from the cross-fostered chimpanzees. He learned to use bowls and spoons as feeding implements, just as Washoe, Moja, Tatu, and Dar used them. He learned to build a sleeping nest with blankets in Washoe's unique way. Washoe builds a nest by taking her blanket and swirling it around herself on the floor, and sometimes she wraps herself in it. Then she collects toys and other objects and places them in her nest. For many months, Loulis simply watched Washoe, or played by himself, although occasionally he helped by giving her a toy. Then, Washoe began to hold Loulis as she built her nest. Eventually, Loulis learned Washoe's nesting methods and when given his blanket at night, he swirled it around himself as she did.

Note that Washoe's tutoring did not always work. Debbi Fouts in her paper presented at this conference reports an instance where Loulis is grooming a cut on Tatu using a stick. Washoe was observed to take the stick away from him, break it, discard it, and then demonstrate how to groom a cut with her lips. Her attempt was in vain because he continued to try to use the stick.

5.6. WASHOE LEARNS FROM THE OTHER CHIMPANZEES

Washoe herself has learned new signs from Moja, Tatu, and Dar. Because the Gardners could not find the BLANKET sign in the sign language manuals then available, Washoe was taught to use the noun/verb COVER for blankets (Gardner, Gardner & Nichols, 1989, pp. 60-61). Later in Reno, Moja, Tatu, and Dar were taught BLANKET, which differs in place, configuration, and movement from COVER. After these younger chimpanzees joined her, Washoe came to use both signs for blankets. From Moja, Washoe acquired a more precise form of the sign APPLE, and used it for apples, only. She continued to use the earlier form of her sign for several different kinds of fruits (cf. B. Gardner & Gardner, 1975, p.261).

TABLE 1. The form of signs in Loulis' vocabulary

SIGN	PLACE	CONFIGURATION	MOVEMENT
ALAN	Temple	Fist	Index edge contacts
APPLE/FRUIT	Cheek	Curved hand palm up	Knuckles rub down
BALL	Fingertips	Both curved hands, facing palm up	Fingertips of opposite hands contact
BANANA	Tip of index, extended from fist	Hooked index, extended from fist	Fingers hook and pull apart
BIB	Chest	Both open hands, palms to signer	Fingertips contact and move down
BIRD	Lips	Pincer hand palm to signer	Index and thumb grasp repeatedly
BLANKET	Underside of chin	Both fists, palms to signer and forearms vertical	Knuckles contact
BOOK	Palms	Both curved hands, palms facing palms	Palms of opposite hands grasp
BRUSH	Arm or part of body to be brushed	Open hand or compact hand, palm to body	Rubs to and fro
CATCH	Palm of curved hand	Fist	Fist contacts with palm of curved hand
CHASE	Wrist of curved hand, forearm extended	Fist	Knuckles contact repeatedly
CLEAN	Palm, open hand, palm up	Fist	Knuckles rub repeatedly in circles
COMB	Temple	Open hand or claw hand palm to signer	Rubs down
COME/ GIMME*	Space in front of signer	Curved hand, pincer hand palm down or palm to side	Arm extended toward addressee or object, then wrist or fingers beckon

DAR	Ear	Index extended from loose fist	Tip of index contacts or rubs forward
DIANA	Nose	Thumb extended from open hand palm to side	Thumb contacts
DIRTY	Underside of chin	Open hand or fist, palm down	Back of wrist contacts repeatedly, often with force so that teeth click together audibly (the audible component is characteristic of Washoe also)
DRINK*	Lips or mouth	Thumb extended from fist	Tip of thumb contacts or inserts
FLOWER/ DEB	Nose	Curved hand, palm to signer	Fingertips contact
FOOD/ EAT	Lips	Curved or pincer hand, palm to signer	Fingers contact
FRIEND	Indexes	Both hooked indexes extended from fists, one palm up and other palm down	Indexes of opposite hand interlock
GO*	Space in front of signer	Index extended from fist or from open hand	Arm extends away from signer, wrist rotates to point index toward location
GOOD*	Lips	Open hand, palm to lips	Palm contacts repeatedly
GUM	Cheek	Index extended from fist	Tip of index contacts then bends and straightens
HAT/ PERSON*	Top of head	Open hand, palm down	Palm contacts repeatedly
HOSE	Space in front of signer	Index extended from fist	Index or hand wiggles up and down
HOT/ COFFEE	Space in front of signer	Open hand, palm to signer	Approaches then turns palm down while extending arm away from signer

HUG*	Upper arms	One or both curved hand(s), forearms crossed	Palms contact
HURRY	Space in front of signer	Spread hand bent at wrist	Shakes
IN	Curved hand, index edge up	Open hand, palm to signer and fingers point down	Fingers insert
ME/ MINE*	Chest	Open hand, palm to signer	Palm contacts repeatedly
MILK	Space in front of signer	Fist, index edge up	Squeezes repeatedly
MORE	Fingertips	Both pincer hands	Fingertips of opposite hands contact repeatedly
NO	Head	N/A	Shakes side to side
NUT*	Teeth	Thumb extended from fist, palm down	Tip of thumb rubs away from signer
OUT	Palm of curved hand, palm to signer, index edges up	Open hand, fingers point down	Back of hand contacts then moves up out of grasp
PAINT	Palm of hand	Open hand, palm to signer	Fingertips contact then fingers bend and straighten
PEEK-A-BOO/ MASK	Both eyes	Open hand or hands, palm to signer	Palm or palms contact
PLEASE	Chest	Open hand, palm to signer	Palm contacts and draws across
ROGER	Ear	Pincer	Index and thumb grasp
SANDWICH	Back of open hand, palm to signer	Curved hand, palm to signer and crossing signer	Palm contacts and hand grasps
SHOE*	On object or space in front of signer	One fist or both fists, side by side and palms to signer	Knuckles contact object or ground

SORRY	Chest	Open hand, palm to signer	Rubs down
SWALLOW	Throat	Index extended from fist or from open hand	Tip of index rubs down
THAT/ THERE	On object or space in front of signer	Index extended from fist	Tip of index contacts object or points toward object or location
TICKLE*	Back of hand, or place on body to be tickled	Hooked index extended from fist	Tip of index rubs to side
TIME	Back of wrist	Hooked index extended from fist	Tip of index contacts sometime repeatedly
TOOTHBRUSH	Lips or teeth	Index extended from fist	Index edge rubs side to side
WANT	Chest	Curved hand, palm to signer	Contacts and rubs down
WASHOE	Contralateral ear	Curved hand	Fingertips rub
YOU	Addressee	Index extended from fist	Index points toward addressee

*Loulis' first ten signs

6. Conclusion

The findings presented here show that Loulis acquired signing and other skills from Washoe and the younger chimpanzees in his community. As in human language acquisition (Moerk, 1976), the chimpanzee mother actively taught her offspring, and the infant actively learned. The laboratory environment provided interesting events and an enriched social atmosphere, "interesting things to talk about ... and good friends to talk to" (R. Gardner & Gardner, 1974). Data collection by trained observers was a matter of course -- as much a part of the routine as serving meals and cleaning. Because of this we have been able to examine the development of social behavior, communication, and other skills in Loulis without disrupting them, and in this way, obtained a comprehensive record of cultural transmission.

7. References

Boesch, C. (1991). Teaching among wild chimpanzees. Animal Behavior, 41, No.3, 530-532.

Fouts, R. S. (1972). Use of guidance in teaching sign language to a chimpanzee. Journal of Comparative and Physiological Psychology, 80, 515-522.

Fouts, R. S. (1973). Acquisition and testing of gestural signs in four young chimpanzees. Science, 180, 978-980.

Fouts, R. S. (1975). Communication with chimpanzees. In G. Kurth & I. Eibl-Eibesfeldt (Eds.), Hominisation und verhalten, (pp. 137-158). Stuttgart: Gustav Fischer Verlag.

Fouts, R. S., Chown, B., & Goodin, L. (1976). Transfer of signed responses in American Sign Language from vocal English stimuli to physical object stimuli by a chimpanzee (Pan). Learning and Motivation, 7, 458-475.

Fouts, R. S., Fouts D. H. & Schoenfeld, D. (1984). Sign language conversational interactions between chimpanzees, Sign Language Studies, 34, pp. 1-12.

Fouts, R. S., Fouts, D. H. & Van Cantfort, T. E. (1989). The infant Loulis learns signs from cross-fostered chimpanzees. In R. A. Gardner, B. T. Gardner & T. E. Van Cantfort (Eds.) Teaching Sign Language to Chimpanzees, (pp. 280-292). Albany: State University of New York Press.

Fouts, R. S., & Goodin, L. (November, 1974). Acquisition of signs in chimpanzees: A comparisons of training methods. Paper presented at the Annual Meeting of the Psychonomic Society, Boston, MA.

Fouts, R. S., Hirsch A. & Fouts D. H. (1982). Cultural transmission of a human language in a chimpanzee mother/infant relationship. In H. E. Fitzgerald, J. A. Mullins & P. Page (Eds.) Psychobiological Perspectives: Child Nurturance,(Vol. 3, pp. 159-193). New York: Plenum Press.

Fouts, R. S., & Rigby, R. (1977). Man-chimpanzee communication. In T. Sebeok (Ed.), How animals communicate, (pp.1034-1054). Bloomington: Indiana University Press.

Fouts, R. S., Shapiro, G., & O'Neil, C. (1978). Studies of linguistic behavior in apes and children. In P. Siple (Ed.), Understanding language through sign language research, (pp. 163-185). New York: Academic Press.

Gardner, B. T., & Gardner, R. A. (1971). Two-way communication with an infant chimpanzee. In A Schrier & F. Stollnitz (Eds.), Behavior of nonhuman primates (Vol. 4, pp. 117-184). New York: Academic Press.

Gardner, B. T., & Gardner, R. A. (1975). Evidence for sentence constituents in the early utterances of child and chimpanzee. Journal of Experimental Psychology: General, 104, 244-267.

Gardner, R. A., & Gardner, B. T. (1974). Teaching sign language to the chimpanzee, Washoe. Bulletin D'Audio Phonologie, 4 (5), 145-173.

Gardner R. A, & Gardner, B. T. (1989). A cross-fostering laboratory. In R. A. Gardner, B. T. Gardner & T. E. Van Cantfort (Eds.) Teaching Sign Language to Chimpanzees, (pp. 1-28). Albany: State University of New York Press.

Gardner R. A, Gardner, B. T., & Nichols, S. G. (1989). The shapes and uses of signs in a cross-fostering laboratory. In R. A. Gardner, B. T. Gardner & T. E. Van Cantfort (Eds.) Teaching Sign Language to Chimpanzees, (pp. 55-180). Albany: State University of New York Press.

Goodall, J. (1986). The chimpanzees of Gombe: Patterns of behavior. Cambridge, MA: Harvard University Press.

Maestas y Moores, J. (1980). Early linguistic environment: Interactions of deaf parents with their infants. Sign Language Studies, 26, 1-13.

Moerk, E. L. (1976). Processes of language teaching and training in the interactions of mother-child dyads. Child Development, 47, 1064-1078.

Rimpau, J. B., Gardner, R. A, & Gardner, B. T. (1989). Expression of person, place, and instrument in ASL utterances of children and chimpanzees. In R. A. Gardner, B. T. Gardner & T. E. Van Cantfort (Eds.) Teaching Sign Language to Chimpanzees, (pp. 240-268). Albany: State University of New York Press.

Schlesinger, H. S., & Meadow, K. P. (1972). Deafness and mental health: A developmental approach. Berkeley: University of California Press.

Shaw, H. (1989). Comprehension of the spoken word and ASL translation by chimpanzees (Pan troglodytes), Friends of Washoe Newsletter, 9, (No.1/2), 8-19.

Stokoe, W., Casterline, D., & Croneberg, C. (1976). A dictionary of American Sign Language. Silver Springs, MD: Linstok Press.

THE USE OF REMOTE VIDEO RECORDINGS TO STUDY THE USE OF AMERICAN SIGN LANGUAGE BY CHIMPANZEES WHEN NO HUMANS ARE PRESENT

DEBORAH H. FOUTS
Chimpanzee and Human Communication Institute
Central Washington University
Ellensburg, Washington 98926
USA

ABSTRACT. The method of remote video recording, with no humans present, was used in studies of chimpanzees Washoe, Moja, Tatu, Dar and Loulis for five consecutive years. We found that chimpanzees sign to each other, in a variety of contexts. The youngest, Loulis, signed more to his male peer Dar than to anyone else. The chimpanzees sign to themselves in the same contexts as humans do. Instances of imaginative play were also recorded.

1. Introduction

Four of the five chimpanzees, Washoe, Moja, Tatu and Dar who now reside at the Central Washington University 's Chimpanzee and Human Communication Institute, were cross-fostered as infants at the University of Nevada, Reno (Gardner & Gardner, 1989, p. 1-28) Loulis, Washoe's adopted son, is the other member of the family. Loulis acquired his signs from Washoe, Moja, Tatu and Dar. Over the years, their behavior has been documented using 16mm film records as well as video recordings.

The four cross-fostered chimpanzees learned signs from friendly human companions. They signed to their human friends, to chimpanzee friends and to cats and dogs and even to toys. (Gardner and Gardner, 1989 p. 1-28). When Loulis joined the family, in 1979, the humans restricted their signing in his presence to see if Loulis would learn signs from his adoptive mother Washoe and the other cross-fosterlings. Washoe and the other chimpanzees were virtually the only source of conversational signing partners. As described in Roger Fouts' presentation at this conference, Loulis learned over 50 signs of ASL from them in the course of 5 years.

We began an observational systematic study of signed interactions between chimpanzees in May, 1981, when Dar and Tatu, each about five years old, came to the Central Washington University institute in Ellensburg to be integrated into the social group which consisted of sixteen-year-old Washoe, her three-year-old adopted son Loulis, and eight-year-old Moja. Dar and Tatu were cross-fostered together with Moja in Reno, but only met Washoe and Loulis in Ellensburg. From May until December, Tatu and Dar lived adjacent to Washoe, Moja and Loulis, but the two groups were separated by a wire mesh gate. In late December, 1981, the gate was opened and from then on the five chimpanzees lived together.

In 1983 we began using the remote video recording technique. Before I present the methods and findings of these studies, it is important to describe what life was like for these five chimpanzees in the Central Washington University institute when they were housed in a compound on the 3rd floor of the psychology building.

R. A. Gardner et al. (eds.), The Ethological Roots of Culture, 271–284.

2. Procedure

Each day in the life of Washoe, Moja, Tatu, Dar and Loulis began around 8:00 A.M., when the human caregivers opened the laboratory and woke the chimpanzees with greetings. When greeting, the humans used both ASL signs and chimpanzee communicative gestures and vocalizations, such as head-bobs, pant-hoots, and bouncing (Goodall, 1986, pp. 144-145). Before breakfast, the chimpanzees were asked to help clean out their enclosures of sleeping blankets, dressup clothes and dinner bowls, spoons, magazines, and tree branches left from the evening before. The chimpanzees were thanked and encouraged for their help.

For breakfast, each chimpanzee was offered a fruit drink called a smoothie, which refers to its texture. They were also offered, depending on the season, fresh, frozen or dried fruit and occasionally an assortment of the three. The fruit smoothie contained a least two varieties of fruit, vitamin E, vitamin C and nutritional yeast, blended to make a tasty drink. Free ranging chimpanzees, eat over 140 different species of plants, as well as a variety of insects, birds' eggs, birds, and small mammals (Goodall, 1986, p. 232) as well as medicinal plants (Wrangham and Goodall, 1989, p. 32-33), and we tried to offer as much variety in the meals we provided. When breakfast was served, all the chimpanzees came into one room and drank smoothies from plastic beakers. During breakfast, the human server often gave the food-grunts that a good meal deserves. With smoothies as well as with the after breakfast fruit, Washoe was served first, as the eldest and most dominant member of the group. This was done out of respect for her social status in the group.

After breakfast, the humans cleaned the chimpanzees' compound, which consisted of four rooms and interconnecting tunnels. The chimpanzees had free access to the entire enclosure while cleaning commenced. Frequently, they chose to interact with the humans who were cleaning.

Interactions ranged from watching the humans at work, to attempts to engage them in games of tickle and chase. The chimpanzees requested DRINK from the nozzle of the hose and often Washoe and Tatu asked to CLEAN. We responded by providing them with a dish of soapy water and a scrub brush. The chimpanzees worked on the floor and inside of the enclosures as the humans cleaned the outside area. Although cleaning was a required daily maintenance activity, it also provided excellent opportunities for the humans and the chimpanzees to interact, and some of the best observations of the chimpanzees as individuals and as a group took place during this task.

Frequently, after cleaning, a snack was served. Sometimes, the chimpanzees requested food or non food treats during cleaning. At other times, we offered a SURPRISE and each of them would respond with joyful and excited hooting. Treat items included toothpaste on toothbrushes, flowers, leaves, snow, ice cubes and water balloons (occasionally filled with fruit juice) as well as food treats. A special treat item was the raisin board, a six-inch-long board with approximately twenty holes drilled through it, which were stuffed with raisins and sometimes marshmallows. This was offered together with a willow or an apple branch that served as a tool to extricate the food, as in termite fishing in the wild (Goodall, 1986, pp. 248-249).

During the time period before lunch, play objects were brought out and offered to the chimpanzees: magazines, hair brushes, plastic mirrors, sunglasses, perfume samples, lipstick samples, paper feed sacks, Halloween masks, shirts, skirts, and so on. Some of the simplest objects were used the most, such as the empty paper sacks that were popular for games of chase. The chimpanzees covered the head with a sack and then engaged in a game similar to the children's game of "blind man's bluff."

At 12:00 lunch was served. Washoe, Moja, Loulis, Tatu and Dar were offered a choice of cooked soups such as split pea, bean, or lentil. These soups were flavored with different vegetables, as well as spices. Other proteins, such as tuna, beef, or chicken were offered occasionally for some variety, but only one of the chimpanzees, Tatu is particularly fond of meat. The lunch meal was served in bowls or cups, with spoons for utensils. If a chimpanzee requested MORE, they were served second helpings. We also provided opportunities for foraging, by putting lettuce leaves mixed with other vegetables on top of the chain link cover of the largest enclosure.

Lunch was followed by quiet activities, among the chimpanzees or interactions with the human caregivers. For quiet activities, the chimpanzees were given drawing materials, as well as small colorful magazines or picture books to look at alone or together with a human caregiver. The caregivers also played active games of tickle and chase with them. Because the humans always remained outside the enclosure, physical contact during these games was minimal and hard plastic toys or long branches were used to tickle and playfully touch the chimpanzees. There are other ways of maintaining contact, and dedicated human play partners frequently used chimpanzee gestures and vocalizations, such as breathy laughter, play-faces, food stomping, and walking with extra large slow movements and so on. Grooming and brushing was another interaction that helped to build good relations between the chimpanzees and their human caregivers. Caregivers could offer their arms and legs for grooming by placing an elbow, an arm, calf of the leg, or sole of the foot next to the chain link enclosure. A good indication of acceptance was when two or even three chimpanzees joined in a grooming session with the human caregiver. Acceptance into the group was important not only in terms of social relations, but also because it led to better opportunities to observe subtleties and individual differences in the behavior of these chimpanzees.

In the late afternoon, the compound was cleaned once more. Again, the chimpanzees were free to be passive observers or to take an active role as helper or play partner. Dinner, served around 4:00 P.M. consisted of rice, cooked cereal, or steamed vegetables, or a combination of cooked grains and cooked vegetables. For variety, we also provided sandwiches, egg salad, tortillas and beans, pizza and occasionally popcorn feasts. After dinner had been served and the enclosure was made tidy, each chimpanzee received a blanket for nesting and sleeping. During some seasons, we provided straw, willow branches, corn stalks and other vegetation for bedding.

In these ways, we attempted to break the routine of an institution, to make each meal an interesting and enjoyable social event, and each day special with many different activities and many opportunities for social interaction. We also celebrated the birthdays of each chimpanzee and the most significant caregivers, as well as most of the holidays on the calendar, which made at least one, and usually more, days of each month extra special. Thus the Christmas celebration started soon after Thanksgiving, when we set up a tree in a room through which a tunnel ran, and began to make garlands of edible ornaments for the tree and the hallways. We always get the tree and decorate it on the weekend following Thanksgiving. By mid-December the halls and tree were decorated with strings of special treats, nuts, popcorn, cranberries, dried fruit, gum and raisins. Throughout this period, the tree was a popular topic of conversation between chimpanzees and their human companions. On Christmas Day, the chimpanzees are given the ornaments to eat and because there are so many, they continued to receive the treats daily until New Year' Day. In 1986, the first snowstorm of the season came on the day after the Thanksgiving celebration. Thanksgiving was celebrated with the chimpanzees and the human caregivers, before the students left for home. The holiday meal was a favorite of Tatu's, roast turkey, BIRD MEAT as well as POTATO SWEET, BERRY SAUCE and PUMPKIN SWEET. The following morning the laboratory seemed especially quiet, with the snow on the ground, visible from the laboratory windows. As was usual during college vacations, the students were away, and only the Foutses were in the lab. Perhaps all of this reminded Tatu of Christmas past, or perhaps she thought the Foutses needed to be reminded to set up a tree. During the cleaning routine, she followed them around signing SWEET SWEET SWEET TREE. Deborah Fouts answered Tatu, YOU MUST WAIT FOR SANTA.

3. Observer Training

There were strict prerequisites and requirements as well as a training regimen for prospective participants in this research. Before being admitted to the training program, prospective participants (for the most part CWU students) completed at least one university level class in ASL. The training program began with a three-month introductory course, which included reading books and journal articles pertinent to this field, as well as viewing films and video tapes of captive and wild chimpanzees.

Next, the participants were trained to recognize specific chimpanzee behaviors, both verbal and nonverbal, as well as the individual traits and habits of the chimpanzees in this group. Their training in ASL was extended to include the criteria of form and usage that had to be met for a gesture to be recorded as a sign and the PCM method of describing signs (Stoke, 1965). A PCM description specifies the place where the sign is made, the hand configuration used to make the sign, and the movement involved in making the sign. Trainees had to master a manual of descriptive terms and abbreviations for nonverbal behavior as well. The majority of the 338 terms in this 40-page manual are based on Goodall's (1968) classification of the facial expressions, vocalizations, postures and actions of wild chimpanzees. All of the trainees had to pass a formal one-hour test on such terms as play-face, pant-hoot, bipedal swagger, and back-hand thump.

After this, each trainee viewed video-taped interactions of the chimpanzees, together with an experienced observer. The experienced observer, a research participant who had at least two years experience with ASL and with these five chimpanzees, identified each behavior as specific examples appeared on the video screen. The trainees recorded observations of taped interactions in longhand, and then practiced identifying signs and transcribing nonverbal behavior into the abbreviations of the manual of nonverbal behavior (NVB).

Finally, each trainee recorded ongoing chimpanzee interactions together with the experienced observer and their records were compared for inter observer reliability. Problem areas were reviewed and there was further individualized instruction until the inter observer agreement was greater than 80%. Only then were the new research participants scheduled for data collection on their own.

Nonverbal behavior, such as the backhand thump or the bipedal swagger, was described with terms from the NVB manual which included 49 terms that referred to facial expressions and vocalizations, 30 that referred to postures, and 117 that referred to actions. An English word gloss was used to describe such gestures that qualified as a sign. The specific word such as FOOD, HUG, HURRY, or COME corresponded to a distinctive combination of place, hand configuration, and movement.

4. Study 1: Conversations Between Chimpanzees

4.1 VARIABLES IN CONVERSATIONAL INTERACTIONS: ADDRESSEE, CONTEXT, DURATION

All interactions that included signing were classified as conversations, and the first chimpanzee to use signs within an interaction was classified as the initiator of that conversation. The addressee was determined on the basis of body and head orientation.

Thus, if Loulis was signing while his body and his head or his head alone was directed toward Dar, then Dar was considered to be Loulis' addressee.

Interactions were classified into one of six behavior contexts, of: reassurance, grooming, play, discipline, eating/feeding, and social interaction. (Fouts & Fouts, 1989 p.299).

Signed conversations consisted of one or more signed utterances and two or more turns. All of the signed conversations included some communicative behaviors in addition to signs (Fouts, Fouts, & Schoenfeld, 1984). For example, the following is a conversation that occurred during a play interaction between Loulis and Dar:

Turn 1: Utterance 1: Loulis solicited a water balloon from Dar by signing HURRY HURRY while holding his hand toward the balloon.
Turn 2: Dar moved away from Loulis.
Turn 3: Utterance 2: Loulis signed WANT to Dar.
Turn 4: Dar again moved away
Turn 5: Utterance 3: Loulis signed HURRY HURRY GIMME to Dar.
Turn 6: Dar terminated the interaction by putting the balloon into his mouth and turned away from Loulis who also withdrew. (Later, in reference to the same piece of balloon, Loulis signed GUM GUM, placing the sign on Dar's cheek, and then signed GUM on his own cheek. Dar continued to chew on the balloon and Loulis did not get his share.)

This conversation has six turns and three signed utterances HURRY, WANT, and HURRY GIMME. Both in Turn 1 and Turn 5, HURRY was counted as a reiterated sign and not as two separate HURRY signs, since reiteration is a way of modulating semantic meaning (Van Cantfort & Rimpau, 1982).

4.2 SYSTEMATIC LIVE OBSERVATION SAMPLES OF CONVERSATIONS

Prior to the introduction of the remote video recording the chimpanzees' use of signs in the course of the daily routine was recorded over a period of 21 months (5/81-1/83). Comprehensive sampling sessions, 40-minutes in duration, were scheduled five times daily, five days per week (excluding weekends and school holidays). In each sampling session, the behavior and signs that occurred during interactions were recorded by an observer, sometimes by an observer and a trainee, standing just outside the enclosure. The observer assigned each interaction to a context, such as reassurance, play, feeding, mutual grooming, or discipline.

Of course, the chimpanzees also signed at other times and such observations of signing were recorded in the daily log that was kept for the individual chimpanzees. Records of interactions obtained outside of the systematic sampling sessions will be referred to as adventitious. The systematic sampling sessions accounted for 76.4% of the reports that were analyzed, while the adventitious samples accounted for 23.6% .

Daily records showed that the introduction of Dar and Tatu to Washoe, Loulis and Moja produced a dramatic effect on the incidence of signing. In the eight-month period just prior to the introduction of Dar and Tatu (May, 1981-Dec.1981), the monthly totals of conversations between chimpanzees ranged from 12 to 85, with a mean of 37.75.

However, during the 13-month period after Dar and Tatu were allowed physical contact with the other chimpanzees (Jan. 1982-Jan. 1983), the monthly totals for conversations ranged from 118 to 659 with a mean of 378.15.

Before the introduction of Dar and Tatu, Loulis was the primary initiator of signing and he addressed 90% of his conversations to his adopted mother, Washoe. As he grew older and became more familiar with his new peers, Loulis addressed an increasing proportion of signs to Dar, his only male peer, and a decreasing proportion to Washoe, his mother.

During the final ten months of this sample (April, 1982-Jan., 1983), Loulis initiated 1,292 conversations and addressed 54% of these to Washoe, 27% to Dar, but only 10% to Tatu and 9% to Moja. Similarly, Tatu and Dar, closest in age to Loulis, initiated most of the signing addressed to Loulis during the same period.

In this sample, 88% of the conversations occurred in just three contexts: social interaction (39%), reassurance (29%) and play (20%). Only 5% of the conversations occurred in a feeding context (Fouts, et al. 1984).

4.3 REMOTE VIDEO RECORDS OF CONVERSATIONS

4.3.1 *Method* . In addition to live observations, we have found remote videotaping (RVT), without humans present, to be an excellent way to begin to understand how and where the chimpanzees use their signs.

We first used RVT during a 15-day period at the end of July and the beginning of August, 1983 in a study that focused on Loulis' use of signs with his chimpanzee family. During this time video cameras and remote video recorders monitored the chimpanzees' interactions with no humans present (D.Fouts, 1984). There were three video cameras, each focused on a portion of the enclosure. There was some overlap of the areas covered by the three cameras, but only 75% of the total enclosure area was covered. The three cameras were connected to video recorders located in a separate recording room. Every day during the 15-day period, the video cameras were turned on for three 20-minute recording periods between the hours of 9:00 A.M. and 5:00 P.M. Forty-five recording periods were scheduled so that each hour of the day was sampled randomly without replacement either five or six times.

During these recording sessions no one entered the chimpanzee quarters, surrounding hallways, and adjacent rooms. Everyone stayed out of view and kept silent, and all activities in the laboratory that might be a distraction were stopped-even the cooking of meals was forbidden, because the cooking odors as well as the preparation and monitoring of the food might distract the chimpanzees from their interactions.

4.3.2 *Reliability.* To determine inter observer reliability, five of the 45 video-taped samples were scored independently by two experienced observers, who had studied ASL, the NVB manual, and the chimpanzees for two and a half years and five years respectively. An agreement was counted when these two observers agreed on all six aspects of the interactions: (1) context category; (2) initiator; (3) terminator; (4) gloss and form description of signs; (5) the number of signs; and (6) the number of turns. If any of these six aspects of an interaction was scored differently, it was counted as a disagreement. In the five segments, there were forty interactions. For 93% of the interactions, the two observers agreed in their reports for all aspects of the interaction. Disagreements were excluded from the final analysis of the data.

4.3.3 *Conversation results.* The 45 video-taped sessions randomly sampled each hour of the daily routine. The amount of signing varied from sample to sample. As many as 29 different conversations were recorded in a single 20-minute session, but in some sessions the chimpanzees rested or groomed during the entire 20 minutes. The 451 interactions that Loulis initiated accounted for 4 hours, 13 minutes or 28% of the total 45-hour video record. The 54 conversations (interactions containing signs) that Loulis initiated accounted for 43 minutes, 20 seconds.

Table 1 shows that the majority of the interactions (signed conversations and nonsigning interactions) that Loulis initiated were directed to Dar, his male peer. Of the 451 interactions, 40% (181) were addressed to Dar, while between 19% and 21% were addressed to each of the other three chimpanzees, Washoe, Moja, and Tatu. Loulis' preference for Dar as an addressee is even more marked when we consider that of the 54 conversations that he initiated, 50% were addressed to Dar and between 5% and 2.2% were addressed to the other three chimpanzees. Table 2 also shows that, in the context of play. Loulis initiated 4.8 times as many interactions with Dar as with his mother Washoe (101 vs 21), and 11.2 times as many with Dar as with Moja, another female peer (101 vs 9). He only initiated one play interaction with his other female peer, Tatu. In the social context, Loulis initiated a sizable and roughly equivalent number of interactions with each of the other chimpanzees. There were less than ten conversations in four of the six contexts (reassurance, grooming, discipline, and eating/feeding), too few to make comparison of Loulis signing to his mother and to his three peers.

Table 1
Contexts and Addressees for Interactions Initiated by Loulis

Context		Washoe	Moja	Tatu	Dar	Total
		Addressee				
Social	S	3	0	6	10	19
	NS	54	66	53	52	225
Play	S	1	3	0	16	20
	NS	20	6	1	85	112
Grooming	S	0	0	6	1	7
	NS	3	12	18	13	46
Reassurance	S	8	0	0	0	8
	NS	3	0	2	1	6
Discipline	S	0	0	0	0	0
	NS	0	0	0	0	0
Feed/Eat	S	0	0	0	0	0
	NS	3	2	0	3	8
Total	S	12	3	12	27	54
	NS	83	86	74	154	397
Total Interaction		95	89	86	181	451

S= interactions with signing; NS=nonsigning interactions

Table 2
Number of Signs Used by Loulis as a Function of Context and Addressee

	Addressee				
Context	Washoe	Moja	Tatu	Dar	**Total**
Social	6	0	25	68	99
Play	2	9	0	43	54
Grooming	0	0	23	3	26
Reassurance	27	0	0	0	27
Discipline	0	0	0	0	0
Feed/Eat	0	0	0	0	0
Total	35	9	48	114	206

Table 2 shows that Loulis used more signs in his interactions with Dar than with any other chimpanzee; 55% of the 206 signs Loulis used during interactions were addressed to Dar, 23% to Tatu, 17% to Washoe, and 4% to Moja. When we compared these findings to earlier samples of conversations initiated by Loulis, it was seen that the focus of his signing shifted from Washoe to Dar. Goodall has described a similar shift from mother to peers in the social development of wild chimpanzees (1971, p 156-157, 170 & 215; 1975, p. 137; 1986, p. 81& 166). At the time of this video sample Loulis' total vocabulary consisted of 43 signs. Table 2 lists the signs that appeared in the sample and the frequencies of each. Most of Loulis' signs were addressed to one of the other chimpanzees. At other times, his head and body were oriented away from his companions and he appeared to be signing to himself. Table 3 lists separately the signs used in interactions and in private signing.

Table 3
Signs Used by Loulis During Interactions and When Signing to Himself

	Interactions	Private signing
HURRY	134	6
BALL	-	59
ME MINE	17	24
HAT	-	16
GOOD	11	4
COME	13	-
MORE	7	-
DRINK	6	1
WANT	4	-
PEEKABOO	4	-
GUM	3	-
PERSON	3	-
GIMME	2	-
OUT	-	2
SHOE	1	1
FOOD	-	1
MASK	-	1
TICKLE	1	-
Total frequency	206	115

4.3.4 *Findings, other behaviors recorded in the RVT sessions* . Examples of meaningful modulations (Van Cantfort & Rimapu, 1982) of the shapes of the signs appeared in the remote video taping of chimpanzee-to-chimpanzee signing. For example, during an interaction before lunch Tatu used repetition for emphasis together with questioning modulation. Tatu approached Washoe and signed TIME TIME TIME TIME EAT EAT. As she signed, Tatu looked steadily into Washoe's eyes while holding the last sign, EAT.

Signs for actions are placed on the addressee, to indicate agent-of-action. For example, we have recorded Dar making the TICKLE sign on Loulis' hand followed by a tickle game between the two young males. Dar often uses yet another inflected form of TICKLE when he signs TICKLE on one of his toys, and then gives the toy to a human who is expected to tickle him with it. Incidentally, at this writing Dar's favorite toy to be tickled with is still a plastic dinosaur.

During a remote video-taping session Loulis used a social strategy that resembles one commonly used by human children. We recorded an instance of this, following a fight between Loulis and Dar. After the fight subsided, Dar moved away from Loulis, while continuing to scream. At the same time that Dar was screaming, Loulis approached and sat quietly behind Washoe and signed GOOD GOOD GOOD to her. Much like a human child, he seemed to be assuring his mother that he was good. Dar soon learned to use this strategy as well. We have recorded Dar, soon after Washoe entered the room to settle a fight between Loulis and he, throwing himself on the floor near Washoe, signing HUG HUG HUG to her and then screaming as he looked toward Loulis.

In some interactions recorded by remote video tape, signing appeared to reduce aggression. There were several instances in which Loulis was harassing both of his female peers, Moja and Tatu, until either Washoe or Dar distracted Loulis by signing TICKLE, CHASE, GROOM or COME to him. This signed invitation to play or groom was often effective in distracting Loulis from further aggressive behavior. In one such incident, Loulis harassed Moja and Tatu by rushing at them and hitting them both as they sat grooming. Moja and Tatu moved away and began to resume grooming. When Loulis started to charge them again, Dar reached out and touched Loulis on the arm. Loulis ignored this touch, and continued his charge. Moja and Tatu moved away from Loulis once more. Now Washoe reached out and touched Loulis on the leg and Dar signed TICKLE on Loulis' arm. Loulis responded to Dar by turning toward him and wrestling with him. Washoe joined in this game by tickling Loulis with her hand, as Dar wrestled with him. Subsequently, Loulis withdrew from the area where Moja and Tatu were, and no further aggression occurred. Thus, these chimpanzees used signs to initiate a distraction which in turn served to inhibit aggression.

During the remote video-taping sessions we recorded three different instances of Loulis using a stick to groom an infection in Tatu's foot. In each case, Loulis held Tatu's foot and then picked up a willow branch from the floor. Next, Loulis used his teeth to break off a portion of the branch, and chewed the open end of the branch until it was pointed. He inserted the pointed end of the branch in the wound, and moved it around in the wound. If Tatu attempted to remove her foot Loulis sometimes signed GIMME, COME, or HURRY to solicit the wounded foot, or merely grabbed the foot and began again. Washoe seemed to disapproved of this method of grooming. She removed the grooming stick from Loulis each time she saw him use it, or begin to fashion it with his teeth. On two occasions, after removing the stick, Washoe took Tatu's foot and groomed it with her lips and fingers, which is her standard method of grooming and used as well by free-living chimpanzees in Africa (Goodall, 1989). Loulis continues to use both methods to groom wounds of his own and others. Tatu's wound was such an attractive target for grooming that the frequency of grooming interactions among all of the chimpanzees increased during the period we were video taping.

5. Study II: Private Signing by Chimpanzees

5.1. PREVIOUS FINDINGS: RENO LABORATORY AND CWU INSTITUTE

Since the beginnings of Project Washoe in the Reno laboratory, in the mid-1960's instances of private signing have been observed and recorded. When Washoe was young, her private conversations were truly private. If caregivers tried to eavesdrop she would turn away; and if they persisted and continued to try and see what she was signing, she would get up and move to a more secluded location. Washoe leafed through magazines and named pictures of things that she saw or merely signed to herself. She did this while alone in her bedroom, or sometimes she took a magazine up to the top of a 30-foot-high willow tree and signed to herself (Gardner & Gardner, 1989). In later studies, involving all five chimpanzees we recorded private signing. In a study with over 5200 instances of chimpanzee conversations we found that 119 of them were private (Fouts, Fouts, & Schoenfeld, 1984).

5.2. REMOTE VIDEO RECORDS OF PRIVATE SIGNING

5.2.1 *Bodamer's analysis of RVT records of conversations* . Following this analysis of remote video tape recording, which primarily focused on Loulis' conversations with his mother and peers, we collected video records for five consecutive years.

We noted on those tapes, in addition to the chimpanzee-to-chimpanzee sign language conversations, there were also instances of the chimpanzees signing to themselves. Mark Bodamer (1987) analyzed 45 hours of existing tapes that focused on chimpanzee-to-chimpanzee conversations for private signing. He found 90 instances of private signing on these tapes that he could classify into categories used for private speech in human children (Furrow, 1984). Our concern with this sample, was that it was biased against private signing because the tapes were originally made with the purpose of recording chimpanzee-to-chimpanzee conversations. For example, whenever we had a choice between recording two chimpanzees or one chimpanzee we would choose two chimpanzees in hopes of taping chimpanzee-to-chimpanzee conversations. With the help of a National Geographic grant we went on to record another 56 hours of remote video recordings, but this time we used criteria that would increase the likelihood of recording chimpanzees signing to themselves.

5.2.2 *RVT sample focused on private signing* . In the Bodamer et al (Bodamer, Fouts, Fouts & Jensvold, in press) study focusing on private signing we collected 12 minutes of tape per day five days a week for a total of 56 weeks which gave us 56 hours of new tape to analyze. But this time if we had a choice between recording two chimpanzees as opposed to a single chimpanzee, we would choose the later. As compared to the sample used in Mark Bodamer's thesis, this procedure increased the instances of private signing by threefold, and yielded 368 instances for the 56 hours. One of the more common categories of signing used by chimpanzees and humans as well, was referential signing. Recorded examples include Washoe naming the pictures in a magazine, and Dar signing DOG when he noticed a dog running by outside the building. Of the 368 instances of private signing that were found 218 (59%) were categorized as referential. This was the largest category. An utterance of private signing was categorized as informative signing if the chimpanzees signed about something not in their present environment. The instances we have recorded refute the claim that chimpanzees only sign about things in their immediate environment and that for them "out of sight, is out of mind". For example, Tatu sat in the enclosure with a bowl of bean soup, signing MEAT MEAT MEAT. Bean soup is not a favorite food for Tatu, and as was reported earlier, any type of meat is. Again this shows that just like us, the chimpanzees talk about things that are not present. In other examples, Washoe signed about DEBBI even though she was not present, Loulis signed about a DOG that wasn't there and Moja signed about absent BIRDs. We found that this category accounted for 44 of the 368 instances (12%) and was the second largest category observed. This again demonstrates the rich mental life of the chimpanzee. Expressive signing was popular with Washoe. This is a category of signing that is used more commonly by adult humans than by children. It occurs when humans become upset or excited by something. Thus, if you accidentally struck your thumb with a hammer, even though you were alone you might say something very expressive. Or in less emotional situations, when the traffic light changes to red as you enter the intersection and you happen to notice a police car observing you,

you might say something even more expressive. During one RVT session, Washoe was being recorded as she was lying on a bench looking at a magazine. Loulis came running into the room and into the camera frame from an overhead tunnel. He was running very fast and Washoe ignored him. As he ran under the bench where she was lying he reached up and took her magazine and then ran immediately out of the room using the overhead tunnel again. By the time Washoe got to her feet, Loulis was gone and as she walked off, she signed DIRTY DIRTY to herself. Washoe uses the DIRTY sign as an insult. This category accounted for 10 of the 368 instances (3%). Instances of private signing that were categorized as Regulatory, Attentional, Interactional, Instrumental, Describing own Activity, and Imaginary ranged from 4 instances (1%) to 31 instances (8%). Two categories, Self-regulatory and Question were not observed in this study but had been observed in the 90 instances Mark Bodamer analyzed for his thesis. The Imaginary category accounted for 17 instances (5%) and it was this behavior that was examined by Jensvold and Fouts (1993). This study focused on imaginative private signing as well as other imaginative behaviors by chimpanzees.

6. **Study III: Imaginative Play and Signing by Chimpanzees**

Imagination is another of those special behaviors that have been considered unique to our species. In the private signing study, we used definitions of types of imaginative play developed by Matthews (1977) for human children to identify instances of imaginative play and signing in the RVT records and we found that 5% of the utterances were imaginative. For example, rhythmic movements of signs or form alliteration of signs can be considered comparable to vocal singing or poems One example, Loulis played with a block of wood by placing it on his head and then referring to it as a HAT. Another instance was when Moja produced an alliteration by rhyming signs that all used the same initial hand configuration. Moja sat by herself and signed RED RED RED CRY CRY CRY FUNNY and then lay down.

Earlier reports of chimpanzee play revealed that chimpanzees engaged in all six categories of imaginative play identified for children (Hayes, 1952; Goodall, 1986; Gardner & Gardner, 1989; Savage-Rumbaugh, 1988).

Jensvold and Fouts (1993) study expanded on this to include play behaviors as well as signing. Fifteen hours of behavior were recorded over a three week period. Data was collected for one hour per day between 9 a.m. and 3 p.m. for 15 days, Monday through Friday. Eight sessions were in the morning and seven in the afternoon. The cameras were turned on only when at least one chimpanzee was present. Objects were placed in the enclosure each day to simulate a playroom environment.

Inter observer reliability was obtained by having two independent observers with at least one year experience with the chimpanzees, analyze 10% of the data. Three measures were analyzed: identification of the individual chimpanzees, the behavior that occurred, and the occurrence of any signs. Each measure was scored one time every minute. The percent agreement of the scores for each minute were calculated. Inter observer reliability for the data analysis was 98% for identification of the individual chimpanzee, 97% for identification of the behaviors, and 100% for the occurrence of signs. The total reliability for data analysis was 98%. Reliability for the occurrence of imaginative play was 90%. Inter observer reliability for the categorization of instances of imaginative play was 100%.

Jensvold and Fouts (1993) found that chimpanzees when playing with toys would treat them as if they were alive. Here, imagination involved attributing certain properties to situations or things that they do not actually possess. Using the remote video recording technique they recorded six instances of imaginative play during 15 hours of RVT, these fell into two categories, Animation and Substitution. Dar showed a type of imaginative play referred to as Animation when he signed PEEKABOO to a stuffed teddy bear. Animation (Matthews, 1977) is defined as acting toward an object as if it were alive, and includes speaking or signing to an object. In another instance, Dar signed TICKLE to a stuffed bear and then positioned the bear between his body and the wire. Dar often presses his own body against the wire in this same fashion when being tickled by a human friend. Dar also signed TICKLE on a puppet as though it were capable of playing a game of tickle with him. Washoe demonstrated a different instance of Animation by using a chimpanzee killing gesture to "hit away " a stuffed animal. Moja displayed a type of imaginative play referred to as Substitution when she began to treat a purse as if it were a shoe. The recording of six instances in 15 hours is impressive when one considers that in the Fouts, Fouts, and Schoenfeld study (1984) over 5200 instances of chimpanzee signed conversation yielded only 2% (119) that were classified as private signing, and in the private signing study only 17 of the 368 instances (5%) fell in the Imaginative category.

Examples for all six categories of imaginative play have been observed in these and other chimpanzees. The category Attribution of Function is defined by behavior that appears to attach a new functional property to an object although the object still retains its original identity. One example was recorded in our laboratory, Washoe picked up a toothbrush signed HAIRBRUSH and then brushed her hair with it. The toothbrush retained its original identity yet Washoe had given it a new function. The category of Insubstantial Material Attribution is defined by the individual acting as if there is an object there when there is none. A fine example of this category was reported for the chimpanzee, Viki (Hayes 1952 p. 81). Cathy Hayes described Viki behaving as if she had an imaginary pull toy, walking around the house trailing her arm behind her. At times she acted as if it had gotten stuck and made tugging motions that culminated in a jerk and then she went on her way trailing her arm behind.

Another category, Insubstantial Situational Attribution, includes word play, rhymes, and songs (Furrow, 1984). This was demonstrated when Moja in the instance of private signing described earlier signed RED RED RED CRY CRY CRY FUNNY to herself. All three signs were made with the same handshape which is analogous to alliteration rhyming in spoken word and the repetition of the signs added to the song like quality of the utterance. A final category is Character Attribution, a young Washoe was recorded bathing and feeding her dolls. She was taking on the characteristics of her caretakers who normally fed and bathed her (Gardner & Gardner, 1989). These observations indicate that chimpanzees engage in all of the types of imaginative play as human children do. Imaginative play is a behavior that is rarely observed compared to all the other things the chimpanzees do, just as it is with our species.

7. Conclusions

Using remote video recording we have demonstrated that the chimpanzees Washoe, Moja, Tatu, Dar and Loulis sign to each other and to themselves when no humans whatsoever are present. We have demonstrated that chimpanzee-to-chimpanzee signing was rich enough to provide texts that could be analyzed for context, initiator, addressee, number of signs and number of conversational turns. We have demonstrated that the signs used during interactions between chimpanzees were clear enough so those independent observers, fluent in ASL, agreed in their identification of more than 90% of the signs. We have also found that the instances of private signing by chimpanzees can be classified into the eleven categories used to classify private speech in humans. Video records of imaginative play and signing by chimpanzees have been obtained as well.

8. References

Bodamer, M. (1987). Chimpanzees signing to themselves. Unpublished master's thesis, Ellensburg, Washington: Central Washington University.

Bodamer, M., Fouts, R.S., Fouts, D.H., & Abshire, M.L. (1994). Functional analysis of chimpanzee private signing. In Press, Human Evolution.

Boehm, C. (1989). Methods for isolating chimpanzee vocal communication. In P.G. Heltne & L. A. Marquardt (Eds.) Understanding Chimpanzees, pp 38-60. Cambridge, MA: Harvard University Press.

Campbell-Jones, S. (Producer & Director) (1974). The first signs of Washoe. (Film). Boston, MA:WGBH Nova series.

Eibl-Eibesfelt, I. (1970). Ethology, the biology of behavior. New York: Holt, Rinehart & Winston.

Fant, L. (1972). Ameslan: An introduction to American Sign Language. Northridge, CA: Joyce Media, Inc.

Fouts, D.H. (1984). Remote video taping of a juvenile chimpanzee's sign language interactions within his social group. Unpublished master's thesis. Ellensburg, WA: Central Washington University.

Fouts, R.S., Fouts, D.H., & Schoenfeld, D. (1984). Sign language conversational interactions between chimpanzees. Sign Language Studies, 34, 1-12.

Fouts, R.S., Fouts, D.H. & Van Cantfort, T. (1989). The infant Loulis learns signs from cross-fostered chimpanzees. In R. A. Gardner, B.T. Gardner & T. E. Van Cantfort, (Eds.) Teaching Sign Language to Chimpanzees, (pp. 280-292). Albany: State University of New York Press.

Fouts, R.S., & Fouts, D.H. (1989). Loulis in conversation with the cross-fostered chimpanzees. In R.A. Gardner, B.T. Gardner & T.E. Van Cantfort,. (Eds.) Teaching Sign Language to Chimpanzees, (pp. 293-306). Albany: State University of New York Press.

Furrow, D. (1984). Social and private speech at 2 years. Child Development, 55, 355-362.

Gardner, R.A., & Gardner, B.T. (1973). Teaching sign language to the chimpanzee, Washoe. (16-mm sound film). State College, PA: (Psychological Cinema Register).

Gardner, R. A. & Gardner, B.T. (1978). Comparative psychology and language acquisition. Annals of the New York Academy of Sciences, 309, 37-76.

Gardner, R. A. & Gardner, B. T., (1989). A Cross-Fostering Laboratory. In R.A. Gardner, B.T. Gardner & T.E Van Cantfort, (Eds.) Teaching Sign Language to Chimpanzees, (pp.1-28). Albany: State University of New York Press.

Goodall, J. van Lawick (1968). The behavior of free living chimpanzees in the Gombe Stream Reserve. Animal Behavior Monographs, 1, 163-311.

Goodall, J. van Lawick (1971). In the shadow of man. Boston, MA: Houghton Mifflin Company.

Goodall, J. van Lawick (1975). The chimpanzee. In V. Goodall (Ed.), The quest for man, (pp. 131-169). London: Phaidon

Goodall, J. (1986). The chimpanzees of Gombe. Cambridge, MA: Harvard University Press.

Hayes, C. (1952). Ape in our house. New York: Harper & Brothers.

Jensvold, M.L.A., & Fouts, R.S. (1993). Imaginary play in chimpanzees. Human Evolution,. 8, No.3, (pp. 217-227).

Matthews, W.S. (1977). Modes of transformation in the initiation of fantasy play. Developmental Psychology, 13, 212-216.

Stokoe, W.C., Casterline, D., & Croneberg, C.G. (1965). A dictionary of American Sign Language on linguistic principles. Washington, D.C.: Gallaudet College Press.

Savage-Rumbaugh, S., & Mcdonald, D. (1988). Deception and social manipulation in symbol-using apes. In R.W. Byrne, & A. Whiten (Eds.), Machiavellian Intelligence (pp. 224-237). Oxford: University Press.

Van Cantfort, T.E., & Rimpau, J. B. (1982). Sign language studies with children and chimpanzees. Sign Language Studies, 34, 15-72.

Wrangham, R.W. & Goodall, J. (1989). Chimpanzee use of medicinal leaves. In P.G. Heltne & L.A. Marquardt (Eds.), Understanding chimpanzees. Cambridge, MA: Harvard University Press.

IS THERE PRENATAL CULTURE ?

M.C. BUSNEL *
Génétique, Neurogénétique et Comportement.
UFR Biomédicale, Univ. René Descartes. CNRS URA 1294
45, rue des Saints Pères – 75006 PARIS – FRANCE

ABSTRACT

Considering the dictionary's definition of culture as : "the sum of all acquired behavioral forms in a society" one can wonder if the fetus already partakes of culture?

By means of the auditory channel, it has been shown that the fetus is already capable of receiving incoming stimuli from the environment. He has also been demonstrated to be able to discriminate between different incoming acoustic stimuli (between two syllables or two speakers for instance).

Newborns have also shown that they recognize two different languages, provided one is that of their mother.

Not only do fetuses and newborns discriminate, but they also prefer to listen to stimuli they have heard prenatally.

One can thus conclude that the fetus is already a cultured being.

1. Introduction

To answer the question, "Is there prenatal culture?" we should first define "culture" for, although everyone will agree that human beings are cultured (some will even say that

---

* Also, Scientific Consultants, Chemin de la Butte au Diable, 91570 Bièvres, FR.

R. A. Gardner et al. (eds.), The Ethological Roots of Culture, 285–313.

it is the unique characteristic of our species), yet each dictionary, each author, has his own definition. With some variance they all evolve around two ideas, the most elaborate one being that : culture is "the sum of all acquired knowledge about the intellectual and social assets of a human group" or, in a simpler expression, culture is "the sum of all acquired behavioral forms in a society". We do not wish to discuss these definitions. It will be the role of the concluding remarks of this meeting.

However since our aim here is to propose that fetuses are already subjected to a certain form of culture, we would like to add a more basic definition, pertaining to the meaning of culture as "inducing an inert field to become fertile and fruitful".

- Agriculture thus consists in preparing the soil for sowing and making it fit for receiving seeds.
- Aquaculture is : improving the sea or rivers to enhance vegetal and animal growth in their waters.
- And how about a "homoculture" and its first phase: "foetoculture" ?

If indeed we accept that culture is not only growing crops, but preparing the substrate to improve its adaptibility at being fertile, then one could, by extension, say that "foetoculture", consists in preparing the human brain for encoding and transforming incoming stimuli into thoughts. And all the characteristics of agriculture might well apply to "foetoculture", through which the youth of humans learns about the "us et coutumes" of their social group. Thus culture for humans really starts with nourishing the brain with informations to allow the development of an intelligence "crop".

While all would agree that babies receive stimuli from their environment at birth, when they enter their world, it was, up to now, more difficult to conceive that this is possible even during the gestation period, and that the fetus, and even maybe the embryo, is also being prepared for receiving elements from his culture.

Although we will mainly describe the acoustic stimuli that reach the fetus and how he reacts, it is necessary to think of each organism as a whole, and to study simultaneously all the sensory channels and their interactions. One knows little, for instance, of the eventual dominance, in the fetus, of one sensory channel over the others. It has now been proven that the fetus hears, tastes, smells and so on, but does he do so, in the same manner as the newborn (Lawson et Turkewitz, 1980) ?

Recent results demonstrate that the newborn himself perceives in a different manner than older babies (after 3,4 months) (Turkewitz 1971, 1989, Turkewitz & al., 1971–1984, Lewkowicz, 1980–1991, Lewkowicz & Turkewitz, 1981). Since the fetus' sensory system is less developed and less discriminative, it may also be less in–

dividualized. The studies on bird embryos (Gottlieb, 1982) as well as those on newborns and older babies concerning the influence of one sense on the receptive capacities of the others, which were triggered by Turkewitz's hypothesis, will not be described here. But since every sensory input can modulate the results obtained on audition alone, the fact that the ear is but one of the many fetal receptors and that there exists an interaction with all the others, must be kept in mind whenever drawing conclusions from experimental results obtained on a single sensory channel.

We will show, that the fetus not only receives many sensory stimuli, but also memorizes them, and thus acquires specific knowledge about its future environment. For instance the fetus of an African mother, who crushes grains a few hours a day, walks bare footed to the well, and comes back with a heavy jar on her head, will not have the same rhythmic and somesthesic inputs than those given by a city mother who rides the subway, walks on paved roads and climbs stairs, or those of a mother horseback riding for hours on a ranch. If these stimuli are indeed integrated by the fetal brain, they become part of its future culture.

Thus three questions need to be answered :

- Is the fetus receiving and integrating incoming stimuli ? (cf. part. 2)
- Do these stimuli, change the receiving organs?
- Do these stimuli, in any manner, influence later behavior of fetuses or babies ? (cf. part. 3 and 4)

The 2nd question is being extensively studied by those dealing with early plasticity of the nervous system. So much so, that some have felt confident in saying that "the function creates the organ". If "creating" now seems to be too gross a concept, we can at least accept that it modulates the way an organ responds. Afferent inputs cannot be neglected as early factors influencing the functional differentiation of sensory systems. Early bilateral auditory deprivation produces a reduced growth of neurons and dendritic arborization and, as a consequence, higher auditory thresholds and a reduction in the ability to discriminate complex auditory structures. Thus early inputs are of primary importance in the future capacities of newborns and infants. This field has been fully reviewed in the last 10 years (cf for instance : Greenough & Juraska, 1986), but it had to be mentioned considering the constant interaction between stimulus, and sensory organ functioning (cf. also Clopton 1986).

It is well known that all sensory systems do not develop at the same time, nor at the same speed, and that each system has its own type of sensitivity as regards to early stimulation. The exact duration of critical or sensitive periods during development must still be defined in humans for each sensory system. Effects of over stimulation during sensitive periods are usually visible on initially over–produced neural connections, which are either strengthened, when the synapses receive appropriate stimulation, or lost, if no stimulation occurs. Greenough has termed this mechanism of developmental organization as "experience–expectant". The over production of synapses is preprogrammed and independent of external stimuli, while the fine tuning of the system is stimulus dependent.

2. Fetal Reception and Integration of Stimuli

Let us now return to the first question : is the fetus receiving incoming acoustic stimuli? Again it can be subdivided in three parts :

- Is the uterus indeed a noisy environment?
- Is the fetus' auditory system mature enough to allow the integration of acoustic stimuli?
- Does the fetus react to strong and even to weak stimuli ?

Since answering these types of questions has been my main concern in the last 15 years, I will mostly rely on our experimental results (Busnel, Granier–Deferre, Lecanuet). Although references will be made to others, this paper should not be considered as an exhaustive review of the literature in this field.

2.1. INTRAUTERINE SOUND ENVIRONMENT

To measure noise in the uterus, microphones have been inserted, either in its nearest cavity, the vagina or into the cervix (Johansson & al., 1964 ; Bench, 1968; Henshall, 1972 ; Busnel, 1979), or inside the uterus after amniotomy (Walker & al., 1971) and during labor (Murooka & al., 1976; Querleu & al., 1981, 1988a). General conclusions from the studies, using the most precise and refined acoustic equipment and analysers are, that both endogenous and exogenous sounds are present.

The highest component of internal background noise is that of blood flow in the placenta, then comes occasionnal intestinal rumblings, and at a lower intensity, maternal heart beats (Querleu & al. 1981, 1988b, Richards & al., 1992, Abrams & al., 1992).

Most external sounds, are well transmitted *in utero*. But this field is presently being reconsidered. New measuring techniques and models are being developed (Benzaquen & al., 1990, Richards & al., 1992, Abrams & al., 1992, Lecanuet & al. in preparation). All conclude that the internal background noise cannot mask incoming sounds, not even voices, which probably represent the most frequent components of a fetal acoustic environment. Since sound transmission to the uterus seems to be much more efficient than previously measured, uncertainty about fetal audition will now mainly concern the real auditory capacities of the fetus.

2.2. FETAL AUDITORY SYSTEM DEVELOPMENT

Anatomically the different sensory systems in all mammals mature in a similar chronological pattern, from touch, which is the first to develop, to audition and vision which is the last one. But they are functional before having attained their full maturation (sometimes occuring long after birth).

In the human auditory system, the organ of Corti is considered as functional and the first afferent synapses appear around twenty weeks (Pujol and Uziel, 1986). But cochlear maturation is only terminated at 8 months gestational age.

Since the degree of maturity of each part of the auditory system, necessary for an effective functioning, is not yet fully understood, it seems more reliable to depend on fetal reactions (physiological, biochemical, electrophysiological and behavioral), to determine whether the organ is efficient and to what degree.

2.3. METHODS OF MEASURING FETAL REACTION TO ACOUSTIC STIMULI

2.3.1. *Electrophysiological measures.*

Fetal brain stem and auditory evoked potentials were recorded in chronically implanted fetal guinea pigs (Scibetta & Rosen, 1969) or sheep (see review in Woods & Plessinger, 1989). There is a similar chronology when compared to gestational length.

In humans, premature babies tested near birth, show evoked auditory responses starting at 24–25 gestational weeks (GA). But the detectability of major components becomes stable only by 30–32 weeks (Barden & al., 1968 ; Starr & al., 1977 ; Schulman–Galambos & Galambos, 1979 ; Goldstein & al., 1979 ; Krumholz & al., 1985; Salamy & al., 1982 ; Rooteveel & al., 1987 ; Pasman & al., 1991). At 25 weeks, brainstem responses are consistent and reproducible, but with very high hearing level thresholds, which gradually decrease with development ; by the 35th week GA they are only 10–20 dB lower than the adults' threshold. All these measures point to good auditory capacities by 35 weeks.

2.3.2. *Cerebral (14–C) 2–Deoxyglucose uptake.*

This method, which allows the investigation of fetal brain functional activity *in utero* through cerebral glucose utilization (energy metabolism) (for review see Granier–Deferre

& Abrams, 1989), has been used in two animal models, the fetal guinea–pig (Horner & al., 1987) and the fetal sheep (Abrams & al., 1987a,b, 1991, 1992). It has been shown that acoustic stimulation induced marked increase in 2–DG uptake in some auditory structures : the brainstem only in the guinea–pig and all structures, including the auditory cortex, in the sheep. This reflects the increased metabolic activity of cells in response to the auditory stimulation. Each specific structure can be evidenced to react *in utero* to external acoustic stimulation of different frequencies.

2.3.3. *Behavioral responses to acoustic stimuli : fetal heart rate.*

In the newborn, responses to different stimuli are measured with various criteria: startle response, general movements, eye opening, head turning, heart rate, respiratory changes and so on. In the fetus three of those only can be used with accuracy : heart rate, respiration and general or specific startle movements, these being visible and measurable by non invasive techniques.

Computed heart rate variability and reflex movements in response to acoustic stimuli before and right after the stimulus are compared with those occuring without acoustic stimulation. Most studies, in the last twenty years, show that with vibroacoustic stimuli of 105 dB SPL and above, fetal responses are elicited irregularly at 24 weeks gestational age (GA) but become consistent in most subjects, at 28 weeks.

Our group (Busnel & Granier–Deferre, 1983, Granier–Deferre & al., 1985, Lecanuet & al., 1986–1988) stimulated near term fetuses (36–40 weeks) through a loudspeaker placed 20 cm above the mother's abdomen, with different acoustic stimuli: 5 sec. narrow band pink noise centered around specific frequencies, (but filtered below 800 Hz, so as to eliminate skin sensitivity). During the test mothers listen to a constant loud music through earphones so as to be unable to hear the stimulation. The fetal heart rate (FHR) was recorded through a tococardiograph, and computed for data treatment. Fetal movements were videotaped from a real time ultrasound system.

As in the newborn, different states of alertness have been observed. To standardize the experimental procedure, we decided to study the effect of acoustic stimulation in two of these states : one comparable to quiet sleep in newborns, characterized by a low variability of the fetal heart rate (LV), and the other comparable to active sleep in which the FHR is highly variable (HV), called respectively 1F and 2F state by Nijhuis & al., (1982).

Fetal state of alertness (for review see Pillai & James, 1990) as well as type of stimulus influences the fetal heart rate's response to sound stimuli (Schmidt & al., 1985, Lecanuet & al., 1989a ; Visser & al., 1989).

3. Behavioral Responses To Non Significant Stimuli

3.1. SHORT TERM REACTIONS TO STRONG ACOUSTIC STIMULI

The first objective of studying fetal responsiveness to non significant loud stimuli, was to prove that fetal audition was a reality and was quantifiable. The second step was to study response modifications as a function of type of stimulus (intensity, frequency band and duration). The third stage was to measure how and how much the response was modulated by fetal behavioral state of alertness. Finally the response's reliability had to be assessed, for we were attempting to use this reaction to sound as a non–stressful diagnostic test for fetal well–being (Granier–Deferre & al., 1985).

Fetal reaction to strong vibroacoustic stimuli was demonstrated by different authors since Forbes in 1927 and Ray in 1932. But we will here base our reasoning mainly on the research we did in our team. A review of general results in this field can be found in Busnel & al., 1992.

Loud and short (5 sec.) stimuli (at, or above 105 dBL *ex utero,* which correspond, according to frequency, to 90 or 80 dBL *in utero*) produce a high proportion (90%) of short latency FHR accelerations (20–22 beats/min) (Granier–Deferre & al., 1985, Lecanuet & al., 1986). These strong FHR accelerations are usually associated with reflex motor reactions, also triggered by the stimulus : 70% (in high variability FHR), 60% (in low variability FHR). When no motor reactions are produced the mean acceleration is of 12.5 beats/min (Lecanuet & al., 1989a,b), The influence of stimulus characteristics and of the fetal state of alertness have been quantified and are described both by our research team (see in bibliography under : Busnel, or Granier–Deferre, or Lecanuet from 1983 on) and by others (see Bench & Mintz, 1978, Schmidt & al., 1985, Visser & al., 1989, etc.).

3.2. SHORT TERM REACTIONS TO WEAK ACOUSTIC STIMULI

As a first step it was necessary to study reactions to loud stimuli but then it was really more informative to study reactions to sounds occuring in the regular fetal acoustic environment. We therefore decided to test sudden stimuli, but at low intensity : a telephone bell, for instance, and, later, human voices and music.

In the low variability state (LV) strong stimuli produced a mean acceleration of 12.5 beats/min. (when no movements are generated). Lower intensity stimulation produced some accelerations of 10 beats/min. amplitude and many decelerative responses (50 to 18 %) of the same amplitude (– 10 bpm).

Due to their small amplitude, these decelerative responses are more easily recognized and recorded, during episodes of low FHR variability. In newborns, these responses to low intensity speech stimuli are mostly seen in a quiet alert state, but are also observed, during quiet sleep, (Eisenberg & al., 1974).

Not only does the fetus hear and react behaviorally to sounds but has also been shown to habituate to the reception of stimuli.

3.3. SHORT TERM HABITUATION

On repeated presentations of 5 sec. stimuli during active sleep, the percentage of FHR transient accelerative responses, shows no or small non significant decreases, from the first stimulus to the third one, while in quiet sleep it is reduced from 98% responses to the first stimulus, to 38% for the third one (Granier–Deferre & al. 1983). Motor reactions in the same conditions decrease by 23% in HV state and from 56% to 14% in LV state (a decrease of 32%) (Lecanuet & al., 1992).

The number of repetitions necessary for a total waning of the responses depends on fetal state, and also on intensity, frequency and duration of the stimulus. It is variable from one fetus to another, this variability being presumably due to each baby's prior and usual acoustic environment (Lecanuet & al., 1986, 1988, 1989 ; Leader & al., 1982 ; Madison & al., 1986 ; Kisilevski & Muir, 1991).

We will indicate in a later paragraph that this habituation is important for two reasons : first a habituation–dishabituation technique will be used to demonstrate fetal discrimination between two sounds ; second, it proves a certain degree of memorization, perequisite to any learning process.

3.4. LONG TERM HABITUATION

Since the fetus reacts to loud as well as to weak acoustical stimuli, and has short term habituation to these, it was interesting to determine whether there was a long term habituation to some of these stimulation. This would answer our third question : do these stimuli, influence later behavior of fetuses or babies ?

Habituation was studied both for loud, possibly stressful acoustic stimuli, and to lower intensity sounds occuring in the natural environment of the fetus, like heart beats or women's voices.

If gestation is, as we believe it, an adaptation period to the future sensory inputs, prenatal stimulation should have an impact on the sensory responses of the newborn and on its adaptiveness.

Habituation to early strong stimuli which, by their intensity, would be considered as stressful if given at a later time, was demonstrated, both on humans and on animals. One example is detailed below :

Mice fetuses were sujected to a loud 110 dB noise, 3 hours a day during their gestation (Busnel & al., 1983). This animal model is interesting, since the auditory system of mice becomes functional only 9 days after birth. Stimulating during gestation is therefore stimulating a fetus, whose auditory system is immature and theoretically non functional.

By proper crossing genetically deaf dams were able to bear pups who had no hearing defect. They were thus not subjected to a maternal acoustical stress. At weaning (21 days) these stimulated individuals were tested with this same noise and their reactions compared with those of pups which had never heared this noise and with pups which had been subjected to it from 9 to 21 days, that is after the beginning of auditory functioning. Corticosteroid blood levels were measured as an indicator of stress. The pups who had been given the stimulation after 9 days, or who had never received it before, had very high levels of stress, while those subjected to it during gestation, showed little increase of their costicosteroids. Their hearing capacities were normal, and no alterations in those parts of the brain, involved with hearing, (olivary nucleus, auditory cortex, and geniculate bodies) were found (Busnel & al. unpublished results).

These findings, on an animal model, illustrate the possible habituation of fetuses to loud recurrent noises. This is but an example of the many experiments that could be performed on the influences of one or several types of early stimulation, to enhance understanding of their role on development and behavior.

It was therefore fascinating to find out if this was also true in the human fetus and newborn. An early retrospective study on babies near a Japanese airport had convinced Ando and Hattori (1970, 1973, 1977) that pregnant mothers living near the airport, during the first half of their pregnancy, gave birth to babies who were usually not disturbed by overflying planes while the babies of mothers who had moved to the airport's vicinity, during or after that time, awoke and cried at most overflights. This habituation was specific since both groups of babies awoke in a similar way to a musical stimulus at the same intensity. Their observations were confirmed by physiological measures (electroencephalograms and peripheral vessel constrictions), which they found to be different in the two groups of babies. Thus if fetuses can become adapted to loud stimuli*, they should also become acquainted and/or adapted to lower intensity sounds of their normal environment.

4. Behavioral Responses To Significant Stimuli

Since voices are audible *in utero*, and since fetuses react to these one can rightly question the influence of this early audition on habituation, and on neonatal adaptiveness.

4.1. SPECIES CALLS (ANIMALS)

An adaptative advantage has been adequately demonstrated in birds, where the young learn to recognize the species' calls *in ovo*. There is a necessity for prenatal, maternal stimulation in the chick (Guyomarc'h, 1972, 1974 ; Impekoven & Gold, 1973 ; White & Mc Donald, 1983) and duckling (Gottlieb, 1968–1982), for post natal species recognition. Gottlieb extensively studied the modalities of this stimulation in Mallard duck embryos, who learn to recognize *in ovo* their mother's contact–contentment call, and remember it post natally, only if the repetition rate of the given call (recorded and broadcasted on the egg) is in the range of normal calls emitted by the female (2.1 to 5.8 notes/sec.). The best conditioning occurs with such a varied stimulus. It can also be produced with a fixed interval call but only if its tempo resembles the mean interval of normal calls (3.7 notes/sec.). Given either at the fastest rate of the normal range (5.8 notes/sec.) or at the slowest one (2.1 notes/sec.) the stimulus will be as ineffective for the duckling as no prenatal auditory experience at all (Gottlieb, 1980 a, b). In this bird prenatal learning depends therefore both on type of stimulus and on its rhythmicity. Although these experiments concern a species very remote from the human one, their findings must be kept in mind when concluding from habituation tests performed after artificial stimulation.

These results open, in addition, interesting trends of thoughts when studying neurosensory maturation. Prenatal habituation to acoustical stimulation has further been shown in some mammals where the auditory development is prenatal. Vince (1979) demonstrated the effect of such a process, in guinea pigs. The cry of another species,

---

* The difference between loud stimuli to which the fetus becomes adapted and those that will either impair the auditory organ (Lenoir & al.,; 1979), or stress the fetus, and disturb its physiological and psychological growth is not discussed here, but must be kept in mind.

considered as frightening by guinea pigs, is well accepted and produced practically no post natal fright reaction in the young, if it had been emitted repeatedly to the mother, during the gestation period. This habituation may, in fact, be a widespread process since it has likewise been demonstrated, on animals, in other sensory systems (smell and taste). (For review see Smotherman & al. 1988 ; Schaal 1988 ; Schaal & al., 1992).

4.2. NEWBORN PREFERENCE FOR HEART BEAT SOUNDS (HUMAN)

The demonstration of postnatal preferences for prenatal stimuli can be an indication of memorisation of such stimuli and therefore early learning. Some experimental evidences were consequently attempted on human babies.

As a first step, a number of studies tried to determine whether newborns in nurseries were calmed and lulled to sleep by hearing adult heart beat sounds (considered as a reminder of the uterine environment). Salk (1960, 1962) found that newborns, thus stimulated, cried less, fell asleep quicker and gained more weight, on equal food supply, than non stimulated ones. These results were criticized and even refuted (Brackbill et al., 1966, Brackbill, 1970, 1973) and were followed by numerous studies, some showing little or no positive calming effect, others observing appeasement, but not above that of soft music or lullabies.

Be it as it may, this controversy did not deter some, from producing, advertising and selling, recordings of uterine sounds, including heart beats to calm babies. That such a reminder of the fetal acoustic environment might give security in periods of stress, is probable, even though unassailable scientific evidence is not as yet available. But their frequent use in every–day life (for instance to put the baby or infant to sleep every night) appears unadvisable, until more information is acquired on potential effects of such stimulation on neonates' psychological and emotional development. This is a good example of hasty commercial utilization of unconfirmed research results, bringing about an unwaranted public enthusiasm.

For premature babies, who are separated from their mother right at birth, the benefit of hearing uterine sounds might override potential risks, and research in this direction should be intensified. Indeed heart sounds were not proven to be more effective in calming agitated babies than other types of musical stimulation (Busnel & al., 1989, Mosser, 1989) (see below). Anyhow it must never be forgotten that these artificial stimuli are but second rank choices, the mother's real presence being irreplaceable.

4.3. FETAL AND NEWBORN'S REACTIONS TO MUSIC

As an alternative solution the calming effect of different music on premature babies has been studied by measuring heart rate variability during an hour prior to the 10 min. stimulation and an hour after. Computing was effectively done only on the 10 min before, during and after the stimulus. It showed that the tested music will soothe most agitated babies. Traditional folk music from different countries calmed 78% of the tested babies, gregorian chanting 75%, ocean sounds 72%, while heart beat sounds and different types of pulses at the same rhythm had a soothing effect on 61% of them. However none of the tested types of music were as appeasing as those that the mother had extensively listened to, during her pregnancy. (Busnel & al., 1989, Mosser, 1989). Thus certain types of music, presented for a short duration and at specific moments might be considered as advantageous substitutes for intrauterine sound emissions, for the purpose of soothing babies.

Similar results were found on rats raised from birth with 12 hours a day exposure to either Mozart or Schoenberg's music (Cross & al., 1967). Since rats do not hear at birth but only 9 to 10 days later it was legitimate to start at birth only. During testing as adults the rats could choose what music to receive according to the place where they stood in the cage: one side (different at each session) gave them Mozart, the other Schoenberg. Those young rats raised to Mozart did choose Mozart. But neither controls nor Schoenberg raised animals made any significant choice. That rat pups did not attach to Schoenberg's music is still a subject for discussion, and would need further experimentation. But that 28 out of 29 subjects did choose Mozart when they had been exposed to it confirms the finding with babies who were best soothed by the music they had heard during their gestation.

4.4. FETAL RESPONSE TO HUMAN VOICE

Different recordings have been performed so as to determine *in utero* attenuation or distortion of human voices. We will not cover this vast corpus of publications which are still under investigation. It is sufficient for our purpose to know that there is a certain degree of attenuation and therefore of intelligibility of speech which varies as recording and analysing equipment improves. Anyhow many speech characteristics, which are part of the general prosody, are maintained : rhythm, intonation, modulation or melodic contours. These qualities are an important part of the individual recognition of a speaker. Thus whether adults, listening to intra–uterine recordings, can understand 46% of the spoken's words (Versyp, 1985 ; Querleu & al., 1988), (or more, according to recent measures) the significance of this factor is difficult to evaluate since the fetus presumably does not understand words. But it does indicate that adults (and perhaps the fetus) can distinguish on speaker from another by means of some of the features of each voice.

Fetal reaction to human voice was therefore tested (Lecanuet & al., 1986 to 1992). Stimulating near term fetuses during LV states with recorded male and female voices (at 92 dB) elicited a high percentage of small (3,4 beats per minute), short term FHR decelerations. The percentage of infant responses to male and female voices was 71% and 65% respectively, this difference was not significant. Chance FHR deceleration is only 9% (during periods without voice utterances). Some accelerations in FHR are also elicited, but to a lesser degree (Granier–Deferre & al., 1992). The same type of response has been described in the newborn (Brazelton 1979, Eisenberg & al., 1979) but detailed experimental results on this fetal reaction are still scarce.

Through a habituation–dishabituation technique consisting of repeating the same sound till the heart rate reaction had completely waned, and then changing to another sound, it was demonstrated that fetuses could however discriminate between male and female voices (Lecanuet & al., 1992), and one syllable from another (Lecanuet & al., 1987). The two syllables used were *bi* and *ba*, but for acoustic homogeneity we used the double syllable babi, compared to its reverse, biba. There is no ambiguity in this differentiation, the second syllable triggering a new response as soon as it appears.

These types of discriminations, had already been demonstrated on newborns : using synthetic syllables as a stimulus Eisenberg & al. (1974), and Siperstein and Butterfield (1973) indicate that neonates preferred speech to non speech stimuli. However more recent results with "non nutritive sucking techniques" have also been very productive in assessing newborn preferences.

4.5. NEWBORN PREFERENCES FOR PARENTS VOICE

Quite a few scientists detected, by different techniques, a preference for female voices over other voices, in babies a few days old (André–Thomas, 1966, Wolff, 1969, Hammond, 1970, Brazelton & al., 1974, Mills & Melhuis, 1974, Mehler & al., 1978) and even at a few minutes after birth (Feijoo, 1981, Querleu & al., 1988). Female voices usually induce heart rate decelerations while male voices do not (Lipsitt, 1979). Non nutritive sucking is, at this time, considered as one of the best tools for demonstrating such early choices, since newborn babies are capable of learning to modify the rhythm of their sucking bouts, in order to trigger one of two sounds. Using this method, 2 – to

3 – day–old babies demonstrated a preference for heart beat sounds (DeCasper & Sigafoos, 1983) and also for intonated mother's voice (DeCasper & Fifer, 1980). In the next few years, Fifer (1980), Panneton (1985), Moon & Fifer (1990) ; Spence & DeCasper (1982, 1987) have demonstrated a scale of preferences for the newborn : he is usually successful in learning to organize his sucking bouts so as to hear a voice rather than silence, female voice rather than male and mother's voice rather than another female's. Aslin & al. (1983) reviewed the different speech features newborns were capable of discriminating, and Bertoncini & Mehler (1981), Bertoncini (1991), and Moon (1985) considered the phonetic components of speech.

The experiments described in this short review cannot be considered as an irrefutable proof of prenatal learning, since experimental results have not excluded very fast post natal imprinting. Therefore testing preference for mother's voice on premature newborns delivered by cesarian section would be useful, since these mothers, bedridden for 2 or 3 days, can usually during that time neither see, nor talk, to their babies in the premature ward.

In addition if instant post natal learning existed, fathers' voices, which are usually heard by the baby after birth, should be preferred to other male voices. This has not been shown even after 4 hours of post natal experience with their voice (Fifer, 1980 ; DeCasper & Prescott, 1984). It takes approximately two weeks for the father's voice to be preferred to an unknown male voice when he does not talk prenatally to his baby.

A number of other clues are however in favor of the prenatal learning hypothesis:

- Results of research on smell and taste in animals demonstrate a post natal preference for prenatal stimuli (Smotherman, 1982, 1986, Schaal & Orgeur, 1992).

- As mentionned above, Busnel & al. (1989) and Mosser (1989) showed that although playing music to premature babies calms them, the most effectively soothing music is that which the parents have extensively listened to, during the pregnancy period.

- Newborns have been shown not to choose between a lowpass filtered or unfiltered voice of their mother, while they persistently chose the unfiltered version of another women's voice (Spence & DeCasper, 1987). Low pass filtering was considered to resemble voices heard *in utero*. Thus they seemed to recognize their mother's voice whether inside or outside the uterus, but had no interest for a filtered stranger voice never heard during gestation.

- Even though preference for father's voice was not found by Fifer, or DeCasper & Prescott, Feijoo (1981) shows that if the father takes the pain of talking to his fetus a learning of his voice does exist. In a preliminary study (with unfortunatly too few subjects), he asked fathers to say one sentence addressed to the baby loudly and repeatedly, while standing near the mother. Right after birth, this same broadcasted sentence, stopped the baby's cry, made him open his eyes, turn towards the loud speaker, and relax hands and legs. The same sentence, emitted in reverse, had no effect. The result was thus not due to the fathers' voice alone but to the prosody of that particular sentence recognized as familiar.

- One is also supported in this idea of prenatal familiarization and even learning : according to Moon & al. (1991) 2 days old babies prefer their mother's language, be it English or Spanish. Also Mehler & al. (1988) showed discrimination, by 4 days old babies, between two languages if one of them is that of the mother. They do not however discriminate between two different unknown foreign languages. Thus the possibility of recognizing that French is different from Russian, or English from Italian is dependent on one of them having been heard prenatally, since the mother's native language is not necessarily that of the country the baby is born in.

- The final argument in favor of prenatal learning is given by results of DeCasper and Spence (1986), who showed (still with the non nutritive suction technique), that 3 days old newborns prefer hearing a lullaby their mother had read aloud for 6 weeks before birth, rather than an unknown one. The babies had not heard either story since their birth, it could therefore only be prenatal familiarization.

- A difference in reaction was also observed, but this time on the fetus, between a short poem the mother had read aloud twice a day for 6 weeks, and another unfamiliar one, quite different in prosody. Small fetal heart rate deceleration were observed to the familiar story and small non significant HR acceleration to the new one (DeCasper & al. in press 1994). This result is all the more important in that : a) the response is given during the fetal period and b) the difference of reactivity between the two stories did not depend on recognition of mother's voice, since the testing stimulus was a recorded female voice, reading the texts (the same voice for all tested fetuses). Thus it was really the prosody of the poem that was recognized.

All of the above cited experimental results were performed with well defined vocal stimuli, the same one for each subject. We have however lately decided to test whether premature babies would react differently to the affective components of their mother's voice when she speaks directly to them. In this procedure precise recorded stimuli are impossible to use since the mother must be allowed to express her feelings and therefore to say whatever she wishes. Thus it is the general pattern of the stimulus which is analysed, whatever its length or verbal content. It is merely considered as "motherese talk" or "non motherese talk".

A preliminary part of the study tested future mothers talking to their fetus (Busnel unpublished). For each fetus the type of reaction was diverse due to varying fetal state of alertness at the beginning of tests. In all cases (but two) the fetal heart rate reaction was different when the mother spoke to her fetus than when she spoke to the experimenter. But it could always be argued that the mother's emotional state while speaking to her fetus, might influence his reactions. Therefore the experiment was continued with premature babies, (30–36 gestional weeks old and 1 to 10 days after birth). Their heart rate (HR) was followed during successive episodes each separated by a few min. of silence. When the mother talks to her baby (infant directed speech).

The results depend on the baby's state before his mother's arrival :

- If he was calm it made him alert and his HR accelerated.
- If he was agitated or crying it lowered his HR frequency and fluctuations and appeased him.
- If, while standing near him, she speaks to someone else (adult directed speech) there is little or no effect on calm babies but agitated or crying ones are soothed. No awaking or alerting effect were observed.
- When mother left, at the end of her visit, most babies fell asleep with a decelerative trend in HR, while some started crying (Busnel & al. 1992).
- A stranger talking to the baby produces a deceleration or no effect.

These results seem to show that the premature baby and the fetus recognize those characteristics of their mother's voice accompanying addressed speech, and is brought to a state of alertness by this audition, while he is usually calmed by the unaddressed voice of his mother.

If I might be allowed to hypothesize on the meaning of these results, I would say that mother's presence (including her voice) is comforting and reassuring, while the attempts of mothers to establish a true relationship by talking directly to the baby, brings him to a calm alert state during which his heart rate is high but not as high and not as erratic as in crying or moving. Continuation of this study will test differences in these reactions according to the type of feelings the mother has towards her baby (easy or difficult bonding for instance).

5. Conclusion

If we accept the two definitions of culture, given at the beginning of this chapter, it seems obvious from the experimental evidence reported here that the fetus' sensory system is being prepared to receive environmental stimuli, and also, becoming habituated to these. He is in the first stage of integrating his group's culture.

Since the very notion that the fetus might be learning is new, demonstrating possible cognitive capacities was not even considered by the scientific community until very recently. It is therefore not suprising that only the most current research is seeking this type of data. Whatever the stimulus, all results indicate that the fetus, not only integrates its environment but also has discriminative capacities towards different kinds of incoming stimuli. Even more important, are finding proofs of a preference for "known" stimuli over unknown ones.

It is obviously necessary to study large groups of subjects, and to work out the statistical significance of all results, but as a concluding remark I would like to stress that babies, fetuses and even animals are individuals. Some of the most important discoveries in this field might well arise at this point, not from analyzing the most frequent type of response, but from studying what causes the infrequent types, what triggers an unusual reaction from a specific individual. After having discovered the general rules which help in understanding systems, comes the time to observe those who depart from the main stream and why.

For instance, while most premature babies are calmed by the sound of an adults' heart beat, one of our tested babies reacted by great agitation and crying. This baby's mother had had a difficult pregnancy, was often beaten and had received blows on the abdomen (triggering the premature birth). Such negative reactions to an usually positively received stimulus, might open new understanding on the underlying causes for the different types of observed reactions, and on the manner in which the fetus learns: not only passively receiving stimuli, but also integrating them into what will become his personality.

Whether these learning processes are called, habituation, familiarization, imprinting or learning, under its many forms, they imply that the fetus acquires, remembers and integrates whatever sensation he receives *in utero*. The fetus can thus already be considered a member of his culture.

Acknowledgement : the author is grateful to Dr. A.G. Lehmann for her help with this manuscript.

REFERENCES

Abrams, R.M., Hutchinson, A.A., Gerhardt, K.J. (1987a) 'Local cerebral glucose utilization in fetal sheep exposed to noise', Am. J. Obstet. Gynecol. 157, 456–460.

Abrams, R.M. Hutchinson, A.A., Mc Tiernann, M.J. & Merwin, G.E. (1987b) 'Effects of cochlear ablation on local cerebral glucose utilization in fetal sheep'. Am. J. Obstet, Gynecol. 157, 1438–1442.

Abrams, R.M., Gerhardt, E.Jr. & Burchfield, D.J. (1991), 'Effect of behavioral State on cerebral glucose utilization in the fetal sheep'. F.A.S.E.B. abstract 5, A407.

Abrams, R.M., Gerhardt, K.J. & Peters, A.J.M. (1992). 'Transmission of sound and vibration to the fetus. Fetal development meeting : psychophysiological perspectives. Paris.

Ando, Y. & Hattori, H. (1970). 'Effects of intense noise during fetal life upon postnatal adaptability (Statistical study of the reactions of babies to aircraft noise)'. J. Acoust. Soc. Amer. 47, 1128–1130.

Ando, Y., & Hattori, H. (1973). 'Statistical studies on the effects of intense noise during human fetal life'. J. Sound and Vibration, 27, 1, 101–110.

Ando, Y. & Hattori, H. (1977). 'Effects of noise on sleep of babies'. J. Acoust. Soc. Amer. 62, 199–204.

André–Thomas, A.S. (1966). 'Locomotion from prenatal life'. Spastic Soc. Heineman. London.

Aslin, R.N., Pisoni, D.B. & Juczyk, P.W. (1983). 'Auditory development and speech perception in infancy'. In M.M. Haith and J.J. Campos (eds.), Infancy and the Biology of Development. Vol. II. Carmichael's Manual of child Psychology, Fourth Ed. Wiley, New–York, pp. 573–687.

Barden, T.P., Peltzman, P. & Graham, L.T. (1968). 'Human fetal electroencephalographic responses to intra–uterine acoustic signals'. Am. J. Obstet. Gynecol. 100, 1128–1134.

Bench, R.J. (1968). 'Sound transmission to the human fetus through the maternal abdominal wall. J. Genet. Psychol., 113, 1172–1174.

Bench, R.J. & Mintz, D.L. (1978). 'Neonatal habituation and state change'. Quart. J. Exp. Psychol. 30, 355–362.

Benzaquen, S., Gagnon, R., Hunse, C. & Foreman, J. (1990). 'The intrauterine sound environment of the human fetus during labor. Am. J. Obst. Gynecol. 163, 484–490.

Bertoncini, J. & Mehler, J. (1981). 'Quelques remarques à propos de la perception chez le nouveau–né'. Foetologie 31, 387–396.

Bertoncini, J. (1991). 'Initial equipment for speech perception'. In : Langage Bases... Discourse Bases, G. Piéraut, Le Bonniec and M. Dolikskey (eds.) John Benjamin Publishing C°. Amsterdam/Philadelphie.

Brackbill, Y., Adams, G., Crowell, D. & al. (1966). 'Arousal level in neonates and preschool children and continuous auditory stimulation. Exp. Child. Psychol. 4, 178–188.

Brackbill, Y. (1970). 'Acoustic variations and arousal level in infants. Psychophysiol. 6: 517–526.

Brackbill, Y. (1973). Continuous stimulation reduces arousal level : stability of the effect over time. Child Dev. 44 : 43–38.

Brazelton, T.B., Koslowski, B., Main, M. (1974) 'The origins of reciprocity : The early mother – infant interaction'. In : M. Lewis & L.A. Rosenblum (eds.). The effect of the infant on its caregiver. John Wiley & Son New York.

Brazelton, T.B. (1979), 'Behavioral competence of newborn infant'. Seminars in Perinatology, 3, 35–44.

Busnel, M.C. (1979). 'Mesures intravaginales du niveau et des distorsions acoustiques de bruits maternels. Electrodiagn. Ther. 16 : 142.

Busnel, M.C. & Granier–Deferre, C. (1983). 'And what of fetal audition ? In : A. Oliveirio & M. Zappella (eds.). The Behaviour of Human Infants. Plenum Press. N.Y., pp. 93–126.

Busnel, M.C., Molin, D., Dupont, W. & Bourgeois, Ph. (1983). 'Late postnatal effect on corticosterone blood level of an acoustical stimulation previously applied to mice during pre or early postnatal life'. In G. Rossi (ed.), Noise as a Public Health Problem, Edizioni Tecniche a cura del Centro Ricerche e Studi Amplifon, Milano, pp. 549–561.

Busnel, M.C., Lecanuet, J.P., Granier–Deferre, C., & DeCasper, A.J. (1986). 'Perception et acquisition auditives prénatales'. In : H. Bonnet (Ed.) "Les compétences sensorielles du nouveau–né", Arnette, Paris, pp. 37–46.

Busnel, M.C., Granier–Deferre, C., Lecanuet, J.P. & Lehmann, A.G. (1989). 'La sensorialité foetale'. In : S. Lebovici, F. Weil–Halpern (eds.) "Psychopathologie du bébé" ; Paris, PUF, pp. 157–164.

Busnel, M.C., Relier, J.P. & Mosser, C. (1989). 'Effect of acoustic stimulation on premature infants'. Abstr. Internat. Neonat. Intensive Care. Collegium. Nijmeguen. Netherlands.

Busnel, M.C., Neymon V., Relier J.P., de Bethmann O., Fouillot S.P., The premature babie's heart rate react differently to their mother's voice talking to him or to someone else. World Congress of Infant Psychiatry Chicago, 1992.

Busnel, M.C., Granier-Deferre, C. & Lecanuet, J.P. (1992). 'Fetal audition' Developmental psychobiology'. In : G. Turkewitz (ed.). Annals New-York Acad. Sci. 662, pp. 118-136.

Clopton, B.M. (1986). 'Neural correlates of development and plasticity'. In : W.T. Greenough and J.M. Juraska (eds.), Developmental Neuropsychology, pp. 363-386.

Cross, H.A., Halcomb, C.G. & Matter, W.W. (1967). 'Imprinting or exposure learning in rats given early auditory stimulation'. Osychon. Sci. 7 (7), 233-234.

DeCasper, A.J. & Fifer, W.P. (1980). 'Of human bonding : newborns prefer their mother's voice'. Science. 208: 1174-1176.

DeCasper, A.J. & Sigafoos, A.D. (1983). 'The intra uterine heartbeat : a potent reinforcer for newborns'. Inf. Behav. Dev. 6 : 19-25.

DeCasper, A.J. & Prescott, P.A. (1984). 'Human newborns' perception of male voices: preference, discrimination, and reinforcing value'. Dev. psychobiol. 17: 481-491.

DeCasper, A.J. & Spence, M.J. (1986). 'Prenatal maternal speech influences newborn's perception of speech sounds'. Inf. Behav. Devel. 9 : 133-150.

DeCasper, A.J., Lecanuet, J.P., Busnel, M.C., Granier-Deferre, C. & Maugeais, R. (1994). 'Fetuses' reactions to recurrent maternal speech', Inf. behav. and devel. (In press).

Eisenberg, R.B., Marmarou, A. & Giovachino, P. (1974). 'Infant heart rate changes to a synthetic speech sound'. J. Aud. Res. 14 : 21-28.

Feijoo, J. (1981, 1990). 'Le foetus, Pierre et le Loup". In : L'Aube des Sens. E. Herbinet & M.C. Busnel. (Eds.) Stock. Paris, n° 5, pp. 192-209.

Fifer, W.P. (1980). 'Early attachment : maternal voice preference in one– and three–day–old infants'. Ph. Diss. Univ. North Carolina, Greensboro.

Forbes, H.S. & Forbes, H.B. (1927). 'Fetal sense reaction: hearing'. J. Comp. Physiol. Psychol. 7 : 353–355.

Goldstein, P.J., Krumholz, A.K., Felix, J.K., Shannon, D. & Carr, R.F. (1979). 'Brainstem evoked responses in neonates'. Am. J. Obstet. Gynecol. 135, 622–628.

Gottlieb, G. (1968). 'Prenatal behavior in birds'. Quant. Rev. Biol. 43, (2), 148–174.

Gottlieb, G. (1971). 'Development of species identification in birds : An Inquiry into the Prenatal Determinants of Perception. University of Chicago Press, Chicago.

Gottlieb, G. (1980a). 'Development of species identification in ducklings : VI. Specific embryonic experience required to maintain species–typical perception in Peking ducklings'. J. Comp. Physiol. Psychol., 94 : 579–587.

Gottlieb, G. (1980b). 'Development of species identification in duckling : VII Highly specific early experience fosters species–specific perception in wood duckling'. J. Comp. Physiol. Psychol. 94, 1019–1027.

Gottlieb, G. (1982). 'Development of species identification in duckling : IX. The necessity of experiencing normal variations in embryonic auditory stimulation'. Develop. Psychobiol. 15, (6) : 507–517.

Granier–Deferre, C., Lecanuet, J.P., Cohen, H. & Busnel, M.C. (1983). 'Preliminary evidence on fetal auditory habituation'. In : Noise as a Public Health Problem G. Rossi (ed.). Edizioni Technicke a cura del Centro Ricerche a Study Amplifon, Milano, pp. 561–572.

Granier–Deferre, C., Lecanuet, J.P., Cohen, H. & Busnel, M.C. (1985). 'Feasability of prenatal hearing test'. Acta–Oto–Laryngol. Stockholm (Suppl.), 421 : 93–101.

Granier–Deferre, C. & Abrams, R. (1989). 'Effects of sound on fetal cerebral glucose utilization'. Sem. Perinatal. 13 : 371–379.

Granier-Deferre, C., Lecanuet, J.P., Jacquet, A.Y. & Busnel, M.C. (1992). 'Prenatal discrimination of complex auditory stimulation, 8th ICIS. Congress, Miami.

Greenhough, W.T. & Juraska, M.G. (1986). 'Developmental Neuropsychobiology'. Academic Press San-Diego.

Guyomarc'h, J.C. (1972). 'Les bases ontogéniques de l'attractivité du gloussement maternel chez la poule domestique'. Rev. Comp. Animal, 6, 79-94.

Guyomarc'h, J.C. (1974). 'L'empreinte auditive prénatale chez le poussin domestique", Rev. Comp. Animal, 8, 3-6.

Hammond, J. (1970). 'Hearing and response in the newborn'. Dev. Med. Child Neurol. 12 : 3-5.

Henschall, W.R. (1972). 'Intrauterine sound levels'. J. Obstet. Gynecol. 112 : 577-578.

Horner, K., Servières, J. & Granier-Deferre, C. (1987). 'Deoxyglucose demonstration of *in utero* hearing in the guinea-pig fetus'. Hear. Res. 26 : 327-333.

Impekoven, M. & Gold, P.S. (1973). 'Prenatal origins of parent-young interactions in birds - a naturalistic approach'. In : Advances in Ontogeny of Behavior and the Nervous System. Vol. 1. G. Gottlieb (ed.). Academic Press. New York.

Johansson, B., Wedenberg, E. & Westin, B. (1964). 'Measurement of tone response by the human fetus. A preliminary report'. Acta Otolaryngol. 57 : 188-192.

Kisilevski, B.S. & Muir, D.W. (1991). 'Human fetal and subsequent newborn responses to sound and vibration'. Infant Behav. Dev. 14 : 1-26.

Krumholz, A., Felix, J.K., Goldstein, P.J. & McKenzie, E. (1985). 'Maturation of the brain-stem auditory evoked potential in premature infants'. Electr. Clin. Neurophysiol. 62 : 124-134.

Lawson, K. & Turkewitz, G. (1980). 'Intersensory function in newborns : Effect of sound on visual preferences'. Child development, 51, 1295-1298.

Leader, L.R., Baillie, P., Martin, B. & Vermeulen, E. (1982). 'The assessment and significance of habituation to a repeated stimulus by the human fetus'. Early Hum. Dev. 7 : 211–283.

Lecanuet, J.P., Granier–Deferre, C., Cohen, H., Le Houezec, R. & Busnel, M.C. (1986). 'Fetal responses to acoustic stimulation depend on heart rate variability pattern, stimulus intensity and repetition'. Early Hum. Dev. 13 : 269–283.

Lecanuet, J.P., Granier–Deferre, C., DeCasper, A.J., Maugeais, R., Andrieu, A.J. & Busnel, M.C. (1987). 'Perception et discrimination foetale de stimuli langagiers, mise en évidence à partir de la réactivité cardiaque. Résultats préliminaires'. C.R. Acad. Sci. Paris. 305 (III) : 161–164.

Lecanuet, J.P., Granier–Deferre, C. & Busnel, M.C. (1988). 'Fetal cardiac and motor responses to octave–band noises as a function of central frequency, intensity and heart rate variability'. Early Hum. Dev. 18 : 81–93.

Lecanuet, J.P., Granier–Deferre, C. & Busnel, M.C. (1989a). 'Differential fetal auditory reactiveness as a function of stimulus characteristics and state'. Sem. Perinatol. 13 : 421–429.

Lecanuet, J.P., Granier–Deferre, C. & Busnel, M.C. (1989b). 'Sensorialité foetale : Ontogénèse des systèmes sensoriels, conséquences de leur fonctionnement foetal'. In Médecine Périnatale. J.P. Relier, J. Laugier, B.L. Salle. (eds.) : pp. 201–225. Médecine–Sciences. Flammarion. Paris.

Lecanuet, J.P., Granier–Deferre, C., Jacquet, A.Y. & Busnel, M.C. (1992). 'Decelerative cardiac responseness to acoustic stimulation in the near term foetus'. Quart. J. of Exp. Psychol. 44, 279–303

Lenoir, M., Bock, G.R. & Pujol, R. (1979). 'Supranormal susceptibility to acoustic trauma of the rat pup cochlea'. J. Physiol. 75 : 521–524.

Lewkowicz, D.J. (1980). 'Cross modal equivalence in early infancy : auditory visual intensity matching'. Develop. Psychobiol., 16, 547–607.

Lewkowicz, D.J. (1988). 'Sensory dominance in infants : 1. Six-months-old infant's response to auditory visual compounds'. Dev. Psychol. 24 : 155-171.

Lewkowicz, D.J. (1991). 'Development of intersensory functions in human infancy : Auditory-visual interactions'. In M.J.S. Weiss & P.R. Zelazo (eds.) Newborn Attention : Biological constraints and the influence of Experience. Ablex Publish. Corp. Norwood N.J. USA.

Lewkowicz, D.J. & Turkewitz, G. (1981). 'Intersensory Interaction in newborns : Modification of visual preferences following exposure to sound'. Child Dev. 52: 827-832.

Lipsitt, L.P. (1979). 'The pleasures and annoyances of infants : approach and avoidance behavior' In : E. Thoman. Origins of the infant's social responsiveness. Johnson and Johnson Baby producte C°. Pediatric Round Table.Hillsdale, N.J. Erlbaum.

Madison, L.S., Adubato, S.A., Madison, J.K., Nelson, R.B., Anderson, R., Erickson, J., Kuss, L.M. & Goddlin, R.C. (1986). 'Fetal response decrement : True habituation ?' J. of Dev. and Behav. Pediatr. 7 : 14-20.

Mehler, J., Bertoncini, J., Barrière, M., Jassik-Gerschenfeld, D. (1978). 'Infant recognition of mother's voice'. Perception. 7, 491-497.

Mehler, J., Jusczyk, P., Lamberz, G., Halsted, N., Bertoncini, J. & Amiel-Tison, C. (1988). 'A precursor of language acquisition in young infants'. Cognition. 29 : 143-178.

Mills, M., Melhuis, E. (1974). 'Recognition of mothers' voice in early infancy'. Nature, 252, 123, 124.

Moon, C. (1985). 'Phonological universals and syllable discrimination in one-to three-day old infants'. Unpublished doctoral dissertation. Columbia Univ.

Moon, C. & Fifer, W.P. (1990). 'Newborns prefer a prenatal version of mother's voice'. Inf. Behav. Dev. 13 : 530.

Moon, C., Panneton–Cooper, R., Fifer, W.P. (1991). 'Two–day–olds prefer the maternal language'. I.S.D.P. Meeting.

Mosser, C. (1989). 'Effet physiologique des stimulations sonores chez le prématuré'. Ph Diss. Paris.

Murooka, H., Koie, Y. & Suda, D. (1976). 'Analyse des sons intra–utérins et de leurs effets tranquillisants sur le nouveau–né'. J. Gynecol. Obst. Biol. Reprod. 5: 367–376.

Nijhuis, J.G., Prechtl, H.F.R., Martin, C.B. & Bots, R.S.G.M. (1982). 'Are there behavioural states in the human fetus ?' Early Hum. Dev. 6 : 177–195.

Panneton, R.P. (1985). 'Prenatal experience with melodies: Effect on postnatal auditory preference in human newborns'. Unpublished doctoral diss. Univ. of North Carolina, Greensboro.

Pasman, J.W., Roteveel, J.J., De Graaf, R., Maassen, B. & Notermans, S.L.H. (1991). 'Detectability of auditory evoked response components in preterm infants'. Early Hum. Dev. 26 : 129–141.

Pillai, M. & James, D. (1990). 'Behavioural states in normal mature human fetuses'. Arch. Disease in childhood, 65, 39–43.

Pujol, R. & Uziel, A. (1986). 'Auditory development : Peripheral aspects'. In : Handbook of Human Biologic Development. P.F. Timiras & E. Meisami (eds.) C.R.C. Press. Boca Raton. Florida.

Querleu, D., Renard, X. & Versyp, F. (1981). 'Les perceptions auditives du foetus humain'. Med. Hyg., 39 : 2101–2110.

Querleu, D., Renard, X., Versyp, F., Paris–Delrue, L. & Vervoort, P. (1988a). 'Fetal hearing'. Europ. J. Obst. Gynecol. Reprod. Biol. 29 : 191–212.

Querleu, D., Renard, X., Versyp, F., Paris–Delrue, L. & Vervoort, R. (1988b). 'La transmission intra amniotique des voix humaines'. Rev. Fse Gynecol. Obst. 1, 43–50.

Ray, W.S. (1932). 'A preliminary study of fetal conditionning'. Child Dev. 3 : 173–177.

Richard, D.S., Fentzan, B., Gerhardt, K.J., McCann, M.E. & Abrams, R.M. (1992). 'Sound level in the human uterus'. Obst. Gynecol., 80, 186–190.

Rooteveel, J.J., de Graaf, R., Colon, E.J., Stegeman, D.F. & Visco, Y.M., (1987). 'The maturation of the central auditory conduction in preterm infants until three months post term'. II The auditory brainstem response (ABR). Hear. Res. 26 : 21–35.

Rooteveel, J.J., Stegeman, D.F., de Graaf, R., Colon, E.J. & Visco, Y.M. (1987). 'The maturation of the central auditory conduction in preterm infants until three months post term'. III. The middle latency auditory evoked response (MLR). Hear. Res. 27: 245–246.

Rooteveel, J.J., Stegeman, D.F., de Graaf, R., Colon, E.J. & Visco, Y.M. (1987). 'The maturation of the central auditory conduction in preterm infants until three months post term'. V. The auditory cortical response (ACR). Hear. Res. 27 : 95–110.

Salamy, A., Mendelson, T. & Tooley, W.H. (1982). 'Developmental profiles for the brainstem auditory evoked potentials'. Early Hum. Dev. 6 : 331–339.

Salk, L. (1960). 'The effects of the normal heartbeat sound on the behavior of newborn infant : implications for mental health'. World Ment. Health. 12 : 1–8.

Salk, L. (1962). 'Mother's heartbeat as an imprinting stimulus'. Trans. New–York Acad. Sci. ser. 2, 4 : 753–763.

Schaal, B. (1988). 'Discontinuité natale et continuité chimiosensorielle : Modèles animaux et hypothèses pour l'homme'. Ann. Biol. 27 : 1–41.

Schaal, B. & Orgeur, P. (1992). 'Olfaction in utero'. Quart. J. of Exp. Psychol. 44 : 245–278.

Schmidt, W., Boos, R., Gniers, J., Auer, L. & Schulze, S. (1985). Fetal behavioural states and controlled sound stimulation'. Early Hum. Dev. 12 : 145–153.

Schulman-Galambos C. & Galambos, R. (1979). 'Brainstem evoked response audiometry on newborn hearing screening'. Arch. Otolar. 105 : 86-90.

Scibetta, J.J. & Rosen, M.G. (1969). 'Response evoked by sound in the fetal guinea-pig'. Obstet. Gynecol. 33: 830-836.

Siperstein, G.N. & Butterfield, E.D. (1973). 'Neonates prefer vocal instrumental music to noise and vocal music to instrumental music'. Unpub. manus. Univ. Kansas, Med. Center.

Smotherman, W.P. (1982). 'Odor aversion learning by the rat fetus'. Physiol. & Behav. 29, 769-771.

Smotherman, W.P. & Robinson, S.R. (1986). 'Environmental determinants of behavior in the rat fetus'. Anim. Behav. 34, 1859-1873.

Smotherman, W.P. & Robinson, S.R. (1988). 'Dimensions of fetal investigation'. In : Behavior of the fetus. W.P. Smotherman & S.R. Robinson. (eds.) pp. 19-34. Telford Press.

Spence, M.J. & DeCasper, A.J. (1982). 'Human fetuses perceive maternal speech'. International Conference on infant studies, Austin, Texas.

Spence, M.J. & DeCasper, A.J. (1987). 'Prenatal experience with low frequency maternal voice sounds influences neonatal perception of maternal voice samples'. Inf. Behav. Dev. 10 : 133-142.

Starr, A.R., Amlie, N., Martin, W.H. & Sanders, S. (1977). 'Development of auditory function in newborn infants revealed by auditory brainstem potentials'. Pediatrics. 60: 831-839.

Turkewitz, G. (1971). 'Neurobehavioral organization of the human newborn'. In : J. Hellmuth (ed) Experimental infant : Brunner/Mezel New York vol. 2, pp. 24-40.

Turkewitz, G. (1989). 'Intersensorialité chez le nouveau-né'. In : J.P. Relier, J. Laugier, B.L. Salle (eds) : Médecine périnatale. Flammarion. Paris, pp. 226-229.

Turkewitz, G., Moreau, T., Birch, H.G. & Davis, L. (1971). 'Relationships among response in the human newborn : the non-association and non-equivalence among different indicators of responsiveness'. Psychophysiology. 7 : 233-247.

Turkewitz, G., Mc Guire, I. (1978). 'Intersensory functioning during early development'. Int. J. Mental Health, 7, 165-182.

Turkewitz, G., Gardner, J.M., Lewkowicz, D.J. (1983). 'Determinants of infant perception'. In : J. Rosenblatt, R.A. Hinde, C. Beer & M.C. Busnel (eds). Advances in the study of behavior vol. 13, N.Y. Acad. Press, pp. 39-62.

Turkewitz, G., Gardner, J.M., Lewkowics, D.J. (1984). Sensory/perceptual functioning during early infancy : the implications of a quantitative basis for responding. In : G Greenberg and E. Tobach (eds) Behavioral evolution an integrative levels. Hillsdale, NJ : Erlbaum, pp. 167-197.

Versyp, F. (1985). 'Transmission intra-amniotique des sons et des voix humaines'. Thèse Doc. Med. Lille.

Vince, M.A. (1979). 'Postnatal consequences of prenatal sound stimulation in the guinea-pig'. Anim. Behav. 27: 908-918.

Visser, G.H.A., Mulder, H.H., Wit. H.P., Mulder, E.J.H. & Prechtl, H.F.R. (1989). 'Vibro-acoustic stimulation of the human fetus on behavioural state organization'. Early Hum. Dev. 286 : 296-312.

Walker, D.W., Grimwade, J.C. & Wood, C. (1971). 'Intrauterine noise : a component of the fetal environment'. Am. J. Obstet. Gynecol. 109 : 91-95.

White, N.R. & Mac Donald, G.E. (1983). 'Embryonic auditory stimulation and posthatch distress calling in the domestic chicken'. Can. J. Zool., 61, 268-270.

Wolff, P.H. (1969). 'The natural history of crying and other vocalizations in early infancy'. In : Determinants of infant behavior. B.M. Foss. (ed.) 4. Methuen. London.

Woods, J.R. & Plessinger, M.A. (1989). 'Fetal sensory sequencing : Application of evoked potentials in perinatal physiology'. Sem. Perinatol. 13 : 380-392.

THE EARLINESS AND COMPLEXITY OF THE INTERACTION SKILLS AND SOCIAL BEHAVIORS OF THE CHILD WITH ITS PEERS

H. MONTAGNER, B. EPOULET, G. GAUFFIER, R. GOULEVITCH,
N. RAMEL, B. WIAUX, M. TAULE
Unité 70 de l'Institut National de la Santé et de la
Recherche Médicale
Rue du Mas de Prunet 388
34000 Montpellier
France

Research on the perceptual competencies of the human infant has brought to light its extraordinary capacities to interact with and learn from its mother. It is clear that studies in this field allow scientists to better understand the processes and mechanisms which regulate attachment and attunement (Stern,1985) between the infant and its mother, and those which form the roots of emotions, communication and learning. However, the study of the infant's interactions with other human partners, familiar and unfamiliar reveals the capacities of the child to adapt to different social behaviors and build up new communication and learning processes.

The interactions of the child with other children are particularly relevant. Based on the study of friendship, social cohesion and cooperation in groups of children of school age and pre-adolescents, Hartup's work in 1969 and 1970 has given important scope to the researches on social behaviors of children. McGrew and Blurton Jones were also pioneers as they published in 1972 two basic ethological books on children's behaviors with their peers in kindergartens.

Researches have also multiplied from the 1970 in the age classes of 1 to 3 years (Eckerman et al.,1975;Mueller and Lucas,1975 ; Eckerman and Whatley,1977;Mueller and Brenner,1977;Montagner,1978,1988;Mueller,1979; Stambak et al.,1979;Vandell and Mueller,1980;Bronson,1981; Howes, 1980 , 1987;Howes and Rubenstein,1981;Jacobson,1981;Musatti and Panni , 1981 ; Destephano and Mueller,1982;Montagner et al.,1982,1984,1988 ; Musatti , 1986;Bakeman and Adamson,1984;Restoin et al.,1983,1985;Schneider et al., 1989;Verba and Musatti,1989).

Considering the children of less than one year, Vincze (1971) observed mutual smiles and laughing between infants as of the age of 6 months and mutual touching and gazes as of the age of 7 months. Their capacities to focalize their gaze on one another are already present between 2 and 3 months. More recently, Dubon et al.(1981) reported that the orientations of the gaze of infants aged 2 to 3 months can be accompanied by a global motor activity, facial expressions, mimicries and smiles. Similar acts can be manifested in response by the partner. Vandell (1980) observed that as of the age of 6 months most socially directed behaviors between peers are composed of two acts.

However, we lack precise and quantitative data on the interaction processes

R. A. Gardner et al. (eds.), The Ethological Roots of Culture, 315–355.

and social behaviors of the young child with its peers throughout the first two years.

As our multidisciplinary team decided in 1988 to follow up infants from birth to better understand their interactive competencies with their mothers and otherpartners, we have set up two experimental procedures to study the emergence and regulations of the child's interactions with his/her peers during the first and second years of age.

1. The study of the interactive capacities of infants of 4 to 7 months with an infant of the same age

1.1. POPULATION STUDIED

The study was carried out with dyads of infants aged from 4 to 7 months. We will report in this paper the data computed for 16 infants. All were in good health since birth and showed no visible anomalies (Table I).

The sample of infants was made up from the files of the Maternity Hospital. The parents were contacted by telephone when their infant was aged 3 to 4 months. The infant-infant dyads were then formed at random, the two infants and their parents having never met.

1.2. METHOD

1.2.1. *General Principle*. Two infants and their mothers are received in the room where the study is carried out. During 15 to 20 minutes the reception phase allows them to familiarize themselves with this room and the experimental layout. The general principle consists of continuous filming of the infants from the moment their mother seats them in two identical seats, the proximity and orientation of which can be controlled from a distance.

1.2.2. *Experimental Layout*. The experimental layout consists of two seats mounted on rails (Fig. 1). Each mother prepares her own child: after having taken off its coat and shoes, she puts on a waistcoat which is attached at the child's back with velcro. The waiscoat has three straps, two of which are latero-dorsal and fasten behind the seat, the third joins the front and back of the waistcoat between the legs. The particular conception of the seat and waistcoat avoids the sagging or sliding of the head, trunk and pelvis of the infant, laterally or forwards.

1.2.3. *Procedure*. Each experimental session is made of 8 phases of three minutes each which follow one another without interruption after the infants have been seated (Table II). During the first and second phases each infant is face-to-face with its mother (Fig. 1A). The third phase puts the infant in interaction under the same conditions with a student he/she is meeting for the first time. From Phase 4 to Phase 8 the two infants are at first in interaction with each other at a right angle (Fig. 1B), then face-to-face. The last phase (Phase 8) is characterized by giving an identical object to each infant (for more details, see Table II and Montagner et al., 1993). The mother is always present at a distance of approximately 1.50 m from her infant. The same dyads are filmed four times at a one-week interval.

1.2.4. *Analysis of the Data*. In this paper we will give only data on the infant-infant interactions. During a priminary study we quantified 14 categories and 209 items of behavior which we analysed to 1/10 th of a second in order to validate Magnusson THEME programme and reveal patterns, that is to say behaviors linked to each other by a significant critical time interval (Montagner et al., 1990). In the present study we have selected the 8 categories and 25 items which have been the most represented and the easiest to identify(Table III) : six categories are made up of items of interaction, a seventh of general motor movements of the limbs (apparently uncoordinated movements of the arms and legs) and an eight of self-directed behaviors.

The frequency and duration are calculated by infant and by unit of time of 180 seconds, the basic duration of each phase.

1.2.5. *Data Processing* . The comparisons between the frequencies are carried out using the test of Kolmogorov-Smirnov which also allows the verification of the distribution normality (Siegel, 1956). Those between the durations are also carried out usingt the test of Kolmogorov-Smirnov. In this paper we will report only data comptuted durirg the first session.

1.3. RESULTS

1.3.1. *Duration of Interaction With a Peer*. Infants have the capacity to accept, without showing distress and for a period of at least 15 minutes (Phases 4 to 8) an uninterrupted interaction with a peer.

1.3.2. *Hierarchy of Behaviors*. A hierarchy appears to exist between the 8 behavior categories (Fig.2). The orientations of the gaze (Category I) logically have a frequency significantly higher than that of the other behavior categories : $p<.05$ as compared to the facial expressions (Category II), and contacts with the partner (Category VI) ; $p<.01$ as compared to vocalizations (Category III) ; $p<.001$ as compared to the movements of the bust (Category IV), targeted positions and movements of the arms (Category V) and general motor movements of the limbs (Category VII) (For the self-directed behaviors,see below).

The two categories of behaviors which are thereafter the best represented are the facial expressions (Category II) and the contacts with the partner (Category VI) both for their duration and frequency. However,the differences are only significant between : the facial expressions and the targeted positions and movements of the arms (Category V) ($p<.001$); the contacts with the partner and the targeted positions and movements of the arms ($p<.05$ for the frequency and $p<.001$ for the duration).

The vocalizations (Category III) are more frequent than the movements of the bust in the direction of the other infant (Category IV), the targeted positions and movements of the arms (Category V)and the general motor movements of the limbs (Category VII), but the differences are only significant with the targeted positions and movements of the arms($p<.05$ for the frequency and $p<.001$ for the duration).

The duration of the movements of the bust in the direction of the other infant (Category IV) is significantly higher than the duration of the movements of the arms and legs whether these are general motor movements (Category VII) ($p<.05$), or targeted positions and movements of the arms (Category V) ($p<.05$).

The targeted positions and movements of the arms (Category V) have a frequency ($p < .01$) significantly higher than that of the general motor movements of the limbs (Category VII).

The self-directed behaviors (Category VIII)are with the bust movements, the least frequent. The differences are significant with the orientations of the gaze (Category I) ($p < .001$), the facial expressions (Category II) ($p < .001$), the vocalizations (Category III) ($p < .001$), the targeted positions and movements of the arms (Category V), the contacts with the partner (Category VI) ($p < .001$) and the general motor movements of the limbs (Category VII) ($p < .01$). They also have a duration significantly shorter than that of the orientations of the gaze ($p < .001$), the facial expressions ($p < .001$), the vocalizations ($p < .001$), the movements of the bust ($p < .001$) and the contacts with the partner ($p < .001$).

The behavior activity of the infants therefore appears essentially targeted on the infant-partner, even though the two mothers can be seen by their infant during the whole session when he/she turns his/her gaze in their direction. In comparison, the apparently non-targeted behaviors (general motor movements of the arms, hands, legs and feet) and the self-directed behaviors are infrequent and not sustained.

1.3.3. ***Visual attention to the Infant Partner*****.** The infant-partner appears as the privileged target of the visual attention of both children (Fig.3). Our study reveals that 34% (N=2708) of the 7850 behaviors measured are the orientations of the gaze in the direction of the face, hands and gaze (catching of the eyes) or the feet of the other infant (their cumulated duration represents 32.5% of the cumulated duration of the whole of the behaviors).

The frequency and duration of the orientations of the gaze in the direction of the infant-partner (cumulation of the items n°1, 2, 3 and 4 : see Table III) are significantly higher than the frequency and duration of the orientations of the gaze in the direction of its own body (self-directed gazes : item n°5) ($p < .001$), and the frequency and duration of the gazes in other directions, including towards the mother (item n°6) ($p < .01$).

However, the visual attention is not equally divided between the different body areas. The face is the most frequent target (29.06% of all the orientations of the gaze and 14.93% of all the behaviors). The face also has the longest duration of visual attention (33.3% of the cumulated duration of the different orientations of the gaze and 16.73% of the cumulated duration of all the behaviors).

The hands of the partner appear, after the face, as the most attractive target-zone for the gaze of the infant. The orientation of the gaze in the direction of the hands of the infant make up 18.15% of the orientations of the gaze and 9% of the whole of the behaviors (their cumulated duration represents respectively 16.48% and 8.27% of these two sets of behaviors). They occupy the third place in the distribution of the 25 behaviors studied, after the two items "orientations of the gaze in the direction of the face of the other infant" and movements of the oral area". Mutual catching of the eyes occupies the 4th place.

The frequency and duration of the infants' gaze in the direction of their mothers only represent 7 and 8% of all orientations of the gaze. The frequency and duration of the self-directed gazes are weak when compared to the orientations of the gaze in the direction of the different body areas of the infant-partner. The experimental situation does not, therefore, appear to induce withdrawal or "autistic" behaviors during the first session when both children meet each other for the first time.

1.3.4. *Hand and Arm Movements and Partner Body Contacts with the Hand*

The voluntary and organized movements of the arms and hands (items of Category V) and body contacts with the hand and taking the hand of the partner (items of category VI) represent altogether 10.4% of the whole of the behaviors. These behaviors appear targeted, visually guided and precisely controlled.

The coordinated movements of the arms and hands are well mastered in spite of their complexity. In particular : stretching out the arm in the direction of the partner, with the hand in supination or with the hand turning from pronation to supination with movement of the arm.

These data reinforce the hypothesis which we have put forward by defining the organizers of the behaviors in our previous studies, that is to say the motor and/or vocal items which seem to constitute the cores around which or from which other items are "grafted", to "build up" structures which are more and more complex and better adapted to the communication processes and grasping and manipulation of objects (Montagner,1978,1988 Montagner et al.,1988,1990 ; Restoin et al., 1983 , 1985). For example, the behaviors whose motor core appears to be the targeted , organized and visually guided stretching out the arm and hand in the direction of a person or object . That is to say :

1. grasping an object ;
2. the behavior of manual solicitation (hand presented in pronation or supination, or in a lateral or oblique position
3. pointing combined with the arm stretched out in the direction of a person or object ;
4. offering ;
5. the behavior of exploring the body of another person ;
6. the threatening behaviors during which a stretching out of the hand in pronation, lateral or oblique, can be followed by pointing, in any case during the second year.

The analysis of corporal contacts between the two infants shows three types of targeted manual actions :

1. the intentional seeking out with the hand of a contact with the hand of the other infant;
2. the sliding of the hand onto the back of the hand of the partner ;
3. the grasping of its hand. This behavior is sometimes followed by repetitive movements, during which the initiating infant shakes the hand of the other infant.

1.3.5. *Modulated Vocalizations.* One must emphasize that even if they are not frequent (N=405 ; 5.16% of the behaviors measured),the modulated vocalizations are more frequent than the brief vocalizations and occupy the 7th place in the distribution of the 25 items studied. To the naked ear they have phonetic and sequential structures which are diverse and complex. We have, therefore, begun a differential study of the spectral and phonetic structure of these vocalizations, notably their linguistic aspects, in relation to the current modalities of the mother-infant interactions.

1.3.6. *Effects of Different Orientations and Distances of the Infants.*

The control panel allows a study of the regulations of behaviors and interactions due to different orientations and distances of the infants with regard to each other and with regard to their mothers. Phase 4 which brings the infants at a distance of 15 cms, appears as a phase of observation and adptation.

1. it is during this phase (and Phase 5) that the frequency of the orientations of the gaze in the direction of the other infant, is the highest (Fig. 4) ;
2. the frequency and duration of the leaning of the bust in the direction of the other infant are the lowest, as if the infants adopt a position of withdrawal, making observation easier without interactive involvements ;
3. the frequency of the vocalizations is the highest (Fig. 4) ;
4. the frequency of the general motor movements of the upper and lower limbs is the highest ;

In other words during Phase 4, the infants distribute the targets of their gaze, put themselves in a position of "withdrawn" observers, make themselves heard and fidget. They are making themselves confortable.

Phase 5, which puts the infants in contact with each other at a right angle, is that of the "maximal" communication and the sophisticated and multi-modality interactions :

1. the frequency and duration of the self-directed behaviors are weak ;
2. the frequency and duration of the orientations of the gaze, facial expressions and vocalizations are at a high level (Fig. 4) ;
3. the frequency and duration of the targeted movements of the arm and hand in the direction of the infant-partner are much higher than during the other phases, and particularly Phase 4 (Fig. 4) ;
4. the frequency and duration of the corporal contacts with the other child are higher than during any other phase (Fig. 4) .

Thus, during Phase 5, the infants develop to the maximum the different behavior modalities which allow them to attract the attention of the other infant, to enter into corporal contact with him/her and to respond to his/her behaviors.

Phase 8 which is characterized by giving an object to each infant, distinguishes itself clearly from the other phases. One observes a decrease in the frequency and duration of all the behaviors which can play a role in the interactions (orientations of the gaze, in particular in the direction of the other infant ; facial expressions ; vocalizations ; contacts with the partner), with the exception of the movements of the arms and hands in the direction of the other infant (the frequency and duration of which are clearly lower to those of Phase 5). As has been observed in other studies, the manipulation of an object is accompanied by a significant decrease in the behavior activity which allows an interaction with a peer.

Children appear, therefore, capable between 4 and 7 months of adjusting their behavior both to the context and the other child's behavior.

1.3.7. *The Capacity to Imitate* . By imitation, we mean for this population of infants all reproductions of an act of one of the infants by the other infant, either at the same time or with an interval of a few seconds. The reproduction of the act is preceeded by a targeted orientation of the gaze in the direction of the partner. The imitations observed are :

1. rhythmic beating of the feet, either alternating or simultaneous ;
2. simple or repeated beating of the hand(s) either on its own body or on the seat, one of the hands holding an object (Phase 8), or not (Phases 4 to 7) ;
3. movements of the hands and arms in the direction of the back of the seat ;
4. noisy scratching of the waistcoat which each child is wearing ;
5. putting an object in the mouth ;
6. throwing an object ;
7. agitating an object ;
8. sucking the collar of the waistcoat or the end of the strap which passes between the legs ;
9. turning its gaze in the same direction as its partner ;
10. vocalizations having the same phonetics to the naked ear (spectral analysis in process).

1.3.8. *The Capacity to Adjust Behavior to the Partner* . The comparative study of the children's behavior of the same dyads at a one-week interval (Montagner et al., in preparation) reveals that during the second experimental session :

1. most children show behaviors which were not observed during the first session, or the frequency of the behaviors is most often higher . Among them are the most complex, i.e. the extension of the arm in the direction of the infant-partner with the hand in pronation, lateral, oblique or in supination, touching the hand of the other child or taking the other child by the hand (Fig. 5) ;
2. the frequency of some behaviors varies similarly from one phase to the other in both children ;
3. there are no more significant differences between both children either in their behavioral repertoire or in the frequency of most categories of behaviors, which was not observed during the first session .

Everything appears as if, after a first session of familiarization, mutual influences and adjustment phenomena developed between the children of the same dyad.

2. The study of the emergence and development of motor and social skills of children aged from 8 to 11 months at the beginning of the study

2.1. POPULATION STUDIED

This paper will report data on the emergence and development of the different skills and capacities of 17 children (9 boys and 8 girls) aged from 8 to 11 months at the beginning of the study . They include motor skills since their emergence and development appeared to be linked to those of interactive skills and social behaviors of toddlers with peers.

All the children were cared for in a collective day-care center at the end of their mother's maternity leave (i.e. 4 months after the delivery). They have, therefore, been able to familiarize themselves to each other during 4 to 7 months before the beginning of the study. All were in good health since birth and showed no visible anomalies (in other studies we have followed up emotionaly disturbed and mentally disabled children -particularly autistic children- and other ones with cerebral palsy: Montagner et al., in preparation).

2.2. METHOD

2.2.1. ***General Principle*** .Groups of 6 children (3 girls and 3 boys) are received every two weeks (every week in other studies) in a specially equipped setting in our research unit (Fig. 6). When one or two children are absent for health or other reasons the group is filled out by substituting one or two toddlers of the same age from the same day-care population .
The toddlers can move around without limitation of their motor activity, either by their physical environment or by orders or restraints from adults. They are filmed continuously, using 6 cameras which do not leave any blind spots and which run without interruption after having been triggered by sound when the toddlers enter the room.

2.2.2. ***Experimental Setting*** . The room we have fitted out consists mainly of :

1. an open space (o.s. in fig. 6) where children of all ages can at will crawl, move about on all fours, walk or run ;
2. a carousel (R) that can be rotated by the children themselves ;
3. a holed partition (H.P.) with three parts : a central part with a big porthole that allows the children to go from one side to the other and two lateral parts with holes of different forms and sizes ;
4. a corridor (Co) between the partition and a superstructure (Su) with a terminal platform at a height of 1.5 m ;
5. a sloping device (ramp Ra) with a gradient of 45° and transverse bars leading to this platform on which there are two cavities filled with plastic balls ;
6. a spiral staircase (S.S.) of 13 steps leading to a 1.5m high platform on which there are two nests large enough to receive one or two children ;
7. a playpen, a climbing wall (C.W.) and a rope ladder ;
8. small cavities allowing each child to withdraw from the peer group, whether alone or with several children.

This setting was designed based on anthropometric standards taking into account the children's bodily proportions, size, and preliminary observations of children of the same age during unrestricted activity in their family environment and day-care centers. For example, the height of the steps is 10 cms and the width of the spiral staircase varies from 52 to 56 cms according to the level. The configuration of the staircase allows each child to support him/herself at all times, using one or both hands either on the central axis, the wall, the side of the superstructure or the vertical bars which define the outside part from the 9th step on . The configuration of the sloping device also allows each child to support himself/herself by using one or both hands on the wooden banisters.

2.2.3. ***Procedure*** . The observation sessions take place as nearly as possible on the same day of the week, and at the same time of the day for 10 consecutive months. The duration of each session is 30 minutes. Two of the usual day-nursery nurses are permanently present. They sit at the edge of the room and are told to intervene only when they see a child in difficulty.

2.2.4. *Analysis of the Data*. The video films are analysed using grids made up of 250 items which are sometimes regrouped into categories, according to the type of analysis and calculation. From 2 to 4 researchers participate in the collation of occurrences of the different items and advance by consensus.

As all the toddlers did not participate in all the sessions because of absence from the day-care center, and as there are differences of several weeks of age between toddlers of the same group (they are aged from 8 to 11 months during the first session), mean frequencies are calculated per child/session for each month of age, by dividing the number of occurrences of each behavior or category of behaviors by the number of toddlers and the number of sessions at each month of age. The mean frequencies are also calculated in relation to the order of the sessions from the first to the last. Finally, they are calculated for each child, from one month to another and from one session to the next.

As the camera timers simultaneously record a time scale to 1/50 th of a second on the films, the mean duration of each item is also calculated for each month of age and for each child from one session to the next (Montagner et al.,in preparation).

2.2.5. *Data Processing*. As the number of toddlers varies sometimes from one age month to another because of absences and in some cases cancelled sessions (day-care center closed, holidays) the statistical comparisons from one age month to another are carried out using the U test of Mann-Whitney.

2.3. RESULTS

2.3.1. *Locomotor Development*. The temporal evolution of bouts of sitting, walking on all fours, and walking, are the developmental indicators which allow us to define the emergence of different skills and capacities at the end of the first year and during the second year (Fig. 7).

The frequency of bouts of walking on all fours and self-initiated sitting evolves in a comparable way from month 9 to month 19 (the beginning and the end of each bout are noted at the time the child changes his or her mode of locomotion or body position). After a significant increase between 9 and 10 months ($p \leq .01$), there is a progressive decrease from one month to another up to 15 months (there is no significant difference from one month to the next except between 12 and 13 months for the sitting position, a difference significant at the level of $p < .05$). The decrease is accentuated between 15 and 16 months (the difference is significant at the level of $p \leq .01$ both for walking on all fours and sitting). Between 16 and 19 months the differences in frequency of the bouts of walking on all fours and sitting are not significant.

The frequency of the walking bouts increases abruptly between 12 and 13 months (difference significant at the level of $p \leq .001$), then by fits and starts with alternation of large increases (between 14 and 15 months :

difference significant at the level of $p<.01$; between 16 and 17 months: difference significant at the level of $p<.00001$) and of decreases, one of which is very noticeable (between 15 and 16 months : $p<.01$).

Between 10 and 13 months, walking evolves in a reciprocal way from walking on all fours and assuming the sitting position : the frequency of walking bouts increases more and more rapidly; at the same time the frequency of the bouts of walking on all fours and sitting decreases . This reciprocal evolution is found again from the 15th month (14 to 15 in fig. 7) after a phase of comparable frequency of the three posturomotor patterns during the 14th month.

2.3.2. *Establishment of Motor Skills.* The indicators of locomotor development having been established, one can distinguish four phases of establishment of motor skills between 8 and 19 months :

1. The first phase is the emergence of motor skills which allow exploration of the third dimension of space (height and depth).Figure 8 shows that partial ascents of the spiral staircase of 13 steps are already observed at the age of 9 months : these are ascents of 3 to 6 steps by two children (Montagner et al., 1993). The first complete ascent of 13 steps was noted in one of these toddlers between 10 and 11 months. Approximately half of the toddlers (47%) showed that they were capable of climbing from 2 to 9 steps before the age of 12 months (all were able to do so before 16 months) and of climbing all 13 steps at the age of 15 months (all at 17 months).

Figure 9 shows that partial ascent of the ramp Ra can appear between 8 and 9 months (only two children, once each). The range of this emergence is more spread out than for the staircase : approximately half of the children (47%) climbed less than the lower half of the ramp for the first time before 15 months (3 before 12 months and all toddlers at 20 months). This entire sloping device was climbed at 10 months by one child and before 16 months by 58% of the toddlers (by all at 21 months).

If one considers the ascending patterns, the standing position was already observed at 12 months, when the children climbed the steps of the staircase (Montagner et al., 1993). The first partial ascents of the ramp in the standing position were also carried out at 12 months.

2. In the second phase there is consolidation of motor acquisitions (12 to 14 months ; Fig. 8 and 9). The age range of 12 to 14 months is characterized by a slow progression of the frequency of ascents of the staircase and the ramp and by the small number of complete ascents. The standing position is still little used in the ascents and descents. The most frequent posturomotor patterns used by toddlers are : flat on the belly, walking on all fours or on the knees, sitting and progression on the buttocks during the descents. Everything happens as if, by reinforcing their mastering of walking between 12 and 13 months (Fig. 7), the toddlers progressively and "prudently" consolidate their newly acquired motor skills (ascent and descent of the staircase and the ramp) by conserving the four support points (hands and feet) of the beginning of locomotor activity and/or the security of the sitting position.

3. In the third phase we find diversification of the posturomotor patterns during ascents and descents of the staircase and ramp and growing predominance of the standing position (14 to 18 months ; Fig. 8 and 9). The age of 14 to 18 months is characterized by the accelerated development

of the motor skills which allow the "conquest" of the third dimension of space, and the emergence of new patterns of ascent and descent.

The frequency of all ascents (partial and complete) of the staircase and the ramp increases in a significant way ($p \leq .05$) between 14 and 15 months. There is also a significant increase in the frequency of ascents and descents of the staircase in a standing position, most often in combination with other patterns ($p \leq .05$) (Montagner et al., 1993).

The frequency of complete ascents of the staircase and of the number of of steps ascended and descended increases significantly between 15 and 16 months ($p \leq .05$) (Fig. 8). Between 15 and 16 months there is also a significant increase in frequency of ascents of the ramp in the standing position in relation to other patterns ($p \leq .01$) (Montagner et al.,1993). This is due to a sharp increase in frequency of the pattern "bringing feet together" (the support foot on the preceding level is brought to the level of the "attacking" foot), and to a doubled representation of the pattern "legs-alternating position" (the alternation of the legs is that of walking) (Montagner et al.,in preparation). Concurrently, walking on all fours decreases significantly (Fig. 7). During descent of the ramp, the frequency of the positions of flat on the belly and walking on all fours decreases (Montagner et al., 1993) at the same time that the frequency of the pattern "progression on the buttocks" increases.

Between 16 and 17 months, there is a significant increase in frequency of partial and complete ascents of the staircase (Fig. 8), a significant increase in frequency of the standing position during ascents and descents of the staircase, and a significant increase in frequency of standing positions during descent of the ramp (Montagner et al.,1993).

Between 17 and 18 months there is a significant increase in frequency of complete ascents of the staircase and the ramp (Fig. 8 and fig. 9), a significant increase in frequency of the pattern " legs-alternating position" in ascents of the ramp and a significant increase in frequency of the pattern "bringing feet together" in the standing position during descent of the ramp (Montagner et al., in preparation).

The age range of 14 to 18 months is, therefore, characterized by :

... mastering of ascent of the staircase by all the toddlers of our sample, in particular of all 13 steps (the age limit was 17 months for only one child) ;

... growing frequency of partial and complete ascents of the staircase ;

... growing frequency of complete ascents of the ramp ;

... diversification of posturomotor patterns during ascents and descents ;

... growing frequency of ascents and descents of the staircase and the ramp in the standing position ;

... growing predominance of locomotion in the standing position, with alternation of the feet as in walking during ascents and descents of the staircase and ascents of the ramp.

This progression of motor skills is accompanied by a progressive decrease in mean duration of ascents and descents of the staircase and the ramp (Montagner et al., in preparation).

4. In the fourth phase there is "expansion" of "motor conquests", interactions, and social behaviors with peers. As of 18 to 20 months, three phenomena modified the evolution of motor processes in most children :

... among the toddlers there was an increase in frequency and duration of interactions and complex social behaviors here and there along the holed partition, in the corridor, and around the carousel ;

... as of 18 months the carousel became an increasingly attractive device for the toddlers; they turned it more and more frequently, tried to climb on more and more often, and asked the nursery nurses to carry them on.

... from the 19th month the children spent more time in the pools filled with the plastic balls.

Consequently, there was a decrease between 18 and 19 months in frequency of ascents of the staircase and the ramp.

In this paper the evolutions and regulations which appear linked to these "expansions" will not be considered.

2.3.3. *Emergence and Evolution of Interactive Skills and Social Behaviors*

We have collected and divided the interactive skills and social behaviors observed in toddlers into two groups:elementary and complex.

The elementary social behaviors are patterns of interaction made up of "morphologically" well defined items of behavior. We have selected four sets of patterns : offering, manual and postural soliciting and pointing of the finger (Montagner, 1978,1988 ; Montagner et al., 1982,1984,1988, 1990 ; Restoin, 1985 ; Restoin et al., 1983,1985) (Fig.10).

The complex social behaviors are those the items and patterns of which cannot be disassociated, which are clearly identifiable in their form or structure, and the function and significance of which can be defined. We have distinguished 4 categories (Fig. 11) : The multimodal interactions from here and there along the holed partition ; the imitations ; the cooperations ; the behaviors of "hide and seek" or their forerunners, and in a more general way, the behaviors of anticipation of the behavior of the partner.

2.3.3.1. Qualitative Observations. The complexity of the children's skills leads to a preliminary description of the sequences and linking of the acts which form the complex social behaviors which we have distinguished above.

Multimodal interactions from here and there along the holed partition form one set of complex social behaviors. We define as multimodal interactions the feebback made by a behavior and/or vocalization from one of the toddlers which induces a modification in the behavior, and/ or vocalizations of the partner, this modification itself, in turn, inducing a response from the first child. They are made up of items and patterns whose reception and decoding rest on combinations of information through several sensory systems (for example, combination of auditory and visual information or of tactile and visual information).

These interactions allow each child to discover the diversity of the behaviors of his/her partners in contexts other than the family or the day-care center and without limitations of her/his initiatives. Each child behaves as if she/he is in a complex logic of communication, far removed from the simplified and linear pattern "emission - response - modification of the behavior of the emitter ". In fact, the response of most toddlers to the behaviors they see, hear, or feel, from here and there along the holed partition, does not appear to be simply in the form of acts or elementary patterns which are morphologically well structured and codifiable in signals. The mutual responses are most often the linking of acts which overlap and coordinate, and which have complex social functions. A "simple" interaction which begins by the catching of eyes, mimicries, vocalizations, or touching can lead, from the age of 11 months, to imitations, behaviors of cooperation and anticipation of the behavior of the partner. For example, after having crouched or stood up along the partition, with her/his face at the level of one of the openings hidden by a curtain, a child visually explores the other side, shouts or calls, introduces through the opening her/his hand which may or may not hold an object, or agitates the curtain hiding the opening (see fig. 6). These patterns attract the attention of another child on the other side of the partition, who her/himself crouches or stands up, to find her/himself face-to-face with the first child. After having caught her/his eyes the second toddler touches or takes her/his hand, shakes the curtain or tries to unstick it, passes a ball through the opening and drops it. The first toddler picks up the ball and offers it to his /her partner, at the same time imitating the other's cries or calls, then coordinates his/her acts with the other's to unstick the curtain. The second toddler continues with behaviors and vocalizations which are similar or different. Several interactions can follow with or without a pause.

The exchanges appear so attractive that certain children may, between 15 and 20 months, participate in up to 10 successive interactions with the same partner or two different partners. These interactions may be followed by behaviors of mutual search and of anticipation of the behavior of the partner (each toddler goes to one of the ends of the partition, pauses, adopts a waiting position with the head leaning forward in the direction of the partner.

Imitative behaviors are a second set of complex social behaviors. In this category we regroup the reproduction of gestures (for example, offering, solicitations, pointing, banging on the partition) and the reproduction of activities (agitating, sticking and unstick a curtain ; passing a curtain or ball through one of the openings of the partition; carrying a cushion and placing it in the porthole or carousel; or turning the carousel). However, we do not prejudge the meaning of the items, patterns and/or the exchange itself for the initiator or the imitator. The imitations may be observed in any area of the room. The interface made up of the partition between the corridor and the "outside" world particularly favors the imitations, since approximately 70% are observed among children around the openings, the porthole and in the corridor.

Cooperative behaviors are another set of complex social behaviors, and we have distinguished two principal forms of cooperation. The first is the children's coordination of their own acts with those of other children in activities modifying or moving objects (unsticking the curtains; transporting cushions to the porthole or carousel ; turning the carousel, etc). The second is the complemention of one child's acts with those of another (pointing out a curtain to the partner after having tried to unstick it, the partner thereafter trying to take off the curtain ; pointing out the carousel, after having tried to turn it, the partner then trying to turn it; trying without success to push a cushion through the porthole, then pointing it out to another child who lifts it up, carries it to the porthole and pushes it through).

The fourth set of complex social behaviors consists of anticipatory behaviors. Different forms of anticipation of the behavior of another child take place around and here and there along the holed partition and the carousel. For example, two toddlers separate after having developed interactions between each other in the porthole or one of the other openings of the partition. The toddlers disappear from each other's sight, pause, adopt a waiting attitude and a behavior of visual, postural or motor research of the partner at the extremity of the partition, that is to say at the end of the corridor from where the partner "must" come out. One of the children goes then to meet the other child, or, having heard or seen his/her partner, the child turns and moves away either walking or on all fours, laughing, vocalizing or shouting, pauses and looks behind as if to assure him/herself that the other child is following.

2.3.3.2. Quantitive data. We have distinguished two phases in the temporal evolution of interactive processes and social behaviors (Fig. 10 and fig.11).

The first is a phase of emergence of interactive and social behaviors. Figures 10 and 11 show that the elementary and complex social behaviors brought to light all emerge before 12 months, whether they involve one or several toddlers. Among the complex social behaviors, interactions from here and there along the partition and imitations of a gesture appear the earliest. More than 50% of the children followed showed one or more interactions in or around the partition before 12 months (all at 13 1/2 months). This capacity appears in close relation with the emergence and rapid development of the child's ability as of 9 or 10 months, to stand up against the partition and to move about in the standing position whilst supporting him/herself with one or both hands on the partition. More than half of the toddlers are also capable of imitating a gesture before 12 months ; this comprises offering, manual or postural solicitations, pointing and tapping the partition. These results are in agreement with those of our previous studies (Montagner, 1988 ; Montagner et al.,1988 ; Restoin, 1985 ; Restoin et al.,1988).

Imitations of activities, cooperative behaviors, and anticipatory behaviors were each observed initially between 10 1/2 months and 12 months in only 1 to 3 children, according to the category but before 15 months in the majority.

The second phase is characterized by "accelerated" development of interactive skills and social behaviors. Figure 11 shows that, between 12 and 15 months (the consolidation phase for motor skills), the frequency of three categories of complex social behaviors increases greatly, either in an almost linear way (the interactions from here to there along the partition: difference between 13 and 15 months significant at $p < .01$; imitations: difference between 14 and 15 months significant at $p < .01$), or slowly and irregularly (the behaviors of cooperation). The most irregular evolution is that of behaviors of anticipation (there is a rapid increase between 13 and 15 months).

The age range from 15 to 19 months appears to be a consolidation phase for these behaviors with, however, an insignificant decrease between 15 and 16 months and between 17 and 18 months. Significant increases are found between 16 and 17 months for the behaviors of cooperation ($p < .05$) and between 18 and 19 months for the four categories of complex social behaviors ($p < .05$ or $< .01$, according to the category).

Between 14 and 18 months forms of cooperation for which the acts of the two children are complementary in the realization of one activity are observed. For example :

1. at 14 months and 2 weeks a girl indicated (arm extended and finger pointing) a curtain, which she had been unable to unstick, after having looked at the face of a boy who had succeeded. This 15-month-old boy then unstuck the curtain ;

2. at 17 months and 2 weeks a boy indicated a cushion which he had placed on the floor after having carried it to the carousel. He was unable to lift the cushion to a height of 60 cm in order to tip it into the carousel. A boy of 18 months who was observing him then lifted the cushion and tipped it into the rourdabout.

The same type of interaction was observed at the porthole between two children of 17 months, where one had tried unsuccessfully to tip a cushion through the porthole.

Finally, we obtained data also on the temporal evolution of aggressive behaviors. Figure 10 shows that from 11 to 18 months (with the exception of the 13th month) the mean frequency of aggressive behaviors (blows dealt with or without an object, biting, hair-pulling) is 2 to 5 times less than the frequency of elementary social behaviors (offering, manual and postural solicitations, pointing). Comparison with figure 11 reveals that as of 11 months this is 2 to 5 times lower than the frequency of the interactions around the partition, imitations and cooperative behaviors.

If one summed up all episodes of the elementary and complex social behaviors, their frequency, at 11 months and every subsequent month, is significantly higher than the frequency of aggressive behaviors (the difference is significant at $p < .01$ at least).

3. Conclusions and discussion

The fundamental principle of our research is to create conditions which "liberate" the child's behavioral expression in a specific interactive context .

In the first study two infants of the same age were seated by their mothers in seats which secure a stable support of the infant's head and prevent any tipping or slaggling of the trunk. Consequently, as in the Grenier's study (1981), the visual attention, motor skills and interaction capacities appear "liberated".

In the second study, toddlers have the opportunity to move freely from the ages of 8 to 11 months with a group of peers in a specially equipped setting without other constraints than those which are linked to developmental processes. These ethological conditions "liberate" hidden skills and capacities which children improve when they spend time regularly in the setting.

Both studies reveal that young children show an unexpected wide range of skills. Their motor skills and social behaviors appear more precocious and complex than previously reported by scientific publications, clinical observations and developmental scales.

These phenomena appear to be based on the precocious, synchronized and interactive development of five types of abilities and drives :

The first of these is ability shown by children to develop, at least from the 4th month, a targeted and sustained visual attention with regard to the face, hands and gaze of another child.

In the first study with 4 to 7 month-old infants in a stable seated position, everything happens as if visual and relatively brief verification of the maternal presence from time to time (7% of the frequency and 8% of the duration of all the orientations of the gaze) is sufficient to secure the infants and "liberate" their abilities of visual exploration and attention in relation to an infant-partner whom they meet for the first time. The frequency and duration of the orientations of their gaze in the direction of the infant-partner represent respectively 69.72% and 64.78% of the frequency and duration of all the orientations of the gaze.

The mobility and sustained orientations of the gaze of the partner, as well as the mobility of its oral area (and also the hands : see further) probably play an essential role in the attraction of the gaze. The past and current interactions with the mother (and other familiar persons). notably during feeding, could play an important role in the "selection" of these behaviors and in the channeling of the visual attention of the infant. In fact, as has been underlined, the gaze and facial expressions, appear to be particularly efficient "filters" of information and emotions (or affects) in the mother-baby relationship (Brazelton, 1975,1980,1982; Brazelton et al.,1974,1975;Carpenter,1975;Tronick et al.,1978,1980,1982; Stern, 1985,1989).

In the second study, the visual attention of the toddlers in unrestricted activity in the specially equipped setting is solicited at all times by the moving about and activities of their peers especially along the holed partition and in the parallel corridor. The openings and porthole are used more and more often from the 9th month to observe the "outside" world or the corridor depending on which side the children find themselves, putting their head or body through and introducing balls or curtains.The partition is a device which gives each child the opportunity to focus his/her gaze and attention in relation to the other children and allows him/her to develop various and complex social behaviors.

A second type is the "driving force" which "pushes" the young child to interact with his/her peers. In the first study the two infants of 4months who have been placed by theirs mothers in experimental seats, behave as if they have a strong "driving force" to interact with each other. This is expressed not only by targeted and sustained gazes but also by coordinated movements of the arms and hands, projection of the bust and head and modulated vocalizations in the direction of the infant-partner, taking of his/her hand, and manual games with him/her.

In the second study, when children are in free activity, the "driving force"for the peers is shown by the high frequency of visual search for the other(s), calling, and interactions around the holed partition. It is also expressed by the activities reproduced by the peers :

1. following a toddler who is moving about on all fours or bouncing on the buttocks, walking, running, or climbing ;
2. initiating and accepting face-to-face situations ;
3. picking up, moving or manipulating an object after another child has done so ;
4. an abrupt change of behavior or activity after having observed or

heard another child ;

5. complex cooperative behaviors ;
6. anticipation of the behavior of another child.

The third type is the capacity of mastering as of the age of 4 months coordinated and targeted movements of the arms and hands in the direction of the infant-partner. In the first study one can stress in particular the stretching out of the arm in the direction of the partner, with the hand in supination or with the hand turning from pronation to supination with the movement of the arm. These behaviors have the same structure as those of the acts of manual solicitation which is observed in free activity, sometimes towards the end of the first year and usually during the second year (Montagner, 1978,1988 ; Montagner et al., 1988,1990).

These data confirm the hypothesis which we have put forward by defining behavioral organizers, that is to say the motor and/or vocal items which seem to constitute the cores around which or from which other items are "grafted", to "build up" structures which are more and more complex and better adpated to the communication processes and relation with an object (Montagner,1988 ; Montagner et al., 1988,1990). The visual information appear to play an essential role in the ontogeny of these processes. In particular, we have proposed as an example, the behaviors whose motor core appears to be the organized and visually adjusted stretching out of the arm in the direction of a person or object, with the hand in pronation, lateral or in supination. That is to say :

1. grasping an object ;
2. the behavior of manual solicitation (hand presented in pronation, in a lateral or oblique position, or in supination) ;
3. offering ;
4. the behavior of exploring the body of another infant ;
5. the conflict behaviors during which a stretching out of the hand in pronation, lateral or oblique, can be followed by pointing.

In the second study, the capacity of the child to master coordinated and targeted movements of the arms and hands allows him/her before the end of the first year, not only to display elementary social behaviors (offering, soliciting, pointing), but also complex social patterns. For example, as of 9 or 10 months visually guided stretching out of the arm appears to be the core of many interactions and exchanges of objects through the holes of the partition and the porthole. Between 14 and 18 months stretching out of the arm and pointing are basic patterns of cooperation for which the acts of two interacting children are complementary in the realization of one activity.

The fourth type is the preference shown by the young child for affiliative and cooperative behaviors with peers, during the first two years. In the conditions of our studies self-centered behaviors and those of aggressions are always infrequent and of short duration.

The fifth and final type consists of the self-learning skills and learning abilities with and by peers. In the first study the 4 to 7 month-old infants show remarkable skills to learn how to adjust their behaviors to those of the infant-partner, in particular during the second experimental session.

There is evidence for the first time that mutual learning processes do exist between two infants aged as of 4 months at least, without any intervention of or mediation by their mothers. They are influencing each other as independent persons with interaction and feedback as if they have a "driving force" to pick up the information from the infant-partner and combine this with the feedback due to their own behavior.

In the second study carried on the toddlers' behavior in free activity, self-learning skills also appear to be permanently combined with learning with and by others.

By walking, running, moving around the partition, climbing and descending etc., the children discover their capacities to control and coordinate their actions and rhythm better and better during locomotion and the succeeding activities.

By either accompanying or following their peers in their activities children acquire in their contact new motor schemes, activities, patterns and rhythms.

It is between 12 and 14-15 months that the children develop a group of skills which allow them to better control their new motor abilities and to better orientate or direct their behaviors with regard to their peers, in relation to the contexts and areas.

Between 14 and 18 months the children develop acute capacities of adjustment and control both in motor skills and systems of interaction with their peers. They appear capable of new acquisitions by combining experience gained from their initiatives and from information gained continuously from the behavior and activities of other children.

The self-learning and learning during the interactions with and by different partners appear, therefore, closely linked to the mastering of motor, interactive and cognitive processes (the characteristics, properties and functions of the objects are discovered in the different dimensions of space and through the interactions with peers).

3.1. FIRST STUDY : SPECIAL QUESTIONS AND COMMENTS

Our study of the interactive capacities of infants of 4 to 7 months with an infant of the same age shows that there already exists between the age of 4 and 7 months intentional gestures the target of which is a partner of the same age (and perhaps intentional vocalizations). One can question the underlying cognitive, emotional and relational processes of these phenomena. Are they "simply" patterns which show a visual and manual exploration, allowing each child to discover a new partner in a new situation? Are they organized attempts to activate or reactivate the attention and the behavior of the other infant? Are they attempts to test the efficiency and perhaps the meaning of its own behaviors, by inducing the responses of the other infant and reproducing its behaviors? Are they "tests" of its own emotions by a manifestation of behaviors and vocalizations which usually allow expression of an emotional state when the infant is in an interactive situation with its mother and other familiar people? Are these attempts to "test" the emotional reactivity and interactive abilities of the other infant? Is this a "test" of its piloting, imitation and attunement abilities with a new partner?

In any case, the results obtained lead to multiple interrogations as to the theoretical assumptions which make the relationship between the infant and its mother the only source of the child's affective, emotional and relational development, and those which limit the ontogeny and relations of the cognitive processes only to the scope of the knowledge of objects, in their physical, spatial and temporal characteristics. Our study shows that in fact when the infant's attention is not focalized on an event which is the giving of an object to him/her its interaction processes with an infant of the same age, develop at the same time on a basis of:

1. reciprocal visual attention ;
2. an intake of information;
3. repeated attempts to attract the attention of the other infant and induce its responses ;
4. the adjustment of its gaze and gestures with regard to the other infant and maintaining the interactive links.

3.2. SECOND STUDY: SPECIAL QUESTIONS AND COMMENTS

The study on the emergence and development of motor, interactive and social behaviors of children aged from 8 to 11 months at the beginning of the study puts new light on the ontogeny of the behaviors which feed the behavioral rules of the child with partners of the same age as if they were shared by all and as is they were valid absolutely and, therefore, as if they constituted one of the bases of human drives. It sheds light on which moment(s) of their development, and how, children proceed to orient, organize and regulate their behavior in relation to the context and behavior of peers, whatever the restraints inherent to their development. It shows how children behave to change the direction of, induce, and direct the attention to the behavior and activity of other children.

More generally, the influence of interactive and nonrelated partners in the development of the child's communication patterns, social behaviors and cognitive abilities, can be compared to the conclusions reached by Gardner, Gardner and Van Cantfort (1989) in their investigations on teaching sign language to chimpanzees : "The chimpanzees by definition can have no human language organ, and therefore their use of sign language signs in context and appropriately must result from the cognitive abilities of the chimpanzee plus consistent interaction with a human being using a linguistic system in the visual-gestural mode? The possibility then arises there is a continuum in nature - even primate nature : what the human infant also experiences between birth and later childhood may result from human (and primate) cognitive abilities plus interaction with others, and not from a genetic black box"

Our data are in agreement with the work of Hartup (1967,1970,1978,1983) who was the pioneer of the study of interaction systems between children. Hartup claimed that a group of peers has an essential influence on the social and affective development of the child. Consequently, it appears important to provide new settings where young children can discover and improve their hidden skills and capacities which are the foundations of more complex motor, communication and cognitive processes. At the same time, it is fundamentally better to understand the ontogeny of the processes of attachment and interaction of the young child with his/her mother and the others members of his/her family, who obviously play an essential role in the development of the child (Bowlby,1958,1960, 1969, 1973,1979 ; Ainsworth,1973,1974,1979 ; Brazelton, 1974,1975,1980, 1982; Tronick et al., 1978,1980,1982 ; Lebovici,1983 ; Stern,1985, 1989).

In fact, in the light of our studies, it appears necessary to develop combined studies of the motor, perceptual, affective, interactive and cognitive development of the child with regard to his/her human environment and also his/her physical and spatial environment.
As it was stated by Gardner and Gardner (1989) "It is a challenge for developmental psychologists to discover what aspects of rearing are influencial".

4. Summary

In this paper new data are reported on the earliness and complexity of the child's motor, interactive capacities and social behaviors with peers.
The first part deals with a research on the abilities and interactive behaviors of 16 infants aged from 4 to 7 months, with an infant of the same age. These infants are studied in dyads from the moment they are seated by their mother in two experimental seats. The position and distance of the seats are regulated at distance according to a protocole made up of 8 phases of 3 minutes each. The mothers of the two infants are present and can be seen by their infant during the experimental sessions.

The principal results can be summed as follows:
1. The infant-partner is the privileged target of the gaze of the other infant. The frequency and duration of the orientations of the gaze in the direction of one or another of its body areas are significantly higher than the frequency and duration of the self-centered gazes (very weak) and the frequency and duration of the orientations of the gaze in other directions, including towards their mothers. These behaviors between the infants do not decrease in frequency or duration as long as the infants do not receive an object.
2. The face and hands of the infant-partner are the most frequent targets of the gaze; they also have the highest cumulated duration. The mutual catching of the eyes is also well represented, since this occupies the fifth place in the 25 items of behavior which were measured.
3. The majority of the best represented items are the interactive behaviors with the other infant.
4.The manual behaviors already appear well organized, targeted, adjusted to the target and visually controlled.
5. Amongst the other complex behaviors one can observe diverse forms of imitation and modulated vocalizations which are rich in phonemes.
6. Most interactive behaviors are significantly influenced by the human context (the different positions of the infants with regard to each other) and giving an object to each infant (last phase of the protocol).
7. Infants have well-developed capacities to imitate each other.
8. They are able to adjust their behavior to the other infant's behavior in particular during the second experimental session, one week after the first session.

The second part ot the paper is devoted to a study on the emergence and development between the 9th and 19th months of motor, interactive and social behaviors of 17 children with peers in a milieu which allows the "conquest" of the third dimension of space without limitation of their motor activity. Their motor skills and social behaviors appear more precocious and complex than previously indicated by scientific publications, clinical observations and developmental scales.
1. The motor skills which allow the "conquest" of climbing structures (ascents of a spiral staircase and a ramp), are already observed before the age of 12 months in some children (ascents of the ramp) or nearly half of them (ascents of the staircase). Complex social behaviors (multimodal interactions from here and there along a holed partition; imitations; cooperations; anticipation of the behavior of the partner) also emerge before 12 months.
2. The age range from 12 to 14-15 months is characterized by consolidation of the motor skills which allow the "conquest" of the third dimension of the space, and a rapid increase of the imitations and interactions around the holed partition.

3. The age range from 14 to 18 months is characterized by diversification of the posturomotor patterns during ascents and descents of staircase and ramp and the growing predominance of standing position. This is also a phase of significant increase in frequency of imitations and cooperations, and of emergence of complex forms of cooperation.
4. The frequency of aggression is very weak at all ages.

Table I. Characteristics of the infants making up the 8 dyads of the sample;par.: parents.

Dyads	DURATION OF PREGNANCY	APGAR TEST AT 5 minutes	BIRTH WEIGHT	SIZE AT BIRTH	RANK ORDER IN THE FAMILY	TYPE OF FAMILY	AGE OF ASSISTED SITTING POS.	AGE OF NON-ASSISTED SITTING POS.
M. ♂ 21 months 21 days	39 weeks	10/10	2.9 kg	48 cm	2nd	both par.	5 months	7 months
P ♂ 4 months 9 days	40 weeks	10/10	3.6 kg	51 cm	2nd	both par.	3 months	5 months
J. ♂ 4 months 6 days	40 weeks	10/10	3.3 kg	51 cm	3rd	both par.	4 months	7 months
A ♀ 4 months 10 days	40 weeks	10/10	3.5 kg	51 cm	2nd	both par.	4 months	6 months
B. ♂ 6 months 5 days	37 weeks	10/10	2.7 kg	48.5 cm	1st	one par.	2 months	6 months
D. ♀ 6 months 16 days	37 weeks	10/10	2.9 kg	49 cm	2 nd	both par.	6 months	10 months
E. ♂ 5 months 28 days	40 weeks	10/10	3.2 kg	49 cm	2 nd	both par.	4 months	6 months
L. ♀ 6 months 2 days	38 weeks	10/10	3.0 kg	51 cm	1 st	both par.	4 months	7 months
J. ♂ 5 months 13 days	40 weeks	10/10	3.0 kg	49 cm	1 st	both par.	3 months	5 months
H. ♂ 6 months 6 days	40 weeks	10/10	2.4 kg	47 cm	1 st	both par.	4 months	6 months
G. ♀ 6 months 2 days	39 weeks	10/10	2.3 kg	46 cm	1 st	both par.	2.5 months	4.5 months
P. ♂ 5 months	40 weeks	9/10	3.2 kg	48 cm	2 nd	both par.	?	?
C. ♀ 4 months	40 weeks	10/10	3.7 kg	52 cm	1 st	both par.	?	6 months
O. ♀ 4 months 13 days	39 weeks	10/10	2.9 kg	47 cm	3 rd	both par.	3 months	7 months
V. ♀ 7 months 3 days	40 weeks	10/10	3.2 kg	47 cm	1 st	both par.	4 months	5 months
N. ♀ 6 months 1 day	40 weeks	10/10	3.9 kg	51 cm	1 st	both par.	?	7.5 months

Table II. Chronology and duration of the 8 procedure phases of each experimental session. The duration of Phases 4 and 5 is prolonged to 6 minutes when children are engaged in an active interaction without pause. The duration of Phase 8 is reduced to 1.50 or 2 minutes when one child shows signs of discomfort.

CHRONOLOGY OF THE PHASES	HUMAN CONTEXT AND ORIENTATIONS OF THE SEATS	DURATION OF THE PHASES
1	The two infants are face-to-face with their mother,without any intermediary object.	3 min.
2	The two infants are face-to-face with their mother who gives them 3 objects,one after the other. Each object for one minute.	3 min.
3	The mothers are replaced by two students. The conditions are the same as in Phase 2.	3 min.
4	The seats are brought together at a distance of 15 cm from each other. The two infants are at a right angle to each other.	3 or 6 min.
5	The seats are brought together until contact is made. The infants keep the same position as in 4 (Fig. 1B).	3 or 6 min.
6	The seats are face-to-face in contact with each other.	3 min.
7	The seats remain in the same position with the back lent forwards.	3 min.
8	The seats remain in the same position.Each infant receives an identical object.	3 min.

Table III. The 8 categories and 25 items of behavior which have been studied. POS.: positions ;MOV.: movements.

I. ORIENTATIONS OF THE GAZE	II. FACIAL EXPRESSIONS	III. VOCALIZATIONS	IV. MOVEMENTS OF THE BUST	V. TARGETED POS. AND MOV. OF THE ARMS	VI. CONTACTS WITH THE PARTNER	VII. GENERAL MOTOR MOV. OF THE LIMBS	VIII. SELF-DIRECTED BEHAVIORS
1.gaze oriented to the face of the other child 2.gaze oriented to the hands of the other child 3.gaze oriented to the feet of the other child 4.mutual catching of the eyes 5.self-directed gazes 6.gaze oriented in another direction including that of its mother	7.smiles 8.face pulling 9.movements of the oral area (-opening of the mouth ; -sucking ; -chewing ; -protrusion of the mouth and tongue)	10.modulated vocalizations 11.brief vocalizations (-shrieking ; exclamations) 12.groaning and grumbling	13.leaning the bust forward in the direction of the other child	Arm stretched out in the direction of the other child: 14.with the hand in pronation ; 15.with the hand lateral or oblique ; 16.with the hand in supination . Arm flexed in the direction of the other child: 17.with the hand in pronation ; 18.with the hand lateral or oblique ; 19.with the hand in supination .	20.hand brought into contact with the other child . 21.foot brought into contact with the other child . 22.take the hand of the other child .	23.movements of the arms and hands . 24.movements of the legs feet .	25.:-finger in the mouth,nose or ear ; -to take hold or rub its hands or feet ; -scratching -to put an object in its mouth .

A

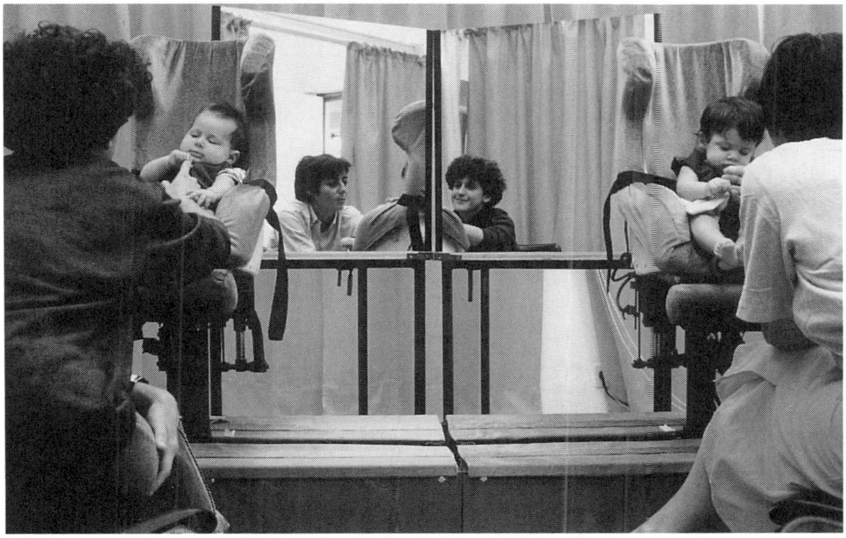

Fig.1 Two phases of an experimental session : A (phase 1) : the two infants are face-to-face with their mothers without any intermediary object . B (phase 5) : the seats are brought together until contact is made (the infants are at right angle to each other).

B

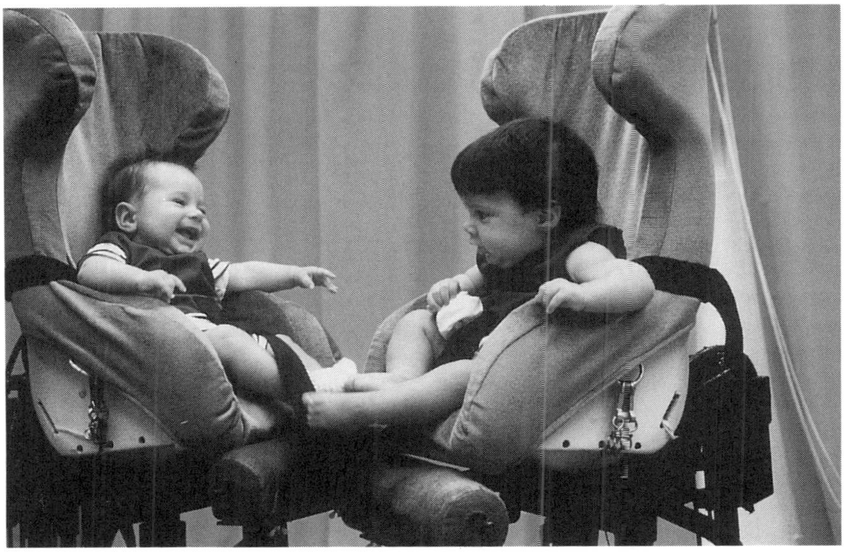

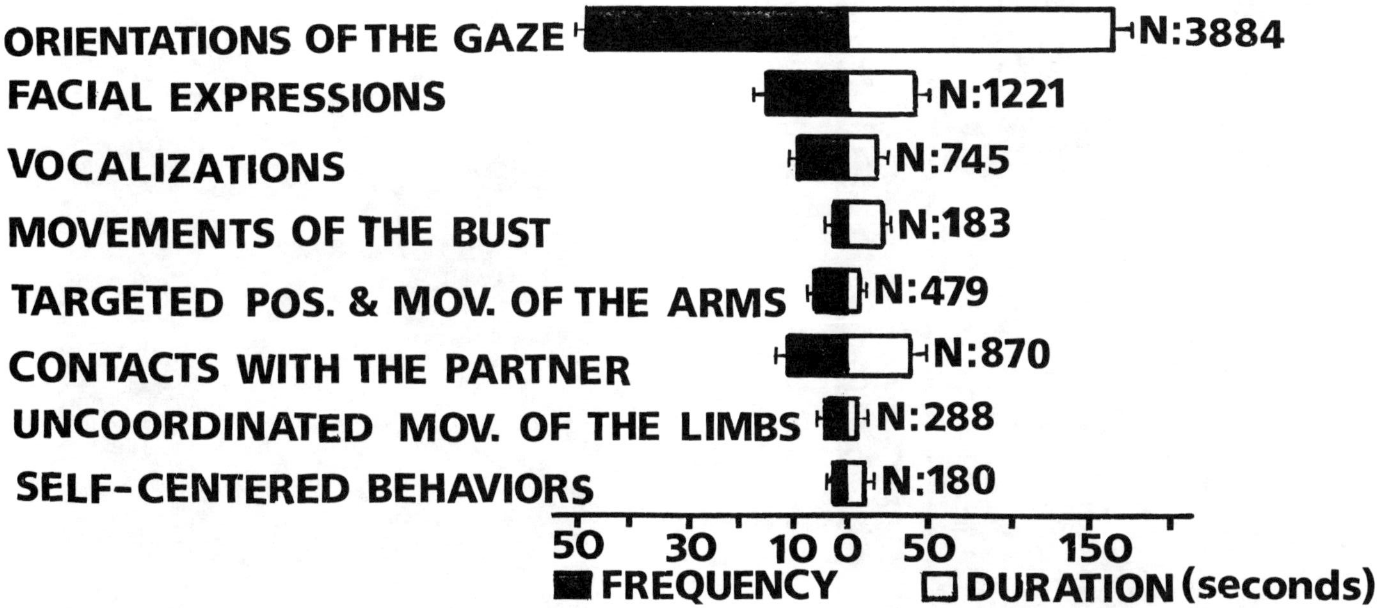

Fig.2. MEAN FREQUENCY AND DURATION OF THE 8 CATEGORIES OF BEHAVIORS (see Table III) FOR ALL THE EXPERIMENTAL SESSIONS, THE DATA OF THE 16 INFANTS CUMULATED.N = TOTAL NUMBER OF OCCURRENCES. EACH MEAN IS GIVEN WITH ITS STANDARD ERROR. CATEGORY I (ORIENTATIONS OF THE GAZE) DOES NOT INCLUDE THE ORIENTATIONS OF THE GAZE IN THE DIRECTION OF THE OBJECT GIVEN TO EACH CHILD AT THE BEGINNING OF PHASE 8.

ORIENTATIONS OF THE GAZE

1. gaze oriented to the face of the other child. N = 1130.
2. gaze oriented to the hands of the other child. N = 705.
3. gaze oriented to the feet of the other child. N = 239.
4. mutual catching of the eyes. N = 634.
5. self-directed gazes. N = 246.
6. gaze oriented in another direction . N = 930.

FACIAL EXPRESSIONS

7. smiles. N = 193.
8. face pulling N = 185.
9. movements of the oral area. N = 837.

VOCALIZATIONS

10. modulated vocalizations. N = 405.
11. brief vocalizations. N = 185.
12. grumbling and moaning. N = 155.

MOVEMENTS OF THE BUST

13. leaning the bust forward in the direction of the other child. N = 183.

TARGETED POSITIONS AND MOVEMENTS OF THE ARMS

Arm stretched out in the direction of the other child:

14. with the hand in pronation ; N = 175.
15. with the hand lateral or oblique ; N = 153
16. with the hand in supination . N = 26.

Arm flexed in the direction of the other child:

17. with the hand in pronation ; N = 52.
18. with the hand lateral or oblique ; N = 44
19. with the hand in supination . N = 29.

CONTACTS WITH THE PARTNER

20. hand brought into contact with the other child . N = 266.
21. foot brought into contact with the other child . N = 528.
22. take the hand of the other child . N = 76.

UNCOORDINATED MOVEMENTS OF THE LIMB

23. movements of the arms and hands . N = 199.
24. movements of the feet and legs . N = 89.

SELF-CENTERED BEHAVIORS N = 180.

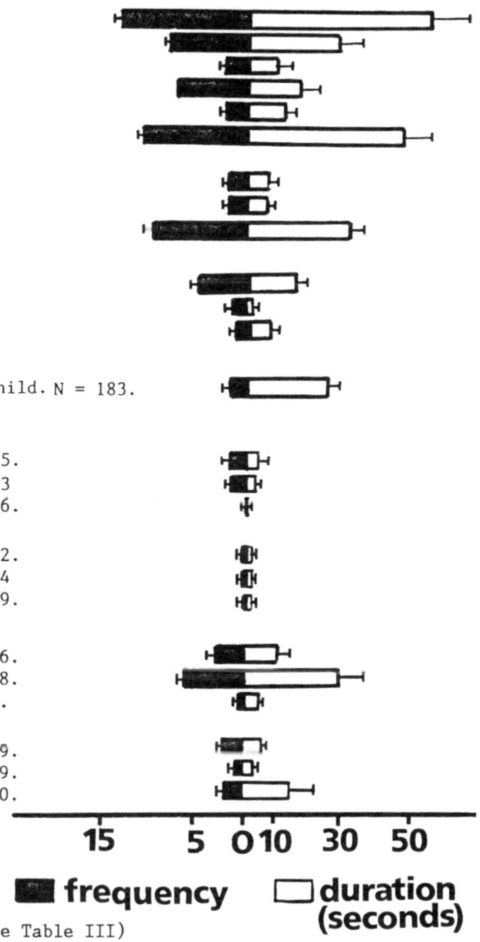

Fig.3. MEAN FREQUENCY AND DURATION OF THE 25 ITEMS OF BEHAVIOR (see Table III) FOR ALL THE EXPERIMENTAL SESSIONS, THE DATA OF THE 16 INFANTS CUMULATED. N = TOTAL NUMBER OF OCCURRENCES. EACH MEAN IS GIVEN WITH ITS STANDARD ERROR.

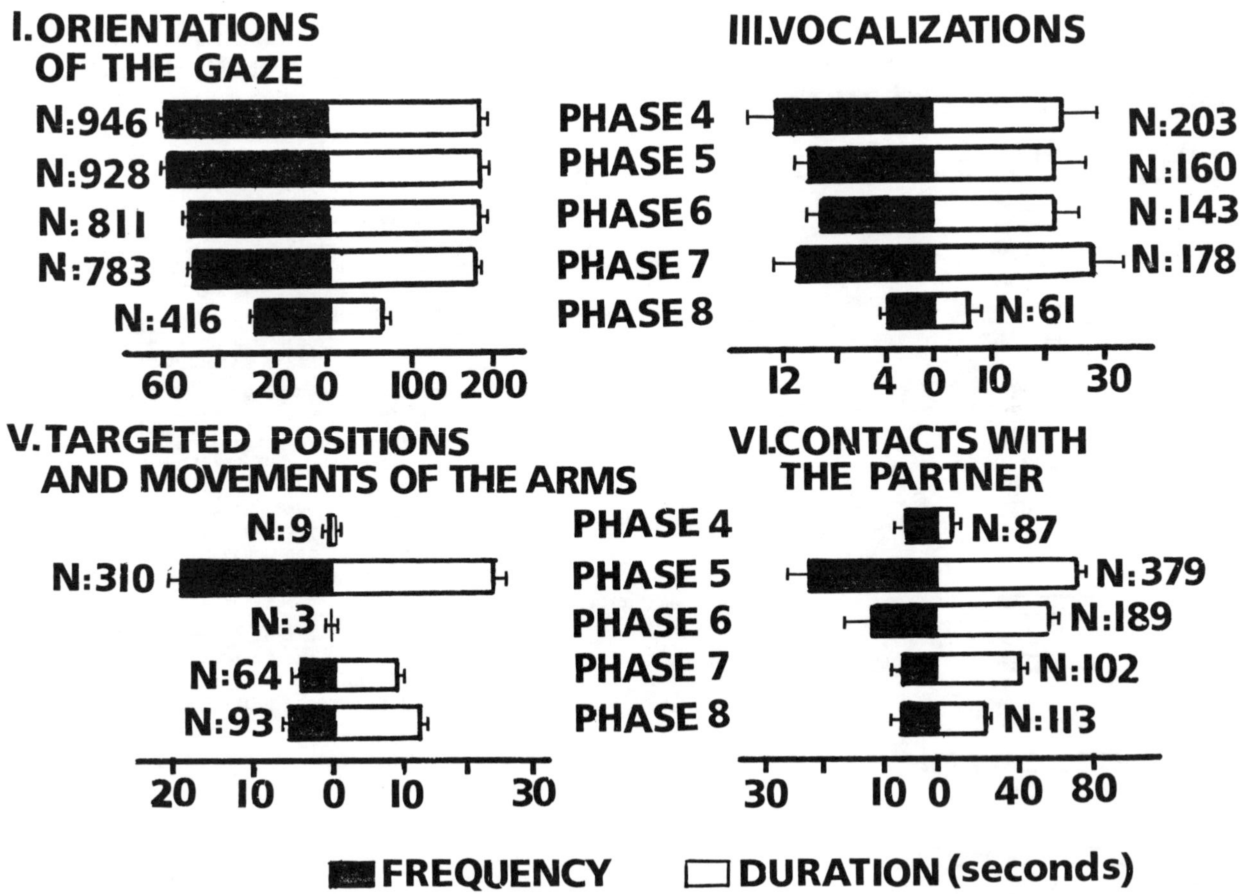

Fig.4. COMPARISON BETWEEN THE 5 PHASES OF THE INFANT-INFANT INTERACTIONS WITH REGARD TO THE MEAN FREQUENCY AND DURATION OF ALL THE ITEMS BELONGING TO BEHAVIORAL CATEGORIES I,III, V AND VI,ALL DATA OF THE 16 INFANTS CUMULATED. EACH MEAN IS GIVEN WITH ITS STANDARD ERROR.

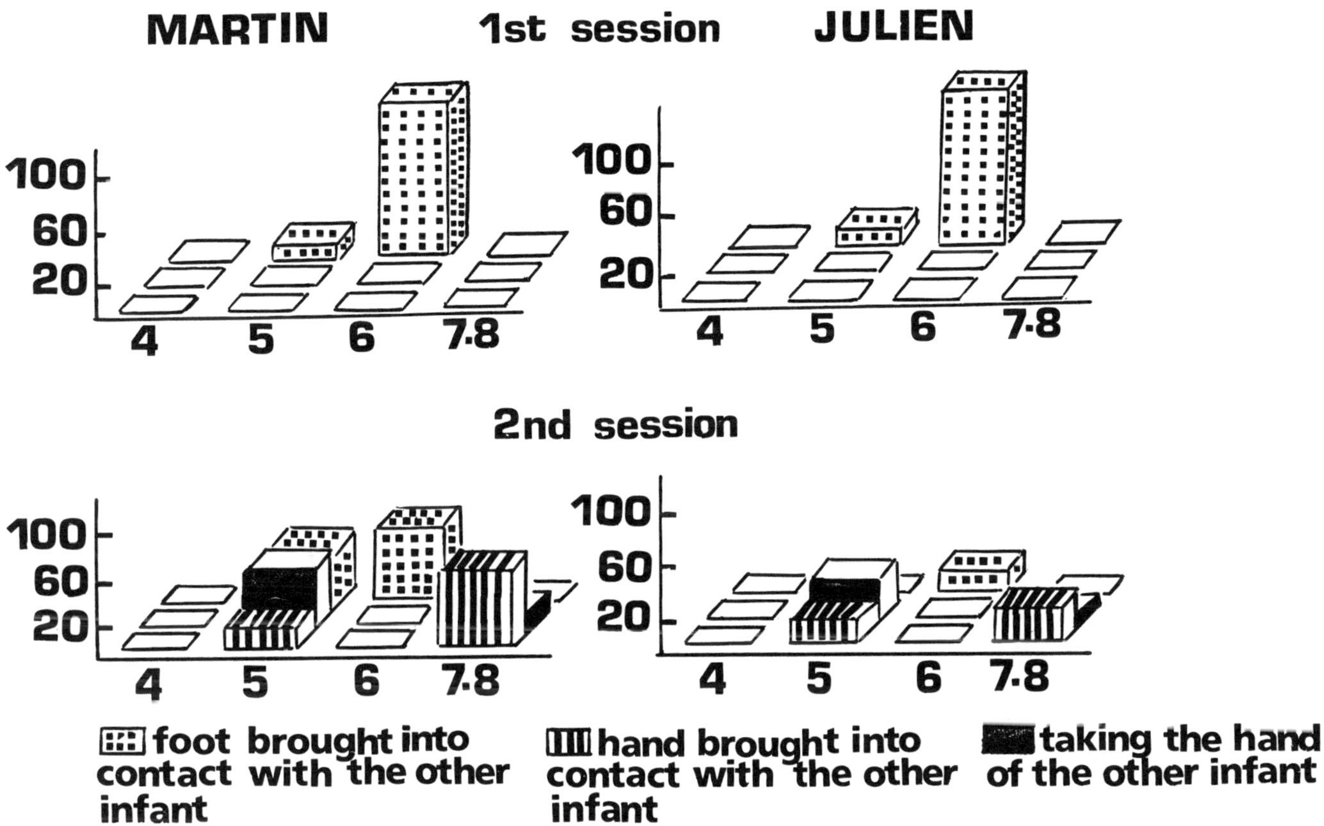

Fig. 5. NUMBER OF OCCURRENCES IN TWO 4 MONTH-OLD INFANTS (MARTIN AND JULIEN) OF THE THREE ITEMS BELONGING TO BEHAVIORAL CATEGORY VI (CONTACTS WITH THE PARTNER) FROM PHASE 4 TROUGH THE CUMULATED PHASES 7 AND 8 DURING THE FIRST AND SECOND EXPERIMENTAL SESSIONS AT A ONE WEEK INTERVAL.

Fig.6 General (A) and detailed views (B to F) of the specially equipped setting in which children are continuously filmed every week or two weeks throughout one or two years during sessions of 30 minutes each . O.S. (A) : open space where children can crawl,move about on all fours, walk or run ; H.P. (A,D,E,F): holed partition with a porthole in the middle of the central panel and openings of various forms and sizes on the two lateral parts ; Co (A,F) : corridor separating the partition from the superstructure Su (on the top two cavities are filled with plastic balls) ; R (A) : carousel ; S.S. (A,C,D) : spiral staircase allowing access to a platform with two "nests" ; Ra (A,B) : behind the railings of the superstructure a ramp allows access to the cavities filled with plastic balls ; C.W. (D,F) : climbing wall ; R.L. (A) : rope ladder.

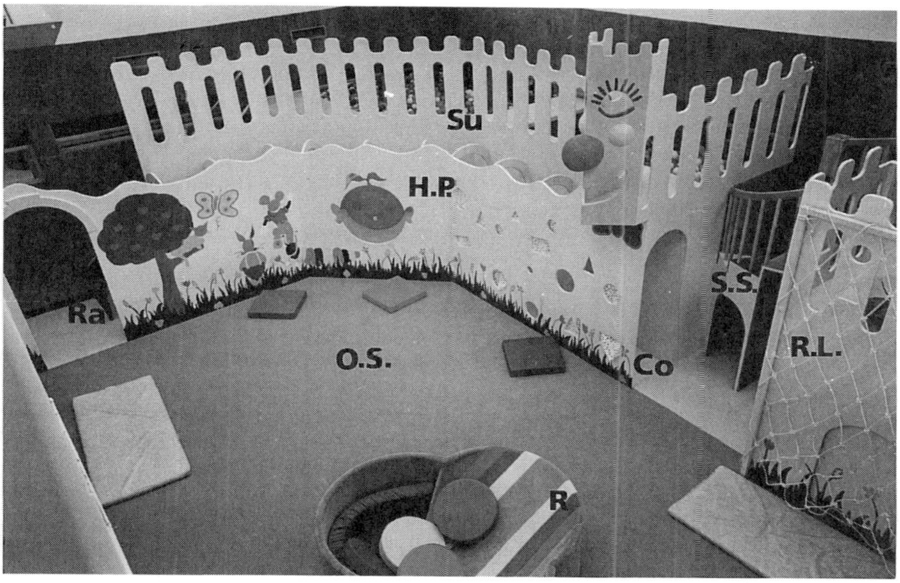

A

B(Ra)

C (s.s.)

D

E

F

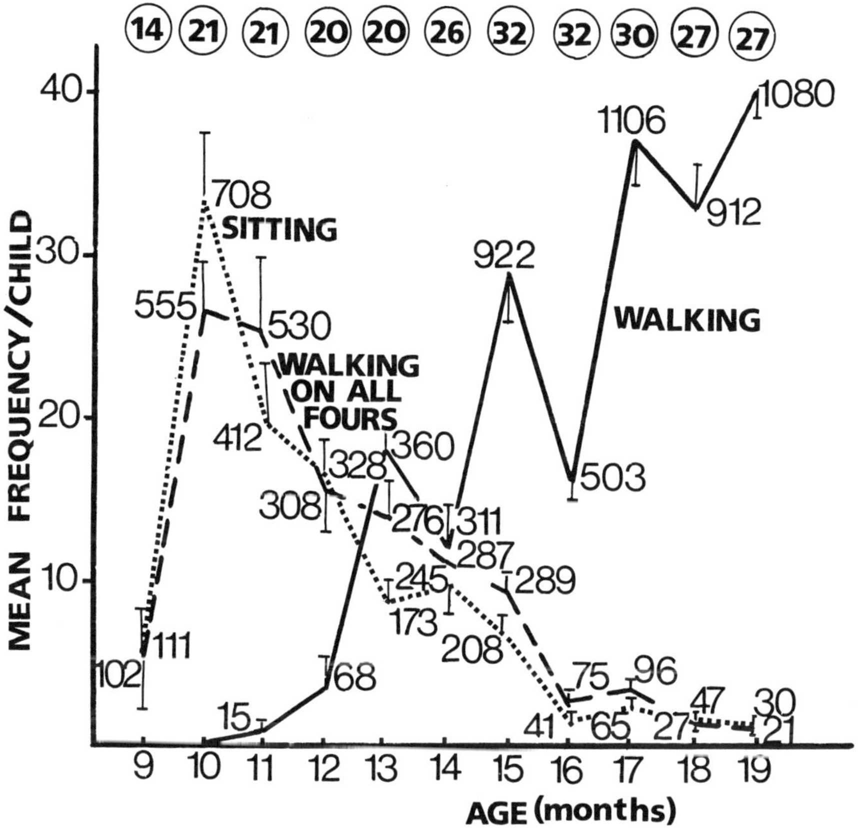

Fig.7. MEAN FREQUENCY (per child) OF SELF-INITIATED SITTING, WALKING ON ALL FOURS AND WALKING OF THE 17 CHILDREN OF THE SAMPLE BETWEEN THE 9th MONTH (9 : DATA COLLECTED BETWEEN THE AGE OF 8 MONTHS AND 8 DAYS AND 9 MONTHS AND 7 DAYS) AND THE 19th MONTH (19 : DATA COLLECTED BETWEEN THE AGE OF 18 MONTHS AND 8 DAYS AND 19 MONTHS AND 7 DAYS). EACH MEAN IS CALCULATED BY DIVIDING THE NUMBER OF OCCURRENCES OBSERVED EACH MONTH FOR EACH PATTERN (FIGURES ACCORDING TO EACH MEAN) BY THE NUMBER OF CHILDREN AND SESSIONS AT EACH MONTH (CIRCLED FIGURES AT THE TOP). THE MEANS ARE REPRESENTED WITH THE STANDARD ERROR.

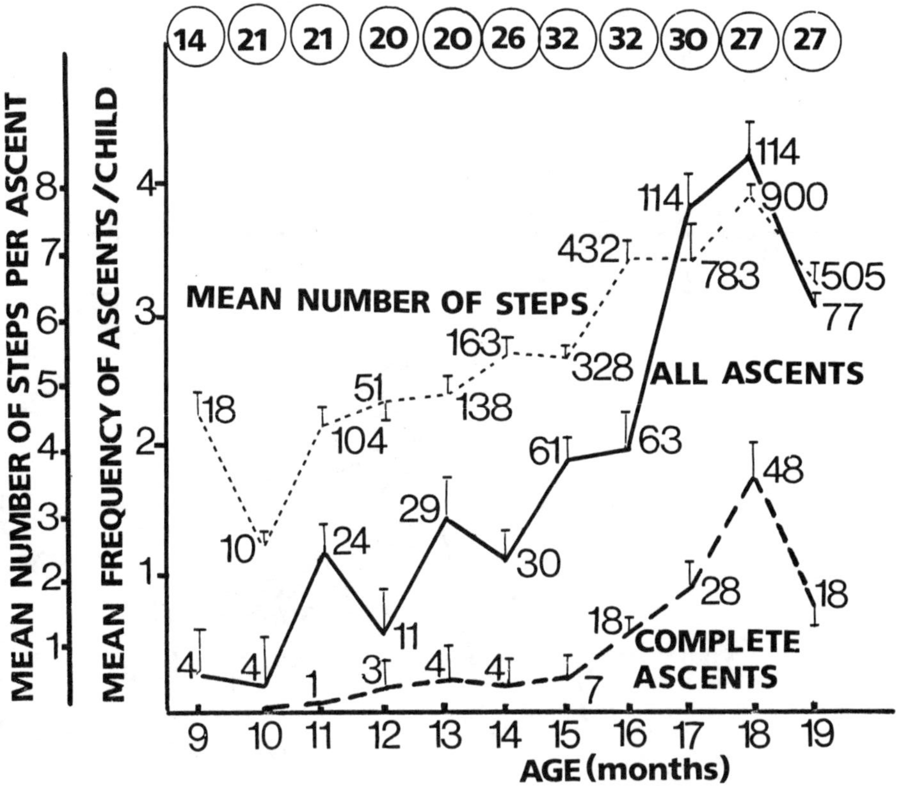

Fig. 8. EVOLUTION BETWEEN THE 9th AND 19th MONTHS OF THE MEAN NUMBER OF STEPS OF THE SPIRAL STAIRCASE WHICH ARE CLIMBED, PER ASCENT, BY THE 17 CHILDREN OF THE SAMPLE, OF THE MEAN FREQUENCY (per child) OF ALL THE ASCENTS (the partial ascents of a minimum of 2 steps and the "complete" ascents, that is to say from 10 to 13 steps) AND THE "COMPLETE" ASCENTS (we have summed up the ascents from the 10th step, that is to say when the children can stand up by holding the banister of the staircase and at the same time discover the platform and the two nursery nurses at the bottom).

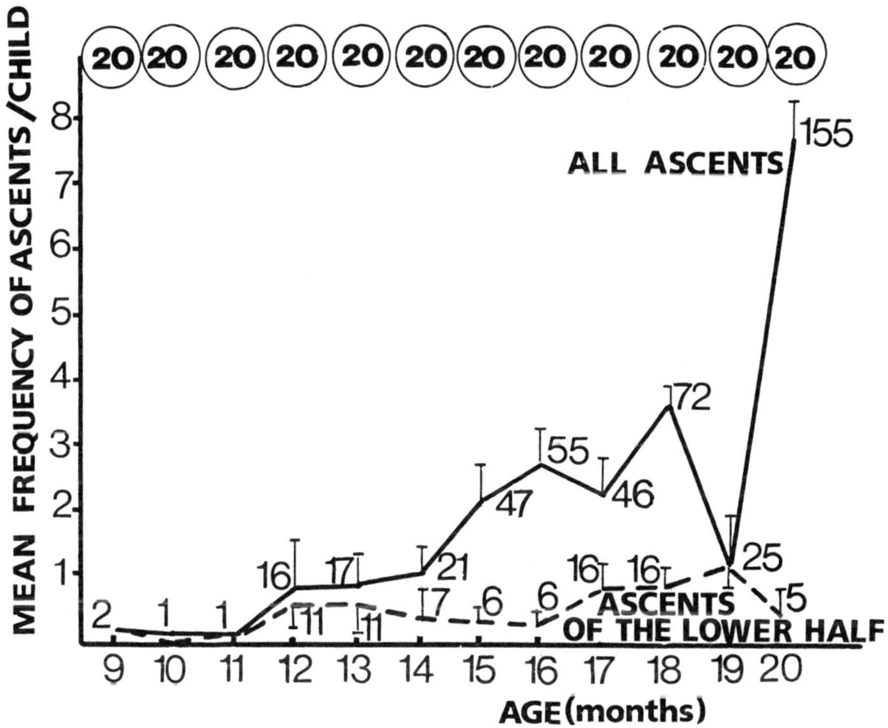

Fig.9. EVOLUTION BETWEEN THE 9th AND 20th MONTHS OF THE MEAN FREQUENCY (per child) OF THE ASCENTS OF THE LOWER HALF AND OF ALL THE ASCENTS (partial and complete) OF THE RAMP, BY THE 17 CHILDREN OF THE SAMPLE. THE NUMBER OF "CHILDREN-SESSIONS" WAS MADE UP TO 20 FOR TECHNICAL REASONS (unusable video-cassettes, non-functioning of the camera covering the ascent of the ramp during several sessions). THE RESULTS ARE GIVEN FOR THE 20th MONTH TO SHOW THE EVOLUTION OF FREQUENCY OF THE ASCENTS AT THIS AGE COMPARED TO THE PRECEDING MONTHS.

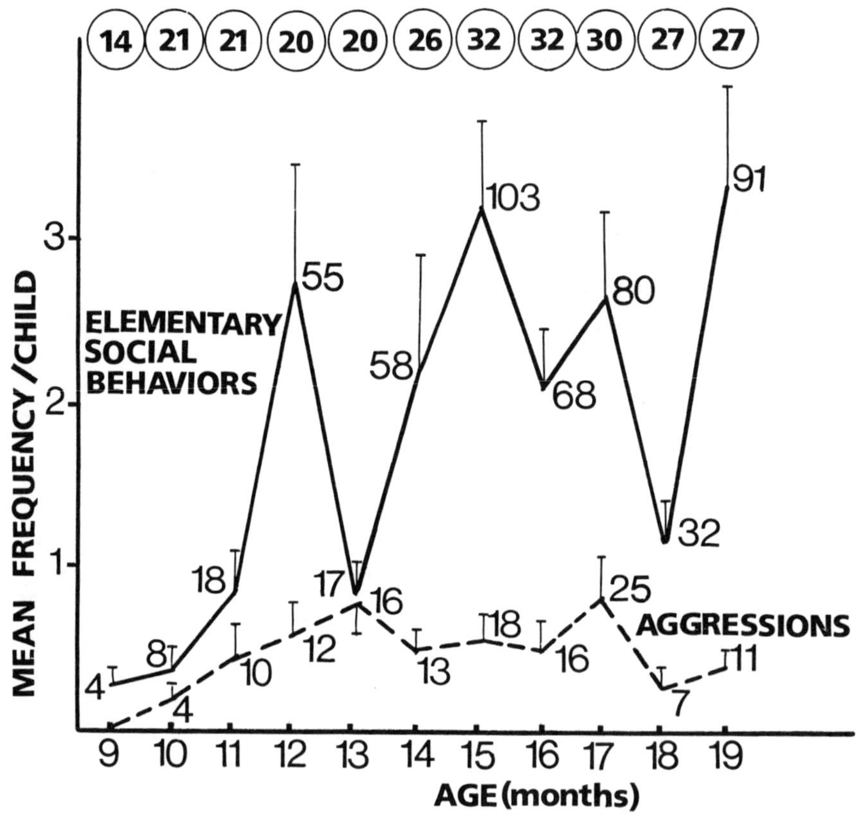

Fig.10. EVOLUTION BETWEEN THE 9th AND 19th MONTHS OF THE MEAN FREQUENCY (per child) OF THE 4 ELEMENTARY SOCIAL BEHAVIORS (offering, manual and postural solicitations, finger pointing) AND AGGRESSIONS (blows dealt with or without an object, biting, hair-pulling) IN THE 17 CHILDREN OF THE SAMPLE.

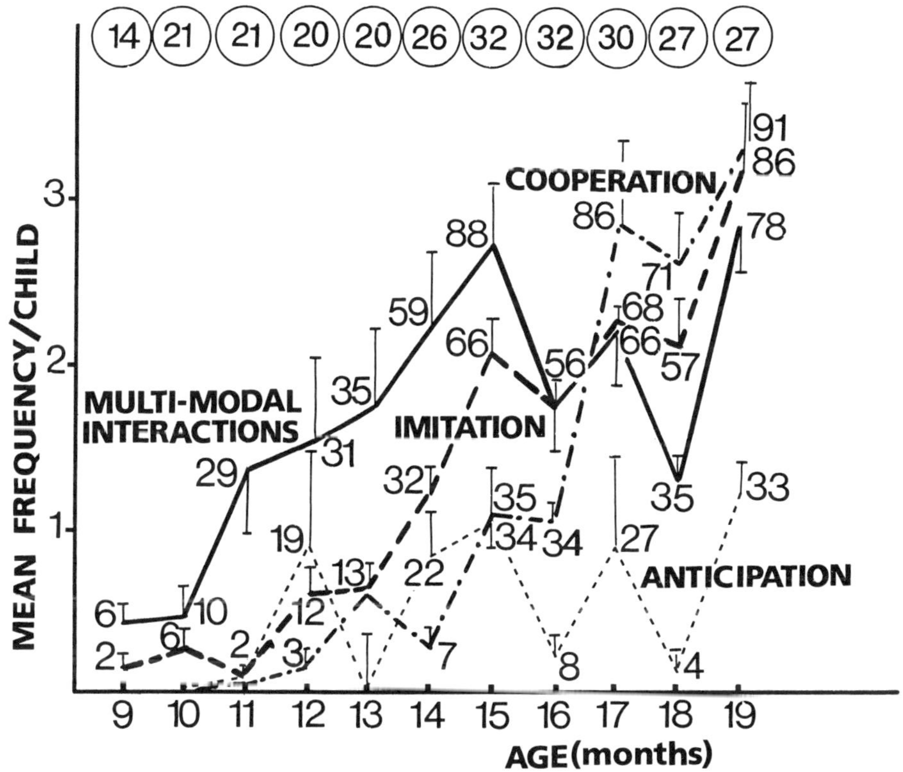

Fig.11. EVOLUTION BETWEEN THE 9th MONTH AND 19th MONTH OF THE MEAN FREQUENCY (per child) OF THE 4 COMPLEX SOCIAL BEHAVIORS (MULTI-MODAL INTERACTIONS FROM HERE AND THERE ALONG THE HOLED PARTITION ; IMITATIONS ; COOPERATIONS ; BEHAVIORS OF ANTICIPATION OF THE BEHAVIOR OF THE CHILD-PARTNER)IN THE 17 CHILDREN OF THE SAMPLE.

BIBLIOGRAPHY

AINSWORTH, M.D.S., 1973 The development of infant-mother attachment. In B.M. CALDWELL and H.N. RICCIUTI (eds), Review of child development research, 3, 1-94, Chicago : University of Chicago Press.

AINSWORTH, M.D.S., 1974 The secure base, New-York:Johns Hopkins University.

AINSWORTH, M.D.S., 1979 Attachment as related to mother-infant interaction, In J.S. ROSENBLATT, R.A. HINDE, C.BEER and M.C. BUSNEL (eds), Advances in the study of behavior , 9, New-York and london:Acad.Press.

AINSWORTH, M.D.S., 1979 Infant-mother attachment, American Psychologist, 34, 932-937.

BAKEMAN, R. and ADAMSON, L.B., 1984 Coordination attention to people and objects in mother-infant and peer-infant interactions, Child development, 55, 1278-1289.

BLURTON JONES, N.G., 1972 Ethological studies of child behavior, London: Cambridge University Press.

BRAZELTON, T.B., 1975 Early mother-infant reciprocity in parent-infant interaction, Amsterdam:Elsevier.

BRAZELTON, T.B., 1980 Neonatal assessment, In S.I.GREENSPAN and G.POLLOCK (eds), The course of life, I, 780-786, U.S. Department of health and human services, DHHS Publications.

BRAZELTON, T.B., 1982 Joint regulation of neonate-parent behavior, In E. Z. TRONICK (ed), Social interchange in infancy , 7-23 , Baltimore : University Park Press.

BRAZELTON, T.B., KOSLOWSKI, B. and MAIN, M., 1974 The origins of reciprocity : The early mother-infant interaction, In M. LEWIS and L.A. ROSENBLUM (eds), The effects of the infant on its caregiver, 49-76, New-York and London:Wiley and sons.

BRAZELTON,T.B., TRONICK, E., ADAMSON, L., ALS, H. and WISE,S., 1975 Early mother-infant reciprocity, In Parent-infant interaction, Ciba Foundation Symposium, 33, 137-154, Amsterdam:ASP Elsevier.

BRONSON, W.C., 1981 Toddlers' behavior with agemates : Issues of interaction, cognition and affect, Norwood (NJ):Alex Publ. Corp.

CARPENTER, G., 1975 Mother's face and the newborn, In R. LEWIN (ed), Child alive, London:Temple Smith.

DE STEPHANO , C.T. and MUELLER, E., 1982 Environmental determinants of peer social activity in 18-month-old males , Infant behavior and development, 5, 175-183.

DUBON, D.C., JOSSE,D. and LEZINE,I., 1981 Evolution des échanges entre enfants au cours des deux premières années de la vie, Neuropsychiatrie de l'enfance, 29, 273-290.

ECKERMAN,C.O., WHATLEY,L.J. and KUTZ,S.L., 1975 Growth of social play with peers during the second year of life, Developmental Psychology 11, 42-49.

ECKERMAN,C.O. and WATHLEY,L.J., 1977 Toys and social interaction between infant peers, Child development, 48, 1645-1656.

GARDNER, R.A. and GARDNER, B.T., 1989 Prelinguistic development of children and chimpanzees, Human evolution, 4, 433-460.

GARDNER, R.A., GARDNER B.T. and VAN CANTFORT,T.E., 1989 Teaching sign language to chimpanzees, Albany (N.Y.):State University of New-York Press.

GRENIER,A., 1981 La motricité libérée par fixation manuelle de la nuque au cours des premières années de vie, Arch. Fr. Péciat.,38, 557-561.

HARTUP, W.W., 1970 Peer interaction and social organization , In P.H. MUSSEN (ed), Carmichael's manual of child psychology, 2, 253 - 268, New-York:Wiley.

HARTUP,W.W., GLAZER,J.A. and CHARLESWORTH,R.,1967 Peer reinforcement and sociometric status, Child development, 38, 1017-1024.

HOWES,C., 1980 Peer play scale as an index of complexity of peer interaction, Developmental psychology, 16, 371-372.

HOWES,C., 1987 Social competency with peers : Contributions from child care, Early childhood research quaterly, 2, 155-167.

HOWES,C. and RUBENSTEIN,J.L., 1981 Toddler peer behavior in two types of day care, Infant behavior and development, 4, 387-393.

JACOBSON,J.L., 1981 The role of inanimate objects in early peer interaction, Child development, 52, 618-626.

LEBOVICI, S., 1983 Le nourrisson, la mère et le psychanalyste, Paris:Le Centurion.

McGREW,W.C., 1972 An ethological study of children's behavior, New-York and London.

MONTAGNER,H., 1978 L'enfant et la communication, Paris: Stock/Pernoud.

MONTAGNER,H., 1988 L'attachement, les débuts de la tendresse, Paris:Ed. Odile Jacob.

MONTAGNER,H., 1993 L'enfant acteur de son développement, Paris: Stock/ Pernoud.

MONTAGNER,H., RESTOIN,A., RODRIGUEZ,D., ULLMANN,V., VIALA, M. and LAURENT,D., 1988 Aspects fonctionnels et ontogénétiques des interactions de l'enfant avec ses pairs au cours des trois premières années, Psychiatrie de l'enfant, 31, 173-278.

MONTAGNER,H., RESTOIN,A., RODRIGUEZ D., ULLMANN,V., VIALA,M.and LAURENT D., 1988 Social interactions of young children and their modifications in relation to environmental factors, In M.R.A. CHANCE (ed), Social fabrics of the mind, 237-259, London and Hillsdale:Erlbaum.

MONTAGNER,H.,MAGNUSSON,M.,CASAGRANDE,C., RESTOIN,A., BEL,J.P., NGUYEN M.HOANG,P., RUIZ,V.,DELCOURT,S., GAUFFIER,G. and EPOULET,B., 1990 Une nouvelle méthode pour l'étude des organisateurs de comportements et des systèmes d'interaction du jeune enfant. Les premières données, Psychiatrie de l'enfant, 33, 391-456.

MONTAGNER,H., RESTOIN,A. and HENRY,J.C., 1982 Biological defence rhythms, stress and communication behavior, In W.W. HARTUP (ed.), Review of child development research, 6, 291-319, Chicago : University of Chicago Press.

MONTAGNER,H., RESTOIN,A.,ULLMANN,V., RODRIGUEZ,D., GODARD,D. and VIALA,M. 1984 Development of early peer interaction, In W. DOISE and A. PALMONARI (eds), Social interaction in individual development, 25 - 41, Cambridge and London:Cambridge University Press.

MONTAGNER,H.,GAUFFIER,G.,EPOULET,B., RESTOIN,A., GOULEVITCH,R., TAULE, M. and WIAUX,B., 1993 Alternative child care in France : Advances in the study of motor, interactive, and social behaviors of young children in settings allowing them to move freely in a group of peers, Pediatrics, 91, 253-263.

MONTAGNER,H., RUIZ,V., RAMEL,N., RESTOIN,A., MERTZIANIDOU,V. and GAUFFIER G., 1993 Les capacités interactives d'enfants de 4 à 7 mois avec un enfant du même âge, Psychiatrie de l'enfant, 36, 489-536.

MUELLER,E., 1979 Toddlers and toys : An autonomous social system, In M. LEWIS and L.A. ROSENBLUM (eds), The child and its family, New-York: Plenum.

MUELLER,E. and BRENNER,J., 1977 The origins of social skills and interaction among playgroup toddlers, Child development, 48, 854-861.

MUELLER,E. and LUCAS,T.A., 1975 A developmental analysis of peer interactions among toddlers, In M.LEWIS and L.ROSENBLUM (eds) , Friendship and peer relations, 223-258, New-York:Wiley.

MUSATTI,T., 1986 Early peer relations : The perspective of Piaget and Vygotski, In E.MUELLER and C.COOPER (eds), Process and outcome in peer relationships, 25-54, New-York:Acad.Press.

MUSATTI,T. and PANNI,S., 1981 Social behavior and interaction among daycare center toddlers, Early child development and care, 7, 5-27.

RESTOIN,A., MONTAGNER,H., RODRIGUEZ,D., GIRARDOT,J.J., ULLMANN,V., CASAGRANDE,C. and TALPAIN,B., 1983 Données nouvelles sur le développement chronologique des systèmes de communication du jeune enfant avec ses pairs, Recherches de psychologie sociale, 5, 31-56.

RESTOIN,A., MONTAGNER,H., RODRIGUEZ,D., GIRARDOT,J.J., LAURENT,D., KONTAR F., ULLMANN ,V., CASAGRANDE,C. and TALPAIN,B., 1985 Chronologie des comportements de communication et profils de comportement chez le jeune enfant, In R.E. TREMBLAY, M.A. PROVOST and F.F. STRAYER (eds), Ethologie et développement de l'enfant, 93-130, Paris:Stock/Pernoud.

SCHNEIDER, B.H., ATTILI,G., NADEL, J., and WEISSBERG,R.P., 1989 Social competence in developmental perspective, 51, Series D : Behavioural and social sciences, NATO ASI Series, Dordrecht, Boston , London : Kluwer Acad. Publ.

SIEGEL,S., 1956 Non-parametric statistics, New-York and London:Mc Graw Hill.

SPITZ,R.A., 1968 De la naissance à la parole, Paris:P.U.F.

STAMBAK,M., BONICA,L., MAISONNET,R., RAYNA,S. and VERBA,M., 1979 Modalités d'échanges entre enfants de moins de deux ans, Cahiers du C.R. E.S.A.S., 19, 3-103.

STERN,D., 1985 Affect attunement, In J.D. CALL, E.GALENSON and R.L. TYSON (eds), Frontiers in infant psychiatry, 2, New-York:Basic Books.

STERN,D., 1989 Le monde interpersonnel du nourrisson. Une perspective psychanalytique et développementale, Paris : P.U.F.

TRONICK,E., ALS,H., ADAMSON, L., WISE,S. and BRAZELTON,T.B., 1978 The infant's response to entrapment between contradictory messages in face-to-face interaction, Journal of the american academy of child psychiatry, 17, 1-13.

TRONICK,E., ALS, H. and BRAZELTON,T.B., 1980 Monadic phases : A structural descriptive analysis of infant-mother face-to-face interaction Merril-Palmer Quaterly, 26, 3-23.

TRONICK,E.,RICKS,M. and COHN,J., 1982 Maternal and infant affective exchanges : Patterns of adaptation,In T.FIELD and A.VOGEL(eds),Emotion and interaction: Normal and high risk children, 83 - 101, Hillsdale: Erlbaum.

VANDELL,D., 1980 Peer interaction in the first year of life, Child development, 51, 481-488

VANDELL,D. and MUELLER,E., 1980 Peer play and friendships during the first two years, In H.C. FOOT, A.J. CHAPMAN and J.R. SMITH (eds) , Friendship during the first two years, New-York:Wiley.

VERBA,M. and MUSATTI,T., 1989 Minor phenomena and major process of interaction with objects and peers in day-care centres,European journal of psychology of education, 4, 215-227.

VINCZE,M., 1971 The social contacts of infants and young children reared together, Early child development and care, 1, 99-109.

LEARNING BY INSTINCTS, DEVELOPMENTAL TRANSITIONS, AND THE ROOTS OF CULTURE IN INFANCY

FRANS X. PLOOIJ and HETTY H.C. VAN DE RIJT-PLOOIJ

Department of Psychology
University of Groningen
Grote Kruisstraat 2/1
9712 TS Groningen
The Netherlands

Department of Pedagogical Sciences
University of Amsterdam
IJsbaanpad 9
1076 CV Amsterdam
The Netherlands

KEYWORDS/ABSTRACT : Learning instincts / Developmental transitions / Mother-infant relationships / Roots of culture / Infancy / Chimpanzee / Man.
Knowledge about the continuity in the ontogenetic processes underlying behaviour in apes is needed in the discussion of culture in non-human species. Both free-living chimpanzee and human infants were studied. 'Perception Control Theory' depicts the build-up over development of a non-linear hierarchy of negative feedback control systems. Each newly emerging level goes together with a new type of perception, learning instinct and consummatory stimuli. As a consequence, infantile regression periods are found, followed by mother-infant conflict and a developmental transition to greater independence. A zone of next development is entered and in ritualized exchanges with the mother the roots of culture are formed.

1. Introduction

The thesis presented in this paper is the following:
During human infancy, the roots of culture are shaped both by a number of learning instincts and by the elaboration of the developmental transitions that are brought about by those learning instincts.

Learning instincts constrain the learning, and thus the culture, within certain limits. They determine the types of things or events the organism pays attention to, or not. With the emergence of each new learning instinct a new 'type of learning' comes into being. It is only fair that species who depend heavily on learning in their life span strategy, at least know what to attend to and from what age onwards (Gould & Marler, 1987). Otherwise they cannot possibly make order out of chaos. We believe these instincts are universal, at least for man, and determine the so-called 'zone of next development' at various ages during infancy.

The other shaping force of culture is the way in which this zone is entered and occupied. This varies greatly and is dependent on the efforts of the developing organism on the one hand and the opportunities it gets or is given on the other hand. These opportunities can be influenced to a large extent by the parent(s). They have an educational task. This causes the immense indivi-

R. A. Gardner et al. (eds.), The Ethological Roots of Culture, 357–373.

dual and cultural differences observed all over the world.

This thesis is going to be illustrated and elaborated by presenting the research we have carried out over the past twenty years on both free-living chimpanzee and human infants. The chimpanzee research is reported as well, because, first of all it inspired the human research and, second, it shows there is quite some continuity in the ontogenetic processes underlying behaviour in both chimpanzees and man. And it is this kind of knowledge that is lacking according to the researchers who are critical of the notion of culture in non-human species (Galef, 1992).

2. Chimpanzees

2.1. Study one

Originally, in 1971, the first author set out to study the ontogenetic development of the organization underlying behaviour in the free-living chimpanzee infants in the Gombe National Park, Tanzania, East Africa. Van Hooff (1971) had just published his dissertation including a structural analysis of the social behaviour of a group of captive, adult chimpanzees. Multivariate analysis of the temporal relationships showed that behavioural elements can be placed in a hierarchical structure. The simple idea was to study how this hierarchical organization emerged in the course of ontogeny. At that time, very little detailed, analytical research had been done on that topic in the domain of animal behaviour (Hinde, 1970). But the few studies that had been carried out supported the idea of an ascending type of development, which proceeds through the emergence of isolated units followed by an ascending integration into a hierarchy.

However, insurmountable problems were met. In the above approach the behaviour repertoire is supposed to be composed of a number of discrete patterns of behaviour. Systems are envisaged to control a group of functionally related activities (McFarland, 1971, pp. 205-206) and discrete patterns of behaviour are used in the search for the existence of the integrating systems (Baerends, 1956, 1976; Hinde, 1953). The difficulty in finding what systems control the behaviour of babies younger than 5 months was that these discrete patterns of behaviour were found not yet to exist. This finding was supported by others in the literature. For instance, Welker (1971) noted that the criteria for classification of behaviour "are derived primarily from observations of already organized behaviour patterns of mature, experienced animals, and consequently are not applicable to the developing patterns characteristic of neonatal and inexperienced animals." Later, Condon (1977, 1979) reached the same conclusion from his frame-to-frame analysis of human neonates.

Another way to establish the existence of such integrating systems is the search for consummatory stimuli. It was argued at the time that it is only logical that a hierarchical integration concerns, among other things, the ontogeny of movement-regulation leading up to the ability to execute discrete behaviour patterns, if the hierarchical integration starts before the age at which discrete behaviour patterns emerge. In such movement-regulation, feedback through proprioception as well as other perception (of the results of movement outside the body) had been found to play an important role in primates (Marsden, Merton, Morton & Adam, 1978; Polit & Bizzi, 1979), in mice (Fentress, 1976), and some bird species (Baerends, 1956, 1970; Bastock, Morris & Moynihan, 1953). In such feedback, a deviation from set-values plays a crucial role (Baerends, 1976). As soon as the perceived stimuli conform to the set-values, all activity aimed at restoring those values stops. Such stimuli have been called "consummatory stimuli" (Bastock et al., 1953). Thus, systems may also be thought of as controlling set-values in the perception, effectuated by the behaviour, as a result of which consummatory stimuli can be observed. Consequently, these consummatory stimuli can be of help in finding the integrating systems.

The next task would be to find how far these systems could be considered to be part of different levels in a hierarchical structure causing the consummatory stimuli to function as "hierarchically nested stopping rules," as Dawkins (1976) called it. In the end we wanted to know how such a structure develops.

The search for consummatory stimuli is not easy. Hofer (1978) speaks of hidden, regulatory processes. As long as one does not know what an organism controls, chaos seems to be present in the observed behaviour. As soon as one does know, order appears in the same overt behaviour (Golani, 1976).

Therefore, one needs a lead to detect the consummatory stimuli. We found Powers' (1973) functional model of the human nervous system very useful as such a lead. It would take too much time to describe this model here in detail. Generally speaking, Powers' hypothetical construct stands for the way in which the organization of the human nervous system functions. It provides a functional block diagram of how overt behaviour is regulated. This regulation of overt behaviour is thought to result from the control of perception; it is continuously aimed at reaching or maintaining certain perceptual set-values. These set-values can belong to different levels in nonlinear hierarchical organization and higher-order set-values are able to adjust lower-order ones. Speaking more specifically, the hypothetical construct consists of hierarchy of negative feedback control systems in which each higher-order system controls a number of lower-order systems (Figure 1). To make a long story short, each level of control systems in the hierarchy goes together with a certain type of perception and, consequently, a certain type of learning and a certain class of consummatory stimuli.

How then, one can ask, would such a functional model of the adult nervous system help in finding out what type of control systems underlie behaviour and what type of perception is controlled at a particular age during early development? After all, the information about types of perception is derived from a hypothetical construct and the observer only has a recording of overt behaviour available. What overt behaviour should be observed before one can conclude that a particular type of perception is under control? The first author (Plooij, 1980, 1984) has used three procedures in making such decisions.

One way is to study reactions to disturbances. An organism can be said to control a variable with respect to a reference condition, if every disturbance tending to cause a deviation calls forth a behaviour which results in opposition to the disturbance. In the type of field study which we have conducted it is impossible to apply disturbances and look for the countering behaviour. What can be done, however, is to look for naturally occurring disturbances, and to observe the behaviour following such disturbances. Everyday life is full of such natural experiments.

A second way to decide what type of perception is under control concerns the speed of control systems. First order control systems are very fast. About 0.1 sec or even less will be the least significant time interval. The higher the order of control, the slower the system. This allows all orders of control to work simultaneously. Furthermore, control systems can oscillate when they become unstable. The higher the order of control the lower its frequency of oscillation must be. Thus, the frequency of this oscillation provides information about the order of control involved. A few examples should clarify this. "Clonus" oscillations result from unstable first-order control

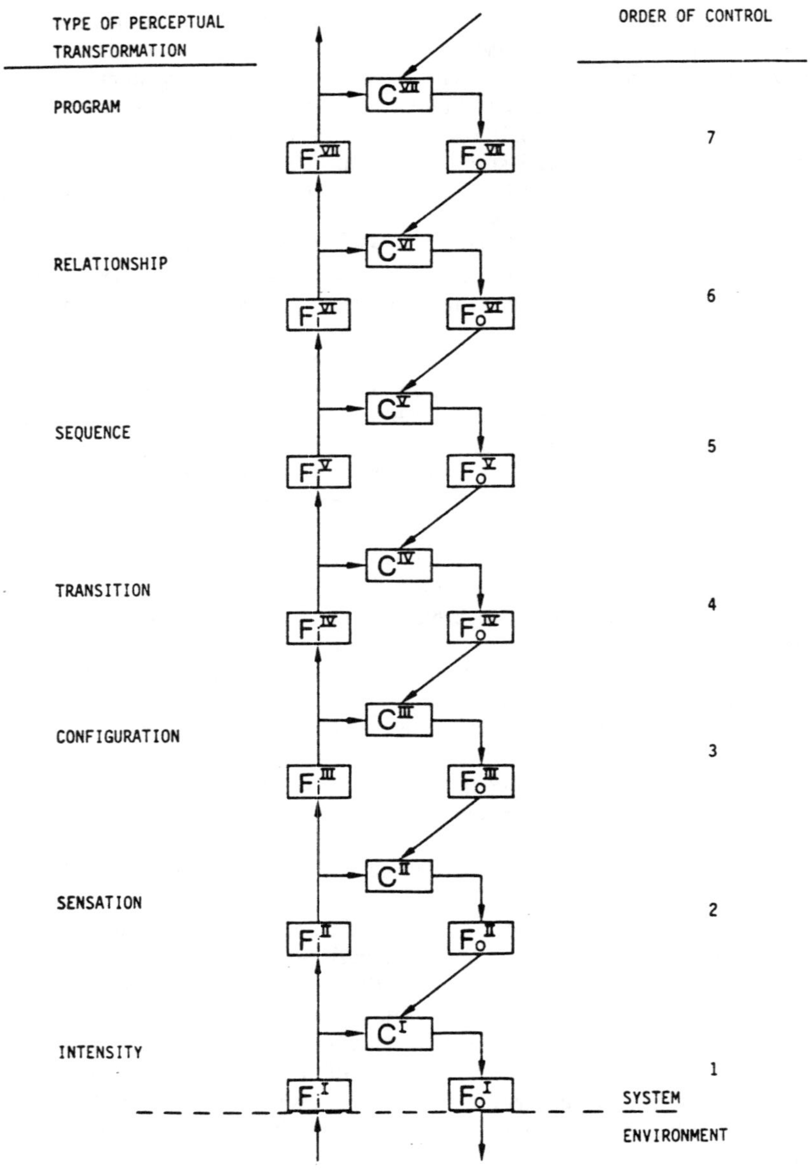

Figure 1. Hierarchy of negative feedback control systems (from Plooij, 1984)

systems when muscles exert too much effort. They oscillate at roughly 10 Hz. Several types of "tremors", such as the ones of Parkinsonianism, oscillate at approximately 3 Hz and witness second-order instability. Finally, over-correction such as over- and undershooting the target while reaching out for something results from an unstable third-order control system.

A third way to decide what type of perception is controlled is concerned with the degree of variability. The disappearance of rigidity in the control of a certain variable indicates that the order of control involved has changed from being highest to being one but highest. This follows logically from the hypothetical construct: as long as a certain control system is the highest in the hierarchy there are no higher order control systems superimposed on it which adjust the reference signal and variation is absent. As soon as these higher-order control systems become operative, the reference signal starts being adjusted and variability in behaviour is observed. This was found to be true for cormorants, a bird species studied by Kortlandt (1955). He concluded that this "zone of variability in behaviour ascends in the same degree as does the progress in maturation."

So, apart from observing what behaviours are shown at a particular age, it is equally if not more important to realize that behaviour is the control of perception, to have knowledge of the types of perception that play a role, and to study certain qualities of the overt behaviour such as speed of oscillation and degree of variability.

With this model in mind it was possible to make a lot of sense of our observational field data. The first author detected five clusters of changes in the behaviour and perception of the chimpanzee babies and infants during the first year of life (Plooij, 1980, 1984). Each of these clusters tended to centre around a certain age and could be understood in terms of a new type of control systems being superposed onto the already existing hierarchy of control systems. The following statements summarize these findings.

2.1.1. *The chimpanzee neonate has a nervous system with no more than two orders of control.* What evidence supports this statement?

First of all, the neonatal perception is still close to a total unity. This is indicated by the fact that a very wide range of stimuli in several sense modalities can elicit the vocalizations staccato and uh-grunt. Furthermore, the production of these vocalizations is simply related to a disturbance, a sudden intensity-change, regardless of the quality of the stimulation. For instance, a sudden change in light-intensity by a cloud moving in front of the sun and the cloud's shadow racing over the ground reaching the baby. But also a sudden sound, such as the breaking of a tree-branch, a sudden thunder, or breaking wind by the mother. This intensity-determined way of reacting to all kinds of disturbances indicates that higher than second order control systems are not yet active. The perceptual signals which feed into higher than second order input functions encode more than intensity alone.

Second order systems control sensations. And the behaviour of chimpanzee neonates is very much ruled by sensations. For instance thermoregulation, which plays a role in the comfort-contact-search and in rooting towards a nipple.

The fact that higher than second-order control systems are not yet active can also be concluded by looking at qualities of the neonate's behaviour through two of the three procedures mentioned earlier.

Lack of variability in the so-called "comfort-contact-search" is the first indication that no more than two orders of control are functional. This search has very much to do with, among other things, temperature regulation. As long as the temperature is off-target, that is discrepant from an optimal temperature, the contact-comfort search is activated. Only when this optimal temperature is present, the search stops. This implies that there is one fixed target value or reference value in this thermoregulation. Consequently, there are no higher than second order control systems varying this reference value.

The speed with which the neonate's head oscillates from one side to the other during rooting provides the second indication that no more than two orders of control are functional. This occurs with a frequency of 2-3 Hz (Prechtl, 1958). If this frequency is compared with the frequency of a clonus or a tremor, rooting may be considered to result from an unstable second-order control system.

2.1.2. *Around the age of two months a cluster of changes is observed which leads to the conclusion that a third order of control has become operative.* This conclusion is based on the following evidence.

The speed changed with which the head oscillates during rooting. Rooting was replaced by the head-turning-response. One turn of the head from one side to the other lasts 2 seconds instead of 1/3 to 1/2 sec. The oscillation has become slower, thus a higher order control system must have become operative.

The chimpanzee baby of two months and older does not whimper anymore when ventro-ventral contact is broken. Thus variability has appeared in the comfort-contact-search. This variability in second-order-control indicates that a higher order control system must have become operative, allowing the second-order-reference signal to vary.

Third, a shift occurred from a staccato-production ruled by mere intensity to a selective responsiveness to more "specific" stimulation. This is shown in figure 2. The shift is quite dramatic and steep after the age of two months. The "specific" stimulation was the sight or the calls of other chimpanzees. This implies the ability to perceive visual and acoustic patterns, and that is among other things what the control of configurations is all about, as indicated in the model at third order control. In addition, a sudden interest in the face of the mother and in visual patterns in general indicates the ability to perceive visual configurations. A sudden occupation with bodily configurations was witnessed by the start of making all sorts of faces, moving mouth parts, the newly gained control over various body positions, and moving the hands to and from in front of the eyes.

Again, the quality of the behaviour gives us an extra clue as to what order of control has become operative after two months. At the age of eight weeks the baby was observed to grasp something for the first time. This occurred with an enormous overshoot, also called purpose-tremor. Such an overcorrection is typical for a third order control system. The fact that fourth order of control is not yet active may be deducted from the quality of another behaviour. At the age of seven weeks a baby was observed to scratch itself for the first time, but in a very wooden way : the whole hand was alternately flexing and stretching. Furthermore, scratching was directed only at the body spot where the hand happened to be, even when an itch was in a totally different spot. This example shows that the baby of two months is not yet able to bring its hand to a certain body spot and that it fails to make smooth transitions between one arm-hand configuration and the other. Thus, fourth-order control of transitions is still absent.

The cluster of changes around two months also had its consequences in the social realm. The mother drastically stopped with carrying and supporting the baby. This is shown in figure 3. This is understandable if one realizes that the baby is now able to perceive and control bodily configu-

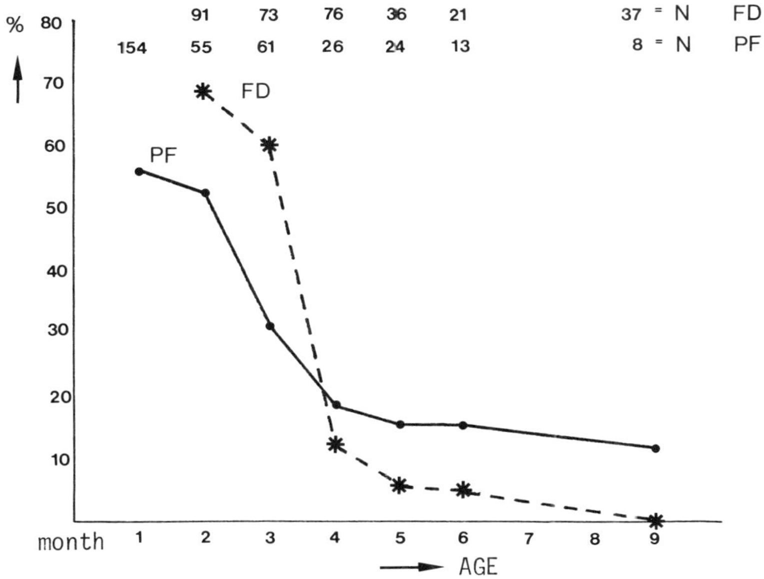

Figure 2. The percentage of staccato's and uh-grunts given as a reaction to "non-specific" stimulation for the two babies PF and FD (from Plooij, 1984)

rations. This includes also the ability to support its own body-weight by maintaining the clinging position while hanging under the belly of the mother. This is something the baby was not able to do before the age of two months. We observed a natural experiment which cruelly proved the latter statement. An adult female chimpanzee lost the use of one arm after an epidemic disease. She had raised two infants successfully before. Two more infants, that were born after her arm was paralyzed, both died in the first few months, because the mother was not able to support them properly. As a consequence the infants frequently fell down, off the body of the mother. When she was in a tree, this was lethal.

Having illustrated the way of reasoning elaborately using this first cluster of changes around two months, the remaining clusters of changes will be dealt with faster and more sketchy.

2.1.3. *Fourth-order control of transitions seems to emerge between the ages of 3-4 months.* All movements were much more smooth. The purpose tremor and other jerky movements were not observed any more. The evidence for the emergence of this order of control is still shaky and more research is needed.

2.1.4. *Towards the age of five months a fifth order of control emerges.* For the first time in the development of the babies it is easy to observe discrete motor-patterns such as walking quadrupe-

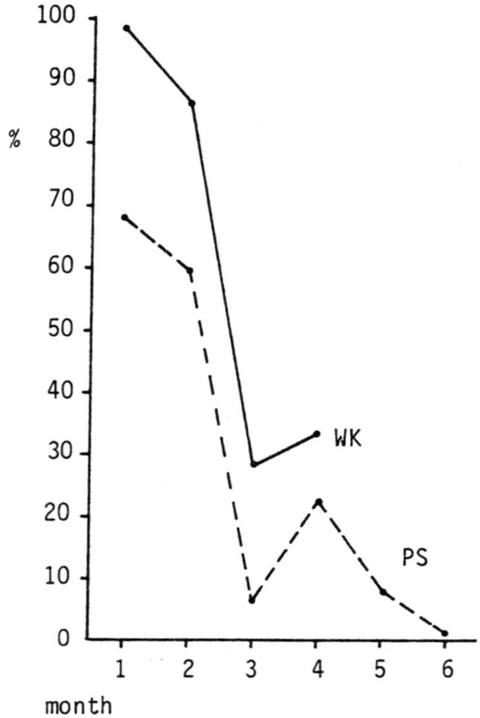

Figure 3. The decrease in the % of carrying and supporting over age (from Van de Rijt-Plooij and Plooij, 1987)

dally and climbing. Fifth-order reference signals control short, familiar sequences or "events". Sequences can occur in all modalities. As for proprioceptions, sequences of transitions in posture are subject to fifth order control. Apart from walking and climbing other discrete motor-patterns such as picking a fruit from the tree are observed.

In the social realm the first severe conflicts and forms of weaning are observed between mother and infant. Access to the belly and to the nipple is a little bit restricted for the first time in the baby's life. The babies do not give in easily and sometimes the mothers have to use force to pull their baby off the belly, off the nipple and onto their back. For the first time in their life the babies are mainly responsible for maintaining ventro-ventral contact. This is depicted quantitatively in figure 4.

2.1.5. *Between 7-9 months quite a large cluster of changes was observed which suggests that yet another order of control has emerged having to do with relationships.* During this time the number of discrete motor-patterns increased remarkably. This suggests that the fifth-order control

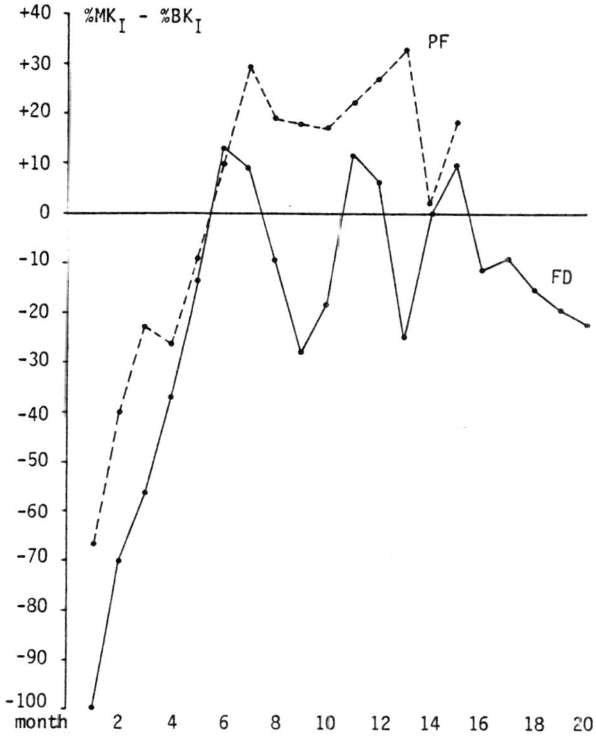

Figure 4. Baby's role in ventro-ventral contact (%MK$_i$-%BK$_i$) over age (from Van de Rijt-Plooij and Plooij, 1987)

of short, familiar sequences or events has come under the control of systems one step higher in the hierarchy, causing a great deal of variability.

The infants start placing objects against certain body spots frequently. Object *on top of* head, object *into* neck-pocket, object *against* belly, etc. The infant could not have performed these behaviours without the ability to perceive and control a relationship between an object and a certain body spot.

In month eight there is a rather abrupt change in the way the infant makes excursions away from and back to the mother. These excursions became more frequent, but shorter in duration and the infant stayed closer to the mother than at an earlier age. The infant showed concern over the distance to its mother by whimpering when this distance was increased beyond the infant's control. The distance between mother and infant is also a relationship.

Six anomalous motor-patterns ("Branch-swaying," "Gymnastics," "Pirouetting," "Slap-stamping," "Stamping," and "Somersaulting;" for definitions see Plooij, 1984) were observed for the first time, being display-patterns and expressing conflict between a tendency to approach and a tendency to withdraw. This conflict could not have arisen if the infant would not yet have been able to perceive and control a distance relationship.

The social gesture "begging" appeared. Also socio-sexual behaviours such as "presenting," "mounting" and "thrusting" emerged around the age of nine months. These behaviours also demand the ability to relate certain body parts to the other individual or certain body parts of that individual.

2.1.6. *And finally, as far as the first year is concerned, towards the end of the first year another order of control appears to become operative. This gives the ability to perceive and control programs.* These are lists of relationships between events and lower order perceptions, brought about one after the other. Examples for humans are: eating porridge with a spoon, or vacuum cleaning. In chimpanzees the first imperfect forms of termite-fishing, nestbuilding, collecting food, and other functional behaviour sequences appeared. Seventh order control of programs is related to the idea of "syntax" of behaviour as suggested by Lashley (1951).

The sixth and seventh order of control also had their consequences in the social realm. Up till the age of seven months the infant was still ventral to its mother most of the time. Thereafter, however, shifts were observed towards predominantly using space more distant from the mother. Figure 5 shows the rankorder of the amount of time spent in each of six distance categories for one mother-infant pair Fifi and Freud. The distribution of the top ranks over age is indicated by the solid black line.

Between month seven and eight we found a drop in the amount of time spent in ventro-ventral contact. This drop was associated with a shift from mainly staying in ventro-ventral contact to mainly staying "in contact", as shown in this figure.

Between month eleven and twelve a drop in the amount of time spent in *body* contact was found. After this drop, time spent in contact remained at about 50%.

This drop in time spent in body contact was associated with a shift from mainly staying "in contact" to mainly staying "out of contact", but "within arm's reach".

2.1.7. *Yet another order of control, the eighth order, may emerge around 1½ years.* To support this statement we only have data on aspects concerning the mother-infant interaction. A second drop in the amount of time spent in body contact was found between month 18 and 19. After this drop, time spent in contact remained at about 30%. This second drop in time spent in body contact was associated with a shift from mainly staying "within arm's reach" to mainly staying within 1.5-5 meters away from mother.

2.1.8. *Summarizing,* the application of Powers' model helped to understand what new types of perception and control of that perception become operative at what age during early development.

This understanding, in its turn, helped to integrate a lot of isolated data.

Not only our own field-data can be nicely integrated in the framework of this model, but also many neurophysiological and neuro-ethological data, such as those reported by Cools (1985) or Teitelbaum & Pellis (1992). One small piece of these neurophysiological data is interesting to mention here in the context of this paper. During recovery from brain manipulation, various authors report the same sequence of types of perception becoming operative in half an hour, just like a film of development that has been speeded up.

distance categories / month	ventral	not-ventral	in contact	within arm's reach	between 1.5–5m	between 5–15m	
1	1						
2	1						
3	1		2				
4	1	3	2				
5	1	4	3	2	5		
6	1	4	2	3	4		
7	1		2	2	4		
							A
8	2	4	1	3	4		
9	3		1	2	4		
10	1	5	3	2	4		
11	2	4	1	3	5		
							1
12	3	5	2	1	3	6	
13	2	5	3	1	3	6	
14	3	5	2	1	4		
15	2	5	4	1	3		
16	4	6	2	1	3	5	
17	1	5	4	2	3	6	
18	3	5	4	1	1	6	
							2
19	4	5	5	2	1	2	
20	4	5	3	2	1	6	

A=drop in ventro-ventral contact
1=first drop in contact
2=second drop in contact

Figure 5. Rankorder of the amount of time spent in each of six distance-categories for one baby (from Van de Rijt-Plooij & Plooij, 1987)

2.2. Study two

Another collection of data that could be nicely integrated through the model was a study on the quality of the chimpanzee mother-infant relationship and on the growing independence of the infant within this relationship (Van de Rijt-Plooij & Plooij, 1987, 1988).

The striking findings in that study were that the growing independence occurs jumpwise instead of gradually, that the jumps towards greater independence are preceded by mother-infant conflict which in its turn may be preceded by regression in the infant, and that this sequence of regression-conflict-jump started at exactly the same ages at which the new control systems emerged.

For instance, the ability to perceive and control so called "configurations" emerged at two months of age. Usually we think of visual configurations when we use this word. However, with the word configurations and all the other words describing various types of perception, very abstract "invariances" in our perception can also be involved. For instance, proprioceptive configurations are involved in the control of body posture. The link with the mother-infant data is that after the age of two months the infant was able to cling to the mother and the mother stopped abruptly with supporting and carrying the infant.

At the age of 4.5-5 months the ability to perceive and control so called "events", that is short and familiar sequences, emerged. Walking is an example of a sequence of body postures and transitions between such postures. Shortly thereafter, the mother was aggressive to her infant when it came ventral or on the nipple. Now it was perfectly capable of climbing dorsal itself or of climbing towards and picking food itself.

At the age of seven months the ability to perceive and control relationships emerged. With relationships is meant such straightforward examples as *in, out of, under, on top of, behind.* Also the distance between two objects or between mother and infant is such a relationship. And at seven months of age we observed the infant to be concerned over the distance to its mother for the first time in its life.

At the age of ten months the ability to perceive and control socalled "programs" emerged. Also at ten months, the first regressive period was observed and one month later mother and infant had conflict over body contact.

Thus, we found that the new potentials created by the maturation of new control systems in the nervous system are not restricted to the behaviour and perception of single individuals, but affect also behaviour in the social realm.

We think that the maturation of the new types of control systems only create new potentials or, in other words, learning instincts. Actual reorganization of the infant's overt behaviour and the actual learning of new skills depend on the interaction with the situational-social context.

So, the infant may never develop certain skills if it does not get the opportunity or if it is not forced to do so. For instance, we presented evidence that, depending on the caretaking environment, chimpanzee babies in captivity show a delay of many months in taking their first steps, because they were not forced to do so.

Maternal conflict behaviour can be viewed as stimulating the infant to develop new skills with regard to independence from the mother to the fullest extent. By demanding more from the infant as soon as it is able to, the mother pushes the infant so to speak to the edge of its abilities. In other words, in each period of conflict, mother contributed by demanding that the infant reorganizes its behaviour according to its new potentials. Furthermore, during the conflict periods, the mothers never rejected the infant as a whole, but only certain aspects of its behaviour. Thus, mother-infant conflict and even maternal aggression has its positive effects in that it enhances development.

This effect is very much enlarged if it is followed by teaching. For example, after the emergence of the perception and control of so called events during the fifth month of life, the mothers were aggressive to their infants when these tried to come ventral or on the nipple, as was just mentioned. One mother FF was observed to offer her infant FD an alternative for being pushed away from the ventral side before travelling. She taught her infant the behaviours that were to serve as 'signals' for climbing dorsal. The following was observed:

FD was walking on a rock (about 50cm high) while FF was sitting out of contact but within arm's reach. Then FF stood up slowly, turned her back towards him and approached him while flexing her knees slightly and looking back at him with her lowered back closest to him. FD did not cling immediately and FF waited motionless while looking at him. Finally, he clung and FF walked a few paces to turn around and come back to the same rock to return him onto the rock. Over and over she started travel in this way and when FD did not walk onto the rock himself, she placed him there. Finally, FD clung immediately whenever FF 'signalled' by flexing her knees and looking back.

Next, FF started the whole procedure again, with FD on the ground. She lowered her back by crouching onto the ground in front of him while looking back at him over her shoulder. Whenever FD seemed not to be looking at FF she would wait and, ultimately, gain his attention by touching him. When touched in this way he always responded by clinging immediately. 'Signalling' in this context was also repeated several times and gradually slight variations were made.

FF was very patient with FD. She sat down with him and waited until he tottered off himself. A baby of this age can hardly walk and FF had to wait for several minutes before FD was far enough away to make it worth her while to signal again. On the first day, FF was seen to use different combinations of the following signals: looking at, reaching for, leg bending, grunting, lowering back. Moreover, she used these signals in a variety of situations: when being on the ground on the same height, when being in a tree, when being with other individuals, and when being alone. FD was seen to agree to these signals by climbing dorsal himself. This apparently tedious 'teaching' and 'learning' of signals had advantages to both partners. Having learnt to understand signals, FD showed that he accepted the dorsal position and FF no longer needed to be aggressive to prevent ventro-ventral contact (Van de Rijt-Plooij & Plooij, 1987).

During the remainder of the first year of life, the development of communication in chimpanzee infants shows the same stages of acquisition at the same ages as in human infants. The "begging" gesture emerges at nine months and definitely is a deictic gesture. The other three gestures initiating grooming, tickling and approach emerge towards twelve months and are referential gestures (referring to complex schemes of action and derived from ritualized exchanges with the mother). By twelve months of age, they are on the verge of symbol use (Plooij, 1978). It can be argued that the ages and stages of acquisition of signs and finally symbols can be understood as consequences of the newly emerging types of learning that were described earlier (Plooij, 1987).

It is in the daily customs and ritualized exchanges with the mother that culture is born. And the types of learning, or learning instincts, are at the root of them.

3. Man

Inspired by the findings in the chimpanzees, we set out to explore the possibility that development in human infants proceeds in the same way. We decided to focus first on the infantile regression or relapses or set-backs that were associated with the clusters of changes in the chimpanzees that witnessed the emergence of a new type of learning. By doing this we followed, in fact, Spitz (1959) and other people who stress that emotional changes are the indicators of transitions.

In our study we did two things. First of all we asked 15 mothers many questions on a weekly basis. Mothers have been proven to be acute observers of new capacities of their infants, certainly when they are asked very specific questions. And, secondly, we validated these questionnaire data by making home observations in two families on a monthly basis (Van de Rijt-Plooij & Plooij, 1992).

One of the questions we asked the mothers was whether the last week was more difficult or tiresome than the week(s) before. As the null hypothesis one would expect such weeks to be evenly distributed over development. But they are not. A clear striping can be seen in the individual data in figure 6A. If the percentage of mothers reporting regression is plotted over age, a number of peaks can be seen around certain ages in figure 6B.

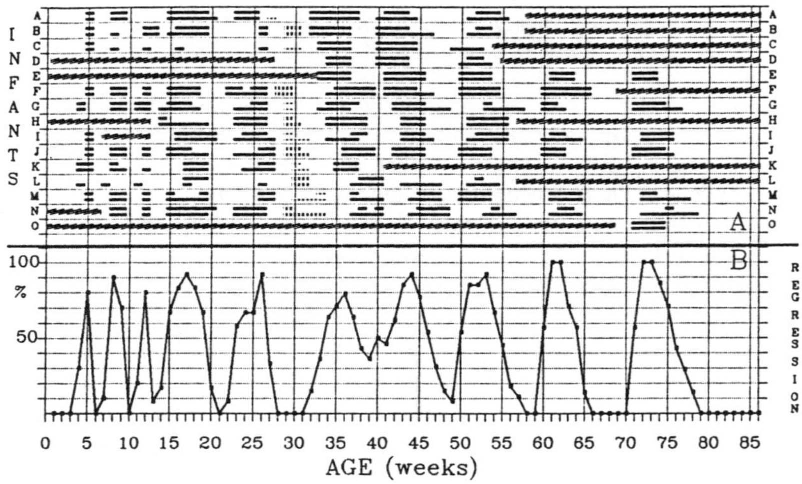

Figure 6. Regressive periods during infant development. A. The pairs of solid, horizontal bars stand for regressive periods. The top bars represent weeks experienced as difficult by the mother. The bottom bars represent concrete regressive infant-behaviours as reported by the mothers. The hatched bars stand for weeks when an infant was not observed. The dotted bars around 30 weeks represent the 'checking back pattern' as reported by Mahler et al., (1975, p. 55). This behaviour was new and not regressive. B. The percentage of mothers that reported concrete, regressive infant behaviours over age (from Van de Rijt-Plooij and Plooij, 1992)

Then we compared these difficult periods with the easy periods in terms of the answers given to the more specific questions in the questionnaire. And it turned out that the difficult periods were characterized by the infants being more irritable, clingy and closer to mother, and sleeping less well. When we scored the weeks as being characterized or not by these behavioural and more objective criteria, the results turned out to be even more striking. These findings were validated by the direct observations in the homes.

Then we focussed our analysis on mother-infant conflict. In Figure 7 we see that periods of mother-infant conflict shortly follow the regression periods. This was tested statistically in the data of individual infants and proven to be significant (Van de Rijt-Plooij & Plooij, 1993).

So the working hypotheses that arose from observations on free-living chimpanzees were confirmed for human mother-infant pairs in the sense that regression periods followed by conflict are found.

The big surprise was the fact that we found no less than ten regression periods! At least three more than the developmental transitions or biobehavioural shifts reported so far in the literature (Fischer & Hogan, 1989). This is understandable, if one realizes that most studies did not sample frequently enough. If one does not sample weekly, a number of periods are missed.

At the moment we are analysing disease data on a weekly basis. What tends to come out is that the percentage of infants being ill rises after the last week of the regression periods. We applied some form of density analysis and a multimodal frequency distribution of diseases of over age is confirmed statistically with nine or more peaks (Plooij, Van de Rijt-Plooij, Van der Stelt, Van Es, & Helmers, submitted for publication). So the disease data pretty well confirm and support the regression and conflict data.

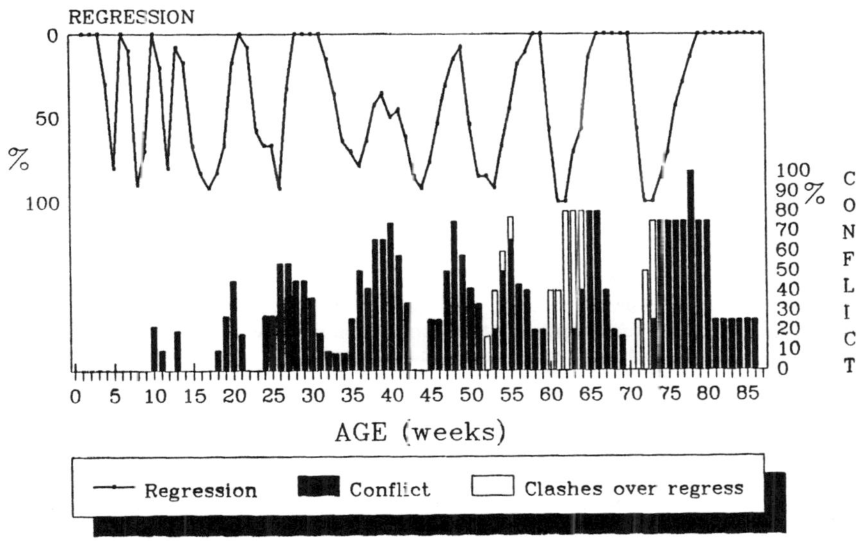

Figure 7. Periods of mother-infant conflict follow periods of regression in the infant. (The data on regression periods are adapted from Van de Rijt-Plooij and Plooij, 1992). The open bars represent the % of mother-infant pairs in which clashes occurred that are initiated by the infant's quarrelsome way of demanding closer contact and more maternal attention (clashes over regression). The closed bars represent the % of mother-infant pairs in which all forms of conflict were initiated by the mother, such as the categories: "annoyance", "promoting progress", and that part of the category "clashes" in which these followed the mother's "promoting progress" (from Van de Rijt-Plooij and Plooij, 1993).

Apparently something fundamental is going on ten times in the first twenty months. We believe this something fundamental is the emergence of new types of control systems with new types of perception and new types of learning or learning instincts. Each time this happens, the infant is off-balance and has to adapt, to reorganize. At first the infant withdraws to a safe base: the mother. Just imagine yourself waking up overnight on a new planet! What would you do? But then, when the mother starts being annoyed and mother-infant conflict follows, the infant is forced to start using the new learning instincts. It enters the zone of next development. In doing so, it shapes the new potential into new skills. This happens in interaction with the mother during daily customs and in ritualized exchanges. It is during these developmental transitions, brought about by the learning instincts, that the roots of culture are formed.

4. Acknowledgements

The studies reported here were financially supported by the Netherlands Foundation for the Advancement of Tropical Research (WOTRO grant no. W84-66), the Dr. J.L. Dobberke Foundation for Comparative Psychology at Amsterdam, and the Netherlands Organization for the Advancement of Pure Research (NWO grant no. 56-83).

5. References

Baerends, G.P. (1956) 'Aufbau des tierischen Verhaltens', in J.-G. Helmcke and G.C. Hirsch (eds.), Handbuch der Zoologie (Vol. 8), Walter de Gruyter, Berlin.

Baerends, G.P. (1970) 'A model of the functional organization of incubation behaviour', Behaviour Supplement 17, 263-312.

Baerends, G.P. (1976) 'The functional organization of behaviour', Animal Behaviour 24, 726-738.

Bastock, M., Morris, D., and Moynihan, M. (1953) 'Some comments on conflict and thwarting in animals', Behaviour 6, 56-84.

Condon, W.S. (1977) 'A primary phase in the organization of infant responding behaviour', in H.R. Schaffer (ed.), Studies in mother-infant interaction, Academic Press, London.

Condon, W.S. (1979) 'Neonatal entrainment and enculturation', in M. Bullowa (ed.), Before speech: The beginning of interpersonal communication, Cambridge University Press, Cambridge, pp. 131-148.

Cools, A.R. (1985) 'Brain and behavior: Hierarchy of feedback systems and control of input', in P.P.G. Bateson and P. Klopfer (ed.), Perspectives in Ethology (Vol. 6), Plenum Press, New York.

Dawkins, R. (1976) 'Hierarchical organization: A candidate principle for ethology', in P.P.G. Bateson and R.A. Hinde (eds.), Growing points in ethology, Cambridge University Press, Cambridge.

Fentress, J.C. (1976) 'Behavioral networks and the simpler systems approach', in J.C. Fentress (ed.), Simpler networks and behavior, Sinauer Associates, Inc., Sunderland, MA.

Fischer, Kurt W., and Hogan, Anne E. (1989) 'The big picture for infant development. Levels and variations', in J.J. Lockman and N.L. Hazen (eds.), Action in social context: Perspectives on early development, Plenum, New York, pp. 275-305.

Galef, B.G. (1992) 'The question of animal culture', Human Nature 3, 157-178.

Golani, I. (1976) 'Homeostatic motor processes in mammalian interactions: A choreography of display', in P.P.G. Bateson and P.H. Klopfer (eds.), Perspectives in ethology (Vol. 2), Plenum, New York.

Gould, J.L., and Marler, P. (1987) 'Learning by instinct', Scientific American, January, 62-73.

Hinde, R.A. (1953) 'Appetitive behaviour, consummatory act, and the hierarchical organization of

behaviour -- with special reference to the great tit (Parus major)', Behaviour 5, 189-224.
Hinde, R.A. (1970) Animal Behavior, McGraw-Hill, New York.
Hofer, M.A. (1978) 'Hidden regulatory processes in early social relationships', in P.P.G. Bateson and P.H. Klopfer (eds.), Perspectives in ethology (Vol. 3): Social behavior, Plenum, New York.
Hooff, J.A.R.A.M. van. (1971) 'Aspecten van het sociale gedrag en de communicatie bij humane en hogere niet-humane primaten' [Dissertation], University of Utrecht, the Netherlands.
Kortlandt, A. (1955) 'Aspects and prospects of the concept of instinct (vicissitudes of the hierarchy theory)', Archives néerlandaises de Zoologie 11, 155-284.
Lashley, K.S. (1951) 'The problem of serial order in behavior', in L.A. Jeffres (ed.), Cerebral mechanisms in behavior: The Hixon symposium, Wiley, New York.
Mahler, M.S., Pine, F., and Bergman, A. (1975) The Psychological birth of the human infant: Symbiosis and individuation, Basic Books, New York.
Marsden, C.D., Merton, P.A., Morton, H.B., and Adam, J.E.R. (1978) 'The role of afferent feedback in the regulation of movement', in D.J. Chivers and J. Herbert (eds.), Recent advances in primatology (Vol. 1): Behaviour, Academic Press, London.
McFarland, D.J. (1971) 'Feedback mechanisms in animal behaviour', Academic Press, London.
Plooij, Frans X. (1978) 'Some basic traits of language in wild chimpanzees?' in A. Lock (ed.), Action, gesture and symbol: The emergence of language, Academic Press, London, pp. 111-131.
Plooij, Frans X. (1980) 'The behavioural development of free-living chimpanzee babies and infants' [Dissertation], University of Groningen, The Netherlands.
Plooij, Frans X. (1984) 'The behavioral development of free-living chimpanzee babies and infants', Monographs on Infancy 3, Ablex, Norwood, NJ, pp. 1-207.
Plooij, Frans X. (1987) 'Infant-ape behavioural development, the control of perception, types of learning and symbolism', in J. Montangero, A. Tryphon, S. Dionnet (eds.), Symbolism and Knowledge, Archives Jean Piaget Foundation, Geneva, pp. 35-64.
Plooij, Frans X., Rijt-Plooij, H.H.C. van de, Stelt, J.M. van der, Es, B. van, and Helmers, R. (submitted) 'Illness peaks during infancy: Nonlinear oscillations in immunity to disease'.
Polit, A., and Bizzi, E. (1979) 'Characteristics of motor programs underlying arm movements in monkeys', Journal of Neurophysiology 42, 183-194.
Powers, W.T. (1973) 'Behavior: The control of perception', Wildwood House, London.
Prechtl, H.F.R. (1958) 'The directed head turning response and allied movements of the human baby', Behaviour 13, 212-242.
Rijt-Plooij, Hedwig H.C. van de, and Plooij, Frans X. (1987) 'Growing independence, conflict and learning in mother-infant relations in free-ranging chimpanzees', Behaviour 101, 1-86.
Rijt-Plooij, Hedwig H.C. van de, and Plooij, Frans X. (1988) 'Mother-infant relations, conflict, stress and illness among free-ranging chimpanzees', Developmental Medicine and Child Neurology 30, 306-315.
Rijt-Plooij, Hedwig H.C. van de, and Plooij, Frans X. (1992) 'Infantile regressions: Disorganization and the onset of transition periods', Journal of Reproductive and Infant Psychology 10, 129-149.
Rijt-Plooij, Hedwig H.C. van de, and Plooij, Frans X. (1993) 'Distinct periods of mother-infant conflict in normal development: Sources of progress and germs of pathology', Journal of Child Psychology and Psychiatry 34, 229-245.
Spitz, René A. (1959) 'A genetic field theory of Ego formation. Its implications for pathology', International Universities Press, New York.
Teitelbaum, Philip, and Pellis, Sergio M. (1992) 'Toward a synthetic physiological psychology', Psychological Science 3, 4-20.
Welker, W.I. (1971) 'Ontogeny of play and exploratory behaviors: A definition of problems and a search for new conceptual solutions', in H. Moltz (ed.), The ontogeny of vertebrate behavior, Academic Press, New York.

AN ETHOLOGICAL PERSPECTIVE ON HUMAN HANDEDNESS

L.F. Marchant & W.C. McGrew
Department of Sociology & Anthropology
Miami University
Oxford, OH 45056, U.S.A.

ABSTRACT

Human handedness is thought to be unique and universal. Critical examination of primary sources challenges both of these conclusions. Data on laterality of function in apes, humankind's closest living relations, indicates hand preference and task specialization, but not yet handedness as crucial studies remain to be done. Ethnographic data across cultures reveal more variance than is usually acknowledged, but existing findings are based on minimal and limited data. Ethological studies of spontaneous behavior in every day life are needed for both human and non-human primates.

R. A. Gardner et al. (eds.), The Ethological Roots of Culture, 375–384.

1. INTRODUCTION

Laterality of function is a basic feature of all bilaterally symmetrical organisms, including Homo sapiens (Coren, 1993). Its most obvious manifestation is in the differential use of the forelimbs, and this has long been linked with asymmetry of cerebral structure (Bradshaw & Rogers, 1993). This congruence of structure and function is often linked to the evolution of humankind's most distinctive and also lateralized trait, language (Corballis, 1991). At the same time, the symbolic significance of Left versus Right in human culture is clear, for example, every time Westerners clasp and shake their right hands in greeting (Needham, 1973). Thus, handedness is an appropriate focus for ethological analysis, from any aspect.

Given that ethology is the biological study of behavior, usually studied observationally in at least naturalistic (if not natural) settings, and given that it is usually comparative (Eibl-Eibesfeldt, 1989), our aim is to apply it accordingly to human hand use. This generates two general topics: uniqueness and universality. We will target each of these in turn, emphasizing methodological strengths and weaknesses, and proposing new knowledge needed to fill existing gaps.

A useful way to visualize the logical possibilities is to generate a 2x2 matrix, using the extent to which hand-use varies within and across task(s) and within and across subject(s). (See Table 1). The two extreme, or homogenous, conditions are straight-forward. If the same hand predominates over most subjects in a population or culture or species and over most tasks performed, then handedness is proven. If hand predominance is shown only for individual subjects on one task, then it is only hand preference. The two heterogenous conditions are more problematical. Consistency across subjects but within only one task is usually also called handedness, but might be better termed task specialization. Finally, consistency across tasks within a subject is stronger than hand preference and so amounts to manual specialization. (Unfortunately, most of these terms have varying and sometimes contradictory meanings in the literature and usually only the first two are commonly used).

2. Uniqueness

The received wisdom in textbooks of psychology is that handedness is unique to the human species, in contrast to all others (e.g. Springer & Deutsch, 1989, p. 251). Mice, cats, monkeys, apes, etc. may or may not show individual hand preferences or (more rarely) manual specialization, but taken as a population or species they are thought to be evenly distributed between left-and right-preferent. Or, in any given study, any or all subjects may be ambipreferent. This random or 50:50 model for non-humans at the population or species level is most forcefully associated, empirically and theoretically, with the late J.M. Warren (1980), who focussed on other primates, especially rhesus monkeys.

This view was challenged in a provocative re-evaluation of existing data by MacNeilage et al. (1987), who revived interest in the issue of species-specificity. They claimed that handedness is wide-spread in primates, especially prosimians, and provided a "postural origins theory" to account for it (MacNeilage et al., 1991). Their assertions spurred further empirical

TABLE 1. Laterality of hand function in terms of tasks and subjects.

		SUBJECT(S)	
		WITHIN	ACROSS
TASK(S)	WITHIN	Hand preference	Task Specialization?
	ACROSS	Manual Specialization?	Handedness

TABLE 2. Variables for analysis of laterality of function (from Marchant & McGrew, 1991)

Variable	Definition
Function	Bilaterally symmetrical organs noted, e.g. hands
Context	Captivity or wild
Sample	Number of subjects in study
Age	Age of subjects
Task	Induced or spontaneous
Number	Number of tasks in study
Trials	Number of trials per subject per task
Complexity	Degree of difficulty or intellectual demand

investigation that strengthened the claim (MacNeilage et al., 1988), so that recently it could be said that gorillas showed "....for the first time in primates, a significant human-like pattern consisting of a greater number of animals which preferred the right hand for all acts...." (MacNeilage et al., 1991, p. 344).

So, who is right, Warren or MacNeilage? The most likely candidates for human-like handedness would seem to be our nearest living relations, the apes (Ward & Hopkins, 1993). We shared a common ancestor with the African Great Apes (Pongidae), bonobo, chimpanzee, and gorilla, perhaps as recently as 5 million years ago, according to molecular anthropological evidence. The likelihood of laterality of hand function is enhanced by findings of the last 10 years that show asymmetry of brain structure (Falk, 1987). In some cases, apparent task specialization (in chest-beating) and asymmetry (in cranial dimensions) have been shown in the same population, for example, the mountain gorillas of the Virunga Volcanoes of Rwanda (cf. McGrew & Marchant, 1993).

Marchant and McGrew (1991) carried out an exhaustive meta-analysis of laterality of function in apes, focussing on methods used. They coded 58 published sources on eight variables: Function, context, sample, age, task, number, trials, complexity. (See Table 2). They found that most studies focussed on hands rather than other organs, captive rather than wild subjects, small rather than large samples of subjects, youngsters rather than adults, spontaneous rather than induced measures, few rather than many tasks and trials, and simple rather than complex tasks. Such are the gaps in the so-far published literature that in only a few studies have hand preference and task specialization been tested, and in no studies have handedness and manual specialization been adequately addressed. Thus, the typical study for assessing laterality was done in captivity with a small number of young chimpanzees; the typical task was one-handed, non-sequential, and gross in movement.

For MacNeilage et al's. (1991) best candidate, Gorilla gorilla, the picture is equally unclear from 22 studies of a total of 227 subjects (McGrew & Marchant, 1993). If one chooses to focus on the most right-sided ("Best Right") results from the array of measures, then gorillas look to be right-handed, but focus on the "Best Left" results makes them out to be left-handed.

In summary, nothing can yet be said confidently about the uniqueness of human handedness. Tentative conclusions suggest: strong hand preference and weak task specialization exist, at least for tool-use tasks in chimpanzees, but there are no compelling signs of manual specialization or handedness.

3. Universality

Laterality of hand function would be species-typical if we found universal patterns across all or most human populations or cultures. (Note that these need not be the same, insofar as biological anthropologists tend to think of the species divided into populations, while socio-cultural anthropologists divide it into cultures). Thus, we might ask if the Yanomami are as predominantly right-handed as Djakartans? In wider comparative exercises, it is two or more species that are contrasted: Are gorillas more right-handed than chimpanzees?

Textbooks make clear the answer: Human beings are right-handed. Buchanan (1862) felt able to state categorically that all nations, races, and tribes preferred to use the right hand to the left. However, quantitative generalizations of extent vary according to source: McManus (1991, p. 251) puts the proportion of left-handers at 8%, Springer & Deutsch (1989, p. 141) estimate 10%, and Corballis (1991, p. 90) has it topping out at 12-13%. All of these figures turn out to be wrong, when reference is made to the primary ethnography from whence the original data came.

Table 3 lists a selected sample of human populations or cultures for which primary data are published. The actual extent of left-hand-use varies from almost nil (5 of 1047 subjects in Katanga) to up to 23% for the Kwakiutl. (Figures of up to 30% left-handedness are reported in secondary sources, e.g. Wile, 1934). The number of subjects tested varies almost 10-fold, from 180 New Guinean highlanders to almost 17,000 Japanese school-children, and the studies span almost 60 years. Even this limited sample suggests that the variation cannot be explained simply as a function of geography, as the California figure is closer to the New Guinea one than to the Canada one. Nor are the differences likely to be a reflection of schooling, as the Sweden figure is more like the Sierra Leone one than the Japanese one. Degree of left-handedness is not correlated with the size of the sample or with the recency of the research. These may be genuine cross-cultural or populational differences that disprove the hypothesized claims of species-typicality, <u>or</u> it may be that confounding factors will otherwise explain the diversity. Methodology is a prime suspect for the latter.

Table 4 presents data on selected aspects of methodology in research on human handedness, from the same studies listed in Table 3. First, few measures or tasks were used to assess hand-use. The greatest number, nine, comes from the most recent data, from New Guinea (and is the only study done by an ethologist, Kevin Connolly).

Second, most subjects were children tested at school. Laterality in the motor skills of immature subjects may not adequately represent the performance of all age-grades in a population, especially adults. (However, there is no apparent correlation here between the age composition of subjects and the extent of left-handedness reported).

Third, all but two of the studies relied upon induced demonstration of requested tasks, usually as part of a standardized protocol. Typically the tester would take each subject through a series of tasks, giving verbal instructions, and then score responses. (<u>No</u> studies reported the hand preference of the testers, but two controlled for the tester's hand-use in administration, Connolly & Bishop, 1992; Rhoads & Damon, 1973). Komai and Fukuoka (1934) used self-report in response to a questionnaire, but <u>no</u> study collected observational data on spontaneous hand-use in everyday activities.

Fourth, the measures or tasks chosen were not a random sample of manual activity. <u>All</u> included use of a pencil or brush, usually in writing. Revealingly, in the Katanga study, after all of the children wrote their names with their right hands, some then erased them with their left! Unexpectedly, the second-most frequent task was the use of scissors, a Western tool of specialized function. (No study reported if the scissors used were biassed or neutral in design

TABLE 3. Primary ethnography on proportion of left- and right-handers across cultures

Culture	N	% Left	% Right	Reference
- Katanga, Zaire	1047	0.5	99.5	Verhaegen & Ntumba, 1964
- Solomon Islands	1352	2.8	97.2	Rhoads & Damon, 1973
- Temne, Sierra Leone	204	3.4	96.6	Dawson, 1972
- Sweden	981	5.4	94.6	Beckman & Elston, 1962
- Jimi, Western Highlands Papua New Guinea	180	8.8	91.1	Connolly & Bishop, 1992
- Berkeley, California, USA	7688	9.6	90.4	Hardyck et al., 1975
- Kyoto, Japan	16,947	11.5	88.5	Komai & Fukuoka, 1934
- Kwakiutl, British Columbia, Canada	180	22.8	77.2	Marrion, 1986

TABLE 4. Methods and subjects of studies of handedness across cultures

Culture	No. of Measures	Age of Subjects	Type of Measure	Measure(s) Used
Katanga	3	children	IT	Erase, scissors, unscrew
Solomon	1	all ages	IT	Trace with brush
Temne	3	adults	IT	Write, scissors, receive object
Sweden	?	?	?	?
New Guinea	9	children adults	IT	Pencil, scissors, hammer, blocks, thread, throw, spoon, match, deal
Berkeley	3	children	IT	Write, scissors, held
Japan	6	children	SR	Write, scissors, pen knife, throw, chop-sticks, match
Kwakiutl	2	all ages	SR, IT	Write

IT = induced task, SR = self report

with regard to handedness). All studies in this limited sample used measures only of object manipulation, and other hand-usage such as gesture was ignored.

Any or all of these four factors present potential problems of validity of measures. (All but one study, Connolly & Bishop, 1992, also present problems of reliability, as they tested each subject only once).

In summary, the only safe generalization about universality that can be made echoes the one put forward by Buchanan (1862), that all known human societies show right-hand predominance for task specialization. Handedness seems likely, but hand preference and manual specialization are largely unexplored. (Recall that hand preference and manual specialization are within subjects - either within or across tasks.) The published literature rarely provides any information about an individual's preferences.

4. Conclusions

In contrast to what textbooks and other secondary sources assert, neither the uniqueness nor the universality of human handedness is proven. This means that more research is needed.

Of primary importance is ethological study of human manual behavior in everyday life. Standard human ethological techniques (Eibl-Eibesfeldt, 1989) should be brought to bear on the widest repertoire of such motor-patterns, going beyond object manipulation to include communicatory gestures, mannerisms, habitual self-directed acts, etc. - both idiosyncratic and conventional. Such research should encompass a wide array of cultures, focussing especially on traditional societies, particularly those in which Western educational practices have not yet been introduced. If these are no longer available for direct observation, then cinematic archives should be explored. The implicit hierarchy of validity is that spontaneous acts are more useful than induced ones, and that induced ones are more useful than verbal self-report of acts.

Similar research is needed on non-human species, especially apes. More field studies are wanted, again on the full array of manual behavior patterns. (This has not yet been done for any species of primate). However, studies of induced laterality of function in other animals also need tightening up in terms of experimental design; this may clarify some of the apparent cross-species contrasts in performance. No study of nonhuman primates matches the methodological rigor of some of the human experimental psychological ones (Connolly & Bishop, 1992; Seltzer et al., 1990). For example, many apes tested in captivity were human-reared, and given their imitational capacities, their lateralized performance might mirror their human models.

Finally, standardized methods of recording both induced and spontaneous hand-use should be devised and implemented systematically for hominoids - apes and humans. Only then will we have a truly comparative ethological perspective on human handedness.

5. Acknowledgements

We thank the organizers of the NATO Advanced Study Institute for their invitation and support. We thank Kevin Connolly for critical comments on the manuscript, although of course all responsibility for any errors remains with us. We thank Cathie Francis, Fay Somerville, and Carol Kist for manuscript preparation.

References

Annett, M. (1985) Left, Right, Hand and Brain: The Right Shift Theory. London: Lawrence Erlbaum Associates.

Beckman, L. & Elston, R. (1962) Data on bilateral variation in man: Handedness, hand clasping and arm folding in Swedes. Human Biology, 34: 99-103.

Bradshaw, J. & Rogers, L. (1993) The Evolution of Lateral Asymmetries. Language, Tool Use, and Intellect. San Diego: Academic Press.

Buchanan, A. (1862) Mechanical theory of the preponderance of the right hand over the left; or, more generally, of the limbs of the right side over the left side of the body. Proceedings of the Philosophical Society of Glasgow, 5: 142-167.

Connolly, K.J. & Bishop, D.V.M. (1992) The measurement of handedness: A cross-cultural comparison of samples from England and Papua New Guinea. Neuropsychologia, 30: 13-26.

Corballis, M.C. (1991) The Lopsided Ape. New York: Oxford University Press.

Coren, S. (1993) The Left-Hander Syndrome. The Causes and Consequences of Left-Handedness. New York: Vintage Books.

Dawson, J.L.M. (1972) Temne-Arunta hand-eye dominance and cognitive style. International Journal of Psychology. 7: 219-233.

Eibl-Eibesfeldt, I. (1989) Human Ethology. New York: Aldine de Gruyter.

Falk, D. (1987) Brain lateralization in primates and its evolution in hominids. Yearbook of Physical Anthropology, 30: 107-125.

Hardyck, C., Golman, R. & Petrinovich, L. (1975) Handedness and sex, race, and age. Human Biology, 47: 369-375.

Komai, T. & Fukoka, G. (1934) A study on the frequency of left-handedness and left-footedness among Japanese school children. Human Biology, 6: 33-42.

MacNeilage, P.F., Studdert-Kennedy, M.G. & Lindblom, B. (1987) Primate handedness reconsidered. Behavioral and Brain Sciences, 10: 247-303.

MacNeilage, P.F., Studdert-Kennedy, M.G. & Lindblom, B. (1988) Primate handedness: A foot in the door. Behavioral and Brain Sciences, 11: 720-746.

MacNeilage, P.F., Studdert-Kennedy, M.G. & Lindblom, B. (1991) Primate handedness: The other theory, the other hand and the other attitude. Behavioral and Brain Sciences, 14: 338-349.

Marchant, L.F. & McGrew, W.C. (1991) Laterality of function in apes: A meta-analysis of methods. Journal of Human Evolution, 21: 425-438.

Marrion, L.V. (1986) Writing-hand differences in Kwakiutls and Caucasians. Perceptual and Motor Skills, 62: 760-762.

McGrew, W.C. & Marchant, L.F. (1993) Are gorillas right-handed or not? Human Evolution, 8: 17-23.

Needham, R. (Ed.) (1973) Right and Left. Essays on Dual Symbolic Classification. Chicago: University of Chicago Press.

Rhoads, J.G. & Damon, A. (1973) Some genetic traits in Soloman Island populations. II. Hand clasping, arm folding, and handedness. American Journal of Physical Anthropology, 39: 179-183.

Seltzer, C., Forsythe, C., and Ward, J.P. (1990) Multiple measures of motor lateralization in human primates (Homo sapiens). Journal of Comparative Psychology, 104: 159-166.

Springer, S.P. & Deutsch, G. (1989) Left Brain, Right Brain. (3rd edition). New York: W.H. Freeman.

Verhaegen, P. & Ntumba, A. (1964) Note on the frequency of left-handedness in African children. Journal of Educational Psychology, 55: 89-90.

Ward, J.P. & Hopkins, W.D. (1993) Primate Laterality. Current Behavioral Evidence of Primate Asymmetrics. New York: Springer-Verlag.

Warren, J.M. (1980) Handedness and laterality in human and other animals. Physiological psychology, 8: 351-359.

Wile, I.S. (1934) Handedness: Right and Left. Boston: Lothrop, Lee and Shepard.

CULTURE AND OLFACTORY COMMUNICATION

M.D. Kirk-Smith
School of Management
University of Ulster
Newtonabbey
Northern Ireland, BT37 OQB

ABSTRACT. Our cultural use of perfume and incense has a basis in our evolutionary biology. Man is highly scented. The axillary glands of human beings are larger and contain more apocrine glands than any other primate, and produce musky steroids for which there are matching olfactory receptors. Pheromonal-like responses to these compounds have been observed. The implications of learning and context on these responses are discussed. Perfumes contain compounds that are sexual attractants in other mammals and which resemble the musky axillary steroids produced by human beings. Examples of the cultural transmission of social responses to these natural and synthetic odours across human cultures present and past are given. Possible evolutionary and cultural reasons for the suppression and adaptation of man's natural chemical communication system are presented.

1. Introduction

The ability of animals to communicate by biological odours has been well-researched in recent years, and odour cues have been shown to be involved in activities including sexual behaviour, parent-offspring interactions, territorial marking and social recognition (for reviews, see Doty, 1976; Brown and Macdonald, 1985). Although the sense of smell has not previously been considered as well developed in primates as other mammals, research has now shown a similar involvement of odour cues (Epple, 1986).

In parallel with this research on other mammals and primates, several authors have proposed that natural odours may have affective consequences on human behaviour. The earliest review of the literature is by Havelock Ellis (1936) although his evidence is mostly anecdotal. More recently, Comfort (1974) drew on circumstantial human and animal experimental evidence, and identified possible active odours and their sites of production.

Although the olfactory sense has not been regarded as playing a significant role in human beings' behaviour (e.g., Schneider, 1974), recent experimental work has identified how human beings are affected by odours and can use them for identification of others (for review, see Schaal and Porter, 1991). This paper examines the evidence for a chemical communication system in man with particular emphasis on the products of the axillary gland, since this is the main scent producing organ in man, and examines how their synthetic analogues in perfumes and incense, have come to be used. The argument developed by the paper is that even though this communication system is biologically well-specified, the combined influence of learning and the social and cultural contexts of human beings ensure that responses to such odours will not be as stereotyped as in other mammals.

R. A. Gardner et al. (eds.), The Ethological Roots of Culture, 385–406.

2. The Components of a Communication System

2.1 SIMILARITIES BETWEEN HUMAN AND OTHER MAMMALS SENSES OF SMELL

It might be thought chemical communication is unlikely in man because of the relative lack of development of the sense of smell compared to other mammals in whom such communication is well established. There are several counters to this argument.

Firstly, a qualitative similarity between the human and rat senses of smell is suggested by the high correlations observed between human psychophysical characterisations of odours and the different mitral cell degeneration patterns in rats exposed to the same odours for long periods (Piggott and Osisiogu, 1977).

Laing (1975) studied the quantitative similarity between the human and rat sense of smell. He found that the difference in sensitivity between rats and humans for non-biological odours was of the same order as the interindividual variability of the human subjects, several humans were even observed to be more sensitive to specific odours than some rats. His results are in accord with threshold studies on rats (Moulton and Eayrs, 1960; Bennett, 1968; and Sakellaris, 1972), rabbits (Moulton, Celebi and Fink, 1970), dogs (Moulton and Eayrs, 1960; Becker, King and Markee, 1962) and pigeons (Henton, 1969), where sensitivities similar to those reported for humans were also obtained.

The information carrying capacity of the olfactory sense in man has also been underestimated (Engen, 1970); human subjects can classify singly presented odours into one of at least 60 categories, rather than the 16 previously thought. This difference appears to be due to better training and the use of more realistic stimuli with higher levels of information (i.e., the use of the complex odours of objects rather than single chemical compounds).

Keverne (1983) has argued that since olfactory information is coded as a pattern that requires a higher order of recognition, animals with a greater neural backup have the potential for more sophisticated pattern recognition, e.g., for complex biological odours. Since the olfactory bulb seems to act mainly as a filter, and decoding of the olfactory message is a more central event, animals with a greater neocortical support system have a great ability to make use of this system. In these terms, primates, including humans, do not have a poorly developed sense of olfactory perception, and Keverne suggests that the ways in which this is employed may be the most evolved of all species.

Thus, man's sense of smell may be neither quantitatively nor qualitatively very different from that of odour communicating mammals, and we may even have the advantage of a greater information processing capability for odours than lower mammals.

2.2 THE PRODUCTION OF AXILLARY ODOURS

2.2.1 *The Axillary Organ.* The axillae are the only human glands recognised by dermatologists (Montagna, 1985), zoologists (Albone, 1984), and chemists (Sastry, Buck, Janak, Dressler, and Preti, 1980) as "scent organs".

Their structural and functional adaptations for odour production suggest their potential relevance for chemical communication. They have three secretory glands; the eccrine (sweat), the apocrine and the sebaceous (Strauss and Ebling, 1970). The mixing of the aqueous eccrine secretion with the sticky apocrine secretion and the continuously excreted sebum helps them to spread over the whole of the axillary skin and hair surfaces.

The axillae are highly vascularised, providing heat for evaporation. Axillary hair enhances diffusion of odour and also provides a very large surface area for bacteria which, with the combination of heat and humidity of the armpit, allows proliferation of selected bacteria. These break down the glandular secretions and so produce both more odour and the characteristic axillary odour.

Although the function of the human apocrine glands is unknown, their secretion is dependent on endocrine activity, which in turn is sensitive to psychological and social events, e.g., being released in emotional stress. Indeed, they are similar in morphology and androgen dependence to the apocrine glands specialised for scent marking in other mammals (Hurley and Shelley, 1960; Montagna, 1962; Strauss and Ebbing, 1970). Apocrine and sebaceous scent glands are widely used for scent marking in mammals, including primates (see Theissen and Price, 1976; and Epple, 1986, for reviews), and are known to produce odours with a communicative function, e.g., in rabbits (Mykytowycz, 1974).

Remarkably, in terms of the numbers and sizes of apocrine and sebaceous glands and their concentration in the axillae, man is the most scented ape of all (Stoddart, 1990). The presence of so many scent glands emphasises that odour is as much a human attribute as any other. As Montagna commented (1985): "Every detail of the axilla is precisely tailored to produce and distribute those substances that give human beings their distinctive odor."

Comfort (1974) draws attention to their position and odour producing properties: "Humans have a complete set of organs which are traditionally described as non-functional, but which seen in any other animal would be recognised as part of a pheromone system. These include the apocrine glands, associated with conspicuous hair tufts, some of which do not produce sweat and most presumably produce some other functioning secretion."

2.2.2 *The Compounds*. Apocrine sweat collected in sterile conditions is odourless, and if incubated at room temperature, develops no odour even after 14 days (Shelley, Hurley, and Nichols, 1953). Bacterial activity generates the typical axillary odour. Thus, an individuals's axillary odours are the result of interactions between genetically influenced factors (e.g., distribution of skin glands, endocrine and metabolic processes) and environmental variables such as diet and the enzymatic breakdown of secretions by microorganisms. Odorous compounds present include the "sweaty" smelling short chain fatty acids (Labows, McGinley, and Klingman, 1982).

Two odorous steroids are also present in male axillary sweat, 5-alpha-androst-16-en-3-one (androstenone) and 5-alpha-androst-16-en-3-ol (androstenol) (Figure 1.). These compounds are synthesised in the testes and transported to the salivary glands via the general body circulation (Claus and Alsing, 1976), and have no known physiological function in man. However, the main source is thought to be due to the breakdown of precursors produced by the apocrine glands by axillary coryneform bacteria (Rennie, Holland, Mallett, Watkins, and Gower, 1988). Steroid production is substantially reduced by treating the axilla with a bacteriocidal agent (Bird and Gower, 1982). These bacteria are more common in men than women, which may explain the presence of the steroids in male axillary sweat.

Androstenone is generally thought to have a predominantly urinous odour and androstenol to have a musky odour (Kloek, 1961) and both have a low threshold of perception for humans (Amoore, Pelosi and Forester, 1977) despite having low volatility. Baker (1974) reports that many authors dwell on the musk-like odour of the human axillae. This muskiness is most likely due to androstenol and androstenone, though both androstenone and androstenol are also found in the urine (Brooksbank, Brown and Gustafsson, 1974; Gower, 1972). Androstenone may be partly responsible for the characteristic smell of male axillary sweat. Untrained women characterise T-shirts bearing a synthetic sweat with an increased concentration of androstenone as more male, independently of the rated strength of the odour (Van Toller, Dodd and Kirk-Smith, 1980).

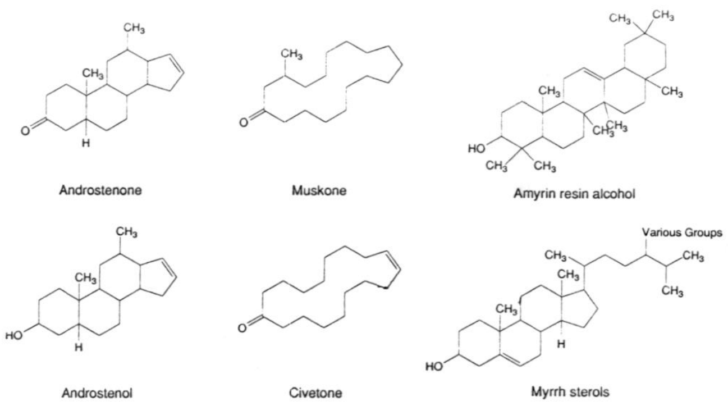

Figure 1: Steroid - like compounds from men, animals and plants

These steroids are active in olfactory communication in another species. Androstenone and androstenol are concentrated in the submaxillary gland of the boar and enter the saliva. The scent of androstenone in the saliva of boars induces an immobilisation reflex and the adoption of a mating posture in oestrous sows (Reed, Melrose and Patterson, 1974). Interestingly, androstenone is also found in truffles, for which sows are used to hunt (Claus, Hoppen, and Karg, 1981). Androstenol enhances the onset of puberty in piglets (Kirkwood, Hughes and Booth, 1983).

The axillary apocrine glands and the production of these odorous steroids develop after puberty, i.e., are under endocrine control (Brooksbank, 1962; Strauss and Ebling, 1970). These odours can therefore be considered secondary sexual characteristics. The above considerations suggest the possibility of a sexual communication function for axillary odours in general, and for androstenol and androstenone in particular.

2.3 ODOUR RECEPTION

Amoore (1970) first suggested that odour perception is effected by odorant molecules interacting with specific receptor sites on the olfactory epithelium. Different types of receptor site would receive the different molecular shapes of different odorant molecules and there would be a limited number of receptor site types specific for an equal number of "primary" odorants.

A family of genes coding for the olfactory receptor protein has now been identified and this appears to be related to the receptors for neurotransmitters and hormones found on other cells in the body, and has a similar associated signal transduction pathway (Buck and Axel, 1991). The number of olfactory receptor types is thought to be large, perhaps over one thousand, and they appear to be associated in sub-families, possibly for specific odour types.

The various primary odours specific for these receptors should exhibit low thresholds of detection compared to similar smelling odours. Additionally, there should occur instances where a specific receptor would be missing (due to a genetic deficiency), in which case the homozygote individual would not be able to detect the corresponding primary odour as other people do, but would detect secondary notes, due to partial interaction of the odour molecule with other sites (parosmia) or nothing at all (specific anosmia) if acuity is low. Intriguingly, the main primary odours identified by these criteria of specific anosmia and low threshold are all produced by the human body; isovaleric acid in vaginal secretions and stale axillary sweat, l-pyrroline in semen and male pubic sweat, trimethylamine in menstrual sweat and androstenone in male axillary sweat and stale urine. Musk (alpha-pentadecalactone) also appears to be a primary odour (Amoore, 1974, 1977).

Human beings are very sensitive to androstenone. Amounts as little as 0.049 μg of androstenone, dried over a 5 cm area of a watch glass, have been detected (Griffiths and Patterson, 1970). It may be significant that women are more sensitive to this male odour than men (Koelega and Koster, 1974). The odour of androstenone is perceived predominantly like urine, perspiration, animal or amine by 40% to 50% of the population. Others perceive the odour as musky or floral, i.e., more pleasant, and a few cannot detect androstenone at all (Amoore, 1977; Beets and Theimer, 1970). In the first group, women perceive androstenone as substantially more unpleasant than men (Griffiths and Patterson, 1970; Gower, Bird, Sharma and House, 1985).

Descriptions of androstenone and androstenol depend on the person and even trained perfumers sometimes find it hard to discriminate a particular steroid as musky or urinous (Kloek, 1961). Androstenone may smell urinous to those with the primary "urinous" receptor. Subjects anosmic to the primary urinous note may perceive the pleasant secondary musky notes through partial interaction of androstenone with the "musky" receptor. Androstenol might fit more closely into the "musky" receptor than the "urinous" receptor, hence giving it its musky odour. Totally anosmic people would lack both types of receptor (Guillot, 1953; Griffiths and Patterson, 1970; Beets and Theimer, 1970).

As mentioned above, the human axillary scent organ is a secondary sexual characteristic, being under the control of gonadal hormones. It begins producing these steroids and the musky odours at puberty. On the reception side, at puberty odour preferences appear to change from sweet and fruity odours towards these same musky and flowery odours (Moncrieff, 1966).

It seems more than a coincidence that man produces these compounds from localised areas of the body (which in animals would be considered sites of pheromone production (Comfort, 1974)) and also has specific receptor sites for these same odorants in his nose, with a correspondingly high sensitivity and a preference towards them at sexual maturity.

2.4 BRAIN CONNECTIONS

The apparent lack of cortical representation of the olfactory sense may suggest that odour is of little importance to man. The main projection areas of the olfactory bulb are to the limbic system. This was previously called the rhinencephalon or "smell brain" due to the olfactory bulb's widespread projections to this area. These projections are to the anterior olfactory nucleus, the olfactory tubercle, the prepyriform cortex and the amygdala. A secondary projection area is the hypothalamus, which is only two synapses from the olfactory bulb. In addition, there are efferents to the olfactory bulb from many of these areas (Heimer, 1975; Alberts, 1974, Powell, Cowan and Raisman, 1965). With such close links, the olfactory bulb might be thought of as part of the limbic system.

These limbic structures influence mood, learning, memory and emotional states (including anxiety, rage, placidity and alertness), visceral and endocrine responses, and behaviour, particularly aggressive, oral and sexual activity. It does not seem possible to define the function of each part of the limbic system because none of it appears to act as a centre for a particular function (Pincus and Tucker, 1974).

The olfactory sense is unique in its connections to the limbic system compared to the other senses, whose main projections are direct to the memory areas in the neocortex via the thalamus. In the olfactory sense the projections to the thalamus are only secondary, via the amygdala and pre-pyriform cortex (Motokizawa, 1974).

This difference may have implications for the perception of odours. Hughes Hendrix, Wetzel and Johnston (1969) measured the electrophysiological responses to odours of human subjects by placing electrodes in their olfactory bulbs. The difference in amplitude recorded when subjects reported detecting and not detecting an odour was often small or non-existent. Thus, the olfactory bulb could still be sending out strong signals, which although not reaching the neocortex and thus awareness, would still affect the primary projection areas, i.e., the limbic structures.

This evidence suggests that olfactory perceptions may be very different, especially in affective associations, from those of the other senses with their greater neocortical representations, and provides a neurophysiological basis for the association between olfaction and sexual functions. In addition, the effects of odour on mood and emotion may not be consciously recognised.

2.5 LEARNING PROCESSES

Odour effects in human young and adults may depend on experience, early and later in life, and the effects of early experience may be retained into adulthood. In any case, the effects of even quite recent experience will rely on long-term memory. Some of these long-term consequences may be experience dependent maturation effects. The presence or absence of odours during development can affect the final organisation of the mitral cells in the rat olfactory bulb, hence influencing the initial processing of olfactory information (Piggott and Osisiogu, 1977).

The effect of odour experience may be very long lasting because odour memory is strikingly persistent, e.g., compared to visual stimuli, especially if the odours have a personal and emotional relevance (Engen, 1977, 1982). Laird (1935) notes the common experience of emotional reactions being evoked by personally meaningful odours, perhaps best exemplified in Proust's "A la Recherche du Temps Perdu", which was inspired by the emotions and memories aroused by the smell of a Madeline sponge in tea. Lawless and Engen (1977) suggest that the olfactory memory acts as part of an alerting mechanism, which sets an affective tone for subsequent behaviour and acts in concert with other sense modalities.

The relative difficulty in labelling odours (Lawless and Engen, 1977) and in verbal association learning (Davis, 1977) suggests odour-dependent behavioural and mood effects would be hard to talk about or introspect on. These results indicate that odour memory is different from the memories of the other senses. This is perhaps a reflection of the lack of cortical representation of the olfactory sense and of the widespread projections of the olfactory tracts to the limbic system.

Regarding the learning process, an odour might gain a signal value as a discriminative stimulus by association with sexual stimuli, that is, the presence of the odour would increase the likelihood of sexual behaviour being reinforced. An odour could also become a conditioned stimulus through pairing with the unconditioned stimuli of the warmth and feel of another's skin, sight of attractive appearance or genital and other erotic sensations. The odour would then be seen as attractive on its own account, or would set up an expectation of the onset of unconditional stimuli.

Learning is also likely to involve some complex of maturational, habituation, reinforcement and conditioning mechanisms. The importance of purely "cognitive" effects on the reactions to odours should not be disregarded either. Hearing other people's opinions and instructions, and seeing other people's behaviour, for instance, may play a subtle part in determining any individual's response to an odour.

Natural or artificial odours may acquire a stimulus value through such complex learning mechanisms. The conditioning and responses to natural odours may be generally more effective than to artificial odours since they are universally present and may exhibit very low olfactory thresholds. The conditioning could be long lasting, especially in emotionally significant circumstances, due to the long term nature of odour memory and its distinctive affective qualities and responses to a stimulus odour could be a change in affective state or mood that is hard to verbalise.

2.6 SUMMARY AND IMPLICATIONS

The reasons given in the foregoing sections for the possibility of odour communication in human beings may be summarised thus:

1. Odour sensitivity and perception in man are similar to those of animals in which odour communication is important.

2. Human beings produce compounds that are similar to, by glands that are similar to, those compounds and glands used for chemical communication by other animals.

3. There appear to be specific, and genetically specified, receptors for these same compounds on the human olfactory epithelium.

4. The widespread and diffuse connections of the olfactory bulb with the limbic system suggest that odour perception is associated with uniquely affective sensory effects.

5. The genetically determined attributes of a specific and sex-dependent odour production and the corresponding highly sensitive and specific reception combined with the underlying brain structure would optimise the likelihood of a long-term, situational specification of responses to the odour.

6. Responses to odours could be hard-to-verbalise changes in mood, affective state or social attitudes, which could have been based on the long-lasting facts of previous emotionally significant experiences.

Androstenone and androstenol are likely to acquire such responses. By being secreted in male axillary sweat, they will be present in human sexual situations, which are highly emotionally significant circumstances and optimally suited for low-level association with these odours. We might expect consequent long-term effects in response to them.

The communication system outlined has a biological basis in the well specified production and reception of the odours. However, the response to these odours may be learned. If learning is involved the inevitable consequence is flexibility - so response to these odours will depend on circumstances and contexts, and will not be easily specified.

3. The Effects of Biological Odours in man

The above analysis suggests that the components of an olfactory communication system exist in man, with both innate and learned aspects. In this section the known influences of biological odours on human behaviour are examined along with the factors which may affect response to them.

3.1 IDENTIFICATION BY ODOUR

In lower mammals, the early olfactory environment has been shown to be important in the development of preferences in rat pups (Marr and Gardner, 1965) and infant spiny mice (Porter and Etscorn, 1976). If odour influenced behaviours may be acquired by habituation or associative conditioning early in life in lower mammals, then there is an increased likelihood of odour effects in higher mammals since the greater development of the neocortex and of neocortical mechanisms implies a lessened dependence on phylogenetically primitive functions, and a greater influence of individual life experiences via learning and memory (Michael, Bonsall and Warner, 1974).

The long-lasting nature of odour memory implies that "odour signatures" may be discriminated for longer periods of time than other phenotypic traits. Neonates may be particularly sensitive to olfactory cues, especially those from lactating females. Discrimination of mothers' individual breast odours is evident as early as 3 days postpartum by their breast-feeding newborns (Schaal, 1988). Since this is several weeks before unambiguous visual recognition of the mother's face (reviewed by Porter, 1991), these olfactory cues might contribute to the initial stages of the developing mother-infant relationship. Two week-old breast fed babies orient towards their mother's axillary odours rather than those of other lactating women (Cernoch and Porter, 1985). Older children of 3 to 5 years old also prefer their mother's axillary odour (presumably the main odour on a worn T-shirt) rather than other women's odour (Schaal, Montagner, Hertling, Bolzoni, Moyse and Quichon, 1980).

The learnt nature of this response is reflected in the ability of human newborns to acquire associations with artificial odours that are similar to those with the mother's own odour. Babies aged 3 to 10 days responded similarly to perfume present on their mother's neck and their mother's breast odour (Schaal, 1986). Such odour contingency can be established very swiftly by associative processes or mere exposure. These new associations need not be linked to feeding. A presentation of a novel odour during a 15 minute massage period is sufficient to bias the preferences in favour of this odour in 4 to 16 hour old neonates (Lott, Sullivan, and McPherson, 1989). A similar shift in olfactory preferences was obtained in 2 day old neonates through exposure to perfumes for the previous 24 hours while laying in their crib (Balogh and Porter, 1986). These studies illustrate how learning mechanisms enable synthetic fragrances to acquire a psychobiological meaning in the same way as body odours.

Interestingly, psychoanalytic writers have long suggested the possibility of a role for odour in human infant psychosexual development (Brill, 1932; Daly and White, 1930; Bieber, 1959; Fitzherbert, 1959; Kalogerakis, 1963). Bieber proposed that the adult sexual odours are the first stimuli that contribute to a two-to-five-year-old child's development of sexual discrimination and organisation, rapidly supplemented by the other sense modalities. If the psychoanalytic concept of repression were valid, olfactory components of sexuality might be operative in adult life but not consciously (Groddeck, 1925).

There is no reason to doubt that the process of learning to discriminate people by odour, with its link to the formation of social attitudes, will be carried through into adulthood, where the relevant odours may typically be a person's body odour, perfume, and other personal fragrances.

The odour of a particular person or type of person will carry a set of associations. These learned associates will set up expectations and emotional responses, and these tendencies will influence the style with which the perceiver reacts to people with that odour in circumstances related to the original experiences. When identity or gender influences behaviour, the olfactory identification will tend to channel the perceiver's response in an individual-relevant or gender-relevant direction.

Thus, all the evidence for olfactory discrimination between people is evidence for olfactory elicitation of attitudes about people. Each olfactorily identified person will have her or his own different salience to the perceiver, who will therefore tend to respond accordingly. Furthermore, one's developing perception of how people should smell will apply similarly to oneself. The elaboration of social awareness and self-image are intimately related.

Surprisingly, studies on adults' abilities to tell gender by odour have provided mixed results. Russell (1976) found that 22 out of 29 subjects correctly identified a T-shirt worn by a man for 24 hours from one worn by a woman. Hold and Schleidt (1977) found that only 16 out of 50 subjects reliably discriminated between 10 T-shirts worn for 7 nights. Some of these results may be due to the high levels of specific anosmia for androstenone (up to 46% in both men and women), since this may be a discriminatory odour (Gower, Bird, Sharma and House, 1985). Non-anosmic women were found to characterise T-shirts bearing a synthetic sweat with an increased concentration of the axillary steroid androstenone as more male, independently of the rated strength of the odour (Van Toller, Dodd and Kirk-Smith, 1980).

However, male odours tend to be described as unpleasant and strong and female odours as pleasant and weak, whether the assignment was correct or not, indicating cultural expectations as to the appropriate odour (Doty, Orndorff, Leyden, and Kligman, 1978b; McBurney, Levine and Cavanaugh, 1977).

A cross-cultural study by Schleidt, Hold and Attili (1981) provides further evidence for cultural influence on odour perception. German women classified their actual partner's odour as pleasant, in Japan and Italy it was classed as unpleasant. However, the supposed partner was classified as pleasant by German and Italian women; but both actual and supposed were rated as unpleasant by Japanese women. This was the only case in which the supposed partner did not smell good.

The Japanese judge personal odours less pleasant than Germans and Italians. There is strong cultural pressure to suppress obvious body odours in Japan, reflected in the old tradition of taking a ritual bath every day, and which may also reflect a more negative attitude towards sexuality compared to Italy and Germany.

3.2 THE EFFECTS OF ANDROSTENONE AND ANDROSTENOL

Many writers have suggested that the sole function of the axillary organ gland is to produce a scent attractive to the opposite sex (e.g., Laycock, 1840; Meisenheimer, 1921; Hagen, 1901; Brody, 1975). Hagen (1901) comments: "In human beings, in whom the body scent is less useful, the sexual attraction seems to be almost the sole effect which still has any significance. In this attractive power the axillary scent in man seems to be superior to the scent of the sexual organ or any other part of the body......the apocrine glands in the axillae, to which the axillary scent is essentially due, develop first a few years before the onset of puberty........ This also supports the existence of sexual significance of the axillary scent". Ellis (1905) and Schiefferdecker (1922) report on the use of axillary odour by men to sexually arouse women, e.g., a young man would carry his handkerchief in his axillae for a time while dancing, and then use it to wipe her perspiring face.

More recently, the search for male olfactory attractants in axillary sweat has centred on the effects of androstenol (musk-scented) and androstenone (smelling like urine), with mixed results across several types of experiments. Subjects breathing androstenol, asked to assess written descriptions of alleged job applicants, rated some candidates differently from controls (Cowley, Johnson and Brooksbank, 1977). In another study, the presence of androstenol at unnoticed levels made women in photographs appear more sexually attractive (Kirk-Smith, Booth, Carroll and Davies, 1978).

Kirk-Smith and Booth (1980) found that women were attracted to sit on an androstenone treated seat in a dentist's waiting room, whereas men avoided it. Gustavson, Dawson and Bonett (1987) have suggested that androstenol may be a male "spacing pheromone", since in the presence of androstenol men's toilet stalls were used less frequently than during baseline periods or during androstenone treatment. Women displayed no difference in choice between androstenol and control periods.

Benton (1982) found that women exposed to androstenol for a month (placed on their upper lips each morning) rated themselves as more submissive than the placebo group in the middle of their menstrual cycle. In contrast, women exposed to androstenone rated themselves as less sexy than control subjects when judging photographs and descriptions of a stimulus male college student, while men's moods were not affected. Neither the male or female judges' ratings of the stimulus male changed (Filsinger, Braun, Monte and Linder, 1984)

In Black and Biron's (1982) study subjects interacted briefly with a confederate wearing androstenol or a control substance. Androstenol treatment had no noticeable influence on ratings of physical attractiveness of the opposite-sex confederate by either males or females. Cowley and Brooksbank (1991) found that women wearing a collar bearing androstenone rated themselves as having significantly more and longer social exchanges with men than control subjects, and it was suggested that androstenol increased female responsiveness.

These experiments form an inconsistent pattern and thus provide little basis for concluding that human males produce odours analogous to the sexual attractants of other species.

3.3 CONTEXTUAL EFFECTS

The lack of clear evidence of general sex attractants in man acting as in other mammals can be explained by the more complex, and verbally mediated, social situation of human beings. If responses to these odours are learnt, we can expect that the learning situations encountered by human beings are likely to be more varied than for animals, with a consequent wider range of responses.

A biological odour is likely to be made up of many components. A particular component may be a constituent of different odour mixtures. For example, similarly smelling musks occur in perfumes and in sweat. If the response is learned, the rest of the mixture would specify whether a perfume or sweat associated response is elicited, i.e., the response is to the whole mixture, not just a particular component, e.g., androstenone, within it.

Generalising this point, the situation, whether perfume or sweat is present, may determine which way the response to the musk-like component goes. Benton (1982) puts the point this way: "In our culture the use of perfumes may lead to the association of musk-like odours with particular situations and expectations. Alternatively, if this odour is produced naturally by the human male, interactions between the sexes may lead to the development of learned associations".

Any examination of the effect of odours on human social behaviour must take into account the complexities of the total setting. The odorous mixture, whether biological or synthetic, is only a part of what determines a response. The odour does indeed carry the meaning of its associate, but only as one of an array of stimuli. What has acquired meaning may not be the odour by itself, but the whole stimulus complex presented to the subject.

This conception of the acquired social effect of an odour is represented in behavioural psychology as the Pavlovian conditioning of the fragrance as just one stimulus element in a compound conditioned stimulus. For example, Asratyan (1961) conditioned a dog to salivate to a tone in one chamber (where it had been paired with food), but to flex a limb to the same tone in a different chamber (where it had been paired with an electric shock). The conditioned stimulus (the tone) by itself determined only when the response would occur, but not what the response would be. The chambers themselves had acquired conditioned properties, i.e., they determined or 'switched' the particular response to the auditory stimulus.

This situation can be generalised to odours. The same odour could cause completely different responses through manipulation of the situation and the state of mind. For instance, a perfume that arouses passion in an intimate setting with a particular person may not have the same effects in a public social context, e.g., at work, where one interacts in non-sexual ways with several others of varying degrees of acquaintance.

Similarly, the influence of a perfume on a new acquaintance will inevitably be different from its effects when repeatedly used in an intimate domestic setting. The response may also be quite individual, e.g., a sexual partner's unique body odour may be arousing because of frequent prior associations with sexual activity, but be of little relevance to others. The fact that any particular component is present, such as androstenone, may not make the body odour any more appealing to others.

In contrast, there may be simple conditioning to the olfactory stimulus. For example, as a child a 55 year-old man was terrified of a female teacher who wore a particular perfume (personal communication). After many years the same perfume still evoked anxiety, despite being encountered repeatedly at social functions and presumably on attractive and friendly women.

Whether compound or simple conditioning occurs, i.e., whether or not one retains the same response to an odour independent of circumstances, will depend, in part, on the salience of the context relative to the salience of the odour stimulus, and the degree of training in discriminating the different circumstances and their perceived consequences. The difficulty in identifying the individual influences of these factors means that the response to a particular odour will be in practice hard or impossible to predict.

The research literature provides clear examples where these factors may explain apparently conflicting effects. For instance, Baron (1983) has shown that when women job applicants wore fragrance, male subjects liked them less, rated them lower in intelligence and friendliness, whereas the opposite was true for female interviewers. Baron gives a plausible cognitive interpretation: "...these findings suggest that males were more strongly affected by scent and other extraneous aspects of the applicants' appearance or grooming than females, and were quite aware of this fact...Males may have realised that such scents would interfere with their ability to serve as an effective interviewer. This, in turn, caused them to experience annoyance or resentment towards the applicants and so to downgrade them on several key dependent measures." In other words, the reduction of attractiveness caused by the perfume can be attributed to crucial aspects of the social setting. This casts in doubt any simplistic notion that perfume can be used as a mere cosmetic or as a signal irrelevant to social context: evaluation for job efficiency is quite different from, say, acting as a host at a party.

These results contrast sharply with Kirk-Smith and Booth's (1991) finding that in the presence of a perfume, both men and women rated photographed men and women as sexier and softer, and rated their mood as becoming more sexy, compared to a no-perfume situation. In this experiment, there was no pressure on subjects to 'perform' as was the case in the Baron experiment. Also, the odour was not shown in a male/female-specific context, since subjects were shown both men and women in the photographs. In other words, Baron gets his results because people have a predisposition to perform in that way in that situation. Similarly, Kirk-Smith and Booth get their results because of the particular social circumstances of their experiment.

An experiment by Rotton, Barry, Frey and Soler (1978) further demonstrates how context influences the response to odour. They hypothesised that the presence of an unpleasant odour (butyric acid) would create a negative mood, which would reduce peoples' rated attractiveness. Each subject rated their liking for the expressed attitudes of a fictitious person, which were chosen to be similar (75%) or dissimilar (25%) to those of the subject. Contrary to expectations, subjects in the presence of the bad odour rated the person as more attractive compared to subjects rating without the odour.

It was then discovered that the experimental subjects had thought that the rated person was also present and suffering the bad odour. When a "fellow sufferer" existed, the malodorous environment increased positive feelings towards them, i.e., a common threat increased the attraction for others. A subsequent study showed that subjects rated people lower when they were perceived not to be present. Thus the same odour can create different responses, in this case the "switch" simply being subjects' differing beliefs about a person's location.

In summary, if a putative human chemical communication system has a learnt component, as outlined above, then in modern human society many factors act together to ensure that responses to biological odours will not be as stereotyped as in other mammals, despite their biologically specified production and reception. Biological odours may or may not be present (through washing or deodorising); synthetic fragrances may replace or mask these natural odours, and may in part mimic them. Since they will be present in different social and emotional situations, and given man's ability to interpret situations in complex ways (e.g., in Rotton et als'. study), the conclusion is that human responses to biological and other odours are likely to be highly variable.

4. The Use of Synthetic Fragrances

We have now reviewed the evidence for the components of a chemical communication system in man, and examined the effects of certain biological natural odours, and the determinants of the responses to them. Human cultures universally use synthetic fragrances, either as perfumes applied to the body, or incense, dispersed by burning. In this section the links between biological odours and such synthetic odours are examined, and how different cultures, past and present, use these fragrances.

4.1 ANIMAL ODOURS IN PERFUMES

Mammalian sex attractants from animal secretions, or more recently their synthetic analogues, have been universally used in perfumes. The most notable of these are musk, extracted from the preputial gland of the Himalayan musk deer, and civet, extracted from the anal gland of the civet cat. In perfumery these animal secretions are considered to be fixatives or "base notes". Their function is to prevent the too-rapid evaporation of the floral notes of a perfume by acting in a similar way to the wax or resin of a flower or leaf.

The odorous constituents of each gland are respectively muscone (3-methyl-1-cyclopentadecanone) from the Himalayan musk deer and the related compound civetone (9-cycloheptadecen-l-one) from the civet cat (Kingston, 1964). In high concentrations their odours resemble the musky/urinous odours of androstenol and androstenone. Figure 1. shows that although they are not steroids, their molecular shapes are remarkably similar. Ohloff, Giersch, Thommen and Willhalm (1983) have shown that the steroid odour (i.e., the musky or urinous note) can be triggered by any molecule having a shape which resembles that of a steroid, whether or not it has steroidal biochemical activity.

The use of musk in perfumes has been reliably traced back over 5500 years to China (Green and Taylor, 1986). The first mention of musk in Europe is attributed to St Jerome in about AD 390, and the earliest reference is in a manuscript, now held by the British Museum, dated 1398. In previous centuries, these compounds could predominate perfumes. The earliest book on perfumes "Les secrets de Maistre Alexis de Piedmont" (1580) gives a scent formulation made up of 20% musk, 8% civet and 12% storax (a musky smelling plan resin). Most modern perfumes also incorporate musk or civet analogues; a recent survey showed 85% female and 94% male fragrances contain musk smelling compounds and 39% and 6% respectively contain civet smelling compounds (Berliner, Jennings-White and Lavker, 1991).

Many writers have long emphasised the impact of musk on sexual behaviour. Fere (1899) considered musk to be more reminiscent of sexual secretion than any other perfume. Hagen (1901) suggests that before the 19th century women used the strongest, most animal, perfumes to emphasise their body odour, as a secondary sexual characteristic. Non-Western cultures have also adopted the odour, in Melanesia, for example, men wear a strong musky aromatic leaf during dances (Davenport, 1965).

Civet has been used in perfumes since Biblical times. At high concentrations it has a urinous/faecal odour, but in very dilute concentrations civet has a more floral-musky odour than musk. Like musk it was also employed as an aphrodisiac with the power to attract the opposite sex. Pratt (1942) quotes the Middle Ages writer Petrus Castellus as saying that civet "will cause so much desire for coitus that she will almost continually wish to make love with her husband." Shakespeare also cites its use. In "Much Ado about Nothing" Don Pedro asks "... he rubs himself with civet. Can you smell him out by that?" To which Claude replies "That's as much to say, the sweet youth's in love".

In summary, people wash off their natural scents and then replace them with the sex attractants of other animals, which resemble human steroidal axillary and urine odours. Apart from musk and civetone, or their analogues, perfumes contain many other compounds that have odours resembling other human, particularly male, body odours, e.g., plant resin alcohols also have a steroidal smell and henna has the strong semen-like odour of henna flowers.

Why are perfumes used? Perfumes are generally marketed as having the ability to enhance sexual attractiveness, reflected in both advertisements and their names, e.g., 'Intimate', Le Temps d'Aimer', Sex Appeal', 'Moment Supreme', L'Aimant', and 'Aphrodisia'. Although perfuming is widely used by both sexes, it seems that in most cultures, including our own, women are more fond of artificial scents than men, perhaps because of an unconscious association of the scent with the distinctly musky male odours they carry.

Jellinek (1954) suggests that these odours are only "stimulating" providing that the mental associations induced by them remain unconscious and that no erotic effect is induced if the resemblance to animal or human body odours is recognised - indeed such a perfume would be rejected. Based on reactions to forty three different odours, Brill (1932) reported that subjects "disliked most" the odour of perspiration.

There is some evidence that this is a culturally learnt reaction to human odours. Other societies do not regard body odour as negatively as ours (see below). Human faeces and sweat smell pleasant to children of three or four years old, and unpleasant afterwards (Stein, Ottenberg and Roulet, 1958). The relative roles of cultural and maturational factors in this change have not been determined, although cultural processes are known to be very powerful by the age of five.

As discussed above, children can be conditioned to associate synthetic fragrances with psychobiological meaning in the same way as body odours. Such acquisition of olfactory attitudes in adulthood has been demonstrated experimentally by Kirk-Smith, Van Toller and Dodd (1983) who showed that a neutral artificial odour paired with an emotional state was able later to evoke that same state. The odour was not noticed by the subjects, which lends some support to Jellinek's suggestion that a fragrance can act unconsciously.

It is also possible that any sexual response to a body odour, albeit unconscious, might be replaced by a more powerful response to synthetic fragrances, perhaps through more frequent pairing in sexual or social sexual situations. Indeed, Kirk-Smith and Booth (1990) found that men and women rated photographed men and women as sexier and softer, and rated their own mood as becoming more sexy, effects not seen when androstenone was used. In each case, the levels of odour used were unnoticeable.

4.2 PLANT ODOURS IN INCENSE

Perfume comes from the Latin 'per fumen' (by smoke). This would be a more suitable name for incense, which is made of fragrant plant resins dispersed into the air by burning. The use of incense for religious ritual is almost universal. Stoddart (1990) identifies the following functions:

1. as a bloodless sacrifice to a deity
2. as a demonifuge, for casting out evil and purification
3. as a sacrifice to the honour of a deceased person
4. as a symbol of honour to the living
5. as an accompaniment to processions, festivities, etc., to establish the correct mood
6. as a refreshing perfume in buildings where people gather.

The earliest recorded use of incense comes from the ancient Chinese, from whom it was taken and used in religious ritual by the Hindus (e.g., the god Siva is offered incense every four hours). It is likely that the ancient Egyptians acquired the incense cult from the Hindu traders around 3600 BC calling frankincense 'sntr' or 'god's odour' (Muller, 1978), believing it to be the sweat of the gods.

The ancient Hebrews used incense from between 300 and 200 BC (Jackson 1909), e.g., Exodus 30:34-36 cites a recipe given to Moses prior to the Exodus; and it was highly valued, c.f. the biblical "gold, frankincense and myrrh". During the first four centuries AD incense was only used by Christians as a fumigant, but became widely used after Constantine the Great inaugurated the Peace of the Church, e.g., the burning of incense at the moment of transubstantiation in the Roman Catholic mass.

There are many plants that could be used in incense, but in practise only between 10 and 20 have been in regular use throughout history. The central ingredients are the five plant resins of myrrh (balsam), frankincense (olibanum), ladanum (onycha), galbanum (asafoetida) and styrax (storax, balm), the barks of cinnamon and cassia and the wood of sandalwood. The use of incense in the Jewish and Christian religions depends upon the first seven ingredients with occasional use being made of sandalwood. Why should such a narrow choice of plants be used?

Major odorous fragrance components of these plant resins are the resin alcohols. These have a four or five fused carbon ring structure and are thought to be derived from phytosterols (Bell and Charlwood, 1980). Figure 1. shows the structural similarity of a typical resin alcohol, amyrin from myrrh, with androstenol and androstenone.

It is possible that resin alcohols are used in incense because they are also molecular analogues of androstenol, just as muscone and civetone in perfume. Significantly, Kloek (1961) noted that subjects more commonly described the smell of androstenol as 'like incense', than musky or like urine, and also related the scents of the major incense ingredients to body odours, i.e., frankincense with sebum of dark-haired subjects, myrrh with sebum of fair-haired subjects, styrax with sweat and skin odour of dark-haired subjects and ladanum with head hair of all hair types.

Thus incense may have an odour that is recognised, possibly unconsciously, as resembling human body odour in general and the steroidal odours implicated in sexual attraction in particular. Stoddart (1990) comments that there can be no odours more able to stimulate the deep emotional levels of the brain than those associated - however distantly and indistinctly - with sexual attraction. This may be a key to the use of incense in the emotionally charged setting of the religious ritual.

4.3 THE CROSS-CULTURAL USE OF FRAGRANCE

Howes and Synnott (1991) have reviewed olfactory traditions across Africa, Oceania and Latin America and identified the uses of the sense of smell as classificatory or dynamic:

The Suya of Brazil use the terms strong-smelling, pungent, bland or rotten to classify animals and plants rather than by morphology or habitat. These same terms are used to classify people, though these change depending on the sex, stage of life cycle and transition through states such as initiation and illness. The Batek Negrito of Malaysia classify virtually everything in their environment by smell, including the sun ("bad, like that of raw meat") and moon ("good, like that of flowers") (Endicott, 1979).

'Dynamic uses' refer to odour in ritual and everyday contexts, with a view to changing the world or returning it to its proper state. For example, to ensure group identity, East African farmers smear themselves with cattle products to smell differently from the neighbouring fishermen. In the Nuba (Sudan), olfactory similarity is socially enforced to group integration (Ebin, 1980); in the Near and Middle East, strangers are welcomed and integrated into the community following several perfuming rituals (Aubaile-Sallenave, 1987). Similar to Western aromatherapy, female Waroa herbalists employ odours to heal, both by curative smells and by creating a pleasant environment, and also use diagnosis by smell in their system of olfactory healing. The Desana exchange ants with a different smell as a ritual to establish exchange relations with other persons and groups. The Temiar use olfactory metaphors to express abstract concepts and values, such as the idea of an 'Odour Soul' (Hall, 1969). In the Amazonian Yanomamo or the New Guinean Kanum-lrebe, rubbing another's body and then sniffing one's fingers are common acts (Ales, 1987, Schulze-Westrum, 1968).

The range of olfactory practices identified is far wider than the idea of "what good is smell for" held by a typical North American or European, and suggest that all these practices are latent in the human condition by their existence in other cultures.

The relative importance of the sense of smell is also reflected in language. For example, in English "to smell" covers both transitive and intransitive meanings. In contrast, the Quechua language used in the central Andes has separate verbs meaning "to smell any smell", "to smell a bad smell", "to give a bad smell to others", "to smell naturally bad", "to leave a good smell", "to come across the remains of a smell", "to let oneself be smelled", etc.. The implication is that smell is more important to this culture than Western culture, where vision predominates (Howes and Classen, 1991).

Hall (1969) suggested that people of different cultures "inhabit different sensory worlds". Howes (1991) confirms that in other cultures smell is of far greater importance in the "ratio of the senses" than in Western cultures. He comments on the apparently universal use of odour in the context of puberty, death and other rites of passage. He suggests this has been weakened in Western society as a result of the declining presence of smells in the wake of the perceptual revolution of the mid-eighteenth century (see next section) and argues that the representation of smell as more individual or "private" than social, more animal than human, more biological than historical is a cultural representation, peculiar to the West, rather than an empirical reality.

However, Western man's use of olfactory cues has been well documented in social contexts. Several authors have discussed the role of real or alleged olfactory cues in racial, social class, or urban/rural differentiation in both past and modern Western societies (Largey and Watson, 1972; Corbin, 1986; Synnott, 1991). Synnott (1991) points out that odours have been used in the West to legitimise inequalities of both class and race, and quotes George Orwell: " The real secret of class distinctions in the West...is summed up in four frightful words...The lower classes smell". John Dollard (1957) notes that the "belief that Negroes have an extremely disagreeable smell to white people" has been used by racists in the USA to justify racial segregation by demonstrating the "impossibility of close association between races".

Although smell is considered the least important sense by Western man, and has received little academic study by sociologists (i.e., Synnott, 1991; Largey and Watson, 1972) and anthropologists, the social significance of odours is evident in the vast industry ($10 billion) involved in the production and removal of odours, e.g., as perfumes, after-shaves, deodorants and air fresheners.

4.4 THE SUPPRESSION OF BODY ODOURS

The universal use of synthetic fragrance over the human body suggests that olfaction is a communicative channel in humans. Rather than mask or be antagonistic to the natural body odours, such fragrances may enhance or complement a person's individual odour (Freytag, 1972). On the other hand, humans also suppress their natural body smell. The use of odour cues in other primates is well documented (e.g., Epple, 1986). Yet in human beings, who have the most developed scent glands of all primates, body odour is regarded as unpleasant and distasteful, with great efforts being expended in its removal. As Stoddart (1990 puts it: "The idea that people should smell of people is deeply distasteful to Westernised man". Doty (1972) quotes a comedian commenting on the number of deodorants available: "What bothers me is that the streets aren't safe, parks are dangerous and the freeways are impossible, but under our arms we have complete protection".

This active suppression of body odours in man can be interpreted from ethological and cultural viewpoints. The examples of the two perspectives presented below are not incompatible, and illustrate how different levels of explanation can account for this modification of the apparent biological adaptations for olfactory communication in man.

4.4.1 *An Ethological Interpretation.* Stoddart (1990) proposes an explanation based on the dual need for early man to live gregariously and to maintain a pair-bond. He suggests that since they lacked the natural weapons to kill the Miocene grassland mammals, man's ancestors had to band together and hunt co-operatively. In addition, the slow growth of the young and the increasing period of juvenile dependency guaranteed the survival of the nuclear family and paternal care as the most efficient means of the enhancement of reproductive success of both sexes.

In other primates, such as gibbons or mandrills, the females advertise oestrus by vaginal odours. In man such advertisement to other males of the group would have threatened the pair bond. Thus vaginal secretions have been rendered meaningless and even unpleasant (Doty, Ford, Preti and Huggins, 1975). Indeed, they do not appear to have any sexual influence in man (Kirk-Smith and Booth, 1991). The whole focus for sexual allurement shifted upwards from the genital region to the upper torso. Natural plant juices may have been used to mask oestrus advertising odours, as well as on those associated with pregnancy and lactation, and so would have created the odour anonymity upon which the pair bonding between male and female depended.

Within the pair-bonding of early man oestrus did not need to be advertised since women had developed extended female sexual receptivity. The odours from the newly developed axillary scent organs were not required for long distance signalling but were part of a highly developed system providing psychological rewards for privatised coitus. The apocrine glands produce musky smelling steroids as part of the psychological reward for pair-bonding sex.

Non-human primates frequently show mutual axillary sniffing; human evolution has simply developed this trait. There is no positive selective advantage for their odours to be perceptible, or liked, by any individual other than the bonded mate. This interpretation is in keeping with the conditioning aspects detailed above. The axillary odour could become a conditioned rewarding stimulus through pairing with unconditioned sexual stimuli, specific and meaningful only for the two partners, and thus help to maintain the pair-bond. Indeed, both sexes have been found to judge their own partner's axillary odour as more pleasant than unpleasant, although cultural differences, possibly related to the reasons for choice of partner and attitudes to sexuality, can affect this (Hold and Schleidt, 1977; Schleidt, Hold and Attili, 1981).

4.4.2 A Cultural Interpretation. In a unique review of the cultural aspects of olfaction, Corbin (1986) provides the following plausible interpretation of the development of suppression and deodorising of body odour in French culture, and by extension, in other European cultures.

He suggests that before the mid-18th century, there was a higher tolerance to malodours. The lack of an organised public health infrastructure meant that all members of society were daily and unavoidably subjected to many stenches, e.g., from excrement, refuse, stagnant water, and unwashed bodies, and these odours were simply accepted as part of life. In the late 17th century excrement was even endowed with therapeutic properties; there was seen to be nothing wrong about using it in aromatic preparations, especially if it came from strong and healthy individuals.

It was commonly thought that the best protection against disease was smelling strongly oneself. Bathing was seen as foolhardy since it removed the dirt that obstructed the pores and thereby gave one a "protective coating" against the penetration of the skin by miasmas. Furthermore, the vigour of the individual was associated with the intensity of his body odour, particularly the "aura seminalis" which was considered vital to sexual attraction. To use excessive amounts of musk and civet based perfumes was seen as a way of protecting oneself and of purifying the air, as well as aiding sexual attraction.

This situation was dramatically changed when scientific discoveries with antiseptics identified infection with organic waste and decomposition and their associated smells. Putrid odours came to signal disease. This led to massive sanitary reform involving the provision of sewers and washing water, the draining of cesspits and the ventilation of hospitals, with a consequent elimination of malodour in both public and private life.

The fear of disease caused the public to be less tolerant of putrid odours, including body odours. Dirt was now seen to obstruct the pores and so favoured the decomposition of substances. It was precisely the absence of strong odour, which was evidence of careful hygiene, that was regarded as the decisive criterion of good taste. The use of powerful masking perfumes now cast doubt upon a person's cleanliness.

Thus, the harmful effects attributed to putrefaction revolutionised attitudes to bodily products and, in a more general way, to all the animal substances used in perfumery, especially musk and civet. Corbin comments that Francis Bacon (1623) had originally noted that the smells of musk and civet were in the products of putrefaction. Doctors began to regard musk and civet as putrid and eminently septic substances, e.g., Virey (1813 cited by Corbin) declared that if human excreta were digested and fermented in a double boiler they acquired a musky odour.

By the mid-19 century doctors unanimously refused to accept the role played by these animal secretions in perfume. Havelock Ellis (1905) considered this discrediting of musk as a major fact in sexological history. In his opinion, and that of Hagen (1901), up to the end of the 18th century women did not use animal odours to mask their own odour but sought them to emphasise it. Light Floral scents such as rose, thyme, violet and rosemary water came to replace animal scents as appropriate for use by women.

Under this interpetation, scientific advances initiated a cultural shift in the perception and significance of odours, leading to a society in which both body odour and the sexual significance of animal odours are suppressed.

5. Conclusion

In conclusion, in man we have evidence of biological adaptation for a chemical communication system. Certain odours appear to be optimised by this system by beingassociated with reproductive activity and individual recognition. Widely varying cultures provide evidence of using these odours, or analogues of them, for this purpose. However, since the system depends on learning, it is also very flexible. From birth, with the identification of the mother's odour there is a life-long process of olfactory enculturation. Different cultures show considerable variation in the significance attached to these and other odours, and they may be used in mixtures for their original and other purposes, e.g., in perfumes and incense. The driving forces for these differences may be found, for example, in environmental determinants that alter the "ratio of senses", or in technically driven social changes. The sense of smell in man is an example of the ethological contention that most forms of behaviour will be either biologically adaptive or will be expressions of evolved tendencies which were adaptive earlier in our evolutionary history.

References

Alberts, J.R. (1974) 'Producing and interpreting experimental olfactory deficits', Physiology and Behaviour 2, 657-670.

Albone, E.S. (1984) Mammalian Semiochemistry, Wiley, Chichester.

Ales, C. (1987) 'Les parfums vegetaux des Yanomaml (Indiens du Venezuela)', in G. Meurgues and F. Aubaile-Sallenave (eds), Parfums de plantes, MNHN, Paris, pp. 242-246.

Amoore, J.E. (1970) Molecular basis of odour, Thomas, Springfield.

Amoore, J.E. (1974) 'Four primary odour modalities of man: experimental evidence and possible significance', in D.A. Denton and J.P. Coughlan (eds.), Olfaction and taste V., Academic Press, New York.

Amoore, J.E., Pelosi, P. and Forester, M.J. (1977) 'Specific anosmias to 5-androst-16-en-3-one and alpha-pentadecalactone: The urinous and musky primary odors', Chem. Senses Flavour 2, 401-425.

Asratyan, E.A. (1961) 'The initiation and localisation of cortical inhibition on the conditioned reflex arc', Annals N.Y.Acad.Sci. 92, 1141-1159.

Aubaile-Sallenave, F. (1987) 'Les parfums dans le monde musulman', G. Meurgues and F. Aubaile-Sallenave (eds.), Parfums de plantes, MNHN, Paris, pp. 165-167.

Baker, J.R. (1974) Race, Oxford Univ. Press, London.

Balogh, R.D. and Porter, R.H. (1986) 'Olfactory preferences resulting from mere exposure in human neonates', Infant Behav. Dev. 9, 395-401.

Baron, R.A. (1983) 'Sweet smell of success? The impact of pleasant artificial scents on evaluations of job applicants', J.Appl.Psychol. 68, 709-713.

Becker, R.F., King, J.E. and Markee, J.E. (1962) 'Studies on olfactory discrimination in dogs', Journal of Comparative and Physiological Psychology 55, 773-780.

Beets, M.G.J. and Theimer, E.T. (1970) 'Odour similarity between structurally unrelated odorants', in G.E.W. Wolstenhome and J. Knight (eds.), Taste and smell in vertebrates, Churchill, London.

Bell, E.A. and Charlwood, B. (1980) Encyclopedia of Plant Physiology. New Series: vol. 8. Secondary Plant Products, Springer Verlag, New York.

Bennett, M.H. (1968) 'The role of the anterior limb of the anterior commissure in olfaction', Physiological Behaviour 3, 507-515.

Benton, D. (1982) 'The influence of androstenol - a putative human pheromone - on mood throughout the human menstrual cycle', Biol.Psychol. 15, 249-256.

Berliner, D., Jennings-White, C. and Lavker, R.M. (1991) 'The Human Skin: Fragrances and Pheromones', J. Steroid Biochem. Molec. Biol. 38(4b), 671-679.

Bieber, J. (1959). Olfaction in sexual development and adult sexual organization. Am. J. Psychother. 13, 851-859.

Bird, S. and Gower, D.G. (1982) 'Axillary 5-androst-16-en-3-one, cholesterol and squalene in men: Preliminary evidence for 5-androst-16-en-3-one being a product of bacterial action', J. Steroid Biochem. 17, 517-522.

Black, S.L. and Biron, C. (1982) 'Androstenol as a human pheromone: No effect on perceived physical attractiveness', Behav. Neural Biol. 34, 326-330.

Brill, A.A. (1932) 'The sense of smell in the neuroses and psychosis', Psychoanal. Quart. 1, 7-42.

Brody, B. (1975) 'The rescual significance of the axillae', Psychiatry, 38, 278-289.

Brooksbank, B.W.L. (1962) 'Urinary excretion of androst-16-en-3-ol. Levels in normal subjects and effects of treatment with trophic hormones', J. Endocrinol. 24, 435-444.

Brooksbank, B.W.L., Brown, R. and Gustafsson, I.A. (1974) 'The detection of 5-androst-16-en-3-ol in human male axillary sweat', Experientia 30, 864-865.

Brown, R.E. and Macdonald, D.W. (eds.) (1985) Social Odors in Mammals, Oxford Univ. Press, Oxford.

Buck, L. and Axel, R. (1991) 'A novel multigene family may encode odorant receptors: a molecular basis for odor recognition', Cell 66, 175-187.

Cernoch, I.M. and Porter, R.H. (1985) 'Recognition of maternal axillary odors by infants', Child Dev. 56, 1593-1598.

Claus, R. and Alsing, W. (1976) 'Occurrence of 5-andros-16-en-3-one, a boar pheromone, in man and its relationship to testosterone', J. Endocrinol. 68, 483-484.

Claus, R., Hoppen, H.O. and Karg, H. (1981) 'The secret of truffles: a steroidal pheromone?', Experientia 37, 1178-1179.

Comfort, A. (1974) 'The likelihood of human pheromones', in M.C. Birch (ed.), Pheromones, North Holland, Amsterdam.

Corbin, A. (1986) The Foul and the Fragrant, Berg Publ., Leamington Spa.

Cowley, J.J. and Brooksbank, B.W.L. (1991) 'Human exposure to putative pheromones and changes in aspects of behaviour', J. Steroid Biochem. Molec. Biol. 39(4b), 647-659.

Cowley, J.J., Johnson, A.L. and Brooksbank, B.W.L. (1977) 'The effect of two odorous compounds on performance in an assessment-of-people test', Psychoneurendocrinology 2, 159-172.

Daly, C.D. and White, R.S. (1930) 'Psychic reactions to olfactory stimuli (a preliminary paper)', Br. J. Med. Psychol. 10, 70-87.

Davenport, W. (1965) 'Sexual patterns and their regulation in a society of the southwest Pacific', in F.A. Beach, (ed.), Sex and Behavior, Wiley, New York, pp. 164-207.

Davis, R.G. (1977) 'Acquisition and retention of verbal associations to olfactory and abstract visual stimuli of varying similarity', J. Exp. Psychology: Human learning and memory 3, 37-51.

Doty, R.L. (1972) 'The role of olfaction in man: Sense or nonsense'. In Bartley, S.H. (ed.), Perception in Everyday Life, Harper and Row, New York.

Doty, R.L., Orndorff, M.M., Leyden, J. and Kligman, A. (1978) 'Communication of gender from human axillary odors: Relationship to perceived intensity and hedonicity', Behav. Biol. 23, 373-380.

Doty, R.L. (ed.) (1976) 'Mammalian Olfaction, Reproductive Processes, and Behavior', Academic Press, New York.

Doty, R.L., Ford, M., Preti, G. and Huggins, G.R. (1975) 'Changes in the intensity and pleasantness of human vaginal odors during the menstrual cycle', Science 190, 1316-1318.

Ebin, V. (1980) 'Les parfums en Afrique Noire', in L.A. Roubin (ed.), Hommes, Parfums et Dieux , Musee de l'Homme, Paris, p.7.

Ellis, H. (1905) Studies in the Psychology of Sex; Sexual Selection in man. Part 11. Smell, Vol. 4, Davis, Philadelphia, Pennsylvania.

Endicott, K. (1979) Batek Negrito Region, Clarendon Press, Oxford.

Engen, T. (1970) 'Man's ability to perceive odours', in J.Johnston, D. Moulton and A. Turk (eds.), Advances in Chemoreception, Vol. 1. Communication by chemical signals, Appleton-Century-Crofts, New York.

Engen, T. (1977) 'Taste and smell', in J.E. Birren and K.W., Schaie (eds.), Handbook of Psychology and Aging, Van Nostrand-Reinhold, New York, pp. 554-561.

Engen, T. (1982) The Perception of Odors, Academic Press, New York.

Epple, G. (1986) 'Communication by chemical signals', in G. Mitchell and J. Erwin. (eds.), Comparative Primate Biology Vol. 2A, Liss, New York, pp. 531-580.

Fere, C. (1899) The Pathology of Emotions (trans. by R. Park), University Press, London.

Filsinger, E.E., Braun, J.J., Monte, W.C. and Linder, D.E. (1984) 'Human responses to pig sex pheromone 5-androst-16-en-3-one', J. Comp. Psychol. 98, 219-222.

Fitzherbert, J. (1959) 'Scent and the sexual object', Br. J. Med. Psychol. 32, 806-809.

Freytag, H. (1972) 'Die Bedeutung des Parfums innerhalb des Adaptations moddells der Kosmetic', J. Soc. Cosmet. Chem. 23, 811-822.

Gower, D.G. (1972) '16-Unsaturated C19 steroids: A review of their chemistry, biochemistry and possible physiological role', J. Steroid Biochem. 3, 45-103,

Gower, D.G., Bird, S., Sharma, P. and House, F.R. (1985) 'Axillary S-androst-16-en-3-one in men and women: Relationships with olfactory acuity to odorous 16-androstenes', Experientia 41, 1134-1136.

Green, M. and Taylor, R. (1986) 'The musk connection', New Scientist 11O, 56-58.

Griffiths, N.M. and Patterson, R.L.S. (1970) 'Human olfactory responses to 5-alpha-androst-16-en-3-one - principal component of boar taint', J.Sci. Food and Agriculture 21, 4-6.

Groddeck, G. (1925) The Unknown Self, Daniel, London.

Guillot, N. (1953) 'Anosmia and parosmia, a few individual cases', Recherches (Paris) 3, 26-31.

Gustavson, A.R., Dawson, M.E. and Bonett, D.G. (1987) 'Androstenol, a putative human pheromone, affects human (Homo sapiens) male choice performance', J. Comp. Psychol. 101, 210-212.

Hagen, A. (1901) Die Sexuelle Osphresiologie, H. Barsdorf verlag, Charlottenburg.

Hall, E.T. (1969) The Hidden Dimension, Doubleday, New York.

Ellis, Havelock (1936) Studies in the psychology of sex, Vol.1, Random House, New York.

Heimer, L. (1975) 'Olfactory projections to the diencephalon', in W.E. Stumpf and L.D. Grant (eds.), Anatomical neuroendocrinology, Karger, Basle.

Henton, W.W. (1969) 'Conditioned suppression to odorous stimuli in pigeons', J. Exp. Analysis of Behaviour 12, 175-185.

Hold, B. and Schleidt, M. (1977) 'The importance of human odour in non-verbal communication', Zeitschrift Tierpsychologie 423, 225-238.

Howes, D. (ed.) (1991) The Varieties of Sensory Experience, Univ.Toronto Press, Toronto.

Howes, D. and Classen, C. (1991) 'Sounding Sensory Profiles', in D. Howes (ed.), The Varieties of Sensory Experience, Univ.Toronto Press, Toronto, pp. 257-288.

Howes, D. and Synnott, A. (1991) 'Anthropology of Odors: Executive Summary', The Fragrance Foundation, New York.

Hughes, J.R., Hendrix, D.E., Wetzel, N. and Johnston, J.W. Jr. (1969) 'Correlations between electrophysiological activity from the human olfactory bulb and the subjective response to odoriferous stimuli', in C. Pfaffmann (ed.), Olfaction and taste III, Rockefeller University Press, New York City.

Hurley, H.J., Jr. and Shelley, W.B. (1960) The Human Apocrine Sweat Gland in Health and Disease, Thomas, Springfield, Illinois.

Jackson, S.M. (ed.) (1909) The New Schaff-Herzog Encyclopedia of Religious Knowledge, Funk and Wagnalls Co., New York.

Jellinek, P. (1954) The Practice of Modern Perfumery, Leonard Hill, London.

Jellinek, P. (1965) Die Psychologischen Grundlagen der Parfumerie, Alfred Hutig Verlag, Heidelburg.

Dollard, J. (1957) Caste and Class in a Southern Town, Doubleday Anchor, New York.

Kalogerakis, M.G. (1963) 'The role of olfaction in sexual development', Psychsom. Med. 25, 420-432.

Keverne, E.B. (1983) 'Chemical communication in primate reproduction', in J.G. Vandenbergh (ed.), Pheromones and Reproduction in Mammals, Academic Press, New York, pp. 79-92.

Kingston, B.H. (1964) 'The chemistry and olfactory properties of musk, civet and castoreum', Proceedings of the 2nd International Congress of Endocrinology, Medica International Congress Service 83, 209-214.

Kirk-Smith, M.D. and Booth, D.A. (1991) 'The Effect of Five Odorants on Mood and the Judgement of others', in D.W. MacDonald, D. Muller-Schwartz and S. Natynzcuk (eds.), Chemical Signals in Vertebrates, OUP, Oxford, pp. 48-54.

Kirk-Smith, M.D. and Booth, D.A. (1980) 'Effects of androstenone on choice of location in other's presence', in H. Van Der Starre (ed.), Olfaction and Taste 7, IRL Press, London, pp. 397-400.

Kirk-Smith, M.D. and Booth, D.A. (1987) 'Chemoreception in human behaviour: Experimental analysis of the social effects of fragrances', Chem. Senses 12, 159-166.

Kirk-Smith, M.D., Booth, D.A., Carroll, D. and Davies, P. (1978) 'Human social attitudes affected by androstenol', Res. Commun. Psychol, Psychiatry Behav. 3, 379-384.

Kirk-Smith, M.D., Van Toller, S. and Dodd, G.H. (1983) 'Unconscious odor conditioning in human subjects', Biol. Psychol. 17, 221-231.

Kirkwood, R.N., Hughes, P.E. and Booth, W.D. (1983) 'The influence of boar related odours on puberty attainment in gilts', Animal Production 36, 131-136.

Kloek, J. (1961) 'The smell of some steroid sex hormones and their metabolites: Reflections and experiments concerning the significance of smell for the mutual relation of the sexes', Psychiatria, Neurologia, Neurochirurgia 64, 309-44.

Koelega, H.S. and Koster, E.P. (1974) 'Some experiments on sex differences in odor perception', Ann. N.Y. Academy of Science 237, 234-46.

Labows, J.N., McGinley, K. and Klingman, A. (1982) 'Perspectives on axillary odor', J. Soc. Cosmet. Chem. 34, 193-202.

Laing, D.G. (1975) 'A comparative study of the olfactory sensitivity of humans and rats', Chemical Senses and Flavour, 1, 257-269.

Laird, D.A. (1935) 'Man's individuality in odor', J. Abnormal and Social Psychology 29, 459-461.
Largey, G.P. and Watson, D.R. (1972) 'The Sociology of Odours', Am.J.Sociology 77(6), 1021-1034.
Lawless, H. and Engen, T. (1977) 'Associations to Odours: Interference, mnemonics and verbal labelling', J. Exp. Psychology 3(1), 52-59.
Laycock, T. (1840) A Treatise on the Nervous Diseases of Women, Longman, London.
Lott, I.T, Sullivan, R.M. and McPherson, D. (1989) 'Associative olfactory learning occurs in the neonate', Neurology 39(53), 1-10.
Marr, J.N. and Gardner, L.E. Jr. (1965) 'Early olfactory experience and later social behaviour in the rat', J. Genetical Psychology 107, 167-174.
McBurney, D.H., Levine, J.M. and Cavanaugh, P.H. (1977) 'Psychophysical and social ratings of human body odour', Personality and Social Psychology Bulletin 3, 135-138.
Meisenheimer, J. (1921) Geschlecht und Geschlechter in Tierreich. vol.I, Die naturlichen Beziehungen, Fischer, Jena.
Michael, R.P., Bonsall, R.W. and Warner, P. (1974) 'Human vaginal secretions: Volatile fatty acid content', Science 186, 1217-1219.
Moncrieff, R.W. (1966) Odour Preferences, Leonard Hill, London.
Montagna, W. (1962) 'The skin of lemurs', Ann. N.Y. Academy of Science 102, 190-209.
Montagna, W. (1963) 'The sebaceous gland in man', in W. Montagna, R.A. Ellis and A.F. Silver (eds.), The Sebaceous Glands, Pergamon Press, Oxford, pp.19-31.
Montagna, W. (1985) 'The evolution of human skin', J. Hum. Evol. 14, 3-22.
Motokizawa, F. (1974) 'Olfactory input to the thalamus; electrophysiological evidence', Brain Research 67, 334-337.
Moulton, D.G. and Eayrs, J.T. (1960) 'Studies in olfactory acuity II. Relative detectability of n-aliphatic alcohols by the rat', Quart. J. Exp. Psychology 12, 99-109.
Moulton, D.G., Celebi, G. and Fink, R.P. (1970) Olfaction in Mammals - Two aspects; proliferation of cells in the olfactory epithelium and sensitivity to odours, in G .E.W. Wolstenholmer and J. Knight (eds.), Taste and Smell in Vertebrates, London : Churchill, 227-250.
Muller, W.W. (1978) 'Weihrauch. Ein Arabisches Produkt, und seine Bedeutung in der Antike', Suppl Band XV der Realencyclopaedie von Pauly-Wissowa, Druckenmuller, Munich.
Mykytowycz, R. (1974) 'Odour in spacing behaviour', in M.C. Birch (ed.), Pheromones, North-Holland, Amsterdam.
Ohloff, G., Giersch, W., Thommen, W. and Willhalm, B. (1983) 'Conformationally controlled odor perception in 'steroid-type' scent molecules', Helvetica Chimica Acta 66, 1343-55.
Piggott, J.R. and Osisiogu, P.T. (1978) 'Similarity between human and animal senses of smell', in J. Le Magnen and P. MacLeod (eds.), Proceedings of the Sixth International Symposium on Olfaction and Taste, I.R.L. Press, London.
Pincus, J.H. and Tucker, G.T. (1974) Behavioural Neurology, Oxford University Press, New York.
Porter, R.H. and Etscorn, F. (1976) 'A sensitive period for the development of olfactory preference in acomys cahirinus', Physiology and Behaviour 17, 127-130.
Porter, R.H. (1991) 'Mutual mother-infant recognition in humans', in P.G. Hepper (ed.), Kin Recognition, Cambridge Univ.Press, Cambridge.
Powell, T.P.S., Cowan, W.M. and Raisman, G. (1965) 'The central olfactory connections', Journal of Anatomy 99, 791-813.
Pratt, J. (1942) 'Notes on the unconscious significance of perfume', Int.J. Psycholanalysis 23, 80-83.
Reed, H.C.B., Melrose, D.R. and Patterson, R.L.S. (1974) 'Androgen steroids as an aid to the detection of oestrus in pig artificial insemination', British Veterinary J. 130, 61-66.

Rennie, P.J., Holland, K.T., Mallett, A.I., Watkins, W.J. and Gower, D.G. (1988) '16-Androstene steroid content of apocrine sweat and microbiology of the human axilla', in W.MacDonald, D.Muller-Schwartz and S.Natynzcuk (eds.), Chemical Signals in Vertebrates, OUP, Oxford

Rotton, J., Barry, T., Frey, J. and Soler, E. (1978) 'Air Pollution and Interpersonal Attraction', J.Appl.Soc.Psychol. 8, 57-71.

Russell, M.J. (1976) 'Human olfactory communication', Nature (London) 260, 520-522.

Sakellaris, P.C. (1972) 'Olfactory thresholds in normal and adrenalectomized rats', Physiology and Behaviour 9, 495-500.

Sastry, S.D., Buck, K.T., Janak, J., Dressler, M. and Preti, G. (1980) 'Volatiles emitted by humans' in G.R. Waller and O.C. Dermer (eds.), Biochemical Applications of Mass Spectrometry, 1st Suppl. Vol., Wiley, New York, pp. 1085-1129.

Schaal, B. and Porter, R.H. (1991) '"Microsomatic Humans" revisited: The generation and perception of chemical signals', Advances in the Studies of Behaviour 20, 135-197.

Schaal, B. (1986) 'Presumed olfactory exchanges between mother and neonate in humans', in J. Le Camus and J. Cosnier (eds.), Ethology and Psychology, Privat-I.E.C., Toulouse, pp. 101-110.

Schaal, B. (1988) 'Olfaction in infants and children: Developmental and functional perspectives' Chem. Senses 13, 145-190.

Schaal, B., Montagner, H., Hertling, E., Bolzoni D., Moyse, A. and Quichon, R. (1980) 'Les stimulations olfactives dans les relations centre l'enfant et la mere' Reprod. Nutr. Dev. 20, 843-858.

Schiefferdecker, P. (1922) 'Die Hautdrusen des Menschen und der Saugetiere' Zoologica (N.Y.) 72, 1-154.

Schleidt, M., Hold, B. and Attili, G. (1981) 'A cross-cultural study on the attitude towards personal odors', J. Chem. Ecol. 7, 19-31.

Schneider, R.A. (1974) 'Newer insights into the role and modifications of olfaction in man through clinical studies', Ann. N.Y. Academy of Science 237, 217-223.

Schulze-Westrum, T. (1968) 'Ergebnisse einer zoologisch-volkerkundlichen Expedition zu den Papuas', Umschau 68, 295-300.

Shelley, W.B., Hurley, H.J. and Nichols, A.C. (1953) 'Axillary odor: Experimental study of the role of bacteria, apocrine sweat and deodorants', Arch. Dermatol. Syphilol. 68, 430-446.

Stein, M., Ottenberg, P. and Roulet, N.A. (1958) 'A study of the development of olfactory preferences', AMA Arch.Neurol.Psychiat. 80, 264-266.

Stoddart, D.M. (1990) The Scented Ape, Cambridge Univ. Press, Cambridge.

Strauss, J.S. and Ebling, F.J. (1970) 'Control and function of skin glands in mammals', Memoirs of the Society for Endocrinology 18, 341-368.

Synnott, A. (1991) 'A Sociology of Smell', Canad.Rev.Soc. and Anth. 28(4), 437-459.

Theissen, D.D. and Price, M. (1976) 'Mammalian scent gland marking and social behaviour', Psychological Bulletin 83. 505-539.

Van Toller, C., Dodd, G.H. and Kirk-Smith, M.D. (1980) 'The effects of odours on human attitudes', SSRC Report (project number HR6090), Social and Engineering Research Council Library, London.

CULTURAL EVOLUTION IN MAN OF POSTURES, GESTURES AND UNVERBALISED SOCIAL RELATIONS

CLAIRE RUSSELL AND W.M.S. RUSSELL
Department of Sociology, University of Reading
Whiteknights, P.O.Box 218, Reading, Berkshire,
RG6 2AA, England

ABSTRACT. Postures, gestures and social relations in man may be transmitted without the use of language. Their spatial distribution may be very stable. As first observed by Herbert Spencer, postures and gestures may change in relation to changes in social organization, they may be transferred from one cultural context to another, and their evolution interacts with that of artefacts. Sign languages may have evolved by the Upper Palaeolithic.

The great achievements of human cultural evolution have been made possible by the use of true language. However, even in literate civilizations, an important part of the culture of a human group is its repertoire of gestures, postures and facial expressions - for instance, 118 such units of behaviour observed by Grant (1969) in Birmingham children. These culture elements, and the situations in which each is used, may be spread, diffused and transmitted as tradition without involving language. 'According to Harald Hoffding (1906), "in a Danish village church the custom of bowing when passing a certain spot in the church wall was maintained into the nineteenth century, but no one knew the reason for this until, on the whitewash being scraped away, a picture of the Madonna was found on the wall", evidently concealed ever since the Reformation in Denmark, three centuries earlier' (Russell, 1981). We once saw, on documentary film, a Japanese mother in traditional dress, her baby on her back, bowing to a Shinto shrine and reaching back to press the baby's head down too (Russell and Russell, 1971).

Even patterns of social organization may be transmitted without language. Adadama village in Nigeria is made up of clearly separated compounds (Harris, 1969). Because a wife lives in her husband's compound, many compounds are self-contained patrilineal groups, each with its compound elder and social rules, and one compound monopolises a priesthood, which thus descends in the male line. Yet the villagers regard their kinship system as exclusively matrilineal, they do not recognize the existence of their patrilineages, and they have no terms in their language to describe them. There is overwhelming evidence for the matrilineal organization of early human societies, and we know of many factors producing a change to patrilineal organization (and none in reverse), so these villagers must have experienced a completely unverbalised change in social organization (Russell and Russell, 1981), such as must often happen among other primates and has actually been observed in Barbary apes (Burton, 1972).

Claire Russell has shown that many aspects of human and higher animal behaviour can fruitfully be considered under the two headings of relaxed and stressful, with gradations and mixtures between them (e.g. Russell and Russell, 1985a). This

R. A. Gardner et al. (eds.), The Ethological Roots of Culture, 407–417.

certainly applies to postures, gestures and social relations. Postures and gestures, besides expressing the feelings of the individual, also act as social signals for other individuals, and both aspects may be relaxed or stressful. Relaxed behaviour is generally friendly, and stems ultimately from relaxed parental behaviour (Russell and Russell, 1985b,1990a). Stressful behaviour is generally competitive, involving attack, escape and aversion, dominance and submission. We have shown that stressful behaviour is produced in response to overpopulation pressure, including spatial crowding; it may also come to be transmitted as stress culture, by stressful conditioning processes applied to the young (Russell and Russell, 1968,1984,1985a,1986a,b).

Under different conditions, the same postures and gestures, and the reactions to them as signals, may be more or less relaxed or stressful. A number of monkey and ape species, for instance, perform incomplete sexual acts, such as mounting, as displacement activities to resolve momentary conflicts, both within and between individuals, when obliged to trespass on each other's social space. Because they can be shown to have no sexual motivation, Claire Russell has called such acts pseudosexual (Russell, 1970,1975; Russell and Russell, 1965a). Kortlandt filmed an attack by chimpanzees on a stuffed leopard suddenly presented to their view (Kortlandt and Bresser, n.d.). During the wild excitement and rushing about, frequent trespass on social space was bound to occur, and between assaults on the leopard there was a great deal of pseudosexual displacement mounting. Plooij (1978) observed the same thing when chimpanzees at Gombe were hunting bush-pigs: the males frequently 'embraced or mounted each other'. In Kortlandt's film the behaviour was clearly relaxed and friendly, as if the chimpanzees were saying: 'Excuse me for jostling you'.

Under crowding stress, however, when conflicts over social space become chronic, the pseudosexual displacement acts of monkeys and apes may be increasingly permeated with attack and escape elements, dominance and submission. Since dominance can be expressed either way, the dominant individual may brutally mount others, or back roughly into them and force them to mount (Russell and Russell, 1968,1985a). So the same act may be more or less relaxed or stressful, according to the conditions.

Alike in monkeys, apes and man, different populations of the same species may have different postures and gestures, or may react to the same posture or gesture signals in different ways (Russell and Russell, 1982,1983). As we observed 28 years ago, 'when beckoning is a summons in Britain, a dismissal in Italy, and an insult in Malawi, cross-purposes readily arise' (Russell, 1964). So far as Britain and Italy are concerned, this statement was amplified and given precision in the course of a remarkable research carried out in 1975-7 by a team led by Desmond Morris (Morris et al., 1979). They visited 40 localities in 25 countries, involving 15 different languages, in Western Europe and the Mediterranean, but also including the Canary Islands. In each locality, they interviewed 30 adult men concerning the use and meaning of 22 gestures made with the head and/or hands. For a given meaning of a given gesture, they could then prepare a histogram of the frequencies in the 40 localities, and finally a map in which different symbols showed where this gesture with this meaning was relatively common, relatively rare, or absent. They found the various gestures and reactions were to a large extent distributed independently of each other, and that the distributions often failed to coincide in any simple way with language or political boundaries. Many intriguing problems were raised, and some of these were elucidated by historical research.

The team found that beckoning, in the sense of summoning, could be performed with the palm up or down. Waving goodbye could be performed with the palm shown, that is, facing the other person, or hidden, that is, facing the waver. Various combinations occurred in different parts of Europe, but it so happened that palm-up beckoning and palm-shown waving were common in northern Europe and rare in Italy, whereas palm-down beckoning and palm-hidden waving were absent in northern Europe and common in Italy. And it so happens that palm-up beckoning looks like palm-hidden waving, and palm-down beckoning looks like palm-shown waving, so north European beckoning resembles Italian waving and vice versa.

This is not the only case where the gestures of different peoples appear to be reversed. During a Balkan war in 1877-8, Russian soldiers serving in Bulgaria had some difficulty in communicating with their Bulgarian comrades. The Russians, like most Europeans, signified 'yes' by nodding and 'no' by shaking their heads. The Bulgarians signified 'yes' by shaking their heads, and 'no' apparently by nodding (Jakobson, 1972). In fact the Bulgarians signifying 'no' were actually tossing their heads upwards rather than nodding them downwards, but when the two movements were performed repeatedly it was easy to confuse them. In Greece and southern Italy, many people signified 'yes' by nodding the head and 'no' by tossing it, with different movements of the eyebrows to prevent confusion between the two.

Morris and his team studied the distribution of head-tossing, with interesting results. They found it was common in Greece, Istanbul, Malta, Sicily and southern Italy, and rare or absent everywhere else in Europe and the Mediterranean. In Italy the border between predominant head-tossing in the south and exclusive head-shaking in the north appeared between Naples and Rome. The team made a detailed study of this border region, from coast to coast, visiting at least 75 places and interviewing more than 750 people. They found two coastal regions where the head-toss was dominant, and a band of country across the peninsula where toss and shake were both frequently used. The western coastal head-toss region gave a remarkably good fit with the area of Greek colonisation between the 7th and 4th centuries B.C., and we know the head-toss is characteristic of both ancient and modern Greece. The eastern coastal head-toss region was occupied during the same centuries by immigrants from across the Adriatic, and was also long under Byzantine influence. The mixed region inland can no doubt be explained by a slow spread of the head-toss from the Greek-influenced coastal regions. A final still more detailed study of the area north of Naples showed an extraordinarily sharp boundary between the rivers Volturno and Garigliano. One could travel from a place where almost everyone used the head-toss to a place where virtually no-one used this gesture in 21 kilometres.

The head-toss study shows that gestures can be transmitted through the generations with great precision for several millennia. But of course postures and gestures do show long-term changes. For a classical study of such long-term change we can turn to a book published exactly one hundred years before that of the Morris team, in 1879: Ceremonial Institutions, being the fourth part, or the first part of the second volume, of Herbert Spencer's monumental Principles of Sociology (Andreski, 1972; Bowle, 1963; our quotations are from the slightly revised second edition, Spencer, 1883). At the outset, Spencer observes that 'the modifications of conduct called "manners" and "behaviour" .... precede human evolution: they are traceable among the higher animals'. He describes how a small dog, on the approach of a large one, 'throws itself on its back with legs in the air .... tacitly saying - "I am conquered, and at your mercy". Clearly then, besides certain modes of behaviour expressing affection, ... there are established certain modes of behaviour expressing subjection'. He thus recognizes that postures and gestures may be relaxed or stressful, and throughout the book he insists that usages of this kind 'are not devised but evolve'. In the chapter on obeisances, he has a full discussion of submissive postures and gestures in man. He begins by remarking that 'probably many readers thought I was putting .... a forced construction' on the behaviour of the small dog. 'They would not have thought so had they known that a parallel mode of behaviour occurs among human beings.' He then quotes from the great African explorer David Livingstone an account of how the Batoka people 'throw themselves on their backs on the ground', and concludes that 'prostration on the back negatives resistance'.

To understand his study of submissive postures and gestures, we need to know how Spencer classified human societies. He first distinguished various levels of complexity. Then, he distinguished between militant and industrial societies. More precisely, since he realised that most societies had both characteristics, he distinguished between mainly militant and mainly industrial societies (Andreski, 1972). The mainly militant type included nearly all complex societies throughout history all over the world. They were societies organized for war, even in peace-time, rigidly hierarchical and oppressively coercive, even productive industry being subject to regimentation. He realised that some of these societies were more extreme than others. He quite rightly singled out the Inca Empire of Peru as a very extreme case. This was one of what Sprague de Camp (1977) has called the watershed empires, which first unified large areas of agriculture based on water control engineering: other examples are the Assyrian Empire and the Ch'in State in China (Russell, 1967; Russell and Russell, 1987a). These empires were totally militaristic and claustrophobically oppressive. At the other extreme were Spencer's mainly industrial societies. These included some rather simple unwarlike societies and also those of Europe, still highly militant in medieval times but mainly industrial by Spencer's time, with Britain the most industrial and least militant of all. These societies were more concerned with production than war; they were relatively decentralised and democratic, with appreciable individual freedom, little hierarchical organization, and social relations based on contract rather than status. Spencer somewhat idealised the modern European societies, and would have been sadly disillusioned about them if he had lived longer into the twentieth century (he died in 1903). If he had read the later works of Malthus, he would have realised that industrialization and the genuine reduction in stress were both due to unprecedently low birth-rates due to later and less frequent marriage, later supplemented by import of resources from and export of people to the rest of the world - in short to reduced population pressure (Russell and Russell, 1980,1987b; Russell, 1983). With renewed population pressure in the twentieth century, Spencer's militant properties returned in these industrial societies. However, he was quite right in maintaining that nineteenth-century European societies were less stressful, militaristic and oppressive than those of other regions or of Europe itself in earlier times, and it is this that makes his study of submissive postures and gestures both interesting and valid. Moreover much of what he observed in European societies is still true there, for they are still benefiting from some of the reductions in stress.

Spencer takes various extreme forms of submissive posture or gesture, and shows they are not present at all in simple industrial societies, and how in modern complex industrial societies they have undergone what he calls abridgement, that is, they have been reduced to slighter and slighter intention movements of the original act. (Some of this abridgement may also be found in the less extreme militant societies.) For instance, he starts with total prostration, as observed in various states of Mexico, Africa, India, south-east Asia and Polynesia. A first abridgement is to kneel and put the head on the ground, which, as he notes, is a necessary intention movement on the way to total prostration. This slightly less extreme submission he finds in various African states, China and medieval Russia. Next comes cowering or crouching, which he finds in Mexico, Melanesia and Polynesia, and which, be it noted, was found by Ewan Grant (1965) as an escape intention movement among schizophrenic women in Birmingham. By a further abridgement, we obtain the kneeling posture, found in Africa, China, Japan, and in medieval Europe, in the posture of feudal homage, to which we shall return. In China and Japan he found a still more abridged form of simply bending one knee, used commonly when meeting superiors in the street. The final abridgement had occurred in Europe, where it still survived as the curtsey in women and the scrape in men, made by a backward sweep of the right foot. These were no longer stressful submissive acts, but relaxed courtesies of greetings between equals; in any case, by Spencer's day they were already rapidly disappearing in Britain. He adds that this whole set of postures is completely absent among the simple unmilitant societies, such as the Andamanese, the Fuegians or the Eskimos. In fine, Spencer showed that patterns of social organization are reflected automatically in postures and gestures, and that changes in these patterns are reflected likewise.

Another expression of this principle was the subject of a two-volume book called The Civilizing Process, by Norbert Elias (1978,1982), first published in German in 1939. His first volume is called The History of Manners, and describes the behaviour normally so classified, chiefly associated with feeding and grooming. He used the deceptively simple technique of juxtaposing passages from books of etiquette written at different dates between the thirteenth and nineteenth centuries in Europe. Paradoxically, these verbal extracts show clearly that gradual changes in attitude and behaviour occurred throughout the period without people being at all clearly aware of what was happening to them. Between relatively barbarous early medieval Europe and highly civilized nineteenth-century Europe, there was a remarkable increase in the refinement of manners. In his second volume, State Formation and Civilization, Elias showed that these changes in behaviour were associated with changes in people's attitudes to each other during the transitions from the feudal system to the court society of absolutist states, and from that to the relatively liberal societies of the nineteenth-century, dominated by the middle classes. It was not that medieval society lacked any concept of manners, but what was done or not done was then justified as showing respect to superiors, whereas nineteenth-century etiquette manuals gave hygienic reasons. As for the increase in refinement, a single example may suffice to illustrate this. In the thirteenth century, it was considered bad manners at meal-times to blow your nose on the table-cloth, for the table-cloth should only be used for wiping your fingers after dipping them in the common sauce-vessel.

An important part in the refinement of manners was played by such artefacts as handkerchiefs and forks, but only when people were ready for them. In the eleventh century a Byzantine princess married a doge of Venice, and brought there little golden forks, which she used at table. The local prelates rebuked her for this sinful luxury, and when the poor girl succumbed to an unpleasant illness, St. Bonventura called it a judgement of God. Yet in the sixteenth century it was the Italians who introduced the fork to western Europe. Once accepted, the artefacts helped to stabilize the changes in behaviour.

This is an example of a principle that acquires great importance in man, namely the interaction between the cultural evolution of behaviour and the cultural evolution of artefacts, called by Alfred J Lotka (1944) exosomatic evolution, or evolution outside the body. This interaction takes many forms. Certain artefacts can have a profound effect on posture. Since ancient times, the people of the western Old World, or at least the richer ones, have sat on chairs, while the people of the Far East have squatted on the floor on mats. Between the 2nd and 12th centuries A.D., the Chinese, alone of all Far Eastern peoples, changed over to sitting on chairs (Fitzgerald, 1965). With this came a number of correlated changes, in costume, furniture, even architecture, and notably in stance and posture, for mat culture and chair culture represent profoundly different ways of life (Russell and Russell, 1982).

Just as organically evolved structures called releasers serve to accentuate signal movements and postures in animals, so culturally evolved releasers, in the form of clothing or ornaments, may serve to accentuate cultural signals in man (Russell and

Russell, 1961; Russell, 1962). One very interesting case of this is the use of costume or armour to give the shoulders an angular look (Leyhausen, 1983). When a prisoner of war in Canada during the Second World War, Paul Leyhausen took full advantage of the presence of a large number of men of varied origins, education and religions to carry out researches in human ethology. Among other things, by means of drawings and questions, he established five characteristics by which the men recognized a threat posture, or what is called in German imponieren. These were: bristling hair, clenched fists, bent elbows, a full dark-coloured beard, and broad shoulders with an angular contour. The last characteristic intrigued Leyhausen, for very strong men do in fact have rounded and not angular shoulders. However, among the large population in the camp, he was able to find some very hairy individuals, and he discovered on their shoulders striking hair-whorls. These were so arranged that, when human beings were as hairy as the great apes, they would have formed angular processes whenever piloerection occurred as an autonomic rage reaction. The angles of armour or costume are therefore simulating these hairy angles. Leyhausen is inclined to regard the releasing mechanism as innate, but since we know behaviour can be transmitted culturally for three centuries, like the Danish bowing to the wall, or nearly thirty centuries, like the head-toss in Italy, there is no reason why it should not be so transmitted for the three thousand or so centuries since the control of fire was firmly established, when, as Claire Russell has pointed out, powerful natural selection must have ensured our final reduction in hairiness (Russell, 1978).

An interesting interaction between clothing and posture appeared in Japan, in the closing years of the Tokugawa Shogunate (May and Annis, 1972). A British sailor called Walter White was on shore leave in Nagasaki in December 1860. He noted that the upper classes wore scarves round their necks, 'each rank wearing the ends a different length - the higher the rank the longer the ends of the scarf. The custom is that when meeting they bow until the ends of the scarf touch the ground, therefore the inferior man having short ends to his scarf has to bow the lower.'

It sometimes happens that a particular gesture is, so to speak, frozen by being carved or moulded as an artefact. Of course this happens in any work of art where postures and gestures are carefully represented. But some gestures are specially modelled by themselves as protective amulets. The Etruscan sculptors were meticulously accurate in showing hand-gestures. In the Villa Giulia at Rome, there is a justly famous Etruscan terracotta sarcophagus from Cerveteri, portraying a husband and wife, and dating to about 530 B.C. (Richardson, 1966). The wife is making a very particular gesture with her right hand, extending the first two fingers and the thumb while bending the other two fingers. There is no doubt that this is a protective or good-luck gesture, to bring blessings on the household. The two fingers extended together may represent the phallus, a protective image against the evil eye, as described by Diodorus Siculus in the 1st century B.C., but certainly going back much earlier (Russell, 1991): it can be traced ultimately to the showing of the penis, sometimes erect, as a displacement activity by male monkeys of several species on the border of a band territory, which becomes an assertion of band territorial integrity (Wickler, 1967; Russell, 1970; Russell and Russell, 1981). However, this hand gesture became an important protective symbol in its own right: a bronze model of a hand showing this gesture was found at Pompeii (Elworthy, 1900, Fig.104). It is mounted on a base, and could have stood somewhere in a house as a protective amulet for the household. But there is a large number of much more remarkable bronze amulets of this kind, dating from the time of the early Roman Emperors, found on Roman sites inside and outside Italy, always in private households, and now scattered over the museums of Europe. Twenty-one of these objects are figured by Frederick Thomas Elworthy (1900), who made a special study of them. They are all models of right hands set on bases, making this same gesture, but also covered with representations of gods and goddesses, often oriental ones, and with animals and objects either associated with particular deities or considered as good-luck charms in their own right. Every specimen has a different set of these mouldings, and they were clearly custom-made for the householders. Elworthy's attention was first attracted to these amulets by a fine specimen in the museum of Cortona, specially adorned with Mercury and his attributes (his Figs. 159 and 160). A specimen from Avenches in Canton Vaud is one of several showing near the base a woman and child, clearly intended to ensure a safe delivery or the health of a child (his Figs. 135 and 136). Finally, a specimen in the museum at Leiden shows near the base a husband and wife with joined hands (his Figs. 149 and 150). Clearly this amulet was for the protection of a married household, and recalls the wife making the gesture on the Etruscan sarcophagus.

Art history has much to offer the student of cultural evolution. We have recently studied the posture of prayer in Christian Europe as shown in works of art (Russell and Russell, 1990b). Many portraits from the fifteenth and sixteenth centuries have familiarized us with the posture of prayer then current in Europe (e.g. Anon., 1964, Fig. 1, p.166; Saunders, 1932, Fig.62; Weightman, 1989, p.199). The praying figure usually kneels, with the hands brought close together, sometimes clasped, sometimes placed palm to palm with fingers together, sometimes with the right thumb hooked over the left hand. The joined hands and the forearms are often turned upwards; sometimes the joined hands are bent upwards at right angles to the forearms, which are held horizontally against the body. The kneeling posture was exosomatically stabilized by manufactured objects, such as the prie-Dieu or the hassock. In brasses as early as the 12th century, the knights lie supine but with their hands together (e.g. Saunders, 1932, Fig. 85). The apposition of the hands seems to be especially characteristic of medieval and later Europe.

The prayer posture of the early Christians was totally different. In about A.D. 198, Tertullian (Apology, 30.4) describes the male Christian praying: he looks up to heaven, with hands outspread and head bare. It was like the pagan prayer posture, except for the bare head (prescribed for males by St. Paul, I Corinthians 11.4): Roman pagans covered theirs. The orant (praying) figure is well known from early Christian (and contemporary pagan) art, from the second to the fifth century (Lowrie, 1906, Figs. 64-6, 85, 92, 144; Grabar, 1968, Figs. 2,13,15,16,59-61,188-93). The praying figure stands, with arms spread wide apart and somewhat forwards, sometimes slightly raised, sometimes bent at the elbow. The hands are open and the fingers often spread apart. This highly relaxed posture looks like an intention movement of embrace, and suggests in particular a child about to embrace an affectionate parent. It is quite different from the crouching posture with hands raised vertically higher than the head, which Grant (1965) observed in schizophrenic patients at Birmingham and classified as an escape posture (no doubt related to the widespread surrender gesture of 'hands up' or 'Hände hoch').

Certain Cretan and Cypriote figures of the later second millennium B.C., with arms raised, have been interpreted by Morris (1985, Figs. 288-9, p.175) as worshippers. At first sight, their posture looks more like Grant's escape posture. But when the Christian orant figure is similarly stereotyped and flattened, as in a fifth-century relief from Coptic Egypt (Stierlin, 1988, Fig.86), there is quite a similarity, so the orant posture may be very old.

In the New Testament, kneeling in prayer (as opposed to in worship) does occur, but only in situations of great intensity (Gethsemane, Luke 22.41; Stephen's martyrdom. Acts 7.60). When the disciples are urged to pray in private, and not like the hypocrites standing in public places, they are not urged to use a different posture (Matthew 6.5-6); and nothing is said in the New Testament about placing the hands together.

When did the posture change? A curious intermediate situation is to be seen on the tympanum of Bourges cathedral (1270-80). The resurrected dead pray standing, but with hands joined palm to palm and bent upwards at right angles to the forearms, which are held horizontally against the body, as is often seen later in kneeling figures. One of the risen dead holds one hand centrally in this position, while with the other he still holds his coffin lid (Aubert, 1964, p.79). In a Byzantine miniature of about A.D. 900, Christ is shown kneeling in prayer in Gethsemane (evidently based on Luke's account), but the hands are clearly separated (Grabar, 1966, Plate 34). The earliest undoubted illustration of figures (praying or worshipping) in the kneeling posture with hands together, that we have so far found, is in the Golden Gospel-Book, commissioned for Speyer Cathedral from the scriptorium of Echternach by the Emperor Heinrich III between 1045 and 1046. The picture shows his parents, Konrad II and Gisela, set beneath the figure of Christ: they kneel with their hands together (Zarnecki, 1975, Fig. 165). The change may therefore have occurred between the fifth and the eleventh centuries, and perhaps nearer the end of that interval.

The timing suggests the new posture was a transfer from another cultural context, the ceremony of feudal homage. The vassal 'puts his hands together and places them, thus joined, between the hands of his lord, - a plain symbol of submission, the significance of which was sometimes further emphasized by a kneeling posture' (Bloch, 1965, Vol. 1, Plate 2). Marc Bloch regarded this as of remote German, presumably pagan, origins, and we may see the change in prayer posture as part of a shift in the centre of gravity of civilization from the Mediterranean to northern Europe.

But the later posture is clearly stressful when compared with the highly relaxed earlier one, and may also reflect the stresses incurred during the violent Dark Ages after the fall of the western Roman Empire. To interpret the act of feudal homage itself, we may return to Spencer. He found the presentation of joined hands, as a submissive gesture, in Africa, Madagascar, Siam and Sumatra. He interpreted it as an intention movement of offering the hands to be tied, as might occur when a man surrendered on the battlefield. He was even able to trace an extreme form in the appalling Inca Empire, where 'a sign of humility was to have the hands bound and a rope round the neck'. He does not seem to have known of the early Christian relaxed orant posture. But he had shown many examples of cultural transfer from obeisances before rulers to obeisances before deities, and he did not miss this one. 'The presentation of the hands joined palm to palm', he wrote, 'once throughout Europe required from an inferior when professing obedience to a superior, is still taught to children as the attitude of prayer.'

In support of this particular transfer, it may be noted that the posture of homage was also transferred to other contexts. Theophilus of Adana was a medieval precursor of Faust who was rescued in the nick of time by the Virgin Mary (Butler, 1979). An illustration of about 1200 shows him kneeling and placing his hands within those of the Devil; and a twelfth-century seal shows a lover in the same posture before his lady (Bloch, 1965, Vol.1, Plates 3 and 4).

One final obvious aspect of the cultural evolution of postures and gestures is the emergence of sign languages. Modes of language other than speech generally evolve under conditions when spoken communication is difficult. Thus whistling languages have been evolved in several places by mountaineers who need to communicate at a distance (Russell and Russell, 1971). Sign languages have been evolved where speech would be drowned by noise, as in sawmills, where speech is impossible, as when teaching the deaf and dumb, where speech is unintelligible, as among the Plains Indians of North America, who used sign language for communication between groups speaking the many different languages of the region, and finally when speech is tabooed, in certain contexts, as among many Australian aborigine tribes, or altogether, as among several orders of monks (Kendon, 1988).

The sign languages of the Australian aborigines are rich and complex: they have been studied in detail by Kendon (1988). The rock art of central Queensland includes stencils of hands 'which appear to have fingers missing or are otherwise distorted' (Kendon). But G.L.Walsh (1979, cited by Kendon) has suggested that these are not, as had been supposed, mutilated hands, but hands making signs: he finds parallels between the stencils and handshapes in existing aborigine sign languages. If he is right this is most interesting, for similar stencilled hands, also interpreted as mutilated, have been found in French and Spanish caves painted during the Upper Palaeolithic (Janssens, 1970, Plates 52-4). It is therefore possible that sign languages evolved very early.

Finally, James Knowlson (1975) has shown that from the seventeenth century onwards, impressed by the success of sign languages among the deaf and dumb, a number of people have recommended sign language as a possible universal language for mankind. But even these optimists can never have dreamed that a sign language developed for use by the deaf and dumb would one day make it possible for human beings to communicate with chimpanzees (Gardner, Gardner and Van Cantfort, 1989).

References

Andreski, S. (ed.) (1972) Herbert Spencer: Structure, Function and Evolution, Nelson, London.

Anon. (1964) Art Treasures of the World, Paul Hamlyn, London.

Aubert, M. (1964) High Gothic Art, Methuen, London.

Bloch, M. (1965) Feudal Society (2 vols., transl. L.A.Manyon), Routledge and Kegan Paul, London.

Bowle, J. (1963) Politics and Opinion in the Nineteenth Century, Jonathan Cape, London.

Burton, F.D. (1972) 'The integration of biology and behaviour in the socialization of Macaca sylvana of Gibraltar', in F.E.Poirier (ed.), Primate Socialization, Random House, New York, pp.29-62.

Butler, E.M. (1979) The Myth of the Magus, Cambridge University Press, Cambridge, 1979.

De Camp, L.S. (1977) The Ancient Engineers, Tandem, London.

Elias, N. (1978) The Civilizing Process: the History of Manners (transl. E.Jephcott), Blackwell, Oxford.

Elias, N. (1982) The Civilizing Process: State Formation and Civilization (transl. E. Jephcott), Blackwell, Oxford.

Elworthy, F.T. (1900) Horns of Honour, John Murray, London.

Gardner, R.A., Gardner, B.T. and Van Cantfort T.E. (eds.) (1989) Teaching Sign Language to Chimpanzees, State University of New York Press, Albany, N.Y.

Grabar, A. (1966) Byzantium, Methuen, London.

Grabar, A. (1968) Christian Iconography, Princeton University Press, Princeton, N.J.

Grant, E.C. (1965) 'An ethological description of some schizophrenic patterns of behaviour', Proceedings of the Leeds Symposium on Behavioural Disorders, 3-14.

Grant, E.C. (1969) 'Human facial expression', Man N.S. 4,525-536.

Harris, R. (1969) 'Unilineal fact or fiction: a further contribution', in M. Douglas and P.M. Kaberry (eds.), Man in Africa, Tavistock, London, pp.137-150.

Hoffding, H. (1906) The Philosophy of Religion, Macmillan, London.

Jakobson, R. (1972) 'Motor signs for "yes" and "no"', Language in Society 1, 91-96.

Janssens, P.A. (1970) Palaeopathology (transl. I. Dequeecker), John Baker, London.

Kendon, A. (1988) Sign Languages of Aboriginal Australia, Cambridge University Press, Cambridge.
Knowlson, J. (1975) Universal Language Schemes in England and France 1600-1800, University of Toronto Press, Toronto.
Kortlandt, A. and Bresser, E.G. (n.d.) Experimentation with Forest-dwelling Chimpanzees in the Congo, 1963, (16 mm film), Stichting Film en Wetenschap, Utrecht.
Leyhausen, P. (1983) 'Kleidung: Schutzhülle, Selbstdarstellung, Ausdrucksmittel', in B.Sitter (ed.), Menschliches Verhalten, Universitätsverlag, Freiburg, Switzerland, pp.137-152.
Lotka, A.J. (1944) 'Evolution and thermodynamics', Science and Society 8, 161-171.
Lowrie, W. (1906) Monuments of the Early Church, Macmillan, London.
May, W.E. and Annis, P.G.W. (eds.) (1972) China Station 1859-1864: the Reminiscences of Walter White, National Maritime Museum, Greenwich, London.
Morris, D. (1985) The Art of Ancient Cyprus, Phaidon, Oxford, in association with Jonathan Cape, London.
Morris, D., Collett, P., Marsh, P. and O'Shaughnessy, M. (1979) Gestures, Jonathan Cape, London.
Plooij, F.X. (1978) 'Tool-use during chimpanzees' bushpig hunt', Carnivore 2, 103-106.
Richardson, E. (1966) Etruscan Sculptures, Collins, London, in association with UNESCO, Paris.
Russell, C. (1970) 'The concept of pseudosex', Guy's Hospital Gazette 84, 241-245.
Russell, C. (1975) 'The chimpanzee carnival: food, space and social behaviour', Biology and Human Affairs 40, 77-100.
Russell, C. (1978) 'Fire control and human evolution', Biology and Human Affairs 43, 14-20.
Russell, C. (1981) 'The environment of ghosts', in H.R.E.Davidson and W.M.S. Russell (eds.), The Folklore of Ghosts, D.S. Brewer, Cambridge, for the Folklore Society, London, pp. 109-137 and 248-253.
Russell, C. and Russell, W.M.S. (1961) Human Behaviour: a New Approach, Deutsch, London.
Russell, C. and Russell, W.M.S. (1968) Violence, Monkeys and Man, Macmillan, London.
Russell, C. and Russell, W.M.S. (1971) 'Language and animal signals', in N. Minnis (ed.), Linguistics at Large, Gollancz, London, pp. 159-194.
Russell, C. and Russell, W.M.S. (1980) 'Scarcities and societal resources', in N. Polunin (ed.), Growth without Ecodisasters?, Macmillan, London, pp. 409-428.
Russell, C. and Russell, W.M.S. (1981) 'Kinship in monkeys and man. 2. The emergence of patrilineal kinship', Social Biology and Human Affairs 46, 37-65.
Russell, C. and Russell, W.M.S. (1982) 'Cultural evolution. 1. Culture, culture transmission and culture change', Social Biology and Human Affairs 47, 17-38.
Russell, C. and Russell, W.M.S. (1983) 'Cultural evolution. 3. The differences between cultures', Social Biology and Human Affairs 48, 135-151.
Russell, C. and Russell, W.M.S. (1984) 'Overpopulation crisis', Social Biology and Human Affairs 49, 23-42.
Russell, C. and Russell, W.M.S. (1985a) 'Conflict activities in monkeys', Social Biology and Human Affairs 50, 26-48.
Russell, C. and Russell, W.M.S. (1985b) 'Sticklebacks and ethology', Behaviour 93, 184-193.

Russell, C. and Russell, W.M.S. (1986a) 'Cultural evolution. 5. The transmission of stress culture', Social Biology and Human Affairs 51, 47-59.

Russell, C. and Russell, W.M.S. (1986b) 'Cultural evolution. 6. The content of stress culture', Social Biology and Human Affairs 51, 60-75.

Russell, C. and Russell, W.M.S. (1987a) 'Orwell, apocalypse and overpopulation', in G.E. Slusser, C. Greenland and E.S. Rabkin (eds.) Storm Warnings: Science Fiction confronts the Future, Southern Illinois University Press, Carbondale and Edwardsville, pp. 159-171 and 255-258.

Russell, C. and Russell, W.M.S. (1987b) 'Cultural evolution. 7. Progress, creativity and renaissance', Social Biology and Human Affairs 52, 101-124.

Russell, C. and Russell, W.M.S. (1990a) 'Cultural evolution. 9. The analogy between organic and cultural evolution' Social Biology and Human Affairs 55, 53-66.

Russell, C. and Russell, W.M.S. (1990b) 'Cultural evolution of behaviour', Netherlands Journal of Zoology 40, 745-762.

Russell, W.M.S. (1962) 'Animals, robots and man; signals and shibboleths', Listener 68, 169-170, 207-208 and 213.

Russell, W.M.S. (1964) 'Violence, monkeys and man', Listener 72, 710-712 and 753-755.

Russell, W.M.S. (1967) Man, Nature and History, Aldus Books, London.

Russell, W.M.S. (1983) 'The palaeodemographic view', in G.D.Hart (ed.) Disease in ancient man, Clarke Irwin, Toronto, pp. 217-253.

Russell, W.M.S. (1991) 'A funny thing happened .... humour in Greek and Roman life, literature and theatre', in G.Bennett (ed.), Spoken in Jest, Sheffield Academic Press, Sheffield, pp. 83-115.

Saunders, O.E. (1932) A History of English Art in the Middle Ages, Clarendon Press, Oxford.

Spencer, H. (1883) Ceremonial Institutions: being Part IV of the Principles of Sociology (the First Portion of Vol.II) (Second edition), Williams and Norgate, London and Edinburgh.

Stierlin, H. (1988) Orient Byzantin, Office du Livre, Fribourg.

Walsh, G.L. (1979) 'Mutilated hands or signal stencils? A consideration of irregular hand stencils from Central Queensland', Australian Archaeology 9, 33-41.

Weightman, C. (1989) Margaret of York, Duchess of Burgundy 1446-1503, Alan Sutton, Gloucester.

Wickler, W. (1967) 'Socio-sexual signals and their intra-specific imitation among Primates', in D.Morris (ed.), Primate Ethology, Weidenfeld and Nicolson, London, pp. 69-147.

Zarnecki, G. (1975) Art of the Medieval World, Prentice-Hall, Englewood Cliffs, N.J., and Harry N. Abrams, New York.

THE EVOLUTIONARY BASES OF INTELLIGENCE

B. CHIARELLI

Istituto di Antropologia
Università di Firenze
via del Proconsolo 12
50122 Firenze, Italy

ABSTRACT. The ecological and adaptive reasons for the evolutionary leaps involved in the encephalic increase, and thus in the evolution of intelligence, from early mammals to Man are analyzed. In particular, an attempt is made to sinthesize the mechanisms by which the environment operates in the transformation of the encephalon, utilizing the plasticity of neurons during childhood. As an example, the origin of language is used as the model for humans.

1. Introduction

When a population (or a species) changes its ecological niche at the genetic level (and consequently also the macro- evolutionary one), it responds by changing some characteristics in order to adapt better to the new environment (Oddum 1971).

In the history of the Mammals and of the Primates, one of the main responses to the change in environment has been the enlargement of the brain with respect to body size.

This change has been defined by H. Jerison (1973) as the "encephalization quotient", expressed by the formula $E=K(\sqrt[3]{P})^2$ where the cube root of body weight squared is the coefficient of the relationship between encephalic weight (E) and body weight (P) and K is the index of cephalization.

In general a larger brain, with respect to body mass, permits greater intelligence and allows a population to deal better with the cognitive demands of the new niche. In living mammals, the value of the encephalization quotient is 0.12 and in Primates is 0.24.

2. From the reptilian to the mammalian stage

During the evolutionary history of interest to us, the first important expansion of the brain occurred at the beginning of the evolution of Mammals, about 220 million years ago, when they separated from the Therapsids (Fig. 1).

Entering a nocturnal niche, in which they slept during the day and fed at night, these primitive mammals, in order to survive, were obliged to integrate visual information with auditory and olfactory information, with a consequent enlargement of the brain with respect to the primitive reptiles and other vertebrates.

In the reptilian brain, the predominant structures were and are those of the corpora striata (Fig. 2), involved primarily in motor functions and those activities which are the basis of social behaviour. To this archaic brain (archipallium) are to be attributed

R. A. Gardner et al. (eds.), The Ethological Roots of Culture, 419–432.

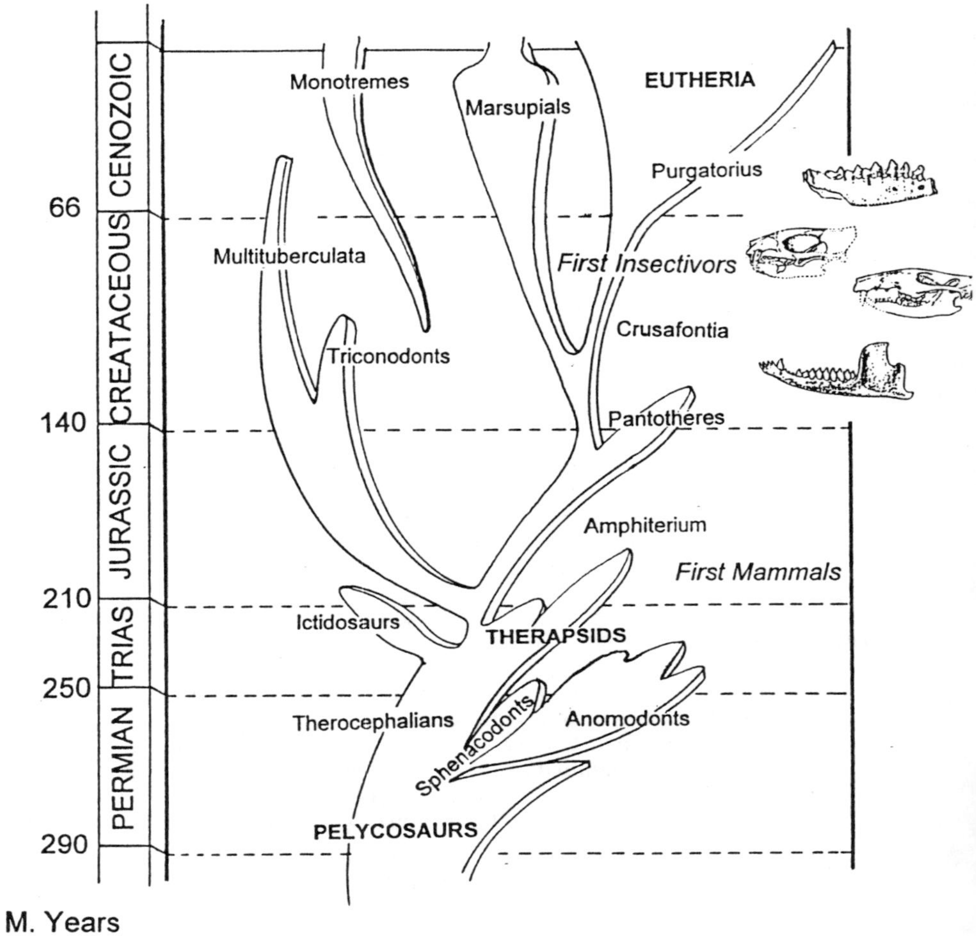

Fig.1. A phylogenetic interpretation of the Eutherian origin

elementary functions which are fundamental for survival, such as the choice and defence of the territory within which the individual carries out vital activities like hunting prey and competing for social rank and for mates. The behavioural sequences produced by the reptilian brain follow a rigid and ritualized code, giving rise to a stream of stereotyped behaviours which are perpetuated in different animal species by means of a genetic substrate which conditions their transmission. These mental codes permit the recognition of similar individuals, and their acceptance into the group, and identify specific sexual characteristics in such a way that a model which repeats even approximately their form and colors is able to elicit complex sequences of behaviour, with attempts at mating or territorial defence. This is the same reason for which the brain, or rather the portion of the archipallium of the brain, of a baboon is stimulated by the sight of a model representing the perineal swelling of an estrous female (Girolami and Bielert 1987) and the human brain is excited by seeing the photograph of a porno star.

In mammals the change of ecological niche which occurred, as mentioned before, 220 million years ago has led to the development of the paleopallium (Fig. 2).

From the earliest phases of embryonic development until birth, the ratio between the volume of nervous tissue and body mass is 6% in mammals. In the postnatal stage this

Reptilian brain (Archeopallium)

Limbic brain (Paleopallium)

Neocortical brain (Neopallium)

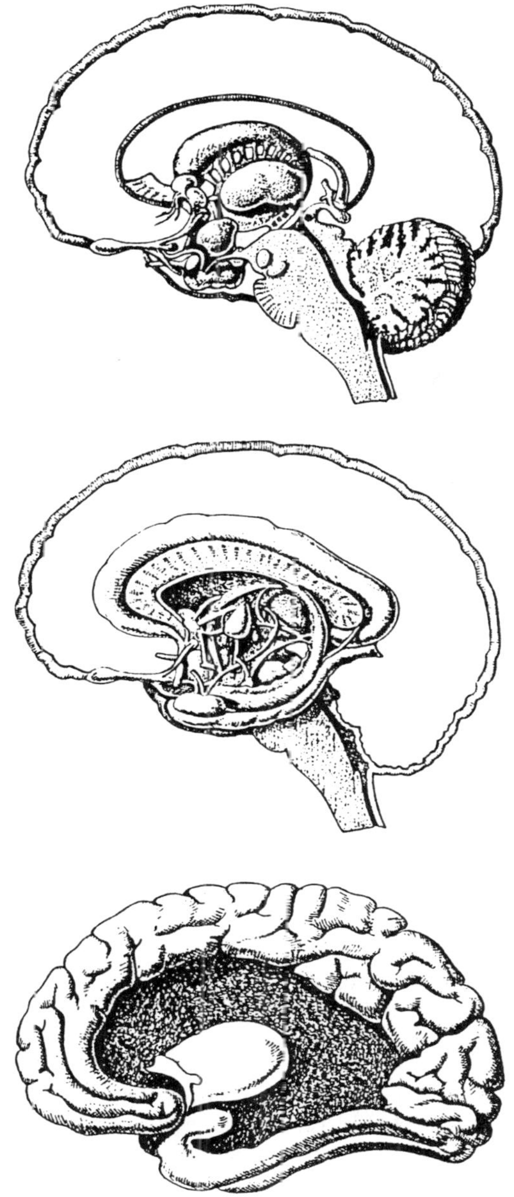

Fig.2. The brain of Man and of the Primates can be roughly subdivided into three organizational stages: the one characteristic of the Reptiles (spinal cord, medulla, mesencephalon and basal ganglia = basic self-preservation processes and instinctive activities); the one characteristic of paleo-mammals (limbic system = viscerosomatic and emotional processes); the one characteristic of neo-mammals (neocortex = processes of learning and complex adaptation).

neural growth stops, or almost does so, while body mass continues to increase; this produces, in adults, ratios which vary from species to species but which are very often markedly less than 6%.

This first expansion of the brain can be defined as the mammalian evolutionary leap. Even though the quantity and location of the genes involved in this phase of transformation from the reptilian to the mammalian stage are beyond our reach, since they occurred in the distant Jurassic, some information will be obtainable once the differential sequences of DNA which control the development of the encephalon in reptiles and mammals are available.

The paleomammalian portion of the human brain was defined by Broca (1878) as the "limbic lobe". Its structures surround the cerebral axis and enclose the reptilian brain (archipallium), covering the corpora striata. To the reptilian type of social behaviour, the limbic brain adds the affective- emotional dimension. In this portion of the encephalon, the information which comes from the external environment is modulated and rendered plastic by the sensations deriving from the visual, auditory, olfactory, taste and cutaneous afferents, as well as by endogenous and endocrine stimuli. With this new encephalic portion, there is an amplification both of the range of cerebral activity and of behaviours which are involved in the preservation of the individual and the species (Fulton 1951; MacLean 1955, 1958; MacLean and Delgado 1953).

In summary, the circuits of the limbic system constitute the anatomical substrate of the emotional sphere which integrates the behavioural aspects of the individual, modulating them by the affective-emotional component.

3. The transition to Primates

The second significant expansion of cerebral mass occurred about 65 million years ago when the phylum Primates originated with *Purgatorius*. The first forms of Primates, following the extinction of the Dinosaurs, again changed niche and became diurnal and arboreal, both to escape competition with other more specialized mammals (Lagomorphs and Rodents) and to supplement a vegetarian diet with insects and larvae which were abundant in the new broad-leaved forests. In order to mate and to raise offspring in an arboreal environment, they were obliged to develop a social structure which demanded greater intelligence; this requirement must have caused a new complex of genetic mutations which produced a new increase in cerebral mass.

Thus the neopallium developed, which is characteristic of the evolutionarily more recent mammals (Fig. 2). It totally envelops the two previous cerebral organizations, reaching its maximum development in Man. The neocortex presents a series of invaginations of its surface which give rise to the cerebral convolutions (which allow the packing within the limited volume of the brain case of a surface area of neocortical tissue of almost 2500 cm^2, of which only one-third is on the exterior and two-thirds is invaginated) (Hoffman 1989).

The neocortex possesses numerous paths of connection, both entering and exiting, with the structures of the two underlying cerebrotypes: the reptilian one (or archipallium) and the limbic one (or paleopallium); however, it does not have direct channels of information with the environment. As a consequence, all the information coming from the external environment, and from the viscera, arrives for evaluation and elaboration by the neocortex by way of the other two cerebrotypes; this information is modulated and colored in particular by the affective-emotional input conferred on it by the limbic brain (or paleopallium).

During the first phases of the evolutionary history of the Primates, there must have occurred a mutation which produced the halving of the body mass of the embryo with respect to the nervous tissue, thus doubling the brain/body ratio. As a result of this mutation, the ratio between cerebral mass and body mass became 12% (Sacher 1982). This allowed the Paleocene Primates to produce smaller offspring with the same degree of neural development as the other mammals (Camperio Ciani *et al.* 1993).

In a subsequent phase, there occurred a reduction in the number of offspring to 1 or 2 per birth, something which, considering the three-dimensional environment of the forest, was indispensable in the early stages of the raising of the young when the mother must carry them constantly on her body.

The diet of these ancient ancestors was almost exclusively vegetarian, sometimes supplemented with insects. The need to capture insects in the trees must have facilitated the development of the dexterity of the anterior limbs, the sense of balance and visual coordination. A few Primates which later became terrestrial, such as the baboons and the chimpanzee, sometimes kill small mammals for food, but meat has never had a substantial role in their diet (Strum 1991).

The evolution of the Primates took place in tropical and subtropical climates with an almost exclusively vegetarian diet. They would not have been able to colonize temperate and cold environments since they would have had little or nothing to eat during winter and spring when there are not enough plants available as food.

When in the middle Miocene, about 16 million years ago, there was again a warmer period, two groups of Primates, *Pliopithecus* and *Dryopithecus* migrated into Eurasia. They survived there until the end of the Miocene when, about 14 million years ago, there was another lowering of the temperature which caused their extinction, except in the tropical zone of southern Asia where the dryopithecines evolved into the Orangutan (Pickford 1986) and the pliopithecines evolved into the Gibbons.

4. The process of sapientization

The third great leap in the development of the brain and intelligence took place, beginning 5-4 million years ago, with the appearance of the australopithecines and of Man. However, underlying this third leap is a long story which must be recounted in detail.

About 15 million years ago in tropical Africa, some Primates differentiated into diverse forms of anthropomorphs. However, around 12 million years ago the ecological niche of the sub-Saharan tropical forest began to undergo profound changes caused by a drier climate (Coppens 1985, Kortlandt A. 1972) as a result of which the major part of the forest was transformed into woodlands with shrubs and grasslands (savanna). As a consequence, some of these anthropomorphs found themselves in a new ecological niche of open savanna and had to adapt to a new environment which required greater intellectual abilities for survival.

Around 5 million years ago, they evolved into the first forms of *Australopithecus*, with erect posture and with a greater encephalization quotient. The australopithecines continued to eat mainly vegetable matter. The wear on their teeth demonstrates that they fed mainly on leaves and fruit (Grine and Kay 1988) but they also ate roots, seeds and insects (Isaac 1978; Parker and Gibson 1979).

During the last 3 million years, the australopithecines have evolved, even if not in strict phylogenetic continuity, into *Homo habilis*, *Homo erectus* and finally *Homo sapiens*. During these phases of evolution, the volume of the brain has increased from 500 cc to 1500 cc. The three-fold increase in brain size in less than 3 million years is the greatest evolutionary change to have occurred in the history of the Mammals!

It was the selective pressure of the new niche which caused this last expansion of the brain just as it previously had imposed the choice of the acquisition of erect posture (Fig. 4). However underlying this increase in cerebral mass was also the highly cooperative social structure which, combined with the freeing of the hands from their function of body support, facilitated the production and use of tools which in turn was made easier by the acquisition of language as a mode of social communication (Chiarelli 1987).

In Man in particular, the neocortex is the site of abstract thought, of logical reasoning, of creative thinking, of the processes of analysis and synthesis, of ideas, of fantasy and of intuition. The control by the neocortex over the other two encephalic portions is carried out by means of symbolic thought and language and has the function of shifting the existential effort from merely biological grounds and from strict dependence on the environment to patterns of autoregulation and of semantic complexity.

The human brain differs from that of animals in its capacity to filter information through the neocortex, for example, sexual stimuli (odors, contacts with the partner, direct and indirect vision of the partner) which in animals instead act immediately and elicit a swift instinctive reaction.

According to MacLean, the human brain is the result of the superimposition of three basic anatomico-structural components, the reptilian (archipallium), the limbic (paleopallium) and the neocortical (neopallium) (Fig. 2). These three cerebrotypes, corresponding to the primitive formation of the brain of reptiles onto which were superimposed, in the course of evolution, the paleocortical structures of the first mammals and later the neocortex of the higher mammals, are different from each other both from the neurochemical and cytoarchitectonic points of view. Therefore in Man, three different brains, each with its own type of intelligence, memory, sense of time and space, motor activity and other functions, come together in a single structure (the "triune brain", MacLean, 1972). Furthermore, these three cerebral entities, even if extensively interconnected and functionally interdependent, are each capable of operating independently from the others (MacLean 1976). As Maclean has written "We are obliged to look at ourselves and the world through the eyes of three quite different mentalities, three interconnected biological computers each of which possesses its own special intelligence, its own subjectivity, its own sense of time and space, its own memory (...). Each of these components corresponds to a fundamental and separate stage of evolution" (1972).

However, the evolutionary stages of this third brain expansion, which began more than 6 million years ago, must be sought in the process of increasing organizational complexity consequent to the acquisition of erect posture and in the influences of life on the savanna on our most ancient ancestors.

In Africa, the australopithecines were replaced by *Homo habilis* and then by *Homo erectus* which continued to live in tropical and subtropical Africa and continued to feed primarily on vegetable matter (Stahl 1989), there being no need in fact to change feeding habits and become carnivorous. The temperatures in sub-Saharan Africa ranged between 20-25°C with a very small seasonal excursion in the course of the year. More than 100 different edible plant species were available throughout the year: berries, bulbs and roots during the wet season and shoots and leaves during the dry season. To enrich the diet, insects such as ants and termites were available.

The change in feeding habits toward carnivory began in the final evolutionary stage of the hominids. Carcasses of large animals, hunted and killed by lions and leopards, were occasionally scavenged and stripped of their meat by these early hominids, but mainly the bones were broken open to extract the marrow. The lions and leopards were not able to extract the marrow from the bones nor the brain from the skull. Thus apart from the occasional prey, the first meat utilized by hominids was that obtained by scavenging the parts of the prey of lions and leopards which were not utilizable by those true carnivores.

Around 1 million years ago, populations of *Homo erectus* began to spread into Eurasia. The problems of survival in the cold and temperate climates of Eurasia and the problems of the long winters were overcome by the use of abundant quantities of animal protein and fat to supplement the diet. Hunting thus became a regular activity, especially during the glaciations of Eurasia, just as it developed in the African populations. The incentive to hunt was certainly a very important stimulus for the development of technology and social cooperation.

The populations of *Homo sapiens* in Eurasia had to resolve problems of survival connected with the low temperatures and the difficulties of the sub-Arctic environment, whereas the populations in Africa evolved in tropical and subtropical environments: two habitats completely different which may have influenced differently the evolution of intellectual characteristics.

5. Quantitative data of the encephalic increase in the hominids

The increase in encephalic volume from the australopithecines to modern humans is the aspect which most interests us. This increase is roughly from 500 to 1500 cc, practically a three-fold increase. This is an evolutionary leap which has occurred over a period of approximately 3 million years, equivalent to a biological time of no more than 150,000 generations (50,000 generations for each 1 million years). Assuming a continuous rate of change, this increase amounts to 500 cc each 50,000 generations or, in other words, to 1 gram per 100 generations or 0.01 gram per generation. In terms of neurons, this is equivalent to about 3 million neurons more every generation, again assuming that the increase has been constant.

This fascinating evolutionary topic has been dealt with by eminent scholars such as Tobias (1968), Falk (1980) and Holloway (1973) to cite only the most famous ones. Among recent Italian authors who have contributed to the subject must first of all be mentioned Raffaello Parenti to whom is due, among other things, the introduction, in a work from 1973, to the English vocabulary (he who knew little or no English) of the term "quantitative sapientization", which has subsequently been taken up by various authors.

More recently the subject has been addressed again in Italy by Roscoe Stanyon and colleagues (1993), who have accumulated the data available in the literature and have analyzed them according to the hypotheses of gradual evolution and of punctuated equilibria. In this synthesis, I use only the raw data presented in that work to report them as coordinates. The growth trend is altogether evident just as it is even more clear on a logarithmic scale (Fig. 3). Particularly interesting is the linearity of the data of *Homo erectus* (HE), archaic *Homo sapiens* (HSA), *Homo sapiens neanderthalensis* (HSN) and modern *Homo sapiens sapiens* (HSS) which has a considerable slope, indicating how the increase of cerebral mass is still occurring.[1]

In presenting these data, the ratio of cerebral mass to body weight has not been considered, nor have the possible sex differences and the exact taxonomic separations; however this approximation does not detract from the general trend of the phenomenon, which is actually emphasized by the fact that in all of these forms the reptilian and paleomammalian brains have the same mass, which means that the increase concerns only the part of the brain related to "sapientization" (the third brain or neopallium).

In my opinion, this clarification is important also for the recent reinterpretations of the cerebral cortex of the neopallium (cerebral isocortex). It has six layers of neurons and a thickness of 300-700 μm in a structural conformation of the columnar type in which the neurons are functionally organized in a vertical way. These columns of neurons constitute the anatomical and functional units on which is based the evolutionary variability of cerebral mass, so that the increase in intellectual capacity of the neocortex can be compared to computer processing units which are added to each other during evolution. The surface area of the human cerebral cortex, with all its convolutions, is about 2430 cm^2, more than 10 times the surface area of the brain of a macaque and more than 100 times that of a rat.

[1] An interesting hypothesis has been recently devised by Leslie C. Aiello (in print in Current Anthropology); the author explains the most recent increase of the human brain as due to the use of cooking the food.

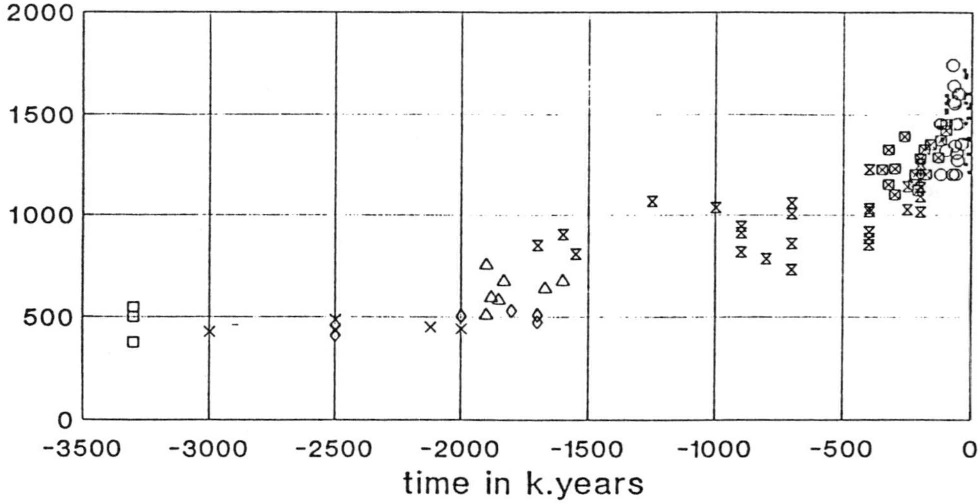

Fig.3. Change in cranial capacity in Hominids

□ *A. afarensis* × *A. africanus* ◇ *A. robustus*
△ *H. habilis* ⧗ *H. erectus* ⊠ *arcaic H. sapiens*
○ *H. sapiens neanderth.* ■ *H. sapiens sapiens*

Since the cortical columns have a similar area in all mammals (even if they vary in length and in density of neurons), the cortical expansion during evolution must have occurred by numerical multiplication of the cortical columns, both by an increase in new columns interposed between the existing ones, and by the multiplication of entire blocks of columns (of whole areas). In fact the replication of blocks might explain the presence in the more-evolved brains of the Primates of a larger quantity of association areas, such as the cortico-visual, somato-sensory, auditory or motor areas, with respect to the insectivores and more primitive mammals. In the cerebral cortex of the spider monkey, there are at least 18 visual association areas, 6 somato-sensory and 4 auditory ones (Diamond et al 1969, Van Essen 1979, Kaas 1989).

6. The possible ecological bases of the diversity in IQ test scores among modern humans

There is another aspect of the evolution of intelligence which must be considered: the different intelligence quotients of the diverse human populations.

The existence of different niches in which the different human populations originated and evolved has stimulated Lynn (1992) to put forward the hypothesis that the Eurasian and African niches exerted different selection pressures on the intellectual development of the Caucasoids and Mongoloids on the one hand and the Congoids and Capoids on the other. In my opinion, more than quantitative differences, there could be different types of intelligence. The hypothesis is based on the theory of Jerison, according to which the cognitive demands produced by a new ecological niche exerted a selection pressure which favoured the increase and differentiation of intelligence during the course of evolution.

Among present-day hunter-gatherers, the quantity of nutrients derived from hunting and from gathering vary according to the latitude (Lee 1968). The populations in tropical and subtropical latitudes are principally gatherers, whereas those in temperate environments practise mainly hunting. The food of populations living in arctic or subarctic habitats is based almost exclusively on hunting and fishing since vegetable matter is virtually impossible to find throughout the whole year, with the exception of some berries available in summer and autumn. In order to hunt large herbivores, these populations need to make a great variety of wood and bone tools, in addition to devising traps and other mechanisms for catching the prey. After killing a large animal, these people must skin and butcher it, dividing it into pieces which can be transported to the home base; necessary for this operation is a large variety of sophisticated cutting instruments. Other prerequisites of effective hunting are a well-organized social group and a sophisticated system of vocal communication which helps to plan and coordinate the hunting activities but which must not interfere with the hunter's vision, which must be concentrated completely on the prey.

Furthermore, the technology for lighting and maintaining a fire must have been much more complicated in Eurasia than in tropical Africa. There would have been the same difficulties in making clothes and in building shelters against climatic adversities. Thus in Eurasia during the different adaptive phases of evolution of the human populations, a greater creative effort was required.

In contrast, for the provision of their daily food, the gatherers of the equatorial and sub-equatorial regions must possess a greater ability for discrimination and a better visual memory in order to remember the different plant species appearing through the seasons. The greater complexity and difficulty of life in the forest must have brought about the development of a heightened ability for the qualitative differentiation of objects and thus a different type of intelligence.

Without wishing to enter into the discussion of Jensen concerning differences in IQ between the diverse human groups, the measurement of which nevertheless is the object of criticism because it tends to measure preferentially only certain types of intellectual abilities, I have no doubt that the qualitative differences in intelligence of the various human populations is based on eco-ethological factors which would have produced genetic differences between populations, just as the type of intelligence is commonly considered to be different between the sexes on account of the unequal division of labor (hunting and gathering) beginning with the most ancient forms of humanity.

7. The Neolithic transition

The Neolithic transition occurred in the Near East, in China and in Central America in an independent manner but at virtually the same time. The beginning of cultivation and the stabilization of settlements was another factor in the enhancement of intellectual abilities. The domestication of plants and animals has involved ever larger permanent groups which have resulted in a general increase in the population. The concentration of people facilitated the development of refined skills and the introduction of innovations. As a result of this new and more active social interaction, intellectual and technological improvements were produced which led to the creation of written language, to a system of mathematics and finally to an understanding of Nature as something separate from ourselves.

Therefore, according to the theory of Jerison, the brain expanded and intelligence developed by means of a constant interaction with the environment.

8. The neuronal bases of cerebral evolution

What are the cellular mechanisms which can explain this interactive evolution? What are the neurophysiological bases for an explanation of this progressive change and enlargement? Is this process still active? What is the influence of the present cultural development on this process?

In recent years, there has been a revaluation of the concept that the structure of nerve cells can be modified during their activity. The influence of the environment on the transformation of the brain is more effective in young individuals in whom neuronal plasticity is greater than in any other postnatal period. Diverse experiments have demonstrated that mice raised in an environment rich in stimuli clearly show an increase in neuroanatomical parameters with respect to those raised in a stimulus-poor environment. Mice exposed at an early age to a complex environment have a heavier brain and a larger ratio of cortical to subcortical weight; also the histochemical parameter and the number of dendritic connections are greater than in individuals maintained in an impoverished environment (Chiarelli 1968).

Similar interactions between neurological development and the type of environmental stimulation have been identified in the Primates. In macaques raised in different environmental conditions, the number and the branching of dendrites of the basal cortex are significantly increased in conditions associated with greater stimulation and with activities which involve leaping, swinging on branches and the handling of objects (Stell and Riesen 1987). Similarly, the development of the visual cortex is affected by the quantity and quality of the light present in the environment where the young macaques were brought up (Riesen 1982, Bianchi 1968).

These researches demonstrate that:

1) the environment can modify the brain in a significant way;
2) the brain of young mammals is anatomically malleable.

In our species the neurological structure remains immature until 12 years of age. Thus it is plausible to imagine that also for Man the environment may play an important role in the modification of neuronal components during the phase of development of the brain. Direct and indirect evidence exists for the neuronal centers of language. We know in fact that if a child is not introduced to a linguistic environment before the age of 8, he will never learn to speak (enfant sauvage); as well, many vocalizations are learned in the early stages of childhood and certain sounds cannot be produced if not learned in infancy (the r for the Japanese and Chinese).

We also know that in the brain of songbirds certain neuronal nuclei involved in sound production are much larger in the males who sing than in the females who only twitter (Mazzi and Fasolo 1977). They must therefore be subjected to the influence of some hormones (especially thyroid hormone and testosterone) which stimulate specific types of cells during ontogenesis. Some sex hormones have in addition a specific importance in the distribution of synapses and in the architecture of the dendrites.

On the basis of these experiments and of other data, Jacobson hypothesized in the 1970's the existence of two types of neurons in the cortex. The first group, formed in the initial stage of development, are strictly influenced by the genome and consist mainly of macroneurons (afferents and efferents). The second group, constituted mainly by interneurons with short axons, instead undergo changes during the entire duration of postnatal development and are less strictly under genetic control. The survival of these neurons is influenced by the functional connections of the synapses which need to be constantly active in order to be fixed (functional confirmation). Marked changes in the form and number of the neurons can occur also in the relatively late stages of ontogenesis of the organism.

The basic program of cortical neuroanatomy is thus genetically defined, but the final organization of the complex of neuronal connections and of their functions is the result of the neuronal activity which occurs during development (Jacobson 1970). In this way, one can seek to explain how new information could have stimulated the increase in brain size during evolution. An individual with more plastic neurons and who managed to acquire more information during the first phases of his life may have survived better and have reproduced more efficiently, in this way giving his genotype more possibility of taking hold and spreading. This is a Darwinian mechanism which leads us to reflect on the origin of the different intellectual abilities of present-day humanity and on the importance of education.

9. Conclusions and consequences

In this attempt to reconstruct the evolution of intelligence, I have sought to demonstrate how this still mysterious mass of protoplasm which dominates the world has progressed through a constant interaction between heredity and the environment. In addition to the discovery and ever more specialized use of technology and language in the earliest hominids, other equally important factors have played a decisive role in the development and evolution of intelligence, such as social structure, the migration to temperate zones, the search for food, the discovery of the cultivation of plants and the raising of livestock, the development of ethical concepts, the knowledge of one's own death, the invention of magic and, if one wishes, of religion.

However, all of these events are nothing but further responses to the necessity of survival and of progress, whose introduction into the human mind comes about not by genetic mechanisms but through education: that is, the facilitated, or better, the conditioned interaction between environmental stimuli and Jacobson's second type of neurons, in a period in which they develop most abundantly and by means of their consequent functional confirmation during the prolonged ontogenetic phase of our species. Heredity intervenes at the level of heterogeneity and individual variability, whereas the different environmental conditions operate by selecting for efficient nervous structures, thus favoring the individuals which possess them.

References

Bianchi, L. (1968) Evolution of the brain in world Primates. In: Taxonomy and phylogeny of Old World Primates with references to the origin of Man. Ed. B. Chiarelli. Turin. Rosenberg e Sellier.

Broca, P. (1878) Anatomie comparée des circonvolutions cérébrales. Le grand lobe limbique dans la série des mammiferes. Revue Anthropol. 1:385.

Camperio-Ciani, A. (1993) Ecobiological aspects of mating system of Primate including Man. Global Bioethics 6:61-74.

Camperio-Ciani, A. and Chiarelli, B. (1993) Socioecologia delle strategie riproduttive nei Primati. Global Bioethics, 5:65-79.

Chiarelli B. (1984) Origine della socialità e della cultura umana. Bari. Laterza.

Chiarelli, B. (1986) Vocalization in non-human primates and the descent of human language. Proceedings International Meeting on Variability and Behavioral Evolution. Accademia dei Lincei. Quad. 259:217-225.

Chiarelli, B. (1987) Coordinamento spaziale, comunicazione gestuale, origine del linguaggio e della lateralizzazione cerebrale nell'uomo. Antropopologia contemporanea 10:151-159.

Diamond, I.T. and Hall, W.C. (1969) Evolution of the neocortex. Science 164:251-261.

Falk D. (1980) Language, handedness, and Primate Brains: did Australopithecus sing? American Anthropology.

Fulton, J.F. (1951) Frontal lobotomy and Affective Behaviour. A neurophysiological Approach. New York. Norton Publ.

Girolami L. and Bielert C. (1987) Female perineal sexual swelling and its effects on male sexual arousal: an apparent sexual releaser in the chacma baboon (Papio ursinus). Int. J. of Primatology 8: 651-661.

Hoffman M.A. (1989) On the evolution and geometry of the brain in Mammals Progress. Neurobiology 32: 137-158.

Grine, F.E. and Kay, R.F. (1988) Early hominid diets from quantitative image analysis of dental microwear. Nature 333:765-768.

Holloway R.L. (1973) Endocranial volumes of Early African Hominids and the role of the brain in human mosaic evolution. J. Human Evolution 2:449-459.

Holloway, R.L. (1978) Problems of Brain Endocast Interpretation and African Hominid Evolution. In: Early Hominids of Africa. Ed. C.J. Jolly. London. Duckworth. Pp 379-401.

Isaac, G. (1978) The food-sharing behavior of Protohuman Hominoids. Scientific American 238(4):90-108.

Kaas, J.H. (1989) Why does the brain have so many visual areas? J. Cogn. Neurosci 1:121-135.

Kortlandt, A. (1972) New Perspectives on Ape and Human Evolution. Amsterdam. Strichting voor Psychobiologie.

Jacobson, M. (1970) Development, Specification and Diversification of Neuronal Connections. In: The neurosciences Second Study Program. Ed. F.O. Smith. New York. Rockfeller University Press. Pp 116-129.

Jerison, H. Evolutionary biology of intelligence: the nature of the problem. In: Intelligence and evolutionary biology. Eds H. Jerison and I. Jerison. Berlin. Springer-Verlag. Pp 1-12.

Lee, R.B. (1968) What hunters do for a living, or how to make out on scarce resources. In: Man in the Hunter. Eds R.B. Lee and I. De Vore. Chicago. Aldine. Pp 30-48.

MacLean, P.D. and Delgado, J.M.R. (1953) Electrical and chemical stimulation on fronto-temporal portion of limbic system in the walking animal. Electroencephal. Clin. Neurophysiology 5:91.

MacLean, P.D. (1955) The limbic system (visceral brain) in relation to central gray and reticulum of the brain stem. Evidence of interdependence in emotional processes. Psychosom. Med. 17: 355.

MacLean, P.D. (1958) The limbic system with respect to self-preservation and the preservation of the species. J. Nerv. Mental Diseases, 127: 1.

MacLean, P.D. (1970) The triune brain, emotion and scientific bias. In: The Neurosciences Second Study Programme. Ed. F.O. Schmitt. New York. The Rockfeller University Press. 366.

MacLean, P.D. (1972) Cerebral evolution and emotional processes: New findings on the striatal complex. Ann N.Y. Acad. Sci. 193:137.

Mac Lean, P.D. (1976) Sensory and perceptive factors in emotional functions of the triune brain. In: Biological Foundations of Psychiatry.

Mazzi, V. and Fasolo, A. (1977a) Evoluzione e determinismo del neurone: dati sperimentali e considerazioni metodologiche. Storia del pensiero scientifico, Juvence 265-294.

Mazzi, V. and Fasolo, A. (1977b) Introduzione alla neurologia comparata dei Vertebrati. Torino. Boringhieri.

Oddum, E.P. (1971) Fundamentals in Ecology. Philadelphia. Saunders.

Parker, S.T. and Gibson, K.R. (1979) A developmental model for the evolution of language and intelligence in early hominids. Behav. Brain Sciences 2:367-408.

Parenti, R. (1973) Quantitative and qualitative trends in human sapientization. Journal Human Evolution 2:499-508.

Passingham, R.E. (1973) Anatomical Differences between the Neocortex of Man and other Primates. Brain and Evolution 7:337-359.

Passingham, R.E. (1979) Brain size and intelligence in man. Brain, Behavior and Evolution 16:253-270.

Piaget, J. and Inhelder, B. (1960) La psicologia della prima infanzia (dalla nascita al settimo anno). In: Trattato di psicologia. Eds D. Ketz and R. Ketz. Torino. Boringhieri.

Piaget, J. (1963) The origins of intelligence in Children. New York. W.W. Norton and Co.

Pickford, M. (1986) Major events in primate paleontology: possible support for climatic forcing models of evolution. Antropologia contemporanea 9:89-94.

Pickford, M. (1988) The evolution of intelligence: a palaeontological perspective. In: Intelligence and evolutionary Biology. Eds H.J. Jerison and I. Jerison. Berlin. Springer-Verlag. Pp 175-198.

Rapoport, S.I. (1990) Integrated phylogeny of the Primate brain, with special reference to human and their disease. Brain Res. Rev. 15:267-294.

Riesen, A.H. (1982) Effects of environments on development in sensory systems. In: Contributions to sensory physiology. Ed. W.D. Neff. New York. Academic. Pp 45-77 vol.6.

Rosenzeweig, M.R., Bennett, E.L. and Diamond, M.C. (1971) Chemical and anatomical plasticity of brain: replications and extensions. In: Macromolecules and Behavior. Ed. J. Gaito. New York. Appleton-Century-Crofts. Pp 205-278.

Sacher G. A. (1982) The role of brain maturation in the evolution of Primates. In: Primate Brain Evolution. Eds E. Armstrong and D. Falk. New York.Plenum Press. Pp 97-112.

Stahl, A.B. (1984) Hominid dietary selection before fire. Current Anthropology 25:151-168.

Stell, M.and Riesen, A. (1987) Effects of early environments on motor cortex neuroanatomical changes following somatosensory experience. Effects of Layer III pyramidal cells in monkey cortex. Behavioral Neurosciences 101:341-346.

Stanyon R., Consigliere S., and Morescalchi M.A. (1993) Cranial capacity in Hominoid Evolution. Human Evolution 8: 205-216.

Strum, S.C. (1981) Processes and products of change: baboon predatory behavior at Gilgil, Kenya. In: Omnivorous Primates. Eds R. Harding and G. Teleki. New York. Columbia University Press.

Tobias P.V. (1971) The brain in hominid evolution. New York. Columbia University Press.

Tobias P.V. (1968) Cranial capacity in anthropoid Apes, Australopithecus and Homo habilis with comments and skewed samples. South African Journal of Science 64:81-91.

Wynn, T. (1981) The intelligence of Olduway Hominids. J. of Human Evolution, 10:529-541.

Van Essen, D.C. (1979) Visual areas of the mammalian cerebral cortex. Ann. Rev. Neuroscience 2:227-263.

EVOLUTION OF HUMAN CULTURE: A COMPOSITE PATTERN

FRANCESCA GIUSTI
Istituto Italiano per gli Studi Filosofici
Palazzo Serra di Cassano
Via Monte di Dio 14
80100 Naples Italy

ABSTRACT This paper outlines and criticizes the hunter-gatherer model that has dominated the study of the evolution of culture since the 60's. This model has been questioned widely on the basis of studies on contemporary foraging societies, on non-human primates and on the basis of archaeological discoveries. In addition, the most profound error of this model for the evolution of human culture was that it was seen as a linear progression instead of a constellation of features combined and recombined in different ways in different species as in different populations within a single species. The behavioural, social and reproductive patterns which formerly seemed to be exclusively human represent an evolving mosaic of labile and reversible features as in the evolving mosaic of the anatomical evolution. The hominid way of life did not develop along a single pathway, but rather along different pathways leading up to our own species.

1. The culture concept in Paleoanthropology

Culture as a distinctive human trait is a very difficult concept to define. For some authors culture is essentially the human aptitude for transmitting knowledge, values and other factors that influence behaviour (Boyd and Richerson 1985). For others, culture is what differentiates one society from another (Hinde, 1987a). It is a complex system embracing technology, knowledge and beliefs. In the paleonthropological context the term culture has been generally used as a synonym for "way of life" and has indicated a complex pattern of behavioural traits which were suggested to be typical of our own species.

My aim is to show that even in this latter meaning of the term, the appearance of a human style of life has been considered as the simultaneous appearance of behavioural traits integrated within a unique evolutionary pattern.

Many models have been proposed in the last few years to account for the appearance of the human way of life. In this paper I will refer in particular to the hunter-gatherer model, not only because it has dominated the study of the evolution of culture since the 60's (Lee and DeVore, 1968), but also because it exploits more than others the notion of culture in the sense of a rigidly structured behavioural pattern. This model began as tentative, but over the years it grew into a doctrine. Even though it has been modified through time, this model has consistently seen as a frame of reference in the studies on human evolution.

The traditional model has proved to be outdated in the light of more recent research; another approach is necessary to account for the evolution of behaviour in our earlier or later ancestors.

R. A. Gardner et al. (eds.), The Ethological Roots of Culture, 433–439.

2. The hunter-gatherer model: its origins and subsequent formulations

This model was first used in the 60's. It was based mainly on field studies on modern hunter-gatherers, but it seemed usefull to combine paleontological and anatomical data with the then current intrepretation from archaeological remains.

Bipedalism and freedom of the upper limbs (the main anatomical changes in human evolution) allowed the male to use deadly weapons for hunting, and allowed the female to care for a more immature infant. This was related to progressive encephalization and the limitation on the size of the pelvis imposed by the upright position (Washburn, 1960; Campbell, 1966, 1974).

A home-base, sexual division of labour and food-sharing were features of modern hunter-gatherers used to reconstruct a unitary model of the primitive life of our ancestors. The human family was an integral part of this model. The pair-bond and the role of the father were the result of economic reciprocity among males and females. Changes in human female sexuality (such as the "loss of estrus", the duplications of sexual signals, the orgasm) were adaptations to ensure an exchange of sexuality and food (Washburn, 1960, 1961; Washburn and Lancaster, 1967).

In subsequent years many aspects of the primitive model were modified. Encephalization was established as a relatively late feature in our evolution and so too were the changes linked to the birth of a more immature infant. In addition, while in the early formulation hunting was the main aspect of the model, in subsequent years much more stress was given to gathering and to the role of the female (See e.g., Leacock, 1978, Dahlberg, 1981). Whether the accent was placed on hunting or gathering the modern hunting-gathering life-style continued to be a powerful framework in which to insert the principal aspects of our evolution.

This model integrated all the principal anatomical and behavioural changes in our evolution: food strategies, patterns of mobility in territories, reproductive strategies, social organization, tecnology and ideology. It depended on the simultaneous appearance of many behavioural traits in a new way of life that persisted for a remarkably long time. Right from the earliest formulations, the hunting style of life was supposed to have appeared in the earliest stage of our evolution maintaining its basic pattern with increasingly complex economic, technical, social and cultural features for 99% of our evolutionary history (Lee and DeVore, 1968).

The model suggested a uniform world totally inhabited by hunter-gatherers that only came apart with the arrival of new climatic conditions. It was only in the postglacial period that new strategies based on food production led to cultural diversification.

3. The rejection of the hunter-gatherer model

New findings and new thinking suggest the following criticism on the hunting model (Giusti, 1994).

It has been known for many years that scavenging rather than hunting was the primary ancestral means of procuring food. However scavenging did not seem to contrast the hunting and gathering hypothesis. Both scavenging and hunting involve food sharing at a home-base. Scavenging could progress smoothly into hunting without a basic change in the cultural structure (Isaac, 1978).

A much more critical problem with the model is the research that questions the existence of the home-base itself. Recent archaeological evidence indicates that the stone heaps previously considered the most ancient shelters were in fact purely accidental (Binford, 1981). Also fierce competition with the carnivores refutes the idea of a home-base as a sure and sheltered refuge near water sources (Binford, 1981, 1983; Potts, 1984, 1988). Both tooth marks and tool marks appear on the same animal bones. This shows that there was close competition between early hominids and large carnivores (Bunn, 1981, 1982; Potts and Shipman, 1981; Bunn and Kroll, 1986; Blumenshine, 1987; Potts, 1984,1988).

Perhaps the hunting model only applies to the latest stages such as the late *Homo erectus* (700,000-400,000 b.p.) (Potts, 1988) or only to modern man of our species. Perhaps, the hunting strategies developed in parallel to agriculture in postglacial conditions (Foley, 1988).

New problems have recently arisen in the anthropological field itself in applying the modern hunter-gatherer model even to the recent stages of our evolution. Recent research indicates that modern hunting-gathering societies do not represent a single category. Simple egalitarian societies, which have been traditionally used for the prehistoric hunting model, are rare. Many hunter-gatherer societies show much more complexity. Some invest labour in more complex equipment, such as traps. Some invest labour in long time food storage. Such activities lead to incipient social stratification (Testard, 1982, 1988; Woodburn, 1982).

The idea that modern hunter-gatherers represent an ancient form may, itself, be an error. Instead, simple egalitarian societies may be the product of social degeneration, as we have only found them encapsulated within other modern societies (Woodburn, 1988).

Recently evidence of the existence of simple hunter-gatherer societies has been questioned. Wilmsen and Denbow (1990) produce archaeological evidence that hunter-gatherer societies have never been self sufficient but have always depended on their agropastoralist neighbours and on some modern economic system. In this revisionist view, "traditionalists" have merely projected their own modern ideals into the evolutionary past.However, Solway and Lee (1990) vigorously defend the evidence for the traditional view. At times this debate has been very heated.

4. Non-human primate studies: behavioural variations

The classical model of hunter-gatherers that has dominated the field since the 60's is therefore in dispute because of recent findings from studies of contemporary foraging societies, as well as from archaeological studies.

Attempts to repair the hunter-gatherer model by minor adjustments fail because the error is profound. The classical hunter-gatherer model is based on a still more classical model of cultural evolution as a linear progression from one type to another. What we see instead, along the pathway of human evolution, are constellations of features combined and recombined in different ways in different species as well as in different populations within a single species.

Therefore let us consider some data from non-human primate research, the other main field of study used to reconstruct the patterns of behaviour of the ancient hominids.

There are many problems in using non-human primates as models for human evolution and we cannot simply extend the behavioural repertoire of some living non-human primates to extinct hominoid forms using referential models (Tooby and DeVore, 1987). However, non-human primate research can be used to extract principles which may be useful in the reconstruction of the behaviour of our ancestors (Hinde, 1987b) In particular chimpanzees still remain the closest model showing at least the minimal capacity of the early hominids (Wynn and McGrew, 1989; McGrew, 1991).

For some time many traits previously considered to be uniquely human have been found in the repertoire of many non-human primates and especially in chimpanzees, our closest relatives. I refer both to aspects of material culture and complex mental abilities. The former (such as tool using in different situations, tool making, techniques to obtain meat etc)., have been widely observed in field research (See e.g. Goodall, 1986; McGrew *et al* 1979); the latter (such as cognitive and communicative abilities) have been studied in laboratory conditions, in crossfostering experiments (Gardner and Gardner,1985, 1991) as well as in the field in communicative and social contexts .

Recent research indicates that traits of material culture show a high degree of variability that some authors relate to different "traditions" between different populations (Goodall, 1970; McGrew *et al* 1979). For example, the most common tools used by wild chimpanzees to obtain termites differ between communities living in different sites as well as between different populations within the same community. Also techniques to process and eat palm nuts and techniques for getting meat differ among different populations (McGrew, 1991).

Such a variability above all may be related to very different habitats in which chimpanzees can live. They vary across a broad spectrum from dry, hot tropical environments (McGrew*et al* ,1981), to most common habitats of savanna woodlands, to rain forests (Boesh and Boesh 1989).However the degree of behavioural variability does not seem to be related simply to different ecological conditions. For example, food preferences and specific food denial are not merely linked to food availability: in places where nuts are present, some chimpanzee populations eat and process them while others ignore them. Moreover there are strong preferences for different kinds of food (McGrew, 1991).Many aspects of material culture may be the result of a free choice and, to some extent, show an arbitrary and random character.

Let us examine another aspect of the complex behaviour of chimpanzees. It often occurs within composite patterns in which each aspect has a positive feedback on the other. Among Tai Forest chimpanzees, collaboration in hunting, food sharing among males, prolonged learning and food sharing between mother and son, using symbolic forms of communications are all aspects that coexist and strengthen each other. These traits often occur in reciprocal association with foresight,the ability to delay action, and social learning (Boesh & Boesh, 1989, 1990; Boesh, 1990; 1991).

Cultural variation is often associated to lability and reversibility of behaviour especially in the case of complex patterns of behaviour. Hunting behaviour in many non-human primates arises under complex environmental conditions such as prey availability, prey size and competition with other predators, under complex social conditions, such as cooperation, synchronization and social learning, and in the presence of a complex social group where different individuals show varying degrees of creativity, foresight and other cognitive skills (Strum, 1987; Strum and Mitchell, 1987).Under these complex conditions there is lability and reversibility. Patterns of behaviour appear, disappear and reappear again.

5. Ancient hominids and different behavioural patterns

The reconstruction of ancient hominid patterns of behaviour has become an ever increasing arduous task in which the concept of culture may be misleading, to such an extent that many authors prefer to speak of the evolution of complex and flexibile behaviour (Foley, 1991). In any case all the data from both evolutionary ecology and archaelogical records seem to concur to show some degree of variability in different species of hominids who adapted to different ecological niches and also in different populations within the same species.

Let us consider the following points:

1) Concerning the most ancient stages of our evolution, in the Plio-Pleistocene scenario many data seem to support a high degree of variability in the models of life of different genera and species. Such variability implies different alimentary strategies, reproductive strategies, growth rates and consequently, different social systems on the whole.

In a highly variable savanna habitat, feeding strategies might have been different, ranging from vegetarian patterns to patterns in which meat consumption prevailed (with a different rate of scavenging or predation). Some variability in stone tool use could have occurred for different species of early hominids according to different ecological niches based either on low quality vegetable food (digging tools of different material for robust australopithecines) or on high quality food, such as meat (stone flake-tools to remove the meat from carcasses for *Homo*). It is also possible that not all populations were necessarily dependent upon the use of stone tools (Toth 1985) According to the data relating to sexual dymorphism also the hominid reproductive strategies varied across a broad-spectrum from harem structures (robust australopithecines) to milder forms of polygyny in multimale groups (earliest forms of genus *Homo*) (Foley, 1987, Foley-Lee, 1989, Kinzey, 1987; Giusti,1992). Recent studies on tooth enamel seem to indicate very fast and ape-like growth rates for the robust australopithecines and slower rates for the genus *Homo* (Benyon and Dean, 1988). In conclusion, the whole life-style might actually have been shaped in very different ways for different kinds of hominids.

3) A high degree of variability is also associated to more recent evolutionary stages. In many cases there is no direct correspondence between modern anatomical traits and “modern”, complex behaviour which is backed up by the increasingly complex archaelogical records. The progressive encephalization of the latest evolutionary stages might have increased the cultural variability over time. At the same time, however, the emergence of a complex language might have determined the conditions for the maintenance and transmission of the behavioural models.

In conclusion, the behavioural, social and reproductive patterns, which formerly seemed to be exclusively human, represent an evolving mosaic of labile and reversible features like in the evolving mosaic of the anatomical evolution (Potts, 1988). The hominid way of life did not develop along a single pathway, but rather along different pathways leading up to our own species.

If we consider cultural variability as a primitive feature, then the spread of a homogeneous lifestyle can be related to the more recent stages in the evolution of the social systems. The relationships between cultural change and cultural stability in the different situations (see e.g. Van der Leeur and Torrence, 1989; Boyd and Richardson, 1985) will be the subject of much discussion in the near future.

References

BEYNON A.D.& DEAN M.C.(1988) "Distinct dental development patterns in early fossil hominids *Nature* **335:** 509-14

BINFORD L. (1981) *Bones: Ancient Men and Modern Myths* Academic Press New York

BINFORD L. (1983) *In Pursuit of the Past. Decoding the Archaeological Record* Thames and Hudson London

BLUMENSCHINE R.J. (1987) "Characteristic of an early hominid scavenging niche" *Current Anthropology* **28** : 383-406

BOESCH C. (1990) "First hunters of the forest" *New Scientist* 19 May 1990: 38-41

BOESCH C. (1991) "Symbolic communication in wild chimpanzees" *Human Evolution* Vol. 6 N.1 : 81-89

BOESCH C. & BOESCH H. (1989) "Hunting behavior of wild Chimpanzees in the Tai National Park" *American Journal of Physical Anthropolgy* **78**: 547-573

BOESCH C. & BOESCH H. (1990) "Tool use and tool making in wild chimpanzees" *Folia Primatologica* **54:** 86-99

BOYD R. & RICHERSON P.J. (1985) *Culture and the Evolutionary Process* The University of Chicago Press, Chicago and London

BUNN H.T. (1981) "Archaelogical evidence for meat-eating by Plio-Pleistocene hominids from Koobi Fora and Olduvai Gorge" *Nature* **291:** 574-577

BUNN H.T. (1982) "Animal bones and archaeological inference" *Science* **215**: 494-495

BUNN H. T. & KROLL E. M.(1986) " Systematic butchery by Plio-Pleistocene hominids at Olduvai Gorge, Tanzania" *Current Anthropology* **27**: 431-452

CAMPBELL, B. (1966;1975) *Human Evolution: an Introduction to Man's Adapatations* Aldine Chicago

DAHLBERG F. (1981) (ed) *Woman the Gatherer* Yale University Press New Haven

FOLEY R. (1987) *Another Unique Species. Patterns in Human Evolutionary Ecology* Longman Essex

FOLEY R. (1988) "Hominids, human and hunter-gatherers: an evolutionary perspective" in INGOLD T., RICHES D. & WOODBURN J. (eds) (1988) *Hunters and Gatherers Vol I-History, Evolution and Social Change* pp: 207-221

FOLEY R. (1991) "How useful is the culture concept in early hominid studies?" in FOLEY R. (1991) (ed) *The Evolution of Human Behaviour* London Unwin Hyman pp. 25-38

FOLEY R. (1991) (ed) *The Evolution of Human Behaviour* London Unwin Hyman

FOLEY R.A. & LEE P.C. (1989) "Finite social space, evolutionary pathways, and reconstructing hominid behavior" *Science* **243:** 901-906

GARDNER R.A. & GARDNER B.T (1991) "A cross-fostering laboratory" in GARDNER R.A. & GARDNER B.T (eds)*Teaching Sign Language to Chimpanzees* State University of New York Press, Albany pp.1-28

GARDNER B.T & GARDNER R.A (1985) "Signs of intelligence in cross-fostered chimpanzees" *Philosophical Transaction Royal Society* **308:** 159-176

GIUSTI F. (1992) "The evolution of human family" paper presented to the Meeting "Human family relationships: evolution biology and culture" (Naples April 1-2) in press.

GIUSTI F. (1994) *La scimmia e il cacciatore. Interpretazioni, modelli sociali e complessità nell'evoluzione umana* Donzelli, Roma

GOODALL J. (1970) "Tool using in primates and other vertebrates" in LEHERMA D.S, HINDE R. & SHAW E. (eds) *Advances in the Study of Behaviour* Academic Press New York, London Vol III pp.195-249

GOODALL J. (1986) *The Chimpanzees of Gombe. Patterns of Behaviour* Harvard University Press, Cambridge

HINDE R. (1987a) *Individuals, Relationships and Culture. Links between Ethology and the Social Sciences.* Cambridge University Press, Cambridge

HINDE R. (1987b) "Can nonhuman primates help us understand human behaviour" in SMUTS B.B., CHENEY D.L., SEYFARTH R.M., WRANGHAM R.W.& STRUHSAKER T.T. (eds) *Primate Socictes* University of Chicago Press Chicago and London pp. 413-420

ISAAC G. (1978) "The food-sharing behaviour of protohumans hominoids" *Scientific American* **238**: 90-108

KINZEY W.G. (1987) "A primate model for human mating system"in KINZEY W.G. (ed.)*The Evolution of Human Behavior: Primate Models* State University of New York Press pp: 105-114

LEACOCK E. (1978) "Women's status in egalitarian society: implications for Social Evolution" *Current Anthropology* **19**: 247-275

LEE R.B & DE VORE I. (eds) (1968) *Man the Hunter* Aldine Publishing Company Chicago

MCGREW W.C. (1991) "Chimpanzee material culture: what are its limits and why?" in FOLEY R. (1991) (ed) *The Evolution of Human Behaviour* cit. pp. 13-24

MCGREW W.C., TUTIN E.G. & BALDWIN P.J. (1979) "Chimpanzees, tools and termites: cross-cultural comparisons of Senegal, Tanzania, and Rio Muni." *Man* **14**: 185-214

MCGREW W.C., BALDWIN P.J. & TUTIN E.G. (1981) " Chimpanzees in a hot, dry and open habitat: Mt Assirik, Senegal West Africa" *Journal of Human Evolution* **10**: 227-244

POTTS R. (1984) "Home bases and early hominids" *American Scientist* **72** : 338-47

POTTS R. (1988) *Early Hominid Activities at Olduvay* Aldine De Gruyter New York

POTTS R. & SHIPMAN P (1981) "Cutmarks made by stone tools on bones from Olduvai Gorge, Tanzania" *Nature* **291**: 577-80

SOLWAY J. S. & LEE R.B. (199O) "Foragers, genuine or spurious? Situating the Kalhari San in history." *Current Anthropology* **31/2**: 109-146

STRUM S.C. (1987) *Almost Human* Random House

STRUM S.C. & MITCHELL W. (1987) "Baboon models and muddles" in KINZEY W.G. (ed.)*The Evolution of Human Behavior: Primate Models* cit. pp: 87-104

TESTARD A. (1982) "The significance of food storage among hunter-gatherers: residence, patterns, population densities, and social inequalities" *Current Anthropology* **23**: 523- 537

TESTARD A. (1988) "Some major problems in the Social Anthropology of hunter-gatheres" *Man* **22**: 287-304

TOOBY J .& DEVORE I. (1987) "The reconstruction of hominid behavioral evolution through strategic modeling" in KINZEY W.G. (ed.)*The Evolution of Human Behavior: Primate Models* cit. pp: 183-237

TOTH N. (1985) "The Oldowan reassessed: a close look at early stone artifact" *Journal of Archaeological Science* **12**: 101-120

VAN DER LEEUR & TORRENCE (eds) 1989) *What's New. A Closer Look at the Process of Innovation* London Unwin Hyman

WASHBURN S..L. (1960) "Tools and Human Evolution" *Scientific American* Sept. 1960

WASHBURN S..L. (1961) *Social Life of Early Man* Aldine Chicago

WASHBURN S. L. & LANCASTER C. S. (1968) "The evolution of Hunting" in Lee R., De Vore I. (eds) *Man the Hunter* Aldine Chicago pp: 293-303

WILMSEN E.N & DENBOW J.R. (1990) "Paradigmatic history of San-speaking peoples and current attempts at revision" *Current Anthropology* **31/9**: 489-507

WOODBURN J. (1982) "Egalitarian Societes" *Man* **17**: 431-451

WOODBURN J. (1988) "African hunter-gatherer social organization: is it best understood as a product of incapsulation?" in INGOLD T., RICHES D., WOODBURN J. (eds) (1988) *Hunters and Gatherers Vol I- History, Evolution and Social Change* Berg Oxford

WYNN T.& MCGREW W.C. (1989) "An ape's view of the Olduwan" *Man* **24**: 383-398

CULTURE AND ITS BIOLOGICAL ORIGINS: A VIEW FROM ETHOLOGY, EPIGENESIS AND DESIGN

Valerius Geist,
Faculty of Environmental Design
The University of Calgary
Calgary, Alberta T2N 1N4 Canada

ABSTRACT After illustrating an example of culture by a large non-primate mammal (mountain sheep), three propositions are advanced based on Geist (1978): 1. The roots of human culture need to be examined within successive adaptive syndromes that adapted hominids to landscapes with increasingly hostile climates. It places into perspective visual and vocal mimicry, and biases in human aesthetics towards enclosed spaces and savannah-like outdoor environments. 2. Our biological capacity for culture is affected by an epigenetic programme of ontogenetic development, which, at the extremes, produces "maintenance" (paedomorphic) and "dispersal" (hypermorphic) phenotypes. This concept ties the evolution of modern humans to culture. 3. Humans, like all vertebrates, possess attention-capturing behaviours or "dominance displays". In humans "dominance displays" use cultural means to attain biological ends. A common denominator to human attention displays or dominance displays is to make "art" of biological process. This, like many features of human behaviour, can also be related to the evolved capacity to make tools, namely the open-ended motivation to perfect designs. Innate biases in human behaviour appear to be at the root of our global ecological crisis.

Key words: human evolution, environment, culture, instinct, phenotype plasticity, landscape, architecture, mimicry.

1.1 HOME RANGE TRADITIONS IN MOUNTAIN SHEEP AS CULTURE. One discovers "culture" in the ecology of large mammals when, in an inadvertent "experiment", their adaptations go astray. Thus domestic goats (Capra hircus), gone feral on the Gulf Islands off southern British Columbia, cling tenaciously to the localities they had been abandoned in by their former owners and fail to exploit abundant, good habitat close by. They use only certain small patches of habitat, sometimes of low quality, while others, equally favourable or even superior, which are separated by only a few hundred yards of dense forest, are not visited. Here a "tradition" (for such is culture), based on the human owner's, not the goats best interests, was adopted by goats and transmitted faithfully for decade after decade of feral life.

In a similar vein one notices that mountain sheep (Ovis dalli, O. canadensis) are quite incapable of extending their range to habitat patches that were once used, but became, somehow, "lost" to sheep. This culture, this tradition seems at first glance to be absurdly confining and non-adaptive. Sheep are loath to cross to uninhabited

R. A. Gardner et al. (eds.), The Ethological Roots of Culture, 441–459.

habitat patches; in one case a Dall's sheep population refused to cross a road ploughed by a bull dozer. There is unused habitat for mountain sheep to exploit, yet they seem to be incapable of the obvious: explore and occupy! Once the surprise over this has passed it is not too difficult to decipher that a tradition of habitat occupation, confers enormous benefits to mountain sheep under natural conditions; today's conditions are not "natural" for mountain sheep, as I shall explain below. The extermination of mountain sheep over much of their former geographic range about a century ago was an inadvertent "experiment" that revealed that habitat occupation in this species is based on a tradition. This tradition is central to understanding much that is puzzling in the social behaviour of mountain sheep, such as the choice of "leaders" by young rams based on horn size, the "homosexual" structure of male-groups, the "slow & gentle" disassociation of short-yearlings from their mothers, and the chaperoning in late spring of these youngsters by barren females (Geist 1971).

"Culture" is no soft option that mountain sheep may or may not exercise, but a dominant factor central to mountain sheep ecology. This species cannot live without culture. Natural selection appears to shape and refine adaptations in sheep that make a specific kind of culture possible, which makes wild sheep the superb, quick learners they are.

1.2 LEADERSHIP

What makes young rams follow large-horned rams? Horn size is a symbol of the ecological success of their bearers. Horn size is either a proxy for successful foraging, and thus of occupation of superior pastures, or of long life and thus of occupation of superior escape terrain - or of both. Horn size increases with the quality and quantity of the summer's food supply, but it also increases with age, that is, the success of the horn-bearer in avoiding being caught by predators. This is, in good part due to living in secure landscapes, such as in steep terrain that precludes effective hunting by predators. These are two excellent habitat options for a young ram to acquire by simply following a large-horned ram.

However, it is not in a large-horned ram's interest to lead potential competitors to his pasture. Therefore, one would expect little tolerance by large-horned rams of their young followers. Conversely, one expects that the young followers, somehow, ingratiate themselves to the large-horned leaders. They do so by playing "estrus female". That is, the behaviour of a young or subordinate male towards a larger, older or more dominant one is a copy of an estrous female's behaviour towards a breeding male. It appears as if female mimicry by young males is used to cancel out rejection by the large male. After all, how can a male reject an estrous female? Thus it is the norm in ram bands for subordinates to play female and for dominants to treat subordinates as females.

The separation of mother and maturing offspring in mountain sheep society is gradual. It is as if this serves to avoid upset and any discontinuity in the young being led by someone experienced. Thus two months before the female gives birth to another offspring, the lamb that followed her begins to periodically follow other sheep in the band. Its mother reacts in no way to this change of behaviour of her child. With every week the child becomes more and more independent in its choice of associates, and returns less and less frequently to follow its mother. It plays with others of its own age, joins a subgroup temporarily that does not contain its mother, follows a passing ram briefly, and plays increasingly frequently with one or two barren females in the nursery group.

As gravid females, inconspicuously, distance themselves from the nursery herd and move into steep cliffs to await birth, their children are found more and more frequently with "old aunts" which actively lead and look after the yearling. That is, a few old females become foci around which the young gather. These old females encourage this cohesion by bleating after youngsters that try to move away on their own, or by grooming young - a behaviour otherwise very rare in sheep - or by enticing them to join in running and bouncing across the slopes and cliffs. Such games belie the age of the old females as they rollick with their little crowd of eager followers. Note, that the year old lambs are kept on as followers. That is, the young are unlikely to be on their own, and - being led - have little chance to leave the secure home range and thus being lost in a sea of unsuitable landscape all around. The old females act as active leaders during a crucial time, namely, when most of the adult females are absent from the home range and they young sheep could be on their own, leaderless and in danger of straying into unsuitable, dangerous land.

Free living bighorn sheep, once they got to know me well due to almost daily contact over the course of two years, not only followed me to obtain salt, but when led to strange places stuck to me closely like "leeches", till I led them back to within a hundred paces of their familiar home range. Only then did they desert me, ran to familiar ground, lolled about and commenced feeding.

Sheep society appears structured as if to insure that the tradition of acquiring and passing on home-range knowledge works as well as possible. This tradition is absolutely vital to the success of individual sheep. The behaviour appears to be "designed" so that individuals may "learn" with precision and high certainty where to go to live, and to pass on, redundantly, that ecological knowledge to the next generation. A young sheep which fails to so learn, which strikes out across the huge expanse of unsuitable, hostile landscapes in search of a place to live, goes to great expenses and risks early death to gain at best no more than can be gained cheaply by simply following an adult (Geist 1971:117-129).

1.3 ECOLOGICAL ORIGIN OF SHEEP HOME-RANGE CULTURE

The origin of the sheep's system of passing on habitat knowledge as a tradition resides in ecological attributes of sheep feeding areas. These are covered predominantly by grasses, sedges or alpine floral mats which are resistant to disturbance. A climax forest that burns returns to climax forest only after many decades of ecological succession of sub-climax plant communities. However, a grassland that burns returns to grassland at once. Grassland is stable, except for changes imposed by climatic change. Grassland persists.

Mountain sheep are glacier followers. As such they must adjust to glacial and inter-glacial conditions over the long stretches of time dictated by the Milankovitch cycle: 60,000-90,000 years of glacial climate and 40,000-10,000 years of inter-glacial climate (Pielou 1991:8). During glacials, landscapes are open and forest is scarce and timberlines low (Guthrie 1990). The glacial conditions in the extreme west of the Yukon and Alaska, in the St. Elias, Wrangle and Alaska Ranges, give us an inkling how sheep lived during glacial times (Geist 1978)

During inter-glacials the open, cold grasslands shrink, fragment and ascend in elevation as temperatures rise and forests encroach on and cover the open spaces. That is, acceptable habitat is reduced to widely dispersed, tiny patches. These tiny patches are "stitched together" by the sheep's tradition of moving in a regular fashion between them. The easiest and safest way to put together a functional home range for a sheep in an inter-glacial is to follow older individuals. For a young sheep to explore is to chance getting lost in the huge space of non-habitat and to chance assembling a mediocre set of seasonal home ranges. Following experienced sheep, which carry proof of success on their body, is thus easier, safer and more profitable in assembling a home range than to strike out on one's own and explore. Thus this tradition of acquiring and passing on home range knowledge assures greater success in survival and reproduction for the individual sheep.

2.1 HOMINID CULTURAL ROOTS

What ecological constellations dictated that learning be vital to reproductive fitness in hominids? Culture and traditions must have been central in the same fundamental, biological sense to hominids as they were to mountain sheep. The following discussion is based largely on Geist (1978), unless referenced otherwise.

Clearly, chimpanzee biology gives some powerful and legitimate insights, legitimate because of our close genetic identity (Miamoto et al. 1987), confirmed by powerful relics of kin-selection in human biology. Unfortunately, kin-selection does not automatically entail ecological factors the adaptation to which requires learning. Thus kin-selection, important as it might be in understanding early hominid evolution, appears not to be relevant by itself in explaining the origins of human culture.

There may be a possible exception, namely the capacity for <u>self-discipline</u> which is so highly developed in humans, as to be virtually pathologic. Is self-discipline essential for culture? If so, then there are at least two circumstances which would severely select for it. Firstly, the self-discipline needed, if wounded, not to run away from one's kin-group when engaged in a territorial fight. Surface damaging weapons, typical in territory defenders, are expected to inflict severe pain, but not to damage internal organs. Consequently, there is a pay-off for self-discipline, that is, for staying with the group in its fight irrespective of the pain, because that enhances the chances of the group retaining its territory, and thus the chances of the wounded individual to survive and reproduce. This selection would occur in a kin-selected, chimp-like hominid ancestor. Any "warfare" requires self-discipline on the part of the fighters, particularly when safety depends on concealment from the enemy.

A second focus of selection for self-discipline would be hominids who were utterly dependent for their reproductive success on hunting. To stalk close to one's prey requires not to flinch when biting insects descend on one's body, or when one stepped inadvertently on a thorn. Any sudden defensive movement on the part of the hunter would send the prey scurrying, making the hunt, and thus food procurement, a failure.

What factors of pre-hominid and hominid biology need close adjustment via learning? Only students of chimpanzee biology can answer that authoritatively. Are the idiosyncratic food habits of chimp populations a clue (Kortland 1985)? Beyond that, according to conventional wisdom, hominid evolution in Africa shaped our cultural roots. This may be true and yet over simplified. The evolution of large mammals suggests that evolutionary lineages generate species as they penetrate landscapes of increasingly inclement, seasonal climates, from the tropics to the Arctic (Geist 1971, 1978, 1987). In a similar vein, hominids passed through several species or adaptive syndromes in their evolution. Within Africa, not one, but several adaptive syndromes can be identified. Within each of these stages culture played different adaptive roles, depending on the life-strategies of each species. This procession of adaptive syndromes, common in ungulates and carnivores, is exceptional in primates. Currently, only the hominids can be shown to have penetrated the barrier posed by a discontinuous vegetation season, to - ultimately - evolve classical Ice Age species that lived alongside woolly mammoths, cave bears and Irish elk in periglacial environments.

Moreover, African hominids apparently speciated at the same time as did species of African bovids, primarily during periods of increasing aridity and possibly cold (Vrba 1985). That is, advances in evolution are related to climatic adversity. Verba (1985) argues for increased hominid speciation during an arid phase 2.0-2.5 million years ago which is associated with the appearance of Homo habilis and robust australopithecines. H. erectus appeared with the first major glaciations about 1.8-1.6 million years ago, but terminates in the large Riss glaciation about 220,000 BP, when H. sapiens makes his appearance. This too was a period of accelerated bovid evolution in Africa.

2.2 HOMINID ADAPTIVE SYNDROMES

The adaptive syndromes leading up to modern man, briefly, can be listed as follows (see Geist 1978):

a) A kin-selected one, based on material resources defense in tropical forests; a chimpanzee-like adaptive syndrome. This is the origin, the basis of our adaptations and various human attributes still can be traced to it (i.e. low birth-weight, long gestation and ontogenetic development, male bonding, humour, in-group/out-group psychology with group defence, withdrawal at night, omnivorous food habits with a craving for plant poisons as flavours, dominance displays with objects external to the body, that is, with artifacts etc.).

b) Opportunistic surface feeder in a weakly seasonal savannah, with security generated by climbing into trees or cliffs, and by selfish-herd formation. This is the australopithecine stage. Some of our attributes are traceable to this stage (i.e. loss of combat-canines, inhibition to bite in combat, upright stance, and probably our great manual dexterity and aesthetics favouring green, open landscapes with trees with low branches).

c) Adapted to discontinuous vegetation season in strongly seasonal, dry steppe; generates security in the absence of escape terrain, deals with climatic vagaries, introduces a significant amount of animal protein into its diet, and begins evolution as a systematic hunter. This is the first "human" adaptive syndrome beginning with Homo habilis and evolving into H. erectus.

d) Advanced hunters, adapted in large part to culturally diverse landscapes, including to cold temperate and periglacial landscapes. Two versions of H. sapiens arise as Ice Age mammals: Neanderthal and modern man.

e) Late glacial and post glacial adaptations to biological and cultural environments, and concomitant differentiation, in part via adaptation to cultural constants.

2.3 AN EXAMPLE OF AN ADAPTIVE SYNDROME

2.3.1 Security - at night How does an adaptive syndrome function? We may use that of hominid adaptations to the open steppe. A vital need for a hominid on a tree-less steppe at night is how to survive this period of intense predator activity. With visual acuity reduced, without potent biological weapons to punish predators or the physical ability to evade them, particularly at night, how does a human protect itself against large predators at night on the ground? Today's casual approach to camping out by native people is deceptive, as they have normally the power to kill carnivores which killed one of the tribe, thereby extinguishing any developing tradition to kill humans by said species of carnivore (Kortland 1980). Also, we can only live where we are able to kill the largest existing carnivore, which is no simple task with primitive weapons. A fauna with very large predators may be immune to human encroachment, as appears to have been North America's Rancholabrean fauna (Geist 1989b).

At the australopithecine stage of life in the savannah there were trees to spent the night in, possibly augmented by platform-like nests which blocked the path of a predator ascending a tree and thus protected the nocturnal occupants of the nest. What can one do on open plains in the absence of trees to escape predation at night?

About 2.5-2.0 million years ago there were in East Africa a diversity of large predators, sabre-toothed cats, leopards, lions, two species of hyenas, hunting dogs and large jackals (Leaky 1979), all of which were potentially dangerous to H. habilis, a small-bodied hominid (Susman and Stern 1982).

The answer appears to be to build a closed over thorn-branch shelter. This is based on the work of Kortland (1980) and on general camping experience. One expands the flimsy, flat, two-dimensional night-nest of the chimp-like ancestor, which is good for one adult, into a sturdy, roofed, three-dimensional structure capable of holding several adults. Such huts may have been built by early H. habilis (Leaky 1979). The walls themselves provide a barrier for carnivores, as demonstrated, for instance by the relative safety of a tent; it is apparently treated by wildlife as a solid structure. The thorns would probably be strong deterrents, as experiments by Kortland (1980) on lions have demonstrated. In addition a human can deliver a powerful jab with a pointed stick to a carnivore through the walls of the enclosure. Thirdly, humans can respond to a disturbance with vocalizations that, in principle, mimic predator defensive sounds. Predators need to avoid contests damaging to their bodies lest they loose competence and are no longer able to hunt successfully. Consequently, being able to adjust one's voice to those of local predators would pay off in increased security. This may be the evolutionary root of vocal mimicry, a unique aptitude of humans, and in refined form absolutely fundamental to language and music.

Security is thus generated in hominids adapted to tree-less steppe by walled, all-enclosing structures, including caves. We shall return to this later.

2.3.2 Obtaining food: tools from hunger. There is a dry and a wet season in the steppe. In the dry season there cannot be much surface foraging, because the vegetation is dead or dormant, and with it the myriads of animal species that inhabit green vegetation mats.

How does a large-bodied hominid then obtain food? How does a female maintain continuous gestation or lactation across seasons of scarcity of surface foods? Small-bodied primates can restrict reproduction to the vegetation season, because of short gestation and lactation periods typical of small body size. Two large hominids emerged during the 2.5-2.0 million year interval of dry climates, one being the robust australopithecine and the other an early Homo. The food habits of the former were related to hard, abrasive plant matter, possibly to seeds. That is, its food habits appears to be an extension of the australopithecine adaptive syndrome of surface feeding. Its built suggests that it climbed for security, which means that it exploited closely all food within secure reach of trees. In addition to robust australopithecines, there were large baboons (Theropithecus) to compete for surface foods. Homo probably approached matters differently, displaced as it was beyond the safety of trees on surfaces closely foraged over by large baboons and robust australopithecines.

Here is one possible solution: one potential food source in the dry season in the steppe are the subsurface stores of plants - bulbs, tubers, corms, roots. However, one needs tools to dig with. Finger nails are not adequate in the hard soil of the steppe. Digging utensils are thus a must to save nails and skin, but above all to save effort and increase foraging efficiency. The most likely source of digging tools is wood from shrubs. Consequently, there is an impetus to develop skills in first choosing and then shaping an adequate digging stick. A vital pay-off in food is thus linked directly to making simple tools. The environment of the dry steppe thus contains the ecological determinants for a culture to create and use tools that are truly life-essential.

However, where would hominids, adapted to life in the dry steppe, have acquired skills and dexterity to shape wood into new shapes as tools and shelter?

Security is probably linked to shaping branches skilfully into structures, both in the australopithecine and the Homo adaptive syndrome. In the australopithecine syndrome building tree platforms (a tree-nest not unlike an eagle's nest) that were sturdy enough to block a climbing predator's advance were postulated. There is no way to test this hypothesis except, possibly, by experiments mimicking what Jim Corbett did when he spent the nights in trees, when he hunted man-eating leopards in India. He had to protect himself against the leopards reaching him in the tree at night, and some leopards did try. The eagle-nest tree-platforms are simply a logical intermediate between the flimsy tree-nest of the chimp and the enclosed thorn-hut of the hominid on the ground. It represents a logical continuity of generating vital skills at shaping wood.

This postulate suggests that australopithecines required a higher level of building skills and dexterity for survival than did chimps. Note a complication: if tree-platforms kept climbing leopards out, how did hominids climb into such nests in the first place? Did they "invent" a hole and plug? Did they place thorn branches strategically to discourage leopards ascending?

The shaping of branches appears to be a prerequisite to the shaping of digging sticks. The improvements in the shaping (design) of such, which results in increased efficiency of uncovering food, depend on personal experience. This experience, transferred from adult to child enhances fitness. Here is a potential for significant pay-off in fitness if the experience of one generation can be transferred to the next. However, such personal experience by one generation is lost on the next generation in the absence of at least rudimentary visual mimicry. A rudimentary ability at visual mimicry, however, is already found in chimps (Kortland 1984). Visual mimicry (imitation) is one fundamental requisite without which culture cannot arise. However, while imitation is highly rewarding under "maintenance" conditions (see below), innovation is usually not, and one expects a paranoid fear of innovation, and therefore, resistance to novelty by individuals under maintenance conditions. The genetic selection for visual mimicry could therefore be rooted in the gain in efficiency of food acquisition through the ability of young to copy successful parents.

Stored plant food is a rich source of carbohydrates, minerals and moisture. It is low on protein, fats and vitamins. Protein and fats are essential for milk production and tissue growth. The most likely source is in the proteins of large mammals that inhabit the steppe; a seed grinder, of course, would obtain some protein and fats from seeds.

2.3.3 Bonding/Hunting. How does one get at the animal protein? Males, unencumbered by gestation and lactation, could become the instrument of providing essential proteins for their offspring, but this could only happen if the males could maximize their own reproductive fitness. Clearly, one way to do it is by each male "bonding" to a specific female which bears and raises the bonded male's child. This presumes a mechanism for bonding, one that persists through gestation and lactation. One way is permanent bonding via sex, which in humans continues partially through gestation. If so, estrus had to be selected against as it would destroy the bond between mates as the offspring would no longer be reasonably that of the bonded male. Therefore males should not be able to sense estrus - and they do not. Both estrus and the ability of the male to detect such should be victims of selection for mate fidelity.

In short, security adaptations and means of getting food in the dry, sharply seasonal thorn steppe, can generate cultural and biological adaptations that we recognize as those of humans. Furthermore, a drive to improve on hand work is essential to a creature whose very survival, let alone reproductive fitness, depends on design and shaping of tools vital to life.

2.3.4 Onion model/Applied aesthetics. The above implies that younger adaptations cover older ones, and that human adaptations may be represented by an "onion model" with successive layers upon layers of adaptations. One cannot take a given human characteristic and ask "How is it related to our African past"? It may not be. Sometimes we can identify certain human attributes that clearly relate to earlier African adaptive syndromes.

A biologist observing human design will ever so often run into our "Darwinian Cage", particularly today, because of our power to alter environments till they please us, and thus expose deep structures within our psyche. Take Gordon Orians'(1986) hypothesis about universals in human landscape aesthetics. The unnatural, inadvertent "experiment" that revealed a gross lack of sensible adaptation on our part, and provided a clue to Orians, is our power to change landscapes. Why, Orians asked, should humans use extraordinary means to create and maintain wherever they go "savannah"? We destroy a species-rich naive prairie and replace it with a species-poor lawn and well-spaced trees, then use large amounts of energy, water, poisons, labour and wealth to maintain this artificial assemblage against native invaders. Similarly, in a rain forest landscape, we chop down trees, plant lawns and may replace native trees with non-native ornamentals that happen to be "beautiful". Orians identified a part of our "Darwinian Cage" within which our perception dwells, this part attributable to our australopithecine African past.

Orians argued that the image of the lush savannah signalled both food and security. The ground vegetation in a savannah is not only exceptionally productive, but a photosynthetic layer which is entirely within the reach of a hominid. Trees spell security from predators. When testing for "beautiful trees", Orians found that such are trees on which we can reach the first branch. In short, our drive to change landscapes into savannah appears to reside in ancient perceptual adaptations to place us into rich and secure natural surroundings. That is, Orians argues, aesthetics has a biological basis. I argued similarly when examining the biological basis of art (Geist 1978:90-98).

In a similar vein one can examine our infatuation with walled and roofed-over structures, that is, enclosed spaces. We spent over 90 percent of our time in them; we make an art of them (architecture). Ironically, we make a point of signalling that we intend to spend some time out of doors by dressing to a code (hiking attire). We internalize "nature" within our "caves" affluence permitting. With little affluence our piece of internalized nature may be a modest potted plant and pet. With greater affluence it is a fire place, then a swimming pool and a fountains. Are we here unconscious victims of our sense for security? Of security once derived from thorn huts on the African plains?

The analysis of our past adaptive syndromes is something that takes much time and is insightful, but it is only one window on the cultural heritage in our biology. One can take off on many tangents, such as human aggression in which cultural controls distinguishes human aggression from that of other large mammals (Geist 1966). While, like other large mammals armed with very dangerous teeth we have an inhibition to bite, and thus still posess in this instant the biological negative feed-back that holds down biting to tolerable limits as in other mammals, we have adopted cultural controls over our non-biological weapons (club, spear, gun etc). Woe when these cultural controls fracture, when the satisfaction of deep instincts is permitted with weapons not integral to the human body! To be human must one culturally control innate instincts?

This exploration of ecological determinants of the earliest adaptive syndrome of Homo suggests that visual and vocal mimicry, unique adaptations of man and fundamental to all human culture, are rooted in severe selection for improved feeding efficiency and in increased efficiency of security adaptations.

3.1 EPIGENETIC MECHANISMS

Biological inequality arises not only from differences in the genetics of individuals, but also from differences in expression of hereditary factors in different environments. We pay lip service to this old insight, but do not normally integrate it in our thinking. It is the domain of epigenetic (Waddington 1957, 1975; Lovtrup 1974), as well as of the messages unjustly ignored in Goldschmidt (1940). Bruton (1989) edited a recent symposium on this matter. The following is based on my work on how large Ice Age mammals speciate, as well as on the practical applications of this concept (Geist 1971, 1978, 1987, 1989).

There appear to be some underlying rules by which environments express themselves; in particular it appears that genomes are pre-programmed to respond to signals about resource abundance from the environment they receive through the phenotype. Thus genomes appear able to adjust the developing phenotypes for life in resource poor or resource rich environments. The adaptive logic appears as follows: to maximize fitness in resource poor environments an individual ought to be cheap to maintain, to be good at competing for scarce material resources, and to reproduce in an opportunistic fashion as resources become available to do so. Conversely, in an exceptionally resource rich environment, which normally is very rare, but occurs regularly when a species is colonizing a new area previously not occupied by its kind, the phenotype - clearly - need not compete for material resources. However, it needs to be very good at social competition as reproductive opportunities are at a maximum; it must be able to turn resources into offspring. Consequently, the dispersal phenotype is not expected to live long since reproduction is generally destructive of life expectancy. The phenotype, in an environment not filled by its species, should explore so as to spread its genes. Also, during colonization, that is, when dispersing into rich, strange environments, the phenotype should not be phased by novelty nor discouraged by pain and adversity. The two phenotype extremes, labelled maintenance and dispersal phenotypes, are found not only in mammals, but also in humans (Geist 1978:116-144, 1989). Their recognition has profound implications.

Dispersal phenotypes in large mammals are born large, and grow rapidly consuming abundant maternal milk and enjoy exceptional maternal care. They mature sexually rapidly, they grow to large body size, they develop disproportionately well the tissues of low growth-priority, they are socially very active, they play a lot, have a high reproductive rate, are very healthy, but have a short life expectancy. For theoretical reasons we expect them to be innovators, as errors and failures in experimenting are not very costly when resources are abundant, but may be prohibitively so during maintenance conditions when material resources are scarce. Dispersal phenotypes are luxury, high-energy, high-competence phenotypes, as selection favours social skills, effective combat, and learning as a means of adjusting to novelty, diversity and high reproductive opportunity.

Maintenance phenotypes are the obverse of dispersal phenotypes, except that when resource are scarce, life expectancy also shrinks. Maintenance phenotypes live under conditions of "efficiency selection", that is they are selected to do the most with the least amounts of material resources.

The differences between these phenotype extremes are not only morphological, but also psychological; the regimes of natural selection must be entirely different under maintenance and dispersal conditions. It takes at least four generations for the full dispersal phenotype to arise from maintenance phenotypes. This was first discovered by Vogt (1948; 1951) on two deer species, and subsequently labelled "maternal effect" (Denenberg et al., 1962; Denenberg and Rosenberg, 1967; Beach et al., 1982). The link between phenotype and health was made first by Uhazy et al. (1973) in a study of lungworm infections in bighorn sheep. Current evidence suggests that the signal to the genome to switch from maintenance to dispersal conditions is the amount of protein ingested by the female during gestation (Geist 1978:139; Chauvez and Martinez 1979). The evidence is reviewed in some detail in Geist (1978); others have reached similar conclusions (Wellington 1960; Ellenberg, 1978; Peterson, 1978; Balon 1981; Packard and Mech, 1983).

The above is significant because people of the Upper Palaeolithic remained dispersal phenotypes throughout, being large-bodied, healthy as judged by the osteological record, and possessing large brains. The cerebral cortex is a "luxury tissue", one that is large only under conditions of both excellent nutrition and diversity of stimulation that generates competence. In animal populations, individuals are expected to be dispersal phenotypes only during colonization, that is for a short period of time when the species colonizes a region. We expect that after colonization the individuals revert rapidly to maintenance phenotypes. However, this does not happen with people of the Upper Palaeolithic. Consequently, these people must have - culturally - maintained "dispersal" conditions in order to generate dispersal phenotypes. Were humans knowledgeably shaping themselves by shaping the ontogenetic environment? In this context the Venus Cult becomes significant: dispersal phenotypes can only be grown in a milieu exceptionally supportive of the mother, both materially and psychologically. Did men, for the first time in human history, worry deeply about the well-being of women, and thus develop ritual to enhance the well-being of their spouse?

The dispersal phenotype is significant in other ways as well: it indicates that change is most palatable to the dispersal phenotype, not the maintenance phenotype. The former can be controlled rationally without too much recourse to historical means of coercion (which are so prominently displayed in European criminal history). Historically, it suggests that we affect our own evolution not only inadvertently, such as evolved smaller harder teeth in grain eating cultures, or evolved myopia and colour blindness, or light skin in response to clothing, or the Intuits distribution of sweat glands in relation to his parka, but that we shaped ourselves consciously in our own image whenever we consciously adjusted our bodies to become culturally shaped tools for making a living. If so, then normal, unaided body growth does not suffice to meet the needs for successful life and reproduction. One expects a Baldwin effect to take over when we test "shaped bodies" against the environmental demands of life. I suggest the Upper Palaeolithic was one such period when shaping of bodies was done consciously.

* The characteristic body build of the Upper Palaeolithic can be reached today, but only in environments offering exceptionally rich nutrition and supportive of the children.

* The Venus Cult is logical in the above context, since the well-being of women was key to producing infants with dispersal phenotype characteristics.

* The cave paintings and "decorations" on the "batons de commandement" reveal aspects of cultural life, that indicate the obvious: hunters taking migratory big game need to know in seasonal climates how to predict the migratory movements of their prey and how to kill dangerous beasts. The hypothesis of Marshack (1972) that chronologic time was kept with lunar calendars is appealing.

* Sketches of bears facing hunters which have penile erections suggests bragging, of fearlessness and great confidence when actually in mortal danger. The "bird man of Lasceau" may be similarly interpreted.

* Teenagers are not infrequent in graves. Does this mean testing and failure to be competent at adult tasks?

* Was descent counted by the male line in order to reduce the chances of inbreeding depression, so fatal to maximizing physical development?

* There is evidence for long distance travel and exceptional constancy of

culture over millennia, and thus firm belief in cultural values that allowed human life in the periglacial zones to flourish.

This suggests that in the Upper Palaeolithic people maximized individual development and competence. That generates dispersal phenotypes, based on excellent nutritional for mother and child during ontogeny, a highly supportive milieu for mothers to maximize lactation and minimize growth disruptions to children due to stress, but also by keeping track of who reproduces with whom to minimize inbreeding depressions.

3.2 ADAPTATIONS TO ICE AGE ENVIRONMENTS

An understanding of maintenance and dispersal phenotypes makes it possible to discuss the differences between ourselves and the late Homo erectus in a new way. In essence, erectus looks and acts like a classical maintenance phenotype, modern humans like dispersal phenotypes. Modern humans are culturally diverse and their technology changes over time by the inclusion of novelty; the technology of erectus, however, remains of the same type, but is improved by ever more refined workmanship. This is as if classical efficiency selection was at work. Homo sapiens appear to have advanced genetically dispersal type adjustments into adaptations. This suggests that the change from erectus to sapiens about 200,000 years ago happened under very adverse conditions for human life, and it indicates a widespread mortality of erectus type people unable to cope with environmental change. It harmonizes with the scarcity of human fossils in the long, severe Riss glaciation, and explains the ability of sapiens to invade the periglacial environments. While the beginning of the Riss glaciation appears to have been catastrophic for erectus, paradoxically, those that survived to continue the human lineage must have been individuals of excellent development, not starvelings.

The adaptation to periglacial cold environments happened twice, once in the form of Neanderthal man and once as Cro-magnon. The adaptive syndromes were quite different, with Neanderthal man apparently highly specialized biologically as a super predator (Geist 1981), and Cro-magnon adapted culturally to a more clever exploitation of migratory large mammals, primarily reindeer (Geist 1978). Only Cro-magnon is of interest.

The essence of adaptation in Ice Age hunters was to intercept migratory ungulates using chronologic time to predict migrations, then kill in excess, and cure the excess meat into a store for future use. In addition the hunters killed by confronting and luring prey to a group of hunters which then dispatched it with throwing spears. Thus knowledge about the habits of prey, physical competence, and observation skills that included the night sky were at a premium. The great uniformity of the European Upper Palaeolithic culture over the long waxing and the waning of the late Wuerm stadial speaks of great interchange of people over a large geographic area.

However, if dispersal type characteristics were carefully fostered in individuals, and such were tested against the vagaries of the environment, then one expects a genetic selection favouring these environmentally fostered features.

Since modern man appears in Europe after Neanderthal man dies out during a brief, warm stadial, it begs the question where modern humans came from earlier. There are a number of indications that they were once desert adapted people originating probably on the fringes of the Sahara. That is, they were people adapted already to a climatically extreme environment.

4.1 ARTFUL EXTENSIONS & PERFECTION

The dominance displays of large mammals are behaviours that focus attention onto the displaying individual. They include means of pinpointing attention within picture planes which are much the same as used in conventional, occidental art (Geist 1978:90-98), and criteria such as "symmetry", as a proxy for health. Dominance displays are as much a part of a vertebrate's biology as are liver and lungs.

Humans, as expected, also have dominance displays, but they enhance and extend these by artificial means, that is, by using culture to satisfy the biological urge to be viewed. These are complex behaviours, ranging from laughter and humour to the enhancement for show of the body's morphology and behaviours, to the individual exhibiting himself together with cultural symbols of power, superiority and success (Geist 1978:103-115). As indicated before, with increased affluence, for instance, we incorporate more and more "nature" within our "caves" as a symbol of our worth, while the "cave" is crafted by architects to catch attention, reflect our individuality and flatter our vanity.

A principle of displays by humans is to take a biologically essential activity and <u>artfully elaborate</u> it and make it showy. (Occasionally, in our clothing fashions, we artfully hide what we really want to draw attention to). The central concept here is: granted the power to change the environment and ourselves we will shape what we have power to shape to conform to deep-seated "aesthetics" urges or instincts. Thus we make from:

<u>preening</u> - hair styling of head hair and beard, plus olfactory enhancements. We hire specialists to do the job, creating a profession; many professions serve our vanity.

<u>body shelter</u> - the cult of clothing styles and fashions.

<u>eating</u> - cuisine plus the styling of table decor and manners.

<u>reproduction</u> - romance with appropriate etiquette, customs, fashions, but also bordellos and pornography.

<u>exercise</u> - dance, sports and fashions for specific activities and occasions.

<u>security</u> - architecture and landscape architecture. As discussed under the "human" adaptive syndrome, an adaptation to steppe life is to generate security could by making the ape-nest into a roofed structure interwoven by thorns. Thus enclosures may be emotionally satisfying because they pander to our innate demand for security. This demand has been culturally expanded into architecture by the addition of art, that is, the use of perceptual principles to impose beauty and "attention-grabbing" on

buildings.

aggression - fashions in clothing, armour, weapons, decor and ritual. See Otto Koenig's (1970) dissertation on cultural ethology, deals in good part with this phenomenon, and I have used his insights in my 1978 (pp:108-115) discourse on human dominance displays.

communication - music, prose, poetry, drama, books and libraries, electronic media.

actions - tools which are organ extensions and which are personified, improved and decorated.

We beautify our homes for private pleasure or display, paint during free time, carve etc. We are driven to "improve", to create superlatives; our self-esteem is tied into it and we may measure our self-worth by the beauty of our creations. We "improve" on everything we find, including plant and animal species. These are "improved" not only for utility, but also for "show": dogs, roses, cats, larkspur, gold fish, onions, pigeons, African violets, dwarf horses etc. We can be characterised as a species that takes natural things into its hands, and "beautifies" all within our powers to modify, from sand made into sand castles, rocks into sculptures, to forests into artificial savannah. Even mathematicians speak of "elegant" solutions, presumably because their sense of aesthetics is aroused. The rational being Homo thus focuses its attention to be noticed on "art", and spends prodigious energies achieving it.

Humans are species condemned to art. Is the drive to "improve" an inescapable consequence of tool-making and of tool use?

5.1 A FEW THOUGHTS

This symposium is based on an underlying preposition that culture has some biological, that is heritable basis. Konrad Lorenz (1973) analyzed in meticulous detail the evolutionary history of our cognition. He emphasized that an evolutionary analysis, as practised in comparative ethology, was a tool that elucidated innate limitations within the cognitive apparatus through which all our experience must pass. There is, however, we suspect, an additional psychological landscape of innate guides and biases acquired during our evolution. We suspect that our reality is a "Darwinian cage", shaped by natural selection. Therefore, we cannot sense matters that are present and real, but which were not vital in our survival and fitness. We have faith that comparative ethology may throw light on just such structures.

Indeed others have shown such a priori structures, be it Otto Koehler's (1952) "ubenannte Denken", or Noam Chomsky's universal rules of grammar, or our blind obedience of Zipf's law in every day life (Geist 1978:3-10), or the clinging to predictability in thought and experience, labelled "conventional wisdom" by Kenneth Galbraith, an apparent extension of fundamental cognitive processes in vertebrates (Geist 1978:24-40), or Gordon Orian's (1986) view that aesthetics is a genetic guide to environments and situations favourable to survival and fitness, or my analysis of homology in vertebrate dominance displays and human forms of self assertion (Geist

1978:86-115). In all these cases there are powerful parallels between animals and humans. In addition, Maslow's "Hierarchy of Needs", Piaget's developmental processes and the operations of psychological processes as analyzed originally by Sigmund Freud (Babcock 1992), suggest inherited cognitive structures and processes. The recognition that heredity plays a part in perception and fundamental rules of conduct is much more pervasive than above biased sample suggests.

Yet, why bother with it? One answer is that with a global environmental crisis looming up, following innate dictates is part of the problem. Writers have warned us about the global environmental crisis, such Meadows' et al. 1972, Limits to Growth, or the subsequent The World Conservation Strategy (WCS, 1980) issued jointly by the International Union for the Conservation of Nature (IUCN), the World Wildlife Fund (WWF) and the United Nations Environmental Programme (UNEP), the Bruntland Report (1987), the WCS's second version Caring for the Earth (1991), or Myers (1984) remarkable effort centering on Lovelock's GAIA. See also how professions wreștle with sustainability and the difficulty of implementing it, as exemplified in Elder (1991), or in recent textbooks of Ecology, such as Stiling (1992), or in Environmental Sciences, such as Cunningham and Saigo (1990), or in Brown's State of the World reports.

Yet if Gordon Orians is right about our innate landscape aesthetics, and I do not see that he is not, then we sacrifice bio-diversity and ecological integrity for the sake of a ghost from the past, an old echo from the genes, long invalid, that instructs us, unconsciously, in what surroundings we aught to feel "safe" and comfortable. For this obsolete ghost we murder wild, co-adapted life at very great economic cost, lest we live in "ugly" surroundings. And should we follow an urge and let nature take over our private lawn before the house, city bylaws, fully enforced, see to it that nature's reign on our lawn is brief. We are also stuck with the bill for the city crew's efforts in setting back the ecological succession on our very private lawn. In a less trivial sense, when we have the power, we make our environments fit irrational, hidden needs for security, eroticism and also for our thirst for "admiration".

Our accepted means of gaining attention and social standing, of satisfying that animal craving for status, is by publicly consuming burgeoning masses of resources, energy and services. We practice "conspicuous consumption" to gain status and flout it, as the keenly observing economist Thorstein Veblen (1934) labelled this practice. If profit is a means to social recognition, is sustainable development possible in principle? If consumption is to be throttled then, clearly, social recognition must be uncoupled from market values. But if not market values, then what cultural values?

Or take nature/nurture, an old bogy man. Research on the biological roots of culture implies that we do not lack instincts, while the humanities teach that we rise to become worthy due to our control over base motives for the sake of higher ideals. To be human is thus to control one's instincts. Aggression appears to be as instinctive as smiling. A decisive proof of this statement would require experiments unethical on humans. Aggression appears to be within us; it is not planted there by a culture. However, it is only legitimate when under control by culture. Our unique capacity at self discipline, rigorously applied as "education" or "socialization", may serve to contain our "animal urges", to make us competent individuals in society. Whether this is so, is an ancient debate in Western and Chinese philosophy; in this debate the side painting humans as intrinsically good, appears to always win out. It may be that daily reality prevents adherers to the opposite view from dying out. Norms of conduct appear to exist in order to control human instincts.

A deeply frustrating aspects of current occidental society to women is the preferential treatment of male roles. One response is for females to strive to share male roles. An alternative is to share real power, but within different roles based on sex. The old Germanic culture as reflected in the Icelandic sagas gives an example of how real power sharing works. It allowed some matriarchs to be leaders of men in the settlement of Ice Land. In essence, while men acquired the wealth and prestige essential to the management and enjoyment of wealth, women were the managers and guardians of wealth. Since men were absent for years on trips of trade and plunder, women had to manage estates and command over men on estates. They had the right to divorce husbands on the spot and take away their share of wealth. A woman could hardly gain wealth without a husband, and a man could hardly go on trade and war journeys without his estate being in capable hands and actively managed. The roles are different, but the power held by one over the other was substantial.

While addiction to innate tastes and preference is at the root of much of what is problematic currently in occidental society - and I did not even go into population growth -, the neglect in fostering individual development robs us of ability to deal with this crisis. I noted the paradox that it was advancement of dispersal phenotype, not maintenance phenotype features, which marked the beginning of major evolutionary changes. Even passing familiarity with human epigenetic development makes it evident that the future, it there be one, will rest with highly developed, competent people. There are thus many lessons from ethology and biology that still need to be sensibly applied.

LITERATURE CITED

Babcock, C., 1992. Essential Freud. Blackwell, Oxford, UK.

Balon, E.K., 1981. Saltatory processes and altricial to precocial forms in the ontogeny of fishes. Am. Zool. 21:573-596.

Beach, R.S., Gershwin, M.E. and Hurley, L.S., 1982. Gestational zinc deprivation in mice: persistence of immunodeficiency for three generations. Science 218: 469-471.

Brown, R. L., 1994. State of the World. (World Watch Institute) Norton & Co., New york.

Bruntland, G. H. (chair), 1987. Our common future: The world commission on environment and development. Oxford University Press, Oxford.

Bruton, M. N., 1989. (Editor). Alternative Life-history styles of animals. Kluwer Academic Publishers, Dordrecht.

Chauvez, A. and Celia Martinez, 1979. Consequences of insufficient nutrition on child character and behavior. In: D. A. Levitski (Editor) Malnutrition, Environment and Behavior. Cornell University Press, Ithaca, N.Y. p.283-255.

Cunningham, W. P. and Saigo, Barbara W., 1990. Environmental science. A global concern. Wm. C. Brown, Dubuque, Ind.

Denenberg, V.H., Ottinger, D.R. and Stephens, M.W., 1962. Effects of maternal factors upon growth and behavior of the rat. Child Devel. 33:65-71.

Denenberg, V.H., and K.M. Rosenberg., 1967. Nongenetic transmission of information. Nature 216:549-550.

Elder, P. S., 1991. Sustainability. McGill Law Journal 36(3):822-852.

Ellenberg, H., 1978. Zur Populationsoekologie des Rehes (Capreolus capreolus L. Cervidae) in Mitteleuropa. Spixiana.(suppl.2). pp.211.

Geist, V., 1966. The evolution of horn-like organs. Behaviour 27(3/4):175-214.

Geist, V., 1967. A consequence of togetherness. Natural History 76(8):24-31.

Geist, V., 1971. Mountain sheep. University of Chicago Press. Chicago.

Geist, V., 1978. Life Strategies, Human Evolution, Environmental Design. Springer Verlag, New York.

Geist, V., 1981. Neanderthal the hunter. Natural History 90(1):26-36.

Geist, V., 1987. On speciation in Ice Age mammals, with reference to deer and sheep. Canadian Journal of Zoology 65:1067-1084.

Geist, V., 1989a. Environmentally guided phenotype plasticity in mammals and some of its consequences to theoretical and applied biology. In M. N. Bruton (Editor) Alternative Life- history styles of animals. Kluwer Academic Publishers, Dordrecht. pp. 153-176.

Geist, V., 1989b. Did predators keep humans out of North America? In J. Clutton-Brock (Editor) The walking larder. Unwin Hyman, London. pp. 282-294.

Goldschmidt, R. B., 1940. The material basis of evolution. Yale University Press, New Haven.

Guthrie, R. D., 1989. Frozen Fauna of the mammoth steppe. University of Chicago Press, Chicago. 323 pp.

Koehler, O., 1952. Vom Unbenannten Denken. Zool. Anzeiger, Supplement 16:202-211.

Koenig, O., 1970. Kultur und Verhaltensforschung. Deutsches Taschenbuch, Munich.

Kortland, A., 1980. How might early hominids have defended themselves against large predators and food competitors Journal of Human Evolution 9:72-112.

Kortland, A., 1984. Comments on A. B. Stahl: "Hominid dietary selection before fire". Current anthropology 25:158-160.

Leakey, M., 1979. Olduvai Gorge. Collins, London.

Lorenz, K., 1973. Die Rueckseite des Spiegels. Piper, Munich.

Lovtrup, S., 1974. Epigenetic. J. Wiley, London.

Marshack, A., 1972. The roots of civilization. McGraw-Hill, New York.

Meadows, D. H., Meadows, D. L., Randers, J. and Behrens, W. W. III. 1972. Limits to growth. A report for the Club of Rome's project on the predicament of mankind. Universe Books, New York.

Miyamoto, M. M., Slightom, J. L. and Goodman, M., 1987. Phylogenetic relations of humans and African apes from DNA sequences in the psi-eta-globin region. Science 238:369-372.

Myers, N., 1984 (ed.) GAIA. An atlas of planetary management. Anchor Books, Doubleday, New York.

Orians, G.H., 1986. An ecological and evolutionary approach to landscape aesthetics. In: E.C. Penning-Rowell (Editor), Symposium on Meanings and Value in Landscape.

Packard, J.M. and Mech, L.D., 1983. Population regulation in wolves. in F.L. Bunnell, D.S. Eastman and J.M. Peek (Editors), Symposium on Natural Regulation of Wild Populations. Proc. No. 14, Forest, Wildlife and Range Experimental Station, Univ. Idaho, Moscow, Idaho. pp. 151- 174.

Peterson, R.O., 1977. Wolf ecology and prey relationships on Isle Royal. U.S. National Park Service Monograph No.11. Washington, D.C. pp. 210.

Pielou, E. C., 1991. After the Ice Age. University of Chicago Press, Chicago.

Stiling, P., 1992. Introductory Ecology. Prentice Hall, Englewood Cliffs, NJ.

Susman, R. L. and Stern, J. T., 1982. Functional morphology of Homo habilis. Science 217:931-934.

Uhazy, L.S., Holmes, J.C. and Stelfox, J.G., 1973. Lungworms in the Rocky Mountain bighorn sheep of western Canada. Can. J. Zool. 51:817-824.

Veblen, T., 1934. The theory of the leisure class. Modern Library, New York. 404pp.

Vogt, F., 1948. Das Rotwild. Osterreichischer Jagd und Fischerei Verlag. Vienna.

Vogt, F., 1951. Das Rehwild. Osterreichischer Jagd und Fischerei Verlag. Vienna.

Vrba, E. S., 1985. Ecological and adaptive changes associated with early hominid evolution. In Ancestors: the hard evidence. pp. 63-71, A. R. Liss Inc.

Waddington, C. H., 1957. The Strategy of the Gene. Allan and Unwin. London.

Waddington, C. H., 1975. The evolution of an evolutionist. Cornell University Press, Ithaca, New York.

Wellington, W.G., 1960. Qualitative changes in natural populations during changes of abundance. Can. J. Zool. 38:239-314.

CAUSES OF OUR COMPLETE DEPENDENCE ON CULTURE

Pouwel Slurink
Department of Philosophy
University of Nijmegen
Thomas van Aquinostraat 3
Postbox 9108, 6500 HK Nijmegen
The Netherlands

ABSTRACT. Culture or the non-genetic passing on of information between successive generations has been shown to be quite common among different species of animals, especially, but not exclusively, among primates. During the last decades several attempts to reveal the essential characteristics that make us "uniquely human" (tool use, grammar) have been falsified by new discoveries. Nevertheless, the extreme degree to which some characteristics have developed in our species suggest that some force during our evolutionary history has tipped the scales finally in favor of behavioral flexibility and the continual adaptation to new surroundings and new techniques and life styles. An attempt is made to integrate several hypotheses into a plausible account concerning the ecological circumstances in which the peculiarities of our genus and species evolved.

1. Introduction

Philosophers have long believed that man is in some respect a unique species. This belief has seduced them often to reflections which in the light of current ethological research may seem a little bit farcical. Often those reflections betray more about the obstinacy with which we are inclined to make dichotomies and with which we are inclined to make chauvinistic, racist or "speciïstic" (Dawkins, 1976) distinctions between the "in-group" and the "out-group" than that they having anything to say about our place in the world. So we read in a paper about "Die aufrechte Haltung" or "the upright attitude" the following sentences:

> Ein Tier, das sich in der Längsachse seines Körpers fortbewegt, ist immer auf die Dinge zugerichtet. Der Mensch aber, der sich in einer Richtung senkrecht zu seiner eigenen Längsachse vorwärts bewegt, den aufgerichteten Körper parallel zu sich selbst verschiebend, findet sich allen Dingen einsam gegenüber (Straus, 1949: 370).

The writer seems to have forgotten that already many dinosaurs were walking upright (bipedal) and that there is little evidence of their ability to objectivity and critical distance toward the world that surrounded them.

In the course of its still short history, ethology has often revealed the superficiality of such easy characterizations of our own supposed uniquenesses. Ethologists have discovered that man is not the only animal that makes tools and has complex social relations and commitments, that has belligerent inclinations, that recognizes itself in front of a mirror, that can link symbols into meaningful utterances. This did not, however, stop the stream of claims about man's uniqueness. Revisionists are always very

1 "An animal that moves itself along the longitudinal axis of its body is always directed towards the objects. Man, however, who moves himself at right angles to his longitudinal axis, displacing the upright body parallel to himself, finds himself facing all things in solitude."

R. A. Gardner et al. (eds.), The Ethological Roots of Culture, 461–474.

inventive in devising new versions. For example, now that linguists can no longer deny that apes are able to manipulate strings of symbols meaningfully, they have shifted their attention to the ability to structure meaningful utterances by means of some kind of grammar (Bickerton, 1990). This theory was stated only a short time after which the first reports appeared about the linguistic capabilities of Kanzi, a bonobo brought up by Savage-Rumbaugh, who showed a sense for grammar. This should warn us that we cannot be cautious enough in making claims about our uniqueness.

However, the fact that several characteristics of humans are more or less exhibited by other animals throws up the question why we are the only ones who exhibit them to such an extreme degree. Many other animals show "culture" or the non-genetic passing on of information between successive generations (Bonner, 1980; for a more symbol-oriented definition of culture, see Durham, 1991), but we have built cultures around every individual biological function of our body, be it feeding, sexuality, sleep, stools, cleaning, etc.. Above that we have built cultures around every individual social interaction, be it parent-offspring relations, friendships, or relationships between the sexes. Many other animals more or less create their own environment by building nests or even by building dikes (think of beavers), but no other animal lives so continuously in a self-created environment that can be adapted continually to the caprices of climate and seasonality. Many other animals have complicated communication systems, but very few of them rely so totally upon them as we humans do.

From our arrogant man-centered perspective it may seem as if our capabilities to create culture and to live in a self-created environment are inevitable products of evolution. Looking back on the course of evolution we see a gradual increase of representational capacities. Recently the linguist Bickerton has repeated the old belief:

> As long as there are creatures whose motor capacities and environmental conditions allow them to benefit from it, evolution will always favor an increment in representational power ... Thus, given freedom from catastrophe and sufficient time to work in, normal evolutionary processes inevitably bring about the progressive development of representational systems... If no catastrophe had eliminated them, then fifty million years ago there might already have been three-toed lizardlike creatures sitting around wondering how language could have evolved out of dinosaur communication (1990: 103-4).

There are three things that Bickerton doesn't mention however. First, there are many taxa and only in some of them can we indicate temporary processes of cognitive progress. Second, there are many indications that there has not only been one, but many catastrophes during the process of evolution on earth (Müller, 1989) and it could well be that they were not obstacles but *necessary conditions* for further evolution - given the stuffing of all available niches during the more stable periods. Third, cognitive progress is not a goal in itself, but only the consequence of the *temporary* survival value of cognition. The moment stupidity would bestow more fitness upon its possessors, stupidity would be selected for. There is no guarantee that the most intelligent individuals will always outreproduce the lesser ones. (At least among human beings we see that many famous philosophers and scientists died childless.)

Representational progress or progress in the direction of culture is thus not an inevitable consequence of the process of variation and selection which Darwin originally called "descent with modification" and which we are accustomed to call "evolution". Many animals manage quite well to live without an extraordinary intelligence or without extraordinary communicative, reflective or cultural skills and there have to be a series of special selection pressures to explain why we humans have evolved such an extreme degree of curiosity, creativity and reliance on culture. Let us examine, then, whether we can explain why man is apparently the only animal on this planet that seems to be totally trapped in "the cognitive niche" (Tooby & DeVore, 1987). To explain this, we will need a scenario about our evolution, in short an evolutionario (Wrangham, 1987). But as there exist evolutionarios of all sorts and varieties I will impose some extra demands upon it:

1. First, it should be *phylogenetically plausible*; that is: it should be in concert with our best present knowledge about the possible characteristics of the common ancestor of both ourselves and the great apes that are most related to us.
2. Second, it should be *paleontologically adequate*; that is: it should be at least compatible with our best knowledge concerning the paleontological record and concerning palaeoenvironments.
3. Third, it should be *evolutionarily feasible*; that is: compatible with our best current understanding of the principles of evolutionary biology.
4. Fourth, it must exhibit *explanatory specificity*; that is: it should not have the characteristics of an easy explanatory *deus ex machina* that could also be invoked to explain for example, why wolves or dolphins are dependent on culture, whereas in fact they exhibit culture only to a limited degree. It should explain not only why humans have become completely dependent on culture, but *at the same time* explain why other animals *did in fact not.*

Any evolutionario that meets these four requirements will have survived what I will call the "Darwin test" and will therefore be considered to be at least a candidate for truth. Of course, given the continual progress in evolutionary biology and paleontology the Darwin test has to be repeated time and again on each model. Above that it remains an open question whether "having survived the Darwin test" is the same as "being true". Philosophers of science are inclined to entangle themselves on this point in endless debates concerning the validity and legitimacy of the "inference to the best explanation" (in the hope of finding a rational justification for purely pragmatic theory choice), but fortunately I can evade that discussion here legitimately.

Let us start to work with those four requirements, then. First, to make a phylogenetically plausible model we have to begin with a reconstruction of a possible common ancestor that we share with the great apes that are most related to us. After that we will have to look at what we can learn from paleontology about the environments in which human evolution occured. With those clues in hand we can begin to reconstruct an evolutionary sequence in which human characterists emerged. To meet our third requirement we will try to account for the story of human evolution as a multi-stage process in which each step is adaptive in its own right. Finally, it is important that our explanation *specifically* applies only to our branch on the evolutionary tree, because the riddle that we will try to solve is the riddle of our own unique complete dependence on culture. When our explanation presupposes the evolutionary inevitability of cognitive progress the price we would have paid for it would have been too high: we would then have to explain why other animals did not reach our stage.

2. The common ancestor

Recently, several attempts have been made to integrate molecular evidence on the phylogeny of great apes and humans (see Tanner, 1986 for a review) and the results of ethological studies relating to the social structures of homonoids (Wrangham, 1986; Ghiglieri, 1987; Foley & Lee, 1989; McGrew, 1991). Molecular studies of relatedness have led to an evolutionary tree in which humans are closer to African apes than to Asian apes and in which they are also closer to chimpanzees and bonobos than to gorillas. Comparison of these data with those from studies in the social structure of the species concerned shows that the molecular relatedness corresponds in some degree to resemblances in social structures. Wrangham (1986) first concentrated on the resemblances in social structures between the African apes and humans, because the molecular data suggested their relative relatedness. From the correspondences in the behavioural repertoire of these species he concluded that the common ancestor (CA) of all African apes and humans

> "had closed social networks, hostile and male-dominated intergroup relationships with stalk-and-attack interactions, female exogamy and no alliance bonds between females, and males having sexual relationships with more than one female" (Wrangham, 1986: 68).

Ghiglieri (1987) showed that it was still possible to improve upon this model by concentrating on the resemblances between chimpanzees, bonobos and humans (who are also genetically most related). His analyses show that the chimpanzee-bonobo-human clade (the HUCHIBO-clade) is characterized by a unique cooperation of closely related males who cooperatively defend a common territory and who only exhibit moderate sexual dimorphism, "presumably because success in male-male competition hinges on having larger *group* size (of kin-related males) during conflicts rather than larger individual size" (Ghiglieri, 1987: 339).

The general pattern of homonoid evolution that he reconstructs is the following:

- Before the split between African and Asian apes, societies were largely structured around female kin groups like most primate societies: young subdominant males have to leave their native group and to find or found a harem elsewhere. Bonds between kin-related females are strongest. This pattern can still be devised rudimentarily in orangutans, where females stay close to their mothers and young males start to wander and emigrate from their natal ranges looking for estrous females and reproductive opportunities.
- All African apes share traces of a decline of the matriachal core characteristic of other primate-groups: the females in a gorilla harem are generally unrelated to one another and female exogamy is more pronounced than male exogamy, because some adult males stay in their natal groups. In contrast to the HUCHIBOS, gorillas are not territorial, however, and most groups contain only one silverback male, who in agressive encounters has to rely on his extreme body size to defend his harem.
- Between the common ancestor of gorillas and HUCHIBOS and the common ancestor of all HUCHIBOS (a period of about two million years) a new revolution occured: males stayed in their own natal social groups and kin-related males started increasingly to cooperate. This cooperation between males sets HUCHIBOS apart from the other great apes:

> "Unlike gorillas and orangutans, males of the chimpanzee-bonobo-human clade retain their male offspring predominantly, live in closed social groups containing multiple females, mate polygynously, restrict their ranging to a communal territory, are *cooperatively* active in territorial defense, and, apparently, when a neighboring community weakens, the males of some communities make a *concerted* strategic effort to stalk, attack, and kill their rivals as do men (ref. op. cit.) (Ghiglieri, 1987: 346)."

According to Ghiglieri it was this propensity for cooperation and solidarity between kin-related males that could well have been "the critical preadaptation for male cooperation in dangerous scavenging or hunting in the hominid ancestor". Above that, the fusion-fission sociality characteristic of HUCHIBOS may have functioned as "a preadaptation for a division of labor on a daily basis and a diversification of ecological modes" (Ghiglieri, 1987: 347).

It is interesting to speculate about the evolutionary mechanism behind the origin of male HUCHIBO-cooperation. What factor could have promoted the transition from male exogamy towards female exogamy? Anderson (1986) noted a correlation between female transfer and predation rate in primates. It is possible to imagine that a temporarily reduced threat of predators in one of chrono-species ancestral to all great apes was the ultimate cause (Ghiglieri, 1987; Alexander, 1989; however, see Boesch, 1991). In this first stage (of which the orangutan may be a model) this diminished threat may have been a factor in loosening the group structure and thus may have set in the decline of the female-centred kin-groups characteristic of most primates. In this stage solitary females and their offspring may have been an easy target for solitary males (see Wrangham, 1986, on rape in orangutans). A change to a more abundant food-source (as in the gorilla) may have triggered the transition to a one-male group. Instead of being harassed time and again by solitary males, females could thus secure themselves the superior genes of a strong male who at the same time could defend their offspring against foreign males and the threat of an infanticidal take-over (infanticide is a well-known phenomenon in gorillas). In a next stage the competition of such strong males for harems may have started an arms race in which dominant males where forced to rely on the help of kin-related males more and more (in the Virungas 40% of gorilla groups contained more than one adult or nearly adult male, see reference in Ghiglieri, 1987). This cooperation between kin-related males may then have been the reason why the common ancestor of all HUCHIBOS was uniquely preadapted to form hunting bands and to include more meat in its diet.

In any case, we now know quite a lot about the common ancestor of all HUCHIBOS. Chimpanzees, bonobos and humans share some unique behavioural traits, which partially seem to derive from male retention in their natal groups. Chimpanzees are known to hunt cooperatively in some regions (Boesch & Boesch, 1989) and at least a part of their diet consists of meat (there are reports on scavenging (Hasegawa, 1983)). Above that, they engage in cooperative intergroup aggression (Goodall, 1986). Finally (and this could be more a consequence of their omnivorous diet than of their unique social structure) chimpanzees are skilful makers and users of tools (McGrew, 1991; Boesch & Boesch, 1984) who seem to pass on their skills in a cultural way, because different techniques are applied in different regions (McGrew, 1991).

Summarily, we share many ingredients of a cultural way of life with the chimpanzee and thus, likely, with the common HUCHIBO-ancestor. The question, however, is how those ingredients were molded into a recipe for a totally culture-dependent species.

3. The problem of the unknown paleoenvironments and the origin of bipedalism

Now we know so much about the common HUCHIBO-ancestor it would seem that the procedure for the creation of human beings can simply be derived by studying paleoenvironments and the fossil record (Foley, 1987). However, it is not necessarily that simple. The fossil record is very patchy and interrupted indeed and, in principle, it is possible that the main events in hominid evolution have not left any traces. It is well possible that human evolution resembles in this respect the origin of life: it could also be a transition that wipes out its own traces. Although spectacular finds have again and again filled major gaps in the human evolutionary story, there remain gaps that continually inspire speculation about unique episodes that changed the course of our race dramatically.

Given the incompleteness of the paleontological record, it is not surprising to see that we meet several difficulties with the reconstruction of our evolutionary divergence from the common HUCHIBO-ancestor. Firstly, we encounter one striking anomaly when we study the paleontological literature. It seems that the collection of fossils of the oldest fully bipedal hominids (the Hadar hominids, dated approximately 4.0 million year b.p.) show a clear sexual dimorphism more within the range of the gorilla than in that of the chimpanzee. This seems to falsify Ghiglieri's model of the common HUCHIBO-ancestor. Either the similarities between the HUCHIBOS are a product of convergent evolution, or the Hadar-hominids belong to more than one species as suggested by the heterogeneity of the Hadar materials (Schmid, 1989).

Secondly, to begin a reconstruction of the selection pressures responsible for our divergence from the common HUCHIBO-ancestor we need a model about the environment in which the first hominids evolved. At this moment there is no unanimity as to what this environment might have looked like. At least three possibilities are defended:

- A Savanna-like environment: this is the most commonly held model among paleontologists. It is supported by data that show that, as a consequence of a colder and drier climate, tropical forests in Africa have given away to more mosaic environments, with much savannas, at least since 10 million years ago (Foley, 1987). Many dramatic changes in the fauna of this period seem to be a result of this transition (Tobias, 1991).
- A forest-like environment. Recently, new evidence has accumulated on Plio-Pleistocene rainforests in East Africa (Williamson, 1985).
- An environment characterized by the presence of water in the form of marshes and lakes. The pollen obtained at the site where the First Family fossils were recovered (at Hadar) is mainly from bulrushes which invariably inhabit marshes ((LaLumiere, 1991). Most sites where early hominids were recovered are in the close vicinity of lakes or rivers (for example: Tabarin, Hadar, Omo, Koobi Fora, Olduvai) (Foley, 1987: 195).

All three models have inspired different explanations of the origin of human bipedalism.

- Explanations derived from the savanna-model differ most widely. Some suggest that uprightness enabled early hominids to reach upward for food (Jolly, 1970), others, that it increased their visual horizon and enabled them to notice both prey and predators from a bigger distance on the savanna.

Some models stress the possibility of carrying food and tools when walking bipedally (Hewes, 1961, Lovejoy, 1981). Other models concentrate on the supposed energetic efficiency of bipedal travel in a large home range (Rodman & McHenry, 1980; Pickford, 1989). The human ability to run over very large distances and for prolonged periods of times is projected sometimes back far in time: it enables modern hunter-gatherers to run down prey. Further, it is noted that the body surface exposed to sunlight is relatively small in an upright stance (Wheeler, 1984) and that this could have enabled early hominids to forage during the mid-day when all predators were asleep (Foley, 1987).

- Explanations based on the forest model point to some primitive characteristics of the limbs of the early *Australopithecus* (short lower limb, long forelimb, curved toes, long forefoot) which enabled it to climb into trees (Susman, 1986: 86). This ability to climb could at the same time have been a necessary preadaptation for bipedalism.
- An explanation based on an environment with much water is the aquatic ape theory that recently has been enriched with supporting data (Morgan, 1990; Roede *et al.*, 1991). Originally inspired by the striking parallel between the human subcutaneous fat layer and that of different aquatic mammals, most of whom happen to have lost their fur as well, it was expanded by Morgan and Verhaegen into an all-encompassing theory that could not only account for human bipedality (a wading ape adaptation), but also for the human transpiration system (that would amount to spoiling too much water and salt in a savanna-environment), the threefold brain expansion (aquatic mammals have also an enlarged brain), and even the origin of language (possible as a consequence of two adaptations of a diving mammal: a descended larynx to inhale large quantities of air in combination with conscious control of respiration).

At the moment it seems the safest strategy to evade both the savanna-theory and the aquatic ape theories in their most extreme versions (Pickford, 1989; Reynolds, 1991; Tobias, 1991). Certainly some of the criticism leveled by aquatic ape theorists against the savanna hypothesis seems right. Especially Verhaegen's argument that the human cooling system is extremely water and sodium-wasting and therefore unfit for a dry environment seems very convincing (Verhaegen, 1987). On the other hand the danger has to be avoided that a theory is devised which can not be verified but nicely fills all the gaps in the fossil record with explanations for supposed uniquely human characteristics (Lalumiëre, 1991). To explain human tears as a device to get rid of a surplus of salt (Morgan, 1990) seems to me an explanation which could well be too beautiful to be true. In addition, it creates more problems than it solves: why do we so seldom see people weep on the beaches, for example?

If it is true that most sites where hominid fossils are found lay in the vicinity of ancient lakes, marshes or rivers this suggests that early hominids may have preferred border-areas between forests, waters and savannas. Their most favoured biotope could well have included a mix of trees, waters and open country like our favourite parklands nowadays. The fusion-fission social structure and the in part omnivorous possibilities of HUCHIBOS may have been factors enabling early hominids to profit maximally from this mixed environment (one clue is a carapace of a giant turtle found among the hominid deposits in lake Turkana, Reynolds, 1991). At a first stage they may foraged at the border of the savanna and at the edges of marshes and lakes during the day and they may have climbed into the trees during the nights. In later stages, when the climate became cooler and drier during the Pliocene, they may have been forced to rely more and more on the ability to migrate across the savanna's looking for suitable environments.

If it is true that bonobos both live more in forests and are better adapted to water than chimpanzees (Susman, 1986; De Waal, 1988) it may well be worthwile to concentrate more on this species and on the circumstances in which it moves upright. If it is true that Bonobos have been seen catching fish while wading in the water upright (De Waal, 1988) it could well be that the transition to bipedality is less revolutionary than Morgan (1990) suspects. It seems that in Bonobos uprightness is sometimes used to intimidate predators, sometimes it is used to carry fruit and sometimes to catch fish (De Waal, 1988). This could imply that the carrying-hypothesis (Hewes, 1961) and other hypotheses (a wading ape theory, a predator intimidation theory (references in Pickford, 1989) need not be mutually exclusive. Bipedality need not have evolved as the result of one specific selection pressure, but it could have been the best general solution for a multitude of separate problems.

Once a general ability to walk bipedally had evolved it may have had crucial advantages in a savanna environment and may have openened up new strategies, like hunting by running down prey and scanning relatively big home ranges. The savanna-theory explanations of bipedality could thus all be wrong when interpreted as referring to the *causes* of bipedalism, but at the same time they could be true when interpreted as referring to its consequences (Pickford, 1989).

4. Hunting, scavenging, gathering?

With this cautious explanation of human bipedality, however, we did not come close to an explanation of other supposed uniquely human properties. The existence of the robust australopithecines proves that it is a healthy strategy to separate explanations of bipedality and explanations of, for example, the enlarged brain. Clearly, the climatic changes of about 2.5 million years ago, when for the first time an ice-cap covered the poles and the climate in Africa became drier and colder than ever before, coincide with the origin of two new types of hominids (*Paranthropus* and *Homo*), of which only one is characterized by an enlarged brain.

According to Foley (1987) the two different evolutionary trends among hominids can best be explained as different strategies in coping with the problems posed by the dry season. When dry seasons were increasingly characterized by a scarcity of plant food, one trend was the evolution of robust australopithecines who with their heavy teeth could live from coarse, hard and dry fruits and from seeds. Another reaction to the problems posed by a dry environment could be a shift to more meat consumption. During the dry season many savanna animals tend to congregate around the remaining water resources and the early hominids could have used this opportunity. Certainly, the dentition of early *Homo habilis* is compatible with an omnivorous diet of which meat was a part.

Recently, there has been much discussion concerning the question whether the first hominids were scavengers or hunters. The presence of toothmarks of large carnivores on about 11 to 15 % of the bones at the hominid sites at Olduvai George suggest that at least a part of the meat was first used by large carnivores: toothmarks are sometimes found on the more meat-rich parts of the bones, while human cut marks appear at the non-meat-bearing bones (Shipman, cited in Binford, 1986). Tooby and DeVore (1986: 221) have criticized the idea that early hominids were scavengers on the ground that scavenging would be inefficient and dangerous, given the risks of returning predators and the risks that the flesh is poisoined by micro-organisms. Observations on chimpanzees show, however, that in this species scavenging is at least sometimes practised (Hasegawa, 1983). Clearly, it would be wrong to rely totally on *a priori* arguments in excluding scavenging. Early hominids may have had an advantage over other scavengers, because they had tools at their disposal with which they could have opened the thick skin of very large animals (Toth, 1986; Foley, 1987: 22).

All in all, the most accurate description of the original *Homo* survival strategy could well be characterized as ecological opportunism optimized by tool use. The increase in brain size of *Homo habilis* may have been in part a consequence of the general higher encephalization quotients of predators (Jerison, 1973) and of animals living on variable and unpredictable resources (Eisenberg, 1981) and it may at the same time reflect an increase in meat consumption. It would be very naïve to suppose, however, that hominids relied totally on meat, as there is a ceiling to the amount of protein that can be processed by the human liver and kidneys (Speth, 1988). There is more reason to assume that in hominids hunting was a male strategy for obtaining rich quality food with which they could "show off" and which they could exchange for higher status and other favours (Hawkes, 1991). This hypothesis is compatible with the observations of Boesch and Boesch (1989) on chimpanzees in Tai-forest, where only male chimpanzees participate in hunting.

According to Hawkes (1991) men in hunter-gatherer-societies preferentially target resources which are more widely shared, also outside their nuclear families. Their strategy is less directed to a constant provisioning of daily food and more to an occasional big bonanza that can be widely shared and with which they can gain status. When hunting is conceived as a risky male strategy to obtain high quality food, constant gathering of resources by women is just the other side of the coin.

I therefore do not believe that the hunting hypothesis (Tooby & DeVore, 1986) and the gathering hypothesis (Tanner, 1986) are incompatible. Tanner (1986) uses observations by Boesch & Boesch (1984) to show that female chimpanzees both use tools more than males and use them in a more complex way. To me this only proves that the different and compatible food strategies that Hawkes noted by hunter-gatherers and that inspired her to her "showing-off hypothesis" were in principle already there on the level of the common HUCHIBO-ancestor.

It could be this pattern of sexual division of labour, based on preadaptations of the common HUCHIBO-ancestor, that finally led to the characteristic combination of semi-permanent home bases and male parental investment that can explain so many human characteristics.

5. Home bases, male sexual jealousy and parental investment

But what could have triggered a further increase in the development of separate male and female strategies? And how could it have resulted in the establishment of home bases with fathers directing help to their own offspring? Lovejoy (1981) argued that the unique reproductive system of humans, in which females and males form lasting bonds and males provision their own offspring, arose all of a sudden, *ex nihilo*, as a reaction to the extreme "K-selected" reproduction strategy of their ancestors. In chimpanzees females are busy for more than two years with only one child. They always have both to carry the child with them and to provide it with food. It would be more efficient, according to Lovejoy, when the female could stay home with the children and whilewhen daddy would look around for food in the dangerous outside world (and carry it homewards walking bipedally). The time between two births could become smaller and one female could produce more children. Once upon a time apes suddenly discovered this more efficient reproductive strategy and males started to walk upright and returning at home with their freed arms full of food. Once this strategy was discovered it was so succesful that the upright walking apes seized the world from all other ape species.

However, the idea that such an extreme transition in reproductive strategy could arise *ex nihilo* and not in reaction to a major change in the environment is probably very naive (Tooby & DeVore, 1986). The idea that one mutant female would leave her group and start a solitary life is improbable given the evidence on predation of leopards on australopithecines (Brain, 1981). The idea that her partner could leave her alone and still be sure of his paternity is also improbable. In all likelihood such a sudden transition from a chimpanzee-like lifestyle to the twentieth century american way of life would be too rash on all accounts. Above that, even in modern humans monogamy is only practised in some societies and the model of Lovejoy assumes that is was already universal around four million years ago.

To explain the origin of male parental investment in humans we have to ask ourselves, first, why a situation could arise in which females could become increasingly dependent on male provisioning. This situation could be the above mentioned dry season in an increasingly savanna-like Africa. When the home ranges that males had to cross in their search for food became very extensive and when there were at least some periods during the year in which male provisioning became an indispensable extra food source for females and offspring, a situation arose in which both parties could profit from a meeting area to exchange food (Tooby & DeVore, 1986: 224). Such a meeting area need not to have existed in *Homo habilis*, although it is clear that this species carried along with it selected pieces of carcasses which it butchered on selected sites (first interpreted as "home bases" (Isaac, 1978), later as "stone caches" (Potts, 1986)). But such a meeting area certainly may have existed in *Homo erectus* which is suspected to have discovered the use of fire and of which some remains are found in caves (especially Zhoukoudian Locality 1, Klein, 1989: 222). This could explain why the brain size of this species increased during its existence for over a million years: home bases tend to make life more complex.

At this stage a problem arose. When females became increasingly dependent on occasional extra food provisioning by males but males could obtain such extra resources by cooperation only, dominant males could no longer monopolize all females as a matter of course. Young succesful hunters would otherwise abandon the group and certainly some females would follow them. Also, to get as much extra food as possible, females had both to compete increasingly with one another and to ensure themselves part of the booty. They could do so by introducing extra elements of conditionality and exclusivity to sexual relationships which increased the certainty of paternity for males. Of course, this would not lead to a system of pure monogamy, but simply to an increase in special bonds between males and females, ignited by feelings of love and secured by sentiments of jealousy (based on preadaptations like chimpanzee "consortships" as described by Tutin (1975) and Goodall (1986: 453). At least in modern humans this system can explain such characteristics as concealed ovulation, continual receptivity and patterns of reciprocity in courtship and sexual selection (Alexander & Noonan, 1979).

Thus there may exist a relation between the mutual dependence of males and more permanent bonds between males and females which fundamentally distinguishes human permanent pair bonds from those of, for example, gibbons. When fusion-fission sociality gradually led to the evolution of complex social systems around home bases a whole new series of problems and possibilities could arise. Planning and looking forward became increasingly important and relationships of reciprocal altruism became increasingly demanding (Trivers, 1972).

At the home base, the ancestral food-call could in principle develop into a reporting ceremony about the best areas to collective specific food items. According to the linguist Bickerton the home base could thus have functioned as an instigator of language evolution. At least one other species with a "home-base", the honey-bee, has developed a reporting system (Bickerton, 1990: 154). According to Bickerton it was *Homo erectus* that first used a system of "protolanguage", a presyntactical language which allowed a rough classification of food-items and in which information about the environment could be communicated. According to Bickerton protolanguage enhanced the ecological flexibility of *Homo erectus* and was thus responsible for its spread beyond sub-Saharan Africa. The gradual increase of its vocabulary would account for the gradual increase in brain growth in *H. erectus*.

6. Home bases, territoriality, fire, and the ecological dominance barrier

The consequences of the institution of more permanent home bases in human evolution can thus not be underestimated. Home bases may have offered the kind of continuity that is needed for the establishment of more elaborate cultural traditions. Home bases, especially caves, may have functioned as a kind of shield against the dangerous outside world. At least at the stage of Neanderthals this shield was used probably to protect wounded individuals temporarily to give them time to recover (Klein, 1989). In a still later stage (that of *Homo sapiens sapiens)* it could be used as the center of a complex social life in which many individuals beyond the reproductive age survived to pass on their wisdom for the use of the whole community (Klein, 1989: 385) and in which a complex division of labour could arise (Toth, 1986).

Many writers have stressed the crucial importance of the discovery of fire in this transition to more complex societies (Poirier, 1987; Goudsblom, 1989). The discovery of fire (going back probably more than one million years, Brain & Sillen, 1988) enabled the conquering and defending of caves and with that it created a place of relative safety from predators and climatic uncertainties. The discovery of fire may have forced hominids to new levels of planning ahead and self-control. Moreover, it may have been a factor in finally averting the danger of predators. With that, however, it may have given the impetus for new levels of intraspecific conflict (Alexander, 1987; 1989; 1990).

One of the effects of a safe home base at a favourable site must have been that of intensified territoriality (which we have seen to be a characteristic of all HUCHIBOS). Groups which had settled on relative favourable positions must have aroused the jealousy of other groups. The moment that home bases became relative safe ports of refuge for predators they may have become at the same time threatened by neighbouring groups.

We have already seen that intergroup conflict is a characteristic of all HUCHIBOS. There is reason to suppose that levels of intergroup conflict increase the moment that predation by other species decreases (Alexander, 1989). The moment the population levels of a species are not corrected by predators of other species the new limits will be fixed by intraspecific conflicts over territory and resources. Alexander speculates that the "turning point" at which "ecological dominance" is achieved and at which predation pressure is replaced by interspecific "balances of power" already could have been passed by the chimpanzee (Alexander, 1989: 473). However, recent observations on chimpanzees (Boesch & Boesch, 1991) and archeological data on australopithecines (Brain, 1981) suggest that both these species have experienced a considerable predation pressure from leopards.

I suggest therefore, that we turn to the paleontological record to look for data which could reveal a stage of increased intraspecific conflict. Roper (1969) reports many incidents of presumable interpersonal violence in early *Homo sapiens*. Recently it has been shown that a skull from Bodo (probably *Homo erectus)* which exhibits numerous cut marks probably has been deliberately defleshed (White, 1986). According to Klein (1989: 387) modern *Homo sapiens sapiens* shows very little signs of deliberate injuries and is in this comparable to modern hunter-gatherers. However, he does mention "an extraordinary terminal Pleistocene (ca. 14,000-12,000 years-old) cemetry near Jebel Sahaba in Sudanese Nubia where nearly half of the fifty-nine individuals exhumed had either unhealed antemortem skeletal injuries or had stone artifacts lodged in or near their bones" (Klein, 1989: 387). A recent excavation in the Netherlands shows also that the picture of *Homo sapiens sapiens* as a peaceful hunter-gatherer could be wrong indeed (Louwe-Kooijmans, 1990).

One critical factor that needs to be explained is the increased group size in *Homo sapiens sapiens.* Alexander (1979) rightly wonders why human societies tend to be much larger than is efficient for other cooperative hunters. His "balance-of-power-hypothesis" (Alexander, 1979; 1987; 1989; 1990) is essentially an answer to that question. Originally developed as a semi-group-selectionist explanation of the threefold brain enlargement (Bigelow, 1969), it was molded by Alexander into an explanation of increased group sizes and the evolution of self-constraint and internal discipline (Alexander, 1987).

It seems to me that this hypothesis deserves further study and that it specifically should be made concrete by paleontological evidence. Probably the best way in which it can be made more concrete is to interpret it as the hypothesis that competition for the best sites for predator-free home bases played an important role in human evolution and led somewhere between early *Homo sapiens* and modern *Homo sapiens sapiens* to a shift from intragroup conflict to intergroup conflict. The increase in cooperation within the group, the result of the arms race between groups, resulted in unique opportunities for psycho-social and cultural evolution.

Already in Neanderthals there is evidence that individuals could stay at the home base to recover from serious injuries. *Homo sapiens sapiens* however, is characterized by the unique presence of many individuals beyond the reproductive age (Vallois, 1961). Care for older and wounded individuals could be a result of the need to maintain a maximal group size and with that a balance of power with other groups. The resulting presence of three generations and relatively many not directly kin-related individuals at the home bases can possibly account for the explosive effect on cultural evolution.

Another fact to be considered is that anatomically modern *Homo sapiens sapiens* is built much lighter than earlier types. This could in principle be an adaptation which enabled individuals to live in bigger groups. Large individuals need more calories and therefore are more likely to experience intragroup conflict over resources. Because of its modest figure early *Homo sapiens sapiens* could achieve much higher densities than its Neanderthal contemporaries (Klein, 1989: 330). This could be a sign that it was during this period that it became more important to belong to a big group than to be big yourself.

In sum, the moment that the ecological dominance barrier was crossed, balances of power between groups forced groups to become bigger and individuals to become smaller and to use resources more efficiently. The increased group size plus the increased need for high quality nutrients to sustain all group members was probably literally the force that made *Homo sapiens sapiens* so creative: with the appearance of large groups in Europe the appearance of many new types of artefacts has been associated (White, 1989). The exploitation of new foods is proved by the presence of spear-points, spearthrowers, fishing hooks, harpoons, etc. The appearance of body ornaments reveals the emergence of new levels of complexity in social life. It was probably the necessity both to keep track of your own kin-relations and to orient yourself in an increasingly complex system of division of tasks that induced a new consciousness of identity, reflected in body ornaments (White, 1989).

Of course, self-consciousness is certainly not completely unique to our species (for example, De Waal,

1981). However, only in an environment that is continually transformed by cultural innovation based on complex relationships of direct and indirect reciprocal altruism, identity and social role problems can become so chronic that a continual monitoring of the "self" in relation to the group is needed. Only in the context of a society in which each individual counts as indispensable (because of the need to maintain an adequate group size) can individuals become "precious" enough to deserve funeral rites and to allow themselves reflections on their real self, or "soul". What probably is unique to our species is thus our tendency to continually reinterpret our abilities and possibilities within the group and to reshape our identity continually within a cultural context. We are a species which not only adapts culturally inherited techniques in the service of natural goals, and which not only adapts its environment and reshapes it in accordance with its natural desires, but which at the same time also has to continually adapt cultural images of the self to escape or correct an always threatening sense of alienation.

7. Conclusion: speculation, skepticism, or science?

Thus, our unique complete dependence on culture may have been reinforced by our success in conquering other species. Once upon a time ecological factors may have forced bipedal ape-men to rely increasingly on their ability to work with tools and to open up new food sources with them (Toth, 1986). Once this may have brought about new levels of cooperation between males and new forms of exchange between males and females (food for paternal certainty, in short). One day, the divergent food strategies of males and females may have led to a society organised around a home base which functioned as a relatively safe refuge where food could be exchanged and where relationships could be renewed (Tooby & DeVore, 1986). Once upon a time fire was discovered (Brain & Sellen, 1988) and our ancestors started to really frighten their natural enemies. One day our ancestors became so skilful in cornering and killing prey that they became unbeatable, except by conspecifics. One day our ancestors started increasingly to compete with each other for the most favourable sites for home bases and their groups became bigger and bigger to defend the best sites (Alexander, 1979, 1987, 1989).

One day a return to "nature" (whatever it is supposed to be) was no longer possible. Each human society brought with it an accumulation of cultural inherited skills and practices with which it was able to survive and to defend itself against the vicissitudes of life - and without which it could not live anymore. At the same time, each human society was entangled in an arms race against neighbouring societies with slightly different traditions. Within this setting you needed the best ways to find food, to cooperate, to make weapons. Creativity paid. And you could not have too much of it: there was a runaway selection process for cultural skills.

In the competition between societies it was the biggest societies that won again and again by conquering other societies. An individual could thus better be a small or gracile member of a big society than a big member of a small society. But even the more gracile humans that thus evolved had to eat and thus new food sources had to be found, again and again. And again, it paid to be creative.

Of course, this is only a model, a "just-so-story", another variant of "paleofiction". No doubt parts of it will prove not to be resistant to new evidence. But to find new evidence you have to know where to look. And you always need a story first to know where to look and to know how to improve upon your original story or to discover that you have to reject it. Even the skeptic needs a story to have at least something to be doubtful about. Let us beware that we do not make evolutionary stories simple just because it is safe and easy to be a skeptic.

The investigations for this article were supported by the Foundation for Philosophical Research (SWON), which is subsidized by the Netherlands Organization for Scientific Research (NWO).

References

Alexander, R. D. 1979. *Darwinism and Human Affairs*. Seattle: Univ. of Washington Press.
Alexander, R. 1987. *The Biology of Moral Systems.* New York: Aldine de Gruyter.
Alexander, R. 1989. Evolution of the Human Psyche. In: P. Mellars & C. Stringer, *The Human Revolution*, Edinburgh: Edinburgh UP, 455-513.
Anderson, Connie M. 1986. Predation and Primate Evolution. *Primates,* 27: 15-39.
Bickerton, Derek. 1990. *Language and Species.* Chigago & London: The University of Chigago Press.
Boesch, Ch. & H. Boesch. 1984. Possible causes of sex differences in the use of natural hammers by chimpanzees. *Journal of Human Evolution,* 13: 415-440.
Boesch, Ch. & H. Boesch. 1989. Hunting Behavior of Wild Chimpanzees in the Taï National Park. *American Journal of Physical Anthropology,* 78: 547-573.
Bonner, J. T. 1980. *The evolution of culture in animals.* Princeton: Princeton UP.
Byrne, Richard & Andrew Whiten (eds.). 1988. *Machiavellian Intelligence. Social Expertise and the Evolution of Intellect in Monkeys, Apes and Humans.* Oxford.
Brain, C.K. 1981. *The hunter or the hunted? An introduction to African cave taphonomy.* Chigago: University of Chigago Press.
Brain, C.K. & A. Sillen. 1988. Evidence from the Swartkrans cave for the earliest use of fire. *Nature,* 336: 464-466.
Bunn, H.T. & E.M. Kroll. 1986. Systematic Butchery by Plio-Pleistocene Hominids at Olduvai Gorge, Tanzania. *Current Anthropology*, 27: 431-452.
Dawkins, R. 1976. *The Selfish Gene.* Oxford: Oxford UP.
Durham, William H. *Coevolution. Genes, culture, and human diversity.* Stanford: Stanford U.P.
Eisenberg, J. 1981. *The Mammalian Radiations: an Analysis of Trends in Evolution, Adaptation and Behaviour.* London: Athlone Press.
Foley, Robert. 1987. *Another Unique Species. Patterns in human evolutionary ecology.* London: Longman.
Foley, R.A. & P.C. Lee. 1989. Finite Social Space, Evolutionary Pathways, and Reconstructing Hominid Behavior. *Science*, 243: 901-905.
Flinn, M.V. & B.S. Low. 1986. Resource distribution, social competition and mating patterns in human societies. In: D.J. Rubenstein & R.W. Wrangham (eds.), *Ecological Aspects of Social Evolution.* Princeton UP.
Giacobini, Giacomo (ed.). 1989. *Hominidae. Proceedings of the 2nd International Congress of Human Paleontology. Turin, Sept. 28- Oct. 3, 1987.* Milano: Jaca Book.
Ghiglieri, Michael P. 1987. Sociobiology of the great apes and the hominid ancestor. *Journal of Human Evolution,* 16: 319-357.
Goodall, Jane. 1986. *The Chimpanzees of Gombe.* Cambridge & London.
Hawkes, Kirsten. 1991. Showing Off: Tests of an Hypothesis About Men's Foraging Goals. *Ethology and Sociobiology,* 12: 29-54.
Hasegawa, T. *et al.* 1983. New Evidence on scavenging Behavior in Wild Chimpanzees. *Current Anthropology,* 24: 231-232.
Hewes, G.W. 1961. Food transport and the origin of hominid bipedalism. *American Anthropologist*, 63: 687-710.
Jerison, H. J. 1973. *Evolution of the Brain and Intelligence.* New York: Academic Press.
Jolly, C. J. 1970. The seed-eaters: a new model of hominid differentiation based on a baboon analogy. *Man* 5: 6-26.
Kinzey, Warren G. 1986. *The Evolution of Human Behavior: Primate Models.* New York: State University of New York Press.
Klein, Richard. 1989. *The Human Career. Human Biological and Cultural Origins.* Chigago & London: The University of Chigago Press.

LaLumière, Leon P. 1991. The Evolution of the Genus *Homo*: Where it Happened. In: Roede, 1991: 23-35.

Louwe-Kooijmans L.P. 1990. Bronstijdstrijd. Slachtoffers van een oeroorlog. *Natuur en techniek,* 58: 748-759.

Lovejoy, C. Owen 1981. The Origin of Man. *Science,* 211: 341-350.

Morgan, Elaine. 1990. *The Scars of Evolution.* Harmonswoth: Penguin.

McGrew, W.C. 1991. Chimpanzee material culture: what are its limits and why? In: R.A. Foley, *The Origins of Human Behaviour*, London Unwin Hyman, 13-24.

Müller, Richard. 1989. *Nemesis. The Death Star.* London: Guild Publishing.

Pickford, Martin. 1989. Pre-Hominids. In: Giacobini, 1989: 23-33.

Pickford, Martin. 1991. Does the Geological Evidence Support the Aquatic Ape Theory? In: Roede, 1991: 127-131.

Potts, Richard. 1986. Reconstructions of early hominid socioecology: a critique of primate models. In: Kinzey, 1986: 28-47.

Reynolds, Vernon. 1991. Cold or Watery? Hot or Dusty? Our Ancestral Environment and Our Ancestors Themselves: an Overview. In: Roede, 1991: 331-341.

Rodman, P. & H. McHenry. 1980. Bioenergetics and the origin of hominid bipedalism. *American Journal of Physical Anthropology,* 63: 687-710.

Rodseth, et al. 1991. The Human Community as a Primate Society. *Current Anthropology.* 32 (3): 221-254.

Roede, Machteld *et al.* 1991. *The Aquatic Ape: Fact or Fiction?* London: Souvenir Press.

Schmid, Peter. How Different is Lucy? In: Giacobini, 109-118.

Speth, John D. 1989. Early hominid hunting and scavenging: the role of meat as an energy source. *Journal of human evolution,* 18, 329-343.

Strauss, 1949. Die aufrechte Haltung. Eine anthropologische Studie. *Monatsschrift für Psychiatrie und Neurologie,* 117: no. 4/5/6.

Susman, Randall L. 1986. Pygmy chimpanzees and common chimpanzees: models for the behavioral ecology of the earliest hominids. In: Kinzey, 1986: 72-86.

Tanner, Nancy M. 1986. The Chimpanzee Model Revisited and the Gathering Hypothesis. In: Kinzey, 1986: 3-27.

Tobias, P. V. 1991. The Environmental Background of Hominid Emergence and the Appearance of the Genus *Homo*. *Human Evolution,* 6: 129-142.

Tooby, John & Irven DeVore, 1986. The reconstruction of hominid behavioral evolution through strategic modeling. In: Kinzey, 1986: 183-237.

Toth, N. 1986. The First Technology. *Scientific American.*

Trivers, R. L. 1971. The evolution of reciprocal altruism. *Quarterly Review of Biology* 46.

Trivers, R. L. 1985. *Social Evolution.* Menlo Park: Benjamin/Cummins.

Vallois, H.V. 1961. The social life of early man: the evidence of skeletons. In: S. J. Washburn, ed., *The social life of early man,* 214-235. Chigago: Aldine.

Van Schaik, C.P. & J.A.R.A.M. van Hooff. 1983. On the ultimate cause of primate social systems. Behaviour. 85: 91-117.

Verhaegen, M. 1987. Origin of hominid bipedalism. *Nature,* 325: 305-306

Waal, F. de. 1988. *Peacemaking among primates.* Cambridge: Harvard University Press.

Wheeler, P. 1984. The evolution of bipedality and the loss of functional body hair in hominids. *Journal of Human Evolution.* 13: 91-98.

Williamson, P.G. 1985. Evidence for an Early Plio-Pleistocene rainforest expansion in East Africa. *Nature,* 315: 487-489.

White, T.D. 1986. Cut marks on the Bodo cranium: a case of prehistoric defleshing. *American Journal of Physical Anthropology* 69: 503-509.

White, Randall. 1989. Visual Thinking in the Ice Age. Scientific American. July: 74-81.

Wrangham, Richard W. 1979. On the evolution of ape social systems. *Social Science Information,*

18: 355-386.
Wrangham, Richard W. 1986. The significance of African apes for reconstructing human social evolution. In: Kinzey, 1986: 51-71.
Wrangham, Richard W. 1986. Ecology and social relationships in two species of chimpanzee. In: D.I. Rubenstein & R.W. Wrangham, *Ecological Aspects of Social Evolution,* Princeton: Princeton U.P.

List of Contributors

Dr. M.C. Busnel
Genetique, Neurogenetique et Comportement
UFR Biomedicale, Universite Rene Descartes
CNRS URA 1294
45, Rue des Saints Peres
75006 Paris, France

Prof. B. Chiarelli
Institute of Anthropology
Via del Proconsolo, 12
50122 Firenze, Italy

Dr. Roger S. Fouts
Chimpanzee and Human Communication Institute
Central Washington University
Ellensburg, Washington 98926 USA

Mrs. Deborah H. Fouts
Chimpanzee and Human Communication Institute
Central Washington University
Ellensburg, Washington 98926 USA

Dr. B.T. Gardner
Center for Advanced Study/296
University of Nevada
Reno, Nevada 89557 USA

Dr. R.A. Gardner
Center for Advanced Study/296
University of Nevada
Reno, Nevada 89557 USA

Dr. Valerius Geist
Faculty of Environmental Design
University of Calgary
2500 University Drive N.W.
Calgary, Alberta, Canada, T2N 1N4

Dr. Francesca Giusti
Istituto Italiano per gli Studi Filosoficie
Palazzo Sera di Cassano
Via Monte di Dio 14
80100 Naples, Italy

Dr. M.D. Kirk-Smith
Department of Marketing and Organizational Studies
University of Ulster
Newtonabbey Co., Antrim BT37 OQB Northern Ireland

Dr. L.F. Marchant
Department of Sociology & Anthropolgy
Miami University
Oxford, Ohio 45056 USA

Dr. W.C. McGrew
Department of Sociology & Anthropology
Miami University
Oxford, Ohio 45056 USA

Dr. H. Montagner
INSERM
9 Goule de Laval
34790 Grabels France

Dr. Daniel Paquette
Jardin zoologique du Quebec
8191 Avenue du Zoo
Charlesbourg, Quebec, Canada G1G 4G4

Prof. F. Plooij
Department of Psychology
University of Groningen
Grote Kruisstraat
9712 TS Groningen, The Netherlands

Dr. E. Prato Previde
Istituto Di Psicologia della Facolta Medica
Universita degli Studi di Milano
Via Francesco Sforza, 23
20122 Milano, Italy

Dr. Dennis R. Rasmussen
Florida State University
Panama Canal Branch
PSC #02 Box 2663
APO AA 34002, Panama

Dr. W.M.S. Russell
Department of Sociology
University of Reading
Whiteknights, P.O. Box 218
Reading, Berkshire, RG6 2AA, England

Prof. P.J.B. Slater
School of Biological and Medical Sciences
University of St. Andrews
Fife, KY 16 9TS, UK

Dr. Pouwel Slurink
Department of Philosophy
University of Nijmegen
Thomas van Aquinostraat 3
Postbox 9108, 6500 HK Nijmegen
The Netherlands

Dr. M.D. Suboski
Department of Psychology
Queen's University
Kingston, Ontario, Canada, K7L 3N6

Dr. G. Theraulaz
Centre de Recherches en Biologie du Comportement
CNRS - URA 664
Universite Paul Sabatier
118 route de Narbonne
31062 Toulouse, France

Dr. P. Valsecchi
Dipartimento di Biologia e Fisiologia Generali
Universita degli Studi di Parma
Viale delle Scienze
43100 Parma, Italy

Dr. Kunio Watanabe
Primate Research Institute
Kyoto University
Inuyama, Aichi, 484 Japan

Dr. A.C. Zeller
Department of Anthropology
University of Waterloo
Waterloo, Ontario, N2L 3G1 Canada